Smart and Sustainable Applications of Nanocomposites

Rajni Garg
Galgotias College of Engineering and Technology, India

Ansar Anjum
Galgotias College of Engineering and Technology, India

A volume in the Advances in Chemical and Materials Engineering (ACME) Book Series

IGI Global
PUBLISHER of TIMELY KNOWLEDGE

Published in the United States of America by
IGI Global
Engineering Science Reference (an imprint of IGI Global)
701 E. Chocolate Avenue
Hershey PA, USA 17033
Tel: 717-533-8845
Fax: 717-533-8661
E-mail: cust@igi-global.com
Web site: http://www.igi-global.com

Copyright © 2024 by IGI Global. All rights reserved. No part of this publication may be reproduced, stored or distributed in any form or by any means, electronic or mechanical, including photocopying, without written permission from the publisher.
Product or company names used in this set are for identification purposes only. Inclusion of the names of the products or companies does not indicate a claim of ownership by IGI Global of the trademark or registered trademark.

Library of Congress Cataloging-in-Publication Data

Names: Garg, Rajni, 1979- editor. | Anjum, Ansar, 1977- editor.
Title: Smart and sustainable applications of nanocomposites / edited by: Rajni Garg, Ansar Anjum.
Description: Hershey, PA : Engineering Science Reference, [2024] | Includes bibliographical references. | Summary: "The book provides a comprehensive collection of the latest information and state-of-the-art development as well as explore the potential applications of nanotechnology and nanocoating in various industries, such as biosensors, agriculture, pollutants remediation, biomedical sciences, and consumer products"-- Provided by publisher.
Identifiers: LCCN 2023048623 (print) | LCCN 2023048624 (ebook) | ISBN 9798369310946 (hardcover) | ISBN 9798369310953 (ebook)
Subjects: LCSH: Nanocomposites (Materials) | Nanotechnology.
Classification: LCC TA418.9.N35 S88 2024 (print) | LCC TA418.9.N35 (ebook) | DDC 620.1/18--dc23/eng/20240402
LC record available at https://lccn.loc.gov/2023048623
LC ebook record available at https://lccn.loc.gov/2023048624

This book is published in the IGI Global book series Advances in Chemical and Materials Engineering (ACME) (ISSN: 2327-5448; eISSN: 2327-5456)

British Cataloguing in Publication Data
A Cataloguing in Publication record for this book is available from the British Library.

All work contributed to this book is new, previously-unpublished material.
The views expressed in this book are those of the authors, but not necessarily of the publisher.

For electronic access to this publication, please contact: eresources@igi-global.com.

Advances in Chemical and Materials Engineering (ACME) Book Series

ISSN:2327-5448
EISSN:2327-5456

Editor-in-Chief: J. Paulo Davim, University of Aveiro, Portugal

MISSION

The cross disciplinary approach of chemical and materials engineering is rapidly growing as it applies to the study of educational, scientific and industrial research activities by solving complex chemical problems using computational techniques and statistical methods.

The **Advances in Chemical and Materials Engineering (ACME) Book Series** provides research on the recent advances throughout computational and statistical methods of analysis and modeling. This series brings together collaboration between chemists, engineers, statisticians, and computer scientists and offers a wealth of knowledge and useful tools to academics, practitioners, and professionals through high quality publications.

COVERAGE

- Materials to Renewable Energies
- Statistical Methods
- Composites
- Computational methods
- Ceramics and glasses
- Numerical Techniques
- Sustainable Materials
- Fatigue and Creep
- Multifuncional and Smart Materials
- Coatings and surface treatments

IGI Global is currently accepting manuscripts for publication within this series. To submit a proposal for a volume in this series, please contact our Acquisition Editors at Acquisitions@igi-global.com or visit: http://www.igi-global.com/publish/.

The Advances in Chemical and Materials Engineering (ACME) Book Series (ISSN 2327-5448) is published by IGI Global, 701 E. Chocolate Avenue, Hershey, PA 17033-1240, USA, www.igi-global.com. This series is composed of titles available for purchase individually; each title is edited to be contextually exclusive from any other title within the series. For pricing and ordering information please visit http://www.igi-global.com/book-series/advances-chemical-materials-engineering/73687. Postmaster: Send all address changes to above address. Copyright © 2024 IGI Global. All rights, including translation in other languages reserved by the publisher. No part of this series may be reproduced or used in any form or by any means – graphics, electronic, or mechanical, including photocopying, recording, taping, or information and retrieval systems – without written permission from the publisher, except for non commercial, educational use, including classroom teaching purposes. The views expressed in this series are those of the authors, but not necessarily of IGI Global.

Titles in this Series

For a list of additional titles in this series, please visit:
www.igi-global.com/book-series/advances-chemical-materials-engineering/73687

Building a Low-Carbon Society Through Applied Environmental Materials Science
Takaomi Kobayashi (Nagaoka University of Technology, Japan)
Engineering Science Reference • copyright 2024 • 300pp • H/C (ISBN: 9798369300039) • US $255.00 (our price)

FlexPDE and Finite Element Method Applications in Thermal Energy Storage and Cavities
Khashayar Hosseinzadeh (Sharif University of Technology, Iran) Mehdi Mahboobtosi (Babol Noshirvani University of Technology, Iran) and Davood Domairry Ganji (Babol Noshirvani University of Technology, Iran)
Engineering Science Reference • copyright 2024 • 300pp • H/C (ISBN: 9798369309322) • US $225.00 (our price)

Advancements in Powder Metallurgy Processing, Applications, and Properties
Shashanka Rajendrachari (Bartin University, Turkey) and Rayappa Shrinivas Mahale (Jain College of Engineering and Research, India)
Engineering Science Reference • copyright 2024 • 300pp • H/C (ISBN: 9781668493854) • US $275.00 (our price)

Innovative Multifunctional Nanomaterial for Photocatalysis, Sensing, and Imaging
Azad Kumar (Siddharth University, India)
Engineering Science Reference • copyright 2023 • 283pp • H/C (ISBN: 9781668487433) • US $270.00 (our price)

Innovations in Materials Chemistry, Physics, and Engineering Research
Eugene de Silva (Virginia Research Institute, USA) and Pramudi Abeydeera (Polymeric Ltd., USA)
Engineering Science Reference • copyright 2023 • 446pp • H/C (ISBN: 9781668468302) • US $270.00 (our price)

For an entire list of titles in this series, please visit:
www.igi-global.com/book-series/advances-chemical-materials-engineering/73687

IGI Global
PUBLISHER of TIMELY KNOWLEDGE

701 East Chocolate Avenue, Hershey, PA 17033, USA
Tel: 717-533-8845 x100 • Fax: 717-533-8661
E-Mail: cust@igi-global.com • www.igi-global.com

Table of Contents

Detailed Table of Contents

Observing the unique properties of matter at nanoscale, nanoscience provides innovative solutions for energy, environment, agriculture, and manufacturing, and addresses the critical challenges of sustainability. This chapter highlights nanotechnology's pivotal role in advancing sustainability across diverse domains including medicine, consumer products, and energy optimization. Ethical considerations, potential benefits, and future prospects are thoroughly examined, providing a comprehensive overview of nanotechnology's transformative capacity. The rational design of nanomaterials achieves exceptionally high surface areas, offering strategic enhancements. In the era of electronics, nanotechnology accelerates devices, enhances memory chip density, and reduces power consumption, shaping a more sustainable technological future. The chapter serves as a valuable reference for research community, offering insights into applications of nanotechnology in sustainable development circular economy, and aims to deepen understanding and exploration of nanotechnology's potential to drive positive change.

The design of nanocoatings and nanostructured based surfaces stands out in many fields and biomedical applications due to their superior advantages over traditional coatings. With current nanotechnological approaches, properties such as antimicrobial activity, biocompatibility, increased product life-time, quality interface bonding, enhanced chemical and mechanical properties, high hardness, heat insulation, dirt and water repellency, corrosion resistance, improved energy efficiency, and reduced moisture absorption have been significantly improved. In recent years, the use of NPs and nanocoatings in the design and development of many dental materials has become highly popular in dentistry due to their ability to provide unique combinations of properties. This section will primarily focus on nanocoating materials used in dentistry.

The materials produced on nanoscale have a new set of properties. Nanotechnology and nanomaterials have positively affected the human lifestyle. It has an impact on agricultural, energy, petroleum, sports, chemical and catalysis, environmental, paints and coating, medical, beauty, and textile sectors. Feeding the increasing needs of the population has given rise to different innovations in various sectors. The agricultural sector also has benefitted from such innovations. Nanotechnology and nanomaterials have come to the rescue of the farmers. Micronutrients are supplied to the crops through fertilizers. Nanomaterials are explored for fuel borne catalysis for soot prevention. Nanoparticle reinforced tires have greater abrasive resistance and gas permeability. Nanoparticles are used in coating for better gloss, scratch and corrosion resistance, and anti-fog coatings. Nanoparticles and nanomaterials like cubosomes, nanodots, liposomes, dendrimers, nano-emulsions have become common ingredients in cosmetics.

Water is a crucial natural resource for human and ecosystem health. However, rapid industrialization, urban expansion, and exploitation of natural resources have negatively impacted water quality. Effluents, containing various contaminants, negatively impact water quality. Various techniques have been used to treat effluents,

To combat the exponentially increasing environmental challenges, there is a greater demand for eco-friendly and sustainable materials in various industries, including sensor technology. Green nanocomposites, which combine biodegradable polymers with nanoparticles obtained from renewable resources, offer a possible solution to the demand for effective biosensors and environmental sustainability. This chapter assesses current advances in developing green nanocomposites for skin cancer biosensors and energy storage devices such as lithium-ion batteries, supercapacitors, and fuel cells, emphasizing their prospective applications and performance enhancements. This chapter provides an in-depth examination of green nanocomposites as sustainable, efficient, and cost-effective biosensors, highlighting their manufacturing, unique features, and numerous sensing applications, and also explores the inclusion of nanomaterials such as cellulose nanofibers, chitosan, and graphene oxide into biopolymer matrices such as starch, polylactic acid, cellulose, proteins, and chitosan to build biosensors.

This chapter highlights the nanocellose-based composite sensors as powerful tools in multiple sensing/detecting applications and polymeric matrixes. The worldwide need for rapid, precise/accurate, selective/specific detection positions the nanosensors as crucial devices for sensing purposes. The authors discuss the basic sensor conception, a basic architecture applied in different spatial scales and research areas, such as environmental sciences, medicine, and optoelectronics. Chemo- and biosensors are also described in the chapter. The use of nanocellulose as supporting or embedding material in sensors and polymeric matrixes gives the attributes of this noble material, which, combined with active and sensitive materials at the nanoscale, assemble new opportunities in the sensors field.

Nanocomposites have an essential role in a number of different applications these days. Therefore, they have been utilised in the biosensing field. The integration of nanocomposites in different types of biosensors for detection or mentoring of different analytes results in improved performance biosensors. In this chapter, the role of nanocomposites in biosensors will be studied, with a number of examples, of such improvements in these sensors' performance. An overview of the biosensors will be given first, along with key characteristics to quantity the enhancing in their performance. After that, a detailed discussion of nanocomposites as materials and their roles in biosensor development will be included.

Although bio-based nanocomposite materials have various packaging options, the future is still evolving. At this point, multifunctional intelligent packaging will replace simple traditional packaging. As the use of synthetic plastics decreases, so does the demand for biodegradable packaging materials. Because of its potential to alleviate environmental issues associated with conventional packaging materials, biodegradable nanocoating for food packaging has seen a significant increase in interest in recent years. This innovative technology offers several perspectives and benefits, including improved food safety, extended shelf life, reduced waste, and enhanced sustainability. The objectives of this chapter are to highlight the latest advancements and research in the field of biodegradable nanocoating and its potential to revolutionize the food packaging industry as well as the advanced development teams about the benefits and potential applications of biodegradable nanocoating in food packaging.

Preface

The conventional materials that have long served our industrial needs are encountering limitations in meeting the demands of today's dynamic applications. In response, the fields of Nanoscience and Nanotechnology have rapidly evolved, adopting innovative synthesis approaches to create nanocomposites tailored for diverse industrial purposes. The burgeoning interest in nanoscience has propelled research into these nanosized structures, unlocking their immense potential across industries such as construction, textiles, sensors, and food packaging. Their exceptional properties—ranging from mechanical and electrical conductivity to flame retardancy, dielectric behavior, and superior barrier qualities—have positioned nanostructures as game-changers in material science.

Yet, the ever-expanding landscape of industrial requirements poses a new challenge: the continuous need for additional properties and functionalities in nanomaterials. The pursuit of environmentally friendly, sustainable nanocomposites has emerged as a focal point, sparking extensive efforts to fortify materials using fibers, whiskers, or nano clays, resulting in green and renewable nanocomposites.

Recent endeavors have explored alternative materials like ceramics, agricultural waste, silica, and polymer-layered silicate nanocomposites, offering unique physical and chemical properties compared to conventional composites. Nano clays, for instance, have augmented mechanical properties and introduced novel functionalities based on their chemical composition, enriching thermal, electronic, magnetic, and redox properties, among others.

Our book endeavors to encapsulate exhaustive research on bioinspired synthesis of nanocomposites and their multifaceted applications in today's industrial realm. Tailored to meet contemporary industrial needs, it delves into synthesis methodologies and applications while spotlighting recent technological strides aimed at creating proficient applications in specialized domains. With examples and illustrations, our aim is to render the book reader-friendly.

This comprehensive compilation will explore recent research, experiments, and industrial paradigms, offering an in-depth examination of nanotechnology and nanocoatings. It will encompass fundamental aspects of nanotechnology, current status, future prospects, various types of nanocoatings, their properties, and potential applications.

Furthermore, the book aims to present cutting-edge information and developments in nanotechnology and nanocoating applications across diverse industries such as biosensors, agriculture, pollutants remediation, biomedical sciences, and consumer products.

Designed for a general audience, this book is crafted to be accessible to readers unfamiliar with nanotechnology or nanocoating. It will leverage illustrations and diagrams to facilitate comprehension of complex concepts. Its aim is to bridge the gap between nanoscience research and industry applications, serving as an invaluable resource for students, researchers, academicians, and industry professionals alike.

ORGANIZATION OF THE BOOK

Chapter 1: Introduction of Nanotechnology: A Transformative Frontier

Authored by Ansar Anjum, Maumita Das Mukherjee. This chapter delves into the role of nanotechnology in various sectors, highlighting its transformative potential in sustainability, energy, environment, medicine, and manufacturing. It explores ethical considerations and future prospects, focusing on the design of nanomaterials and their impact on electronics, memory chips, and sustainable technological advancements.

Chapter 2: Bioinspired Synthesis of Nanocomposites

Authored by Semiha Ekrikaya, Hikmetnur Danisman, Seda Baktir, Soley Arslan, Ismail Ocsoy. This section emphasizes nanocoating materials in dentistry, discussing the remarkable properties of nanocoatings and nanostructured surfaces in enhancing product lifespan, biocompatibility, antimicrobial activity, and other desirable characteristics crucial in biomedical applications.

Chapter 3: Multifaceted and Diverse Applications of Nanocomposites

Authored by Sunil Kulkarni. This chapter explores the broad spectrum of applications for nanomaterials, focusing on their impact across agriculture, energy, petroleum, environmental remediation, paints, coatings, medical, beauty, and textile sectors. It delves into innovations that cater to increasing population needs in various industries.

Chapter 4: Applications of Nanocomposites in Water Remediation: A Mechanistic Overview

Authored by Rajni Garg, Rishav Garg, Nnabuk Eddy. Highlighting water's criticality, this chapter evaluates the efficacy of nanocomposites in treating water effluents, emphasizing adsorption techniques for wastewater treatment. It explores the advantages of this approach and its potential to alleviate water quality concerns.

Chapter 5: Nanophotocatalyst-Infused Textile Composites for Environmental Remediation

Authored by Vanshika Sharma, Amit Kumar, Priya Jaswal, Chirag Porwal, Akshay Gaur, Rahul Vaish. Focusing on environmental cleaning, this chapter examines textile composites infused with nanophotocatalysts. It elucidates their diverse uses in oil spill cleanup, water purification filters, and air purification masks, emphasizing scalability and durability.

Chapter 6: Characteristics, Applications, and Limitations of Nanocomposites in Biosensing

Authored by Jyoti Bhattacharjee, Debjani Bhakta, Subhasis Roy. Addressing environmental challenges, this chapter explores green nanocomposites' role in biosensors for skin cancer detection and energy storage devices. It assesses their sustainability, cost-effectiveness, and performance enhancements.

Chapter 7: Nanocellulose Composites as Chemo/ Bio Sensing Agents and Polymeric Matrixes

Authored by Paula Tatiana Peña González, Inés Hernández Celi, Franci Nathalie Gómez Jaimes, Carlos Andrés Martínez Bonilla. This chapter highlights nanocellulose-based sensors in various applications and polymeric matrices, discussing their importance in rapid, precise detection across environmental sciences, medicine, and optoelectronics.

Chapter 8: Nanocomposites for Advanced Biosensing Applications

Authored by Nawal Alghamdi. Focusing on biosensors, this chapter examines how nanocomposites elevate biosensors' performance across different analyte detections. It explores the role of nanocomposites in improving sensor functionalities.

Chapter 9: Applications of Bionanocomposites in Food Packaging

Authored by Mukhtar Iderawumi Abdulraheem, Gholamreza Abdi, Nafeesa Farooq Khan, Abdul Wahab, Kasahun Gudeta, Chandan Singh, Oluwadamilola Oluwatoyin Hazzan. Exploring the evolving landscape of biodegradable nanocoating in food packaging, this chapter discusses its potential in enhancing food safety, shelf life, reducing waste, and improving sustainability in the packaging industry.

Chapter 10: Ethical and Environmental Implications Associated with the Application of Nanotechnology

Authored by Nnabuk Eddy, Rajni Garg, Richard Ukpe, Nyeneime Akpanudo, Hillary Abugu, Rishav Garg, Ansar Anjum, Balu Anand. Examining the broader implications of nanotechnology, this chapter delves into its ethical considerations, toxicity concerns, and the need for balanced discourse between its advantages and potential risks. It scrutinizes the ethical application of nanoparticles and their impact on the environment.

IN SUMMARY

In concluding this comprehensive journey through the realms of nanocomposites and nanotechnology, this edited reference book stands as a testament to the expansive potential and transformative power embedded within these fields. The chapters meticulously curated by esteemed experts offer profound insights into the multifaceted applications, innovations, and challenges in nanoscience.

From the foundational exploration of nanotechnology's transformative capacities to the intricate design and application of nanomaterials in various industries, each chapter unravels a spectrum of possibilities. The bioinspired synthesis of nanocomposites, their impact on water remediation, environmental cleanup, biosensing, and their innovative use in food packaging portray a landscape brimming with potential for societal betterment and sustainability.

This collection encapsulates not only the remarkable achievements but also the ethical considerations and environmental implications imperative in steering the course of nanotechnology's advancement. It prompts us to ponder not just the advantages but also the responsibilities inherent in harnessing the power of the nano-scale world.

As editors, we envision this book to serve as a compass guiding researchers, academicians, industry professionals, and enthusiasts through the intricacies and vast applications of nanocomposites and nanotechnology. We hope that the insights shared within these pages inspire further exploration, innovation, and responsible utilization of these groundbreaking technologies for the collective betterment of humanity and our planet.

Warmest regards,

Rajni Garg
Galgotias College of Engineering and Technology, India

Ansar Anjum
Galgotias College of Engineering and Technology, India

Chapter 1
Introduction to Nanotechnology:
Transformative Frontier

Ansar Anjum
Galgotias College of Engineering and Technology, India

Maumita Das
Amity University, Noida, India

Rishav Garg
https://orcid.org/0000-0001-6565-9234
Galgotias College of Engineering and Technology, India

ABSTRACT

Observing the unique properties of matter at nanoscale, nanoscience provides innovative solutions for energy, environment, agriculture, and manufacturing, and addresses the critical challenges of sustainability. This chapter highlights nanotechnology's pivotal role in advancing sustainability across diverse domains including medicine, consumer products, and energy optimization. Ethical considerations, potential benefits, and future prospects are thoroughly examined, providing a comprehensive overview of nanotechnology's transformative capacity. The rational design of nanomaterials achieves exceptionally high surface areas, offering strategic enhancements. In the era of electronics, nanotechnology accelerates devices, enhances memory chip density, and reduces power consumption, shaping a more sustainable technological future. The chapter serves as a valuable reference for research community, offering insights into applications of nanotechnology in sustainable development circular economy, and aims to deepen understanding and exploration of nanotechnology's potential to drive positive change.

DOI: 10.4018/979-8-3693-1094-6.ch001

Copyright © 2024, IGI Global. Copying or distributing in print or electronic forms without written permission of IGI Global is prohibited.

INTRODUCTION

The rapid advancement of technology has led to an increased demand for nanomaterials, particularly in the fields of electronics, medicine, energy conservation, and agriculture. Nanotechnology is the manipulation of matter on an atomic and molecular scale, typically between 1 and 100 nanometers. This means that nanomaterials are about 100,000 times smaller than the width of a human hair. It is a rapidly evolving field capable to transform industries by utilizing the unique properties of materials at the nanoscale. Harnessing matter at the nanoscale opens up a realm of possibilities, offering enhanced efficiency, accessibility, and sustainability. The enlarged surface area of nanoscale materials fosters higher reactivity, enabling more efficient processes and their reduced size translates to lower energy consumption, promoting sustainability. Additionally, their small stature allows nanoscale materials to penetrate areas previously inaccessible, proving advantageous in the medical field, where human anatomy poses limitations for treatment. (Recordati et al., 2016) (Chen et al., 2019) Nanomaterials, with their unique properties, exceptional strength, conductivity, and reactivity have led to significant advancements in technology and preceded for smaller, lighter, and more efficient devices, enabled the development of new medical treatments, targeted drug delivery systems, and environmentally friendly products (Chis et al., 2020), semiconductor technology, material science, electronics, optics, medicine, plastics, and the textile industry. (Iravani et al., 2023) Thus, the technology is rapidly transforming industries and holds immense promise for shaping the future of technology and society. (Kiessling et al., 2010) (Jamaluddin et al., 2021) As nanotechnology continues to evolve, it is poised to revolutionise numerous aspects of human life. Its potential impact is vast and transformative, making it one of the most exciting and promising fields of scientific research today. (Adriana et al., 2020) (Juanola-Feliu et al., 2014)

In the dynamic landscape of modern science and technology, nanomaterials emerge as versatile powerhouses, wielding unparalleled influence across magnetic, electrical, mechanical, and catalytic realms. Nanotechnology, with its multifaceted impact, stands as an irreplaceable force driving innovation in critical sectors ranging from transportation, energy, and agriculture to healthcare, information, and communication as shown in figure 1. ("Titanium Dioxide Nanomaterials," 2012) This transformative influence not only sparks new developments but also propels sustainability into the forefront of progress. (Chee et al. 2022)

Nanomaterials have unique optical properties that can be utilized to develop novel materials and coatings that can protect against UV radiation and prevent photodegradation of various substances. (Hussain et al., 2022) By incorporating nanoparticles into materials such as paints, plastics, and textiles, we can create products that have enhanced durability and resistance to UV-induced degradation.

Figure 1. Applications of nanomaterials in versatile domains

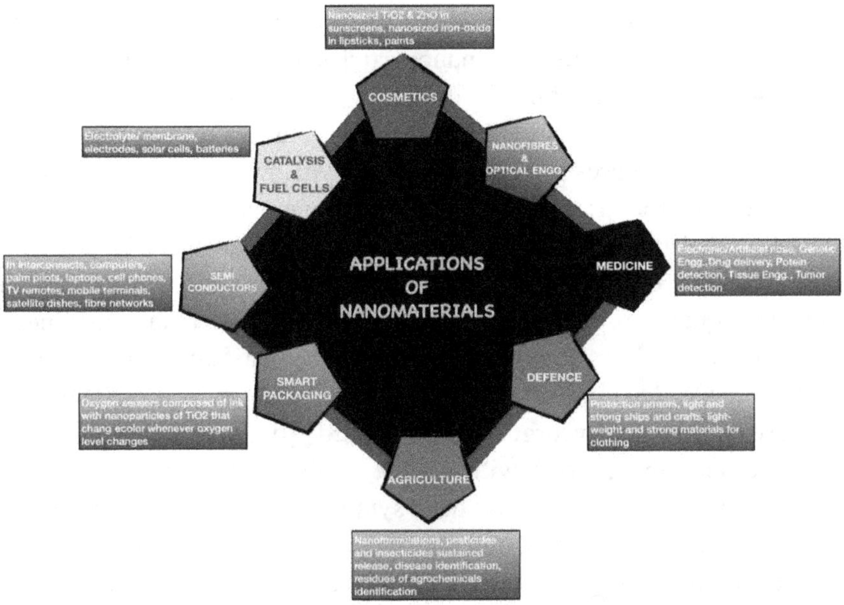

(Rani et al., 2022) Another area of nanotechnology application is in the field of food packaging. Nanomaterials can be used to develop smart and active packaging materials that can extend the shelf life of food products. (González-Pedroza et al., 2023) For example, nanoparticles can be incorporated into packaging films. Furthermore, nanocellulose composites have emerged as promising chemo/biosensing agents and polymeric matrices in the field of nanotechnology. (X. Yang & Wang, 2022) These nanocellulose composites offer a range of applications, including biosensors and medical diagnostics. Nanocellulose, derived from renewable sources, exhibits exceptional mechanical properties, biocompatibility, and biodegradability, making it an ideal candidate for use in medical applications. (Thakur et al., 2021) (X. Wu et al., 2023) In the realm of biosensors, nanocellulose composites have shown great potential for detecting various biological substances.

HISTORICAL AND IMPLICATIVE EXPLORATION OF NANOTECHNOLOGY

Materials behave differently at the nanoscale due to increased surface area, leading to higher reactivity and distinct properties. This has fascinated scientists and engineers, opening avenues for exploration and innovation. Nature depends on nanoscale processes, from simple colloids to complex proteins. Nanoscience and nanotechnologies are not new; size-dependent properties have been exploited for centuries, like using gold and silver nanoparticles as pigments. Nanotechnologies have been used in computer chips for decades.

Nanomaterials have vast potential and can revolutionize many aspects of our lives, from treating diseases to generating energy. However, the use of nanoscale materials dates back centuries, with gold and silver nanoparticles being employed as pigments in stained glass and ceramics since the 10th century AD. Nanotechnology is based on scientific concepts such as interatomic and intermolecular forces, single atom imaging, quantum confinement, self-assembly, molecular machinery, and nano magnetism. It allows for the modification of materials at incredibly small scales, which can lead to improved chemical and physical properties. The scientists' research reveals that the Lycurgus cup is crafted from soda-lime glass, blended with 1% gold and silver, along with a 0.5% addition of manganese. (Barber & Freestone, 1990) Chemists have long been synthesizing polymers, large molecules composed of nanoscale subunits, and nanotechnologies have been instrumental in creating the intricate features on computer chips for the past two decades. (Erhardt, 2003) A substantial portion of nanoscience and numerous nanotechnologies focus on the creation of novel or improved materials. Certain products empowered by nanotechnology have already made their way into the market, experiencing notable commercial success. The windows equipped with self-cleaning capabilities feature a specialized coating of activated TiO_2, engineered to be a mere 15 nm thick. This design imparts exceptional water repellency, allowing rainwater to effortlessly cascade off the surface, carrying away accumulated dirt. In certain sunscreens, nanoparticles play a pivotal role in reflecting and absorbing ultraviolet light. The captivating properties of carbon nanostructures have garnered significant attention and research efforts since their initial discovery in the mid-1980s. The intricacies of nanoscale behaviour unveil both the captivating potential for advancement and the nuanced considerations essential for responsible exploration in scientific and engineering pursuits.(Dubey et al., 2022)

In the medical field, nanotechnology has the potential to revolutionize diagnostics and therapeutics, with applications such as tumor cell detection, infectious disease identification, and targeted drug delivery. Nanotechnology is a promising and emerging technology that combines biotechnology and information technology, and it has already brought about innovative changes in healthcare and biomedical applications.(Jia et al., 2023) (Kontogiannis et al., 2023) Additionally, nanomaterials are being used to develop new drugs and medical devices that can target and treat diseases more effectively. One of the most promising aspects of nanotechnology is its potential to address some of the world's most pressing challenges, such as climate change, technology and disease. They are being developed to create new solar cells and fuel cells that are more efficient and environmentally friendly.(L. Wang et al., 2022) (Banin et al., 2020) Not only this, over the past five years, the field of nanotechnology has witnessed significant advancements in the coatings and biosensors sectors. These innovations have had a profound impact on various industries, including electronics, communication, energy, advanced materials, space technology, and biomedicine. (Ferreira & Filipe, 2018) Nanotechnology has transformed the coatings industry by revolutionizing material development and application. Nanoparticles empower coatings to achieve heightened durability, scratch resistance, and corrosion protection. Notably, self-healing properties have extended the lifespan of coatings, contributing to sustainable solutions. In the biosensors sector, nanotechnology has played a crucial role in crafting highly sensitive and selective devices for medical diagnostics and environmental monitoring. Nano biosensors, equipped with the ability to detect specific molecules with remarkable accuracy, mark a significant advancement. The synergy of nanotechnology and biosensors has given rise to nano-biosensors, ushering in a new era of precision and efficiency in molecular detection. (Omietimi et al., 2023) In the field of biosensors, nanotechnology has enabled the development of highly sensitive and efficient sensing systems. These nano biosensors have become crucial in vital areas such as healthcare, agriculture, genome analysis, food and drink, environmental monitoring, defense, and security.

Nanoscience and nanotechnology have embarked on a captivating odyssey, weaving a tapestry of scientific marvels that have reshaped our understanding of matter and its transformative potential. Each milestone along this remarkable journey has pushed the boundaries of knowledge and redefined the realm of possibilities, underscoring the extraordinary evolution of this field.

In 1857, Michael Faraday while studying “Ruby” gold, a colloidal suspension of gold nanoparticles. He discovered that these tiny particles possessed unique optical and electronic properties, differentiating them from other nanoparticles.

Faraday’s experiments unveiled the incredible ability of gold nanoparticles to produce solutions that changed color under specific lighting conditions. This discovery laid the foundation for further research into the captivating world of nanomaterials and their diverse applications (Figure 2). As of now, several studies have underscored the substantial impact of nanotechnologies on biomedicine, particularly in diagnostics and treatment of human diseases. (Sciau, 2012) Bio-nanotechnology is one of the most captivating fields of nanoscience. Over the past decades, extensive research has investigated the applications of nanotechnology across various biology-related realms, including diagnosis, drug delivery, and molecular imaging. These investigations have yielded impressive outcomes, marking significant strides toward the integration of nanoscience into the forefront of biomedical advancements. (Kinnear et al., 2017)

Figure 2. Diagrammatic representation of timeline of nanotechnology

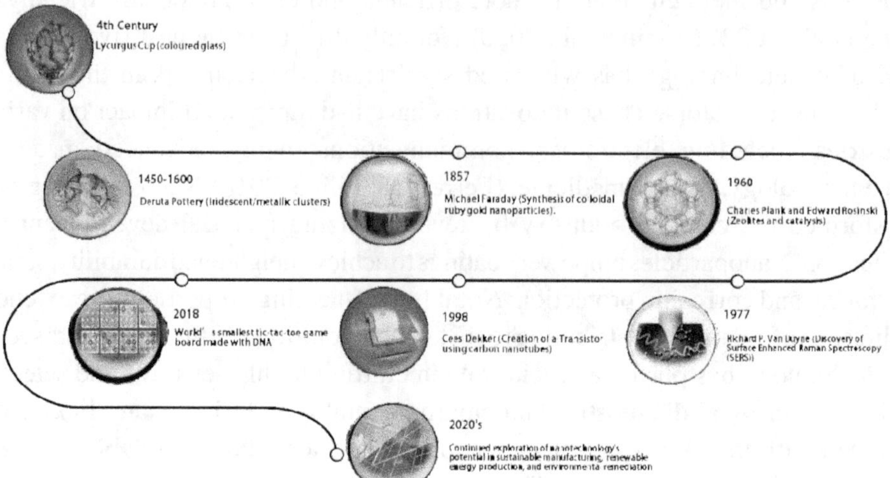

Nanotechnology witnessed remarkable progress from the early conceptualizations of Feynman until a pivotal moment in 1981. This breakthrough came with the invention of a groundbreaking tool – the Scanning Tunneling Microscope (STM) that marked a transformative leap in nanoscience, enabling the visualization and manipulation of matter at the atomic and molecular levels, that led to unprecedented exploration through research. (Binnig et al., 1982)

PROPERTIES OF NANOMATERIALS

The nanoscale particles show the properties of matter dramatically different from their bulk counterparts. With the decrease in size of particles the size-dependent effects come to the fore, leading to captivating phenomena as seen in bulk gold that exhibits a characteristic yellow color, whereas its nanoscale counterpart transforms into a vibrant purple or red. This remarkable shift highlights the profound influence of size on the properties of nanomaterials. Furthermore, by precisely tailoring the size of these tiny materials, scientists have unlocked the potential to fine-tune their properties, opening doors to exciting new avenues in diverse fields.

In the domain of nanomaterial-based detection, contemporary methodologies harness the extraordinary surface-area-to-volume ratio, coupled with the remarkable optical, magnetic, and high sensitivity dielectric properties inherent in nanomaterials. This confluence of properties serves as a foundation for cutting-edge detection techniques. The investigations show a pivotal role of metal nanoparticles in specific biomolecule detection, strategically leveraging the Localized Surface Plasmon Resonance (LSPR) property. This technique relies on the exceptional sensitivity and magnetic characteristic that is intrinsic to metal nanoparticles, enhancing the precision of biomolecular detection.

Diverging from LSPR-centric approaches, metal nanoparticles have lately been in forefront for identification of alterations in DNA sequences. In this innovative method, gold nanoparticles travel through a biological environment, moving towards a crafted sensing platform comprising silica and gold nanoparticles. Within this intricately designed system, the interplay captures the DNA sequence, revealing any mutations present by comparing it with colloidal particles as a reference. This multifaceted approach signifies a substantial advancement in the arena of impactful and precise detection methodologies (Muller et al., 2005).

Electronic and Optoelectronic Properties

The electronic characteristics of nanoparticles are intricately linked to factors like size, surface area, chemical composition, and modifications (Shin et al., 2015). These properties encompass parameters describing the state and behavior of electrons within the material, such as electronic band structure, elucidating their energy and momentum states (Lindsay, 2010). This property of nanoparticles is closely entwined with electric properties like electrical conductivity and dielectric response, as well as optical properties including refractive index, damping constant, absorbance, and more, delineating the material's response to electromagnetic radiation.

Optical properties in nanomaterials, such as absorption, transmission, reflection, and light emission, exhibit dynamic characteristics deviating significantly from those of bulk materials. Manipulating the shape, size, and surface functionality offers a spectrum of optical effects for various applications, achievable through different means based on composition, size, and orientation (Adewuyi & Lau, 2021).

The integration of organic compounds like ligands enhances the diverse surface attributes of nanoparticles, with ligand monomer selection influencing structural features. An illustration of the same is the study conducted on modification of Al_2O_3 with organic ligands, dictating material size through the aggregation process and imparting distinctive electrical properties. The introduction of inorganic compounds into polymer systems further elevates the electrical properties of polymer nanocomposites (Hassan et al., 2019). Notably, the addition of barium titanate with its perovskite nanostructures and piezoelectric properties enhances electrical conductivity and dielectric constant in the polymer matrix.

Gold nanoparticles exhibit a quantum size effect, resulting in discrete electron transition energy levels. For instance, hexanethiol-functionalized nanoparticles

(Au147, d = 1.62 nm) display 15 redox states at room temperature, showcasing molecule-like redox properties. Additionally, the quantized capacitance charging behavior of gold nanoparticles is tunable by external ligands, magnetic fields, and electrolyte ions, presenting appreciable use in electronic devices and electrochemical labels. Besides redox activity, these nanoparticles showcase a surface plasmon band, a result of combined electron vacillation prompted by photons. This property is not present in small nanoparticles (< 2 nm diameter) and large sized gold particles, highlighting distinctive nature of surface plasmon resonance.

The forefront of next-generation display technology is marked by quantum dots, recognized as highly promising optoelectronic materials. These dots exhibit exceptional physical characteristics, being both electroactive (electroluminescent) and photo-active (photoluminescent), positioning them as the cornerstone of forthcoming displays. In comparison to organic luminescent materials employed in OLEDs (organic light-emitting diodes), quantum dot-based materials boast advantages such as lower power consumption, reduced manufacturing costs, extended lifespan, and more vibrant colors.

A noteworthy advantage of quantum dot displays is their versatility in size, with the capability to create displays of various types, including rollable, flexible, and printable variants. Quantum dot displays can be seamlessly deposited on practically any substrate.

Mechanical Properties

Nanomaterials, with their unique size-dependent effects, offer a revolutionary paradigm in mechanical properties. Nanoparticle incorporation refines material grain structure, creating inter/intragranular structures that strengthen grain boundaries, resulting in remarkable improvements. Concrete infused with 3% nano-SiO_2 show significant increase in compressive, bending, and tensile strengths. Beyond these fundamental enhancements, nanomaterials offer tailored properties. By strategically manipulating size, shape, and composition, scientists can design materials for specific applications. (X. Wang et al., 2011) The horizon expands further with advanced coatings, self-cleaning surfaces, and functional materials with integrated sensing and actuation capabilities.

The exploration of nanomechanical properties holds immense promise for advancements in diverse fields. As researchers investigate deeper, even more revolutionary materials can be anticipated with unprecedented capabilities, shaping the future of engineering, medicine, energy, and electronics. (J. L. Wang & Meng, 2014) Nanomaterials, with their incredible size-dependent properties, are revolutionizing the way we think about materials and their performance. These tiny particles, measuring just billionths of a meter, exhibit unique characteristics that lead to enhanced mechanical properties far exceeding those of traditional materials. (Zou et al., 2006)

Three key factors contribute to the exceptional mechanical properties of nanomaterials:

- Volume effects: The high surface area-to-volume ratio of nanoparticles allows for stronger interactions between the particles and the matrix material, leading to improved strength and stiffness.
- Surface effects: The surface chemistry of nanoparticles can be tailored to optimize their interaction with the matrix, further enhancing the overall mechanical performance.
- Quantum effects: When dimensions approach the nanoscale, quantum phenomena come into play, altering the electronic and mechanical properties of the materials in significant ways.

One of the most impactful applications of these remarkable properties is in the realm of composites. By incorporating nanomaterials as fillers into various matrices, researchers are able to create materials with enhanced strength and stiffness. Adding nanoparticles like nano-SiO_2 to concrete significantly improves its compressive strength, bending strength, and splitting tensile strength. (Al Ghabban et al., 2018) Increased ductility and toughness: Nano-oil palm empty fruit string filler, when introduced into kenaf epoxy composites, leads to considerable improvements in tensile strength, elongation at break, and impact strength. Also, nanomaterials can act as reinforcements, effectively preventing microcracks and enhancing the material's ability to withstand wear and tear. (Saba et al., 2016)

These are just a few examples of how nanomaterials are revolutionizing the field of mechanics. With their potential to create lighter, stronger, and more durable materials, they hold immense promise across various industries, including aerospace, automotive, electronics, biomedical. The future of mechanical engineering is undoubtedly intertwined with the ongoing developments in nanotechnology. As research continues, we can expect even more innovative applications that will push the boundaries of what's possible and shape the world around us.

Nanomaterials containing metal nanoparticles exhibit superior fracture toughness and strength compared to pure Al_2O_3. This improvement can be attributed to the "pinning effect" of the nanoparticles. These particles restrict the growth of Al_2O_3 grains, resulting in a finer grain size and ultimately, a stronger and more fracture-resistant nanocomposite. Interestingly, the hardness of these nanocomposites depends on the specific metal incorporated. While nanocomposites with copper show slightly lower hardness than pure Al_2O_3 due to copper's inherent softness, the addition of nickel-cobalt nanoparticles leads to a hardness increase, reflecting the higher hardness of nickel-cobalt. Furthermore, hybrid composites containing both silicon carbide and boron carbide nanoparticles outperform single-reinforced composites. This synergy results in higher Vickers hardness, impact strength, and ultimate tensile strength due to the combined effect of the different nanoparticles. This highlights the potential of tailoring the mechanical properties of nanomaterials by strategically incorporating different types of nanoparticles, opening doors to the development of advanced materials for diverse applications. (Oh et al., 2006)

NANOTECHNOLOGICAL FRONTIERS FOR SENSING INNOVATIONS

Contaminated food leads to annual fatalities of 420,000 and 600 million cases of foodborne infections globally with 30% of these deaths occurring in children aged five or younger due to foodborne diseases. However, the conventional microbial culture methods identify a single specific bacterium but the process is protracted and yields limited data output. In contrast, nanotechnology-driven approaches present a more streamlined solution to detect pathogenic bacteria in intricate food products, delivering heightened sensitivity and specificity in comparison to conventional methods. (Lazcka et al., 2007)(Bhunia, 2008)

Amid the surge in chronic diseases including diabetes and cancer, there is a growing urgency to heighten awareness for the premature detection of these conditions. Biosensors play a pivotal role with their rapid response and heightened sensitivity in the initial stages of such diseases. The creation of nanoscale electrodes, a fundamental aspect of nanofabrication, involves the lithography technique. (Van Gerwen et al., 1998) Initially applied to develop glucose biosensors utilizing glucose oxidase, this technique progressed with the incorporation of platinum nanoparticles onto carbon nanotube layers, enhancing the detection of analytical materials. The sophisticated biosensors are exceptionally proficient not only in detecting glucose beyond blood but also possess applications in immune-sensors. The thin films layered above the sensing surface enable swift and improved identification of antigen–antibody complexes, expanding the sensors' capabilities for efficient analyte recognition. (Pak et al., 2001)

Furthermore, the aptasensor, a sensor wielding DNA or RNA aptamer as a biological recognition element, underwent stringent validation for sensitivity to detect *S. enteritidis and S. typhimurium* through the Polymerase Chain Reaction (PCR) technique. While PCR established a sensitivity limit of for both Salmonella strains, the aptasensor outshone this in practical application, showcasing an elevated sensitivity limit of 101 cfumL^{-1} when exposed to Salmonella-contaminated raw chicken samples. This underscores the unparalleled effectiveness of the electrode paired with an aptamer, surpassing the constraints of the PCR method. Crucially, the developed sensor exhibited exclusive specificity to Salmonella bacteria, presenting no reactivity toward non-Salmonella strains. In 2014, a groundbreaking biosensor emerged as researchers immobilized anti-Escherichia coli antibodies on a single-walled carbon nanotube-based joint biosensor, tailored for the detection of the *E. coli* K-12 strain. This sensor exhibited remarkable specificity and sensitivity, achieving a detection limit within an impressive 5-minute timeframe. (Yamada et al., 2014) Selective aptamer-based colorimetry was employed for the detection of acetamiprid, as demonstrated by Justino et al. in 2017. The detection mechanism involved sensitive target-induced color changes, driven by interparticle plasmon coupling within the gathered AuNPs, resulting in a low detection limit of 5 nM for acetamiprid. In the case of atrazine, an electrochemical method was utilized for detection, while bisphenol A was detected using an optical method. (C. Justino et al., 2017)

Notably, gold nanoparticles, frequently employed in nanotechnology, find stability in their synthesis through surfactants, with the choice of surfactant influencing various properties and preventing nanoparticle aggregation.(Sperling et al., 2008) Characterized by biocompatibility, significant conductivity, and a high surface-to-volume ratio, AuNPs stand out as a focal point in nanotechnology. (Guo & Wang, 2007) The redox activity of gold NPs contributes to their intriguing nature, particularly in improving the responsiveness of electrochemical biosensors to augment their detection capabilities. for foodborne pathogen analysis as shown in Figure 3. Integrating gold NPs onto electrochemical biosensors, coupled with ssDNA complementary to the microbial DNA under scrutiny, improves binding on the transducer surface, thereby amplifying the biosensor's sensitivity. (Y. Zhang et al., 2009) In another application, gold NPs are paired with redox enzymes that precisely catalyze the oxidation/reduction of analyte during reaction.

As the immobilized enzyme-gold NPs interact with the analyte, they elevate the current signal, providing confirmation of detection. Illustrating further innovation, a research team developed an advanced AuNP aptasensor utilizing Surface-enhanced Raman spectroscopy (SERS) for rapid detection of S. aureus and S. typhimurium in a deliberately contaminated pork sample. (H. Zhang et al., 2015)

Carbon quantum dots, minute artificial semiconductor particles boasting dimensions below 10 nanometers, stand as powerful tools in contemporary research of nanoparticles in nanotechnology. Renowned for their intense luminescence, exceptional solubility, and biocompatibility, carbon dots have become indispensable in the scientific realm. (Xu et al., 2004)(Sun et al., 2006) Researchers have harnessed the potential of carbon dots for detecting hazardous microbial cultures and toxins, including the likes of *S. typhimurium* and aflatoxin B1. Wang et al. engineered sophisticated carbon dot aptamer complexes that showcased remarkable prowess in quantitatively identifying *S. typhimurium* within eggshell and tap water

Figure 3. Schematic diagram showing construction of biosensor

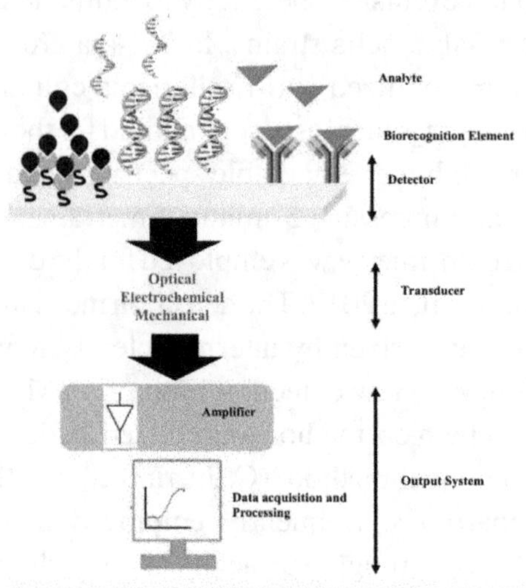

solutions. Operating within a dynamic test range of 103 to 105 cfu mL^{-1}, the limit of detection stood impressively at 50 cfu mL^{-1}, accomplishing this feat in a mere 2 hours (R. Wang et al., 2015). The developed sensor delivered results on par with the standard plate count method, demonstrating its reliability. Trials across diverse food settings have repeatedly affirmed the specificity of microbial identification using biosensors, wherein carbon dots, along with counterparts like carbon nanotubes, and semiconductors, have played pivotal roles. (B. Wang et al., 2016) (Zhao et al., 2009) (Mandal & Parvin, 2011) In 2019, Liu et al. innovatively devised a biosensor based on fluorescence resonance energy transfer, leveraging gold nanorods (NRs) and carbon dots for the identification of lead ions. (Liu et al., 2019)

In recent decades, magnetic nanoparticles have undergone remarkable advancements and found extensive applications in biosensors, either by application on the sensor surface or as labels, offering a promising alternative to fluorescent labels when attached to biomolecules. Their appeal extends to characteristics such as biocompatibility, stability, environmental safety, and cost-effectiveness. Notably, magnetic nanoparticles demonstrate an advantageous magnetic behavior, showing superparamagnetic traits due to a lower number of magnetic domains. This superparamagnetic property, exemplified by Fe_3O_4, contributes to its widespread use in biosensor development, further enhanced by its biocompatibility and ease of preparation. (Rocha-Santos, 2014) Despite its propensity for aggregation in biological liquids due to high surface area and magnetic dipole attraction, functionalization effectively addresses this issue. Various functionalized magnetic nanoparticles,

Figure 4. Schematic representation of application of nanomaterials in sensors [Au = Gold, C = Carbon, Si = Silicon, Mag. Nos = Magnetic nanoparticles, Gox = Graphene oxide, CPs = conducting polymers]

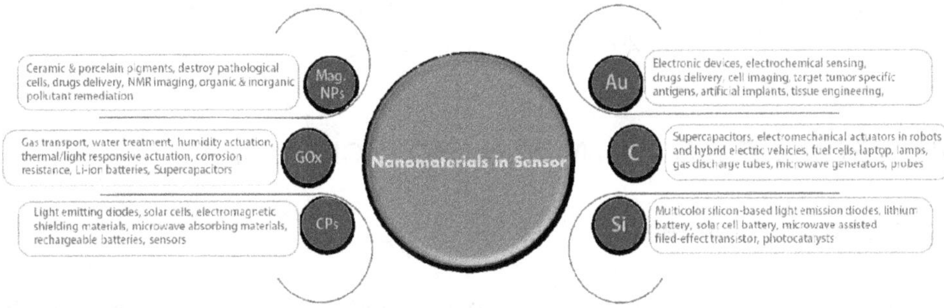

including core-shell Au- Fe_3O_4, core-shell Fe_3O_4-SiO_2, Au- Fe_3O_4 composite NPs, Fe_3O_4-SiO_2/MWCNTs, and Fe_3O_4 on graphene oxide, have been employed to modify carbon electrodes, creating biosensors capable of detecting organophosphorus pesticides. This strategic application of these nanoparticles showcases their versatility and efficacy in developing advanced biosensing technologies as shown in Figure 4.

Various methods have emerged for incorporating magnetic labels into the biosensing of molecules. These techniques employ magnetometers, Hall sensors, and magneto resistive sensors to detect magnetic particles. A particularly successful approach, based on magnetic resonance, utilizes magnetic nanoparticle sensors to accelerate the relaxation rate of surrounding water molecules. Known as Diagnostic Magnetic Resonance (DMR), this method employs two distinct approaches to induce changes in the proton relaxation rates through the binding of molecular targets using affinity molecule-conjugated magnetic nanoparticles (Saha et al., 2012).

Graphene-based nanomaterials, acting as pivotal transducers in biosensors, are at the forefront of cutting-edge research. Widely embraced for their role in converting targeted and receptor molecules into discernible measurements through EDC/NHS chemistry, graphene, in particular, stands as a dominant force in diverse biosensor designs. Its expansive surface area and adeptness at immobilizing diverse molecules, coupled with high electron transmission rates and electrical conductivity, make it an unrivaled choice. (Pumera, 2011) Beyond transduction, these graphene-based materials exhibit prowess as quenchers, crucial in the development of fluorescent transducer-based biosensors. (Rao et al., 2009) Noteworthy variants like reduced graphene oxide (rGO), graphene (G), and graphene oxide (GO) showcase exceptional fluorescent quenching efficiency. The sensor design, intricately woven with graphene elements, profoundly shapes the detection limits for targeted molecules. Meanwhile, the sensitivity and selectivity of biosensors are intricately woven with factors such as bioreceptor presence and the orientation of G, GO, or rGO sheets. Biosensor performance nuances arise from considerations like functional groups, graphene oxidation state, layer count, and the utilization of diverse derivatives, making graphene-based nanomaterials a dynamic frontier in biosensing exploration. (Kasry et al., 2012) (X. Wu et al., 2018) (Batır et al., 2018)

SHAPING THE WORLD WITH NANOTECHNOLOGY

Biomedical Applications

Various classes have been extensively documented in the literature, including but not limited to metal nanoparticles, dendrimers, liposomes, biodegradable polymers, carbon-based nanomaterials, hydrogel nanocomposites, and others. These classes not only support but also expand the applications and services in the medical domain through the principles of nanotechnology. (Jiang et al., 2021) Nanofibers play a crucial role in diverse applications such as wound dressing, medical fabrics, grafts, regenerative medicine, and artificial organ components. Scientists are advancing the field by developing intelligent bandages that are absorbed into the tissue upon wound healing. These intelligent bandages incorporate embedded nanofibers with functionalities such as coagulation, antibiotics, and even sensors for detecting infectious symptoms. (Jackson et al., 2017) An alternative avenue within nanotechnology involves the utilization of nanomachines to prepare for space colonization, aiming to construct structures and establish ecosystems on other planets. Concurrently, researchers are exploring ways to modify human physiology to enhance adaptation to the atmospheric conditions of different worlds. (Shi et al., 2010)

In healthcare, nanomedicine harnesses nanotechnology for the remedy and analysis of various ailments, employing nanoparticles, nanoelectronics biosensors, and molecular nanotechnology. This enables the precise examination of the human body, medications, and medical equipment at the nano level, ensuring a markedly elevated degree of precision in the field of medicine. The healthcare sector is leveraging this technology for diagnostics and advanced medical equipment. The rapid advancements in nanotechnology are leading to the development of novel diagnoses and therapies with improved success rates. Presently, nanomedicine is making strides in the creation of intelligent medications and advancements in cancer treatment. (L. Zhang et al., 2021; Y. Zhang et al., 2009) Drug delivery systems based on nanotechnology typically involve nanoparticles encapsulating one or more therapeutic drugs, capable of binding or scattering within adsorbed polymer matrices. Recent years have witnessed notable advancements in the production of nano-drugs, incorporating imaging, treatments, and diagnostics. The core objectives of nano-drug systems are centered around enhancing the bioavailability of targeted tissue delivery, prolonging the half-life of injectable medications, and facilitating oral administration of medicinal products. Administered at lower doses, nano drugs demonstrate remarkable improvements in pharmacological effects, mitigating health risks and minimizing adverse effects. In the realm of advanced medical methodologies, nanotechnologies exhibit a prowess on par with their reactive counterparts. Tailored wearable monitors hold the potential to transmit intricate data directly to hospital systems, streamlining elderly patient care, particularly in remote locales. Additionally, nanomaterials, once associated with metastatic malignancies, now offer a means to intricately control circulating tumor cells. (Kai, 2006) The technological industry, delving into the realm of individual atoms and molecules, particularly within Nanoscience, unfolds futuristic implications for the realms of science, engineering, and technology. Its profound impact resonates in the meticulous development, characterization, and application of specific atoms and molecules. Across diverse sectors, encompassing many important fields among others, nanotechnology stands as a catalyst for transformative breakthroughs. (Huang et al., 2021)

In the realm of healthcare, the integration of nanotechnology holds tremendous potential for advancing the diagnosis, treatment, and prevention of various illnesses. There is a growing interest among innovators in exploring the health applications of nanotechnology, promising to propel the industry into a new era of growth. Unlike current diagnostic and therapeutic methods reliant on clinical expertise and the analysis of external biometric information, future research is anticipated to be directly informed by nanotechnology data localized at the site of the disease. (Neelakantan P, 2014). This technology empowers our bodies with artificial mechanics, facilitating the development of proactive, preventive, and personalized drugs from within. One prominent example frequently cited is the use of nanomaterials, where the nanometer-scale manipulation of devices and systems opens up numerous opportunities for more precise testing and the diagnosis and treatment of illnesses. (Keller, 2007) (Editorial et al., 2019)

In the realm of nanomedicines, the predominant shape observed in clinical trials or lab-scale studies has been spherical. Nevertheless, the last two decades have witnessed significant strides in our comprehension of inorganic nanoparticle growth and the synthesis of non-spherical shapes. (Xia et al., 2009) This evolution is underpinned by advancements in self-assembly strategies, fostering the creation of anisotropic polymeric nano- and micro-particles. These innovative shapes are now making notable contributions across various domains, including drug delivery, imaging, therapy, and diagnostics. (Williford et al., 2015) The remarkable progress in shaping inorganic nanoparticles has resulted in the generation of diverse forms such as rods, 2D prisms, and branched structures. The conventional method for producing these materials involves wet chemical syntheses, particularly for metals and their oxides. Conversely, polymeric particles often necessitate templates or post-synthesis manipulation to modify their shapes. For a more in-depth exploration of synthetic routes for shape control in both inorganic and organic nanoparticles, we recommend referring to several comprehensive reviews available in the literature. (Tan et al., 2011)

The appealing and often superior physicochemical characteristics of nanoparticles (NPs) stem from their diminutive size, with various shapes frequently influencing these properties positively. However, the impact on material performance when interacting with biological interfaces and entities is not always clear. (Michalet et al., 2005) Nature, offering a diverse array of shapes, includes examples like the spherical HIV virus, rod-shaped tobacco mosaic viruses, and star-shaped bacteria. Bacterial morphology, with its known evolutionary advantages, is shaped by specific mechanisms such as surface interaction, passive diffusion, and active motility. (Dykman & Khlebtsov, 2014) In the context of Darwinian evolution, the form precedes function, driven by random mutations and natural selection. Bacteria and viruses exhibit these diverse shapes for specific and advantageous reasons. As scientists, we have the responsibility to shape future nanoparticles in medicine based on a form that aligns with their intended function. Currently, the predominant spherical shape of the majority of systems is attributed to both the simplicity of synthesis and a limited understanding of their structure-activity relationships. (Wanger et al., 2008)

Another crucial aspect is the heightened affinity of bacteria to substrate, facilitated by its expansive surface areas. Intuitively, filaments offer a larger surface area in contact with a surface compared to spheres, ensuring that detachment at one point does not necessarily result in complete removal. Notably, Streptococcus pneumoniae has demonstrated stronger adhesion to A549 lung epithelial cells in vitro and in vivo when forming chains. This suggests that morphological heterogeneity may play a role in promoting colonization of the upper respiratory tract. Even the typically spherical Staphylococcus aureus has been observed to form filamentous biofilm streamers under flow conditions. These observations underscore two significant evolutionary advantages of filamentous forms: evading elimination by the innate immune system and heightened surface adherence. Both properties hold promising implications for the design of next-generation nanomedicines.(Rodriguez et al., 2012)

The burgeoning field of utilizing plasmonic nanoparticles as the foundation for cutting-edge, innovative, and highly sensitive COVID-19 detection tools is rapidly gaining prominence. These novel biosensors hold the potential to evolve into cost-effective and reliable options for future COVID-19 diagnostics. (Gharpure & Ankamwar, 123 C.E.) Numerous ongoing clinical trials are evaluating real-time biosensor platforms leveraging the unique properties of plasmonic nanoparticles, with gold, copper, and silver nanoparticles serving as fundamental components. ·(Solanki et al., 2023)Diverse nanomaterials are being harnessed to create detection

Table 1. Nanotechnology advancements in medical domain

Application	Characteristics	Features
Wound treatment	Nanoscale Precision	Utilization of Nanogenerators
	Overcoming Natural Barriers	Precision Solutions
	Specialized targeted Medications	Tissue Identification
Antibacterial treatments	development of biocompatible materials	Multidisciplinary Applications
	Nanoscale Agents	improved cleaning of medical instruments
	Comprehensive Therapeutics	Progress in Illness Detection
Minimise damage of healthy cells	Targeting Precision	Targeted Drug Delivery
	Therapeutic Advancements	Improved Patient Results
	Therapeutic Accuracy	Accurate Heat Treatment
	Tumor-Specific Approaches	Tumour-Attracted Nanotubes
Nanomedicine Diagnostics	Precision in Disease Diagnosis	Early Disease Identification
	Early Intervention and Cost-Effective Approaches	Submicrometer Instruments
	Accelerated Drug Development	Quick and Inexpensive Detection
	Mimicking Natural Tissues	Quality of Life Improvement

platforms that offer rapid and precise identification of the SARS-CoV-2 virus. These platforms include colorimetric assays utilizing nanoparticles, cutting-edge optical sensors, nano immune-sensors tailored for COVID-19, innovative immune-chromatographic assays, aptamer-based COVID-19 assays, and bio-barcode assays. (Yakoubi et al., 2023) The various classical associated features and characteristics of nanotechnology for the medical domain are discussed in table 1.

Over the past three years, the impact of nanotechnology, nanoparticles, and nanomedicine in the realm of prevention and the development of COVID-19 vaccines has been extraordinary. (Rauf et al., 2022)The FDA has granted authorization for the use of diverse nano-formulations, encompassing nano-crystals, polymer-based nanoparticles, lipid nanoparticles, as well as non-polymer and metallic, inorganic, and protein-based nanoparticles for COVID-19 vaccine and drug delivery applications. (D. Yang, 2021) (Chauhan et al., 2020) Currently, a range of COVID-19 vaccines has been formulated to combat the spread of the SARS-CoV-2 virus. Nanoparticle-based vaccines are categorised into two types based on their antigen loading protocols. The first type involves antigen-presenting nanoparticles attached to the surface, while the second type encompasses nanoparticles encapsulated onto the vaccine antigen. (Jesús De La et al., 2023) COVID-19 vaccines, including the mRNA vaccine by BioNTech & Pfizer, the lipid-nanoparticles-based vaccines by Moderna (already in clinical use), and experimental nanoparticles-based nasal vaccines, are currently progressing through different phases of clinical trials. (Dosio et al., 2023) (Krishnani et al., 2023)

Personal Protective Equipment (PPE) masks are categorized into three groups based on their filter types: disposable one-use masks, respirator masks, and surgical masks. (Ayan et al., 2023)The ideal mask standards involve the efficient filtration of bio-aerosols while ensuring proper comfort for the wearer. (Mallakpour et al., 2023) When designing a PPE mask, consideration must be given to various factors, including humidity levels, ambient temperature, airflow patterns, and, most importantly, the properties of the material used, as all these elements significantly impact the mask's filtration effectiveness and quality. (López et al., 2023) In the ongoing COVID-19 pandemic, the integration of nanomaterials with selected textiles has provided effective options for the construction of high-quality masks. (Papi et al., 2023)

In conclusion, the future of personalized medicine is intricately linked to the indispensable role that nanomedicine will play, encompassing prediction through monitoring. Nanoscale materials form the foundation for highly sensitive sensors and biomarkers, enabling the early and precise identification of various illnesses. Nanomedicine excels in providing precise mapping of diseases with enhanced targeting and chemical sensitivity. Post-diagnosis, nanomedicine proves to be more efficient in targeting cells while minimizing side effects and preserving healthy cells. Ongoing utilization of products like nano-encapsulated doxorubicin exemplifies these advancements. The evolving landscape includes challenges in drug loading, release, and the continued exploration of metallic nanoparticles for diagnosis and treatment. To harness the full potential, rigorous controls, toxicity evaluations, and comprehensive clinical studies are imperative, ensuring the responsible integration of nanomedicine into future medical practices. Despite its promising potential, careful consideration of risks and thorough evaluation are crucial, emphasizing the need for stringent regulatory measures and extensive clinical assessments before widespread implementation. Moving forward, nanotechnology holds the promise of ground-level issue detection, offering a paradigm shift from reliance on external sensors, medical expertise, and probabilistic diagnostic algorithms. A surgical nanorobot, under the guidance or programming of a human surgeon, has the potential to function as a semi-autonomous on-site surgeon within the human body. It could be introduced through the vascular system or cavities and carry out diverse tasks, including searching for pathology, diagnosing, and correcting lesions through nanomanipulation.(Freitas, 2005) The nanorobot would operate with coordination from an onboard computer and maintain communication with the supervising surgeon through coded ultrasound signals. Exploration into early iterations of cellular nano surgery is currently underway. One instance involves the use of a rapidly vibrating micropipette with a tip diameter of less than 1 μm, operating at 100 Hz. This approach successfully severed dendrites from individual neurons without compromising cell viability. (Kirson & Yaari, 2000) Another method employed femtosecond laser surgery for axotomy in roundworm neurons, leading to functional regeneration of axons. The femtolaser, akin to nanoscissors, vaporizes tissue selectively, leaving adjacent areas unaffected. This precise surgical technique has even been applied to individual chromosomes, showcasing its versatile applications. (Uchugonova et al., 2012)

Application in Consumer Products

The semiconductor industry extensively utilizes nanotechnology in consumer electronics. Nanoscale components, like nanochips, enable the creation of faster and more energy-efficient devices. These advancements contribute to the development of smaller and more powerful gadgets such as smartphones, laptops, and other electronic devices that consumers use on a daily basis. The demand for larger and brighter displays in recent years has led to the increased prevalence of nanoparticles in display technology. These displays, commonly found in computer monitors and televisions, utilize materials like nanocrystalline lead telluride, cadmium sulphide, zinc selenide, and sulphide in the production of light-emitting diodes (LEDs). Additionally, the growing popularity of portable consumer devices such as mobile phones and laptops has created a need for small and lightweight batteries with high capacity. In this context, nanoparticles emerge as the preferred choice for materials used as separator plates in batteries (Teng et al., 2008).

Conventional sunscreens, designed to shield against ultraviolet (UV) rays, exhibit instability with prolonged use. The integration of nanoparticles, specifically titanium dioxide, into sunscreen offers several advantages. Titanium oxide and zinc oxide nanoparticles are chosen in certain formulations for their UV protection capabilities. Notably, both titanium oxide and zinc oxide nanoparticles are transparent to visible light while effectively absorbing and reflecting UV rays. Additionally, certain lipsticks incorporate iron oxide nanoparticles as pigments. (Wiechers & Musee, 2010) Contemporary sun protection products in the cosmetic industry utilize a combination of organic and inorganic sun filters, the selection of which is contingent upon the intended product and its application. Organic sun filters are preferred for their ease of formulation and high efficacy at low concentrations. However, achieving high Sun Protection Factors (SPF) often requires the combination of various sun filters, leading to the necessity of complex formulations. This is especially challenging with solid filters that resist easy solubilization. Some organic sun filters may degrade upon exposure to UV light. In contrast, inorganic sun filters exhibit photostability, and a high SPF can be attained using a single filter, such as titanium dioxide (TiO_2).

Some consumer clothing items are designed with nanotechnology to enhance their properties. Nanoparticles, such as silver nanoparticles, can be embedded in fabrics to impart antibacterial and odor-resistant properties. This is particularly useful in sportswear and undergarments, where maintaining hygiene and preventing odor is crucial. A study reports the development of versatile luminescent textiles by introducing novel iridium-supported silica (Ir-SiO_2) compounds in a coating application for the first time. The treated textiles exhibited promising self-marking attributes and demonstrated antibacterial activity. The effective interaction between Ir complexes and silica nanoparticles facilitated easy detection under UV light, both in textile coating and post-release in the environment. The uniform distribution of this luminescent coating can serve as a valuable tool for wear diagnostics. Furthermore, antibacterial tests revealed compelling light-induced antimicrobial properties against both Gram-positive (*Deinococcus radiodurans*) and Gram-negative (*Escherichia coli*) bacterial systems, likely attributed to the generation of reactive oxygen species (ROS). For practical applications, the Ir-SiO_2 coating, considered as a safety by design solution, can undergo a comprehensive human dermal exposure assessment throughout its entire life cycle, from production to nanoparticle release. (Zanoni et al., 2020) Despite their potential, Ag-doped TiO_2 photocatalysts suffer from instability due to Ag oxidation, limiting their long-term performance. To address this challenge, researchers have explored co-doping with nonmetals like nitrogen, creating synergistic effects. Ag doping further narrowed the band gap, enhancing visible light absorption. This synergistic effect resulted in a stable, efficient photocatalyst. (Bezerra et al., 2017) Furthermore, tri-doped anatase TiO_2 with In, Ag, and N demonstrated improved visible light absorption, highlighting the potential of multi-element doping strategies for further enhancing photocatalytic performance. (Khan et al., 2014)

Enhancing both driving safety and environmental sustainability involves the utilization of numerous nanodevices and advanced materials, as depicted in Figure 5. Autonomous or intelligent driving, prioritizing safety, can be actualized through the incorporation of sensors, radars, real-time processors featuring AI chips, and seamless intervehicle and road-to-vehicle communication. A reduction in the environmental footprint of driving is facilitated by employing high-power semiconductor devices utilizing wide band gap materials, advanced motors incorporating rare earth magnets, next-generation lithium (Li)-ion batteries, and fuel cells powered by green hydrogen. Additionally, the integration of high-strength, lightweight materials like carbon-fiber-reinforced polymers (CFRP) contributes significantly to minimizing the overall environmental impact of vehicular transportation.

Figure 5. Challenges and opportunities in sustainable growth of medicine and health care

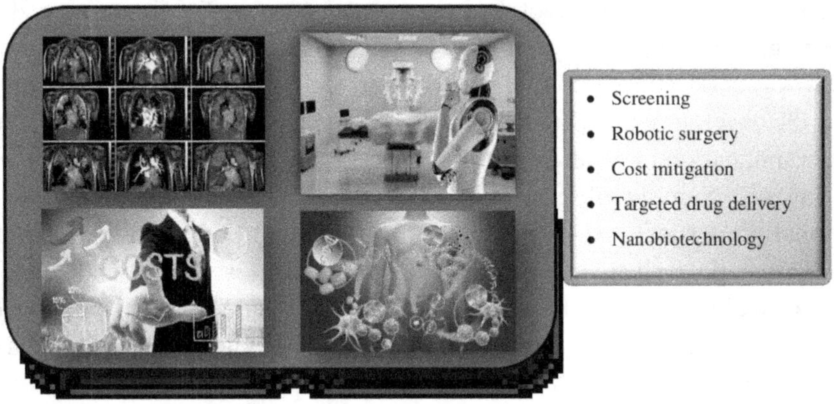

REVOLUTIONIZING NANOMATERIALS FOR A SUSTAINABLE FUTURE

The intersection of nanotechnology with materials science holds tremendous promise to lead into a modern and sustainable world. Nanotechnology has the potential to drive the digital transformation by providing diverse nano-devices to be used such as IoT sensors, autonomous driving vehicles, and smart robots. The fusion of nanotechnology with materials science promises not just a technological evolution but a revolution towards sustainability, addressing urgent global challenges like water purification, CO_2 emissions, and efficient material circulation through recycling.

The pursuit of a sustainable future transcends mere scientific exploration and demands a commitment to responsible innovation. As we unlock the potential to redefine industries from aerospace to medical applications, the lightweight yet robust materials emerge as catalysts for transformative innovation and environmental sustainability. The eco-friendly alternatives, envision water purification processes enhanced by biodegradable polymers reinforced with nanoparticles, showcasing unparalleled filtration and adsorption capabilities. Yet, the scalability, cost-effectiveness, and the potential environmental and health impacts of nanoparticles beckon our rigorous attention. It is in overcoming these challenges that our collective efforts will guide nanocomposites towards a future that is responsible and sustainable. Overcoming these challenges is not just a scientific endeavor; it is the cornerstone for ushering in a revolutionary society, where nanotechnology realizes its full potential for a sustainable and harmonious future.

The lightweight yet robust nature of nanocomposites has reshaped aerospace and automotive manufacturing, addressing challenges like fuel efficiency. In the medical field, nanocomposites play a pivotal role in drug delivery, medical imaging, and tissue engineering, offering precise control over drug release and innovative approaches to tissue regeneration. Additionally, environmental sustainability is championed through the development of biodegradable polymers reinforced with nanoparticles and their application in water purification processes. (Schrand et al., 2010)

Taking an example of graphene, a fundamental building block, with its exceptional properties, opens new vistas for diverse carbon allotropes. Graphene oxide, its versatile derivative has contributed to materials science, nanotechnology, and biomedical research with its hydrophilic nature and enhanced processability and thus, a driving forces of innovation and progress. Graphene oxide actuators, leverage the unique properties of graphene oxide, respond to changes in humidity or temperature by absorbing and releasing water molecules. This process induces controlled mechanical deformation, enabling precise movement ideal for micro- and nano-scale devices such as micro-robots and sensors. Researchers are lately exploring their potential in various applications due to their responsive movement capabilities. The distinctive properties and controllable actuation behavior position these actuators as potential game-changers in fields like robotics, biomedical devices, and micro-electromechanical systems. Despite their promising attributes, researchers are actively working to overcome challenges linked to long-term stability and hysteresis, aiming to unlock the full potential of graphene oxide actuators in practical applications.

Nanotechnology has revolutionized our world, bringing countless benefits to society. From medicine and electronics to energy and environmental sustainability, its impact is undeniable. Moreover, an abundance of undiscovered possibilities awaits exploration, showcasing the expansive potential of this transformative field.

CHALLENGES AND OPPORTUNITIES IN NANOTECHNOLOGY

Nanocomposites despite their myriad advantages, confront certain challenges that necessitate attention and resolution. One prominent challenge revolves around the cost of production. The intricate processes involved in creating nanocomposites render them expensive, thereby limiting their widespread use in various applications. Another crucial hurdle is the reproducibility of these materials. Ensuring consistent properties across different batches of nanocomposites proves challenging, demanding innovative approaches to overcome this variability.

Nanotechnology encounters various technical challenges, shaping the future of revolutionary societal advancements. These include innovations in computing with

IoT edge, AI chips, and quantum devices, ensuring safe and environmentally friendly transportation, advancing nanobiotechnology for health and medical care, integrating service robots into daily life, developing smart materials for sustainable water, air, and materials, and creating energy materials and devices for renewable energy production and conservation.(Payal & Pandey, 2021) A central issue in nanomaterial challenges revolves around the profound impact of defects on performance. Carbon nanotubes, known for their remarkable strength, face potential setbacks due to impurities, discontinuous lengths, defects, and random orientations. Balancing the exploitation of their unique properties while mitigating the influence of defects becomes crucial for their tensile strength.

Cost-effective synthesis of nanomaterials poses a significant hurdle, especially in 2D nanomaterial production where current methods yield subpar quality with inherent defects. Overcoming this challenge requires innovative synthesis methods striking a delicate balance between cost-effectiveness and quality, addressing issues like chiral selectivity, conductivity, and precise control over diameters in carbon nanotube synthesis. These challenges underscore the need for continued research and development to unlock the transformative potential of nanotechnology. Nanotechnology faces problems particularly in pharmaceuticals, due to high costs, uncertainties about the health and environmental impacts of nanoparticles, and concerns regarding their short- and long-term effects on the body (Figure 5).

These factors contribute to investor hesitancy, underscoring the urgent need to address obstacles hindering the effective implementation of nanotherapeutics and the realization of nanotechnology's full potential. (Q. Wu et al., 2020)Challenges in reproducibility, quality control of nanomaterials, scalability, and managing undesired nano-engineering by-products present formidable hurdles, significantly limiting the broader utilization of nanotherapeutics.

Furthermore, the fusion of information and communication technologies, encompassing ubiquitous sensors, IoT, data analytics, machine learning, and AI-enabled interventions, facilitate transition towards supporting a circular economy. While this paradigm shift holds promise, it introduces crucial research challenges for progressing nanotechnologies. The lack of quantification regarding the impacts of circular economy on nanomaterials, manufacturing, and economies underscores the need for standardized measurement methods and best-practice principles. Additionally, developing robust decision support tools is imperative for optimizing product design and manufacturing processes, proactively utilizing AI-based green chemistry to enhance sustainable practices.

CONCLUSION

The challenges inherent in nanocomposites and nanotechnology at large are intricate and multifaceted. However, through persistent research, innovative solutions, and unwavering commitment, the transformative potential of nanomaterials and nanotechnology can be fully harnessed. Anticipating ground-breaking applications, continued advancements in science and technology are poised to reshape industries significantly.

In confronting these challenges, the research community plays a pivotal role in not only identifying the gaps but also in devising innovative solutions. By standardizing measurement methods, establishing best practices, and integrating AI into the decision-making process, nanotechnologies can be harnessed more effectively to contribute to a circular economy, minimizing environmental impact and maximizing sustainability. A continuous stream of research and development endeavours is diligently addressing these obstacles, fostering the emergence of new and inventive applications for these materials.

REFERENCES

Al Ghabban, A., Al Zubaidi, A. B., Jafar, M., & Fakhri, Z. (2018). Effect of Nano SiO2 and Nano CaCO3 on the Mechanical Properties, Durability and flowability of Concrete. *IOP Conference Series. Materials Science and Engineering, 454*(1), 012016. doi:10.1088/1757-899X/454/1/012016

Ayan, S., Aranci-Ciftci, K., Ciftci, F., & Ustundag, C. B. (2023). Nanotechnology and COVID-19: Prevention, diagnosis, vaccine, and treatment strategies. In Frontiers in Materials (Vol. 9). Frontiers Media S.A. doi:10.3389/fmats.2022.1059184

Banin, U., Waiskopf, N., Hammarström, L., Boschloo, G., Freitag, M., Johansson, E. M. J., Sá, J., Tian, H., Johnston, M. B., Herz, L. M., Milot, R. L., Kanatzidis, M. G., Ke, W., Spanopoulos, I., Kohlstedt, K. L., Schatz, G. C., Lewis, N., Meyer, T., Nozik, A. J., & Brudvig, G. W. (2020). Nanotechnology for catalysis and solar energy conversion. In Nanotechnology, 32(4). IOP Publishing Ltd. doi:10.1088/1361-6528/abbce8

Barber, D. J., & Freestone, I. C. (1990). *An Investigation Of The Origin Of The Colour Of The Lycurgus Cup By Analytical Transmission Electron Microscopy (Vol. 32).*

Batır, G. G., Arık, M., Caldıran, Z., Turut, A., & Aydogan, S. (2018). Synthesis and Characterization of Reduced Graphene Oxide/Rhodamine 101 (rGO-Rh101) Nanocomposites and Their Heterojunction Performance in rGO-Rh101/p-Si Device Configuration. *Journal of Electronic Materials*, *47*(1), 329–336. doi:10.1007/s11664-017-5758-4

Bezerra, P. C. S., Cavalcante, R. P., Garcia, A., Wender, H., Martines, M. A. U., Casagrande, G. A., Giménez, J., Marco, P., Oliveira, S. C., & Machulek, A. Jr. (2017). Synthesis, characterization, and photocatalytic activity of pure and N-, B-, or Ag-Doped TiO2. *Journal of the Brazilian Chemical Society*, *28*(9), 1788–1802. doi:10.21577/0103-5053.20170040

Bhunia, A. K. (2008). Biosensors and Bio-Based Methods for the Separation and Detection of Foodborne Pathogens. In Advances in Food and Nutrition Research (Vol. 54, pp. 1–44). Academic Press Inc. doi:10.1016/S1043-4526(07)00001-0

Binnig, G., Rohrer, H., Gerber, C., & Weibel, E. (1982). Tunneling through a controllable vacuum gap. *Applied Physics Letters*, *40*(2), 178–180. doi:10.1063/1.92999

Chauhan, G., Madou, M. J., Kalra, S., Chopra, V., Ghosh, D., & Martinez-Chapa, S. O. (2020). Nanotechnology for COVID-19: Therapeutics and Vaccine Research. *ACS Nano*, *14*(7), 7760–7782. doi:10.1021/acsnano.0c04006 PMID:32571007

Chen, M., He, Y., Ye, Q., Wang, X., & Hu, Y. (2019). Shape-dependent solar thermal conversion properties of plasmonic Au nanoparticles under different light filter conditions. *Solar Energy*, *182*, 340–347. doi:10.1016/j.solener.2019.02.070

Chis, A. A., Dobrea, C., Morgovan, C., Arseniu, A. M., Rus, L. L., Butuca, A., Juncan, A. M., Totan, M., Vonica-Tincu, A. L., Cormos, G., Muntean, A. C., Muresan, M. L., Gligor, F. G., & Frum, A. (2020). Applications and Limitations of Dendrimers in Biomedicine. *Molecules (Basel, Switzerland)*, *25*(17), 3982. doi:10.3390/molecules25173982 PMID:32882920

Chis, A. A., Dobrea, C., Morgovan, C., Arseniu, A. M., Rus, L. L., Butuca, A., Juncan, A. M., Totan, M., Vonica-Tincu, A. L., Cormos, G., Muntean, A. C., Muresan, M. L., Gligor, F. G., & Frum, A. (2020). Applications and Limitations of Dendrimers in Biomedicine. *Molecules (Basel, Switzerland)*, *25*(17), 3982. doi:10.3390/molecules25173982 PMID:32882920

Dosio, F., Huang, Y.-F., Kumar Prabhakar, P., Khurana, N., Vyas, M., Sharma, V., El-Saber Batiha, G., Kaur, H., Singh, J., Kumar, D., Sharma, N., Kaushik, A., & Kumar, R. (2023). Aspects of Nanotechnology for COVID-19 Vaccine Development and Its Delivery Applications. *Pharmaceuticals.* doi:10.3390/pharmaceutics15020451

Dubey, S. K., Dey, A., Singhvi, G., Pandey, M. M., Singh, V., & Kesharwani, P. (2022). Emerging trends of nanotechnology in advanced cosmetics. *Colloids and Surfaces. B, Biointerfaces*, *214*, 112440. doi:10.1016/j.colsurfb.2022.112440 PMID:35344873

Dykman, L. A., & Khlebtsov, N. G. (2014). Uptake of engineered gold nanoparticles into mammalian cells. In Chemical Reviews, 114(2), 1258–1288. doi:10.1021/cr300441a

Editorial, G., Cheon, J., Chan, W., Zuhorn, I., & Authors, C. (2019). The Future of Nanotechnology: Cross-disciplined Progress to Improve Health and Medicine. *Accounts of Chemical Research*, *52*(9), 2405. doi:10.1021/acs.accounts.9b00423 PMID:31525881

Erhardt, D. (2003). NEWS & VIEWS. *Nature*. www.nature.com/naturematerials

Ferreira, M. A. M., & Filipe, J. A. (2018). Nanotechnology Applications – The Future Arrived Suddenly. In Computational Approaches in Biomedical Nano-Engineering (pp. 23–41). Wiley. doi:10.1002/9783527344758.ch2

Freitas, R. A. (2005). Nanotechnology, nanomedicine and nanosurgery. In International Journal of Surgery, 3(4). doi:10.1016/j.ijsu.2005.10.007

Gharpure, S., & Ankamwar, B. (123 C.E.). Use of nanotechnology in combating coronavirus. *3 Biotech, 11*, 358. doi:10.1007/s13205-021-02905-6

González-Pedroza, M. G., Benítez, A. R. T., Navarro-Marchal, S. A., Martínez-Martínez, E., Marchal, J. A., Boulaiz, H., & Morales-Luckie, R. A. (2023). Biogeneration of silver nanoparticles from Cuphea procumbens for biomedical and environmental applications. *Scientific Reports*, *13*(1), 790. doi:10.1038/s41598-022-26818-3 PMID:36646714

Guo, S., & Wang, E. (2007). Synthesis and electrochemical applications of gold nanoparticles. *Analytica Chimica Acta*, *598*(2), 181–192. doi:10.1016/j.aca.2007.07.054 PMID:17719891

Huang, J., Yang, B., Peng, Y., Huang, J., Wong, S. H. D., Bian, L., Zhu, K., Shuai, X., & Han, S. (2021). Nanomedicine-Boosting Tumor Immunogenicity for Enhanced Immunotherapy. In Advanced Functional Materials, 31(21). John Wiley and Sons Inc. doi:10.1002/adfm.202011171

Hussain, A., Attique, F., Naqvi, S. A. R., Ali, A., Ibrahim, M., Hussain, H., Zafar, F., Iqbal, R. S., Ayub, M. A., Assiri, M. A., Imran, M., & Ullah, S. (2022). Nanoformulation of Curcuma longa Root Extract and Evaluation of Its Dissolution Potential. *ACS Omega*. doi:10.1021/acsomega.2c06258 PMID:36643543

Iravani, M., Khalilnezhad, Z., & Khalilnezhad, A. (2023). A review on application of nanoparticles for EOR purposes: History and current challenges. *Journal of Petroleum Exploration and Production Technology*, *13*(4), 959–994. doi:10.1007/s13202-022-01606-x PMID:36644438

Jackson, T. C., Patani, B. O., & Ekpa, D. E. (2017). Nanotechnology in Diagnosis: A Review. *Advances in Nanoparticles*, *06*(03), 93–102. doi:10.4236/anp.2017.63008

Jamaluddin, N. A., Talib, N., & Sani, A. S. A. (2021). Performance comparative of modified jatropha based nanofluids in orthogonal cutting process. *Evergreen*, *8*(2), 461–468. doi:10.5109/4480729

Jesús De La, J., Bayo, T., Ángeles Martín-Lara, M., Calero Hoces, M., Miguel, P., Castillo, S., Pula, H. J., & Zamorano, M. (2023). *Management of Used COVID-19 Personal Protective Equipment: A Bibliometric Analysis and Literature Review*. doi:10.3390/app13042377

Jia, Y., Jiang, Y., He, Y., Zhang, W., Zou, J., Magar, K. T., Boucetta, H., Teng, C., & He, W. (2023). Approved Nanomedicine against Diseases. In Pharmaceutics, 15(3). MDPI. doi:10.3390/pharmaceutics15030774

Jiang, W., Wang, Y., Wargo, J. A., Lang, F. F., & Kim, B. Y. S. (2021). Considerations for designing preclinical cancer immune nanomedicine studies. *Nature Nanotechnology*, *16*(1), 6–15. doi:10.1038/s41565-020-00817-9 PMID:33349682

Juanola-Feliu, E., Miribel-Català, P., Avilés, C., Colomer-Farrarons, J., González-Piñero, M., & Samitier, J. (2014). Design of a Customized Multipurpose Nano-Enabled Implantable System for In-Vivo Theranostics. *Sensors (Basel)*, *14*(10), 19275–19306. doi:10.3390/s141019275 PMID:25325336

Justino, C., Duarte, A., & Rocha-Santos, T. (2017). Recent Progress in Biosensors for Environmental Monitoring: A Review. *Sensors (Basel)*, *17*(12), 2918. doi:10.3390/s17122918 PMID:29244756

Kai, K. (2008). *Nanotechnology and Medical Robotics*. Legal and Ethical Responsibility.

Kasry, A., Ardakani, A. A., Tulevski, G. S., Menges, B., Copel, M., & Vyklicky, L. (2012). Highly efficient fluorescence quenching with graphene. In Journal of Physical Chemistry C, 116(4), 2858–2862. doi:10.1021/jp207972f

Keller, K. H. (2007). Nanotechnology and society. *Journal of Nanoparticle Research*, *9*(1), 5–10. doi:10.1007/s11051-006-9193-3

Khan, M., Cao, W., Li, J., Zaman, M. I., & Manan, A. (2014). Density functional theory calculations for the investigation of (Ag, N) codoping effect on the electronic and optical properties of anatase TiO 2. *International Journal of Modern Physics B*, *28*(18), 1450112. doi:10.1142/S0217979214501124

Kiessling, F., Liu, Z., & Gätjens, J. (2010). Advanced nanomaterials in multimodal imaging: Design, functionalization, and biomedical applications. *Journal of Nanomaterials*, *2010*, 1–15. doi:10.1155/2010/894303

Kinnear, C., Moore, T. L., Rodriguez-Lorenzo, L., Rothen-Rutishauser, B., & Petri-Fink, A. (2017). Form Follows Function: Nanoparticle Shape and Its Implications for Nanomedicine. In Chemical Reviews, 117(17), 11476–11521. American Chemical Society. doi:10.1021/acs.chemrev.7b00194

Kirson, E. D., & Yaari, Y. (2000). A novel technique for micro-dissection of neuronal processes. *Journal of Neuroscience Methods*, *98*(2), 119–122. doi:10.1016/S0165-0270(00)00194-1 PMID:10880825

Kontogiannis, O., Selianitis, D., Lagopati, N., Pippa, N., Pispas, S., & Gazouli, M. (2023). Surfactant and Block Copolymer Nanostructures: From Design and Development to Nanomedicine Preclinical Studies. In Pharmaceutics (Vol. 15, Issue 2). MDPI. doi:10.3390/pharmaceutics15020501

Krishnani, S., Yadav, R., Rishi, N., & Goel, A. (2023). Microbial nanobionics: future perspectives and innovative approach to nanotechnology. *Environmental Applications of Microbial Nanotechnology: Emerging Trends in Environmental Remediation*, 17–32. doi:10.1016/B978-0-323-91744-5.00022-9

Lazcka, O., Del Campo, F. J., & Muñoz, F. X. (2007). Pathogen detection: A perspective of traditional methods and biosensors. In Biosensors and Bioelectronics (Vol. 22, Issue 7, pp. 1205–1217). doi:10.1016/j.bios.2006.06.036

Liu, G., Feng, D. Q., Qian, Y., Wang, W., & Zhu, J. J. (2019). Construction of FRET biosensor for off-on detection of lead ions based on carbon dots and gold nanorods. *Talanta*, *201*, 90–95. doi:10.1016/j.talanta.2019.03.101 PMID:31122466

López, A. D. F., De-la-Torre, G. E., Fernández Severini, M. D., Prieto, G., Brugnoni, L. I., Colombo, C. V., Dioses-Salinas, D. C., Rimondino, G. N., & Spetter, C. V. (2023). Chemical-analytical characterization and leaching of heavy metals associated with nanoparticles and microplastics from commercial face masks and the abundance of personal protective equipment (PPE) waste in three metropolitan cities of South America. *Marine Pollution Bulletin*, *191*, 114997. doi:10.1016/j.marpolbul.2023.114997 PMID:37148588

Mallakpour, S., Behranvand, V., & Hussain, C. M. (2023). Worldwide fight against COVID-19 using nanotechnology, polymer science, and 3D printing technology. *Polymer Bulletin*, *80*(1), 165–183. doi:10.1007/s00289-021-04006-z PMID:35106016

Mandal, T. K., & Parvin, N. (2011). Rapid detection of bacteria by carbon quantum dots. *Journal of Biomedical Nanotechnology*, *7*(6), 846–848. doi:10.1166/jbn.2011.1344 PMID:22416585

Michalet, X., Pinaud, F. F., Bentolila, L. A., Tsay, J. M., Doose, S., Li, J. J., Sundaresan, G., Wu, A. M., Gambhir, S. S., & Weiss, S. (2005). Quantum Dots for Live Cells, in Vivo Imaging, and Diagnostics. *Science Mag.* www.sciencemag.org/cgi/content/full/307/5709/538/DC1

Neelakantan, P. S. (2014). Nanotechnology In Dentistry - What Does The Future Hold In Store? *Dentistry (Sunnyvale, Calif.)*, *04*(02). doi:10.4172/2161-1122.1000198

Oh, S. T., Yoon, S. J., Choa, Y. H., Jeong, Y. K., & Niihara, K. (2006). Wear Behavior of Nano-Sized Metal Particle Dispersed Al_2O_3 Nanocomposites. *Key Engineering Materials*, *317–318*, 369–372. . doi:10.4028/www.scientific.net/KEM.317-318.369

Omietimi, H. B., Afolalu, S. A., Kayode, J. F., Monye, S. I., Lawal, S. L., & Emetere, M. E. (2023). An overview of nanotechnology and its application. *E3S Web of Conferences*, *391*. doi:10.1051/e3sconf/202339101079

Pak, S. C., Penrose, W., & Hesketh, P. J. (2001). An ultrathin platinum film sensor to measure biomolecular binding. In *Biosensors & Bioelectronics, 16*. www.elsevier.com/locate/bios

Papi, M., De Spirito, M., & Palmieri, V. (2023). Nanotechnology in the COVID-19 era: Carbon-based nanomaterials as a promising solution. *Carbon*, *210*, 118058. doi:10.1016/j.carbon.2023.118058 PMID:37151958

Payal, & Pandey, P. (2021). Role of Nanotechnology in Electronics: A Review of Recent Developments and Patents. *Recent Patents on Nanotechnology*, *16*(1), 45–66. doi:10.2174/1872210515666210120114504

Pumera, M. (2011). Graphene in biosensing. *Materials Today*, *14*(7–8), 308–315. doi:10.1016/S1369-7021(11)70160-2

Rani, N., Kumari, K., Sangwan, P., Barala, P., Yadav, J., Vijeta, Rahul, & Hooda, V. (2022). Nano-Iron and Nano-Zinc Induced Growth and Metabolic Changes in Vigna radiata. *Sustainability (Basel)*, *14*(14), 8251. doi:10.3390/su14148251

Rao, C. N. R., Sood, A. K., Subrahmanyam, K. S., & Govindaraj, A. (2009). Graphene: The new two-dimensional nanomaterial. In Angewandte Chemie - International Editioni, 48(42), 7752–7777. doi:10.1002/anie.200901678

Rauf, A., Abu-Izneid, T., Ahmed Khalil, A., Hafeez, N., Olatunde, A., Mominur Rahman, M., Semwal, P., Saleh Al-Awthan, Y., Salem Bahattab, O., Khan, I. N., Arslan Khan, M., & Sharma, R. (2022). Nanoparticles in clinical trials of COVID-19: An update. *International Journal of Surgery*, *104*, 106818. doi:10.1016/j.ijsu.2022.106818 PMID:35953020

Recordati, C., De Maglie, M., Bianchessi, S., Argentiere, S., Cella, C., Mattiello, S., Cubadda, F., Aureli, F., D'Amato, M., Raggi, A., Lenardi, C., Milani, P., & Scanziani, E. (2016). Tissue distribution and acute toxicity of silver after single intravenous administration in mice: Nano-specific and size-dependent effects. *Particle and Fibre Toxicology*, *13*(1), 12. doi:10.1186/s12989-016-0124-x PMID:26926244

Rocha-Santos, T. A. P. (2014). Sensors and biosensors based on magnetic nanoparticles. In *TrAC - Trends in Analytical Chemistry* (Vol. 62, pp. 28–36). Elsevier B.V., doi:10.1016/j.trac.2014.06.016

Rodriguez, J. L., Dalia, A. B., & Weiser, J. N. (2012). Increased chain length promotes pneumococcal adherence and colonization. *Infection and Immunity*, *80*(10), 3454–3459. doi:10.1128/IAI.00587-12 PMID:22825449

Saba, N., Paridah, M. T., Abdan, K., & Ibrahim, N. A. (2016). Effect of oil palm nano filler on mechanical and morphological properties of kenaf reinforced epoxy composites. *Construction & Building Materials*, *123*, 15–26. doi:10.1016/j.conbuildmat.2016.06.131

Saha, K., Agasti, S. S., Kim, C., Li, X., & Rotello, V. M. (2012). Gold nanoparticles in chemical and biological sensing. In Chemical Reviews (Vol. 112, Issue 5, pp. 2739–2779). doi:10.1021/cr2001178

Schrand, A. M., Rahman, M. F., Hussain, S. M., Schlager, J. J., Smith, D. A., & Syed, A. F. (2010). Metal-based nanoparticles and their toxicity assessment. In Wiley Interdisciplinary Reviews: Nanomedicine and Nanobiotechnology (Vol. 2, Issue 5, pp. 544–568). doi:10.1002/wnan.103

Sciau, P. (2012). Nanoparticles in Ancient Materials: The Metallic Lustre Decorations of Medieval Ceramics. In The Delivery of Nanoparticles. InTech. doi:10.5772/34080

Shi, J., Votruba, A. R., Farokhzad, O. C., & Langer, R. (2010). Nanotechnology in drug delivery and tissue engineering: From discovery to applications. In Nano Letters (Vol. 10, Issue 9, pp. 3223–3230). doi:10.1021/nl102184c

Solanki, R., Shankar, A., Modi, U., & Patel, S. (2023). New insights from nanotechnology in SARS-CoV-2 detection, treatment strategy, and prevention. In *Materials Today Chemistry* (Vol. 29). Elsevier Ltd., doi:10.1016/j.mtchem.2023.101478

Sperling, R. A., Gil, P. R., Zhang, F., Zanella, M., & Parak, W. J. (2008). Biological applications of gold nanoparticles. *Chemical Society Reviews*, *37*(9), 1896–1908. doi:10.1039/b712170a PMID:18762838

Sun, Y. P., Zhou, B., Lin, Y., Wang, W., Fernando, K. A. S., Pathak, P., Meziani, M. J., Harruff, B. A., Wang, X., Wang, H., Luo, P. G., Yang, H., Kose, M. E., Chen, B., Veca, L. M., & Xie, S. Y. (2006). Quantum-sized carbon dots for bright and colorful photoluminescence. *Journal of the American Chemical Society*, *128*(24), 7756–7757. doi:10.1021/ja062677d PMID:16771487

Tan, S. J., Campolongo, M. J., Luo, D., & Cheng, W. (2011). Building plasmonic nanostructures with DNA. In Nature Nanotechnology, 6(5), 268–276. Nature Publishing Group. doi:10.1038/nnano.2011.49

Teng, W.-Y., Jeng, S.-C., Kuo, C.-W., Lin, Y.-R., Liao, C.-C., & Chin, W.-K. (2008). Nanoparticles-doped guest-host liquid crystal displays (*Vol. 33*, Issue 15).

Thakur, V., Guleria, A., Kumar, S., Sharma, S., & Singh, K. (2021). Recent advances in nanocellulose processing, functionalization and applications: A review. In Materials Advances (Vol. 2, Issue 6, pp. 1872–1895). Royal Society of Chemistry. doi:10.1039/D1MA00049G

Uchugonova, A., Lessel, M., Nietzsche, S., Zeitz, C., Jacobs, K., Lemke, C., & König, K. (2012). Nanosurgery of cells and chromosomes using near-infrared twelve-femtosecond laser pulses. doi:10.1117/1.JBO.17.10.101502

Van Gerwen, P., Laureyn, W., Laureys, W., Huyberechts, G., Op, M., Beeck, D., Baert, K., Suls, J., Sansen, W., Jacobs, P., Hermans, L., & Mertens, R. (1998). Nanoscaled interdigitated electrode arrays for biochemical sensors. In Sensors and Actuators (Vol. 49). doi:10.1016/S0925-4005(98)00128-2

Wang, B., Chen, Y., Wu, Y., Weng, B., Liu, Y., Lu, Z., Li, C. M., & Yu, C. (2016). Aptamer induced assembly of fluorescent nitrogen-doped carbon dots on gold nanoparticles for sensitive detection of AFB1. *Biosensors & Bioelectronics*, *78*, 23–30. doi:10.1016/j.bios.2015.11.015 PMID:26584079

Wang, J. L., & Meng, L. J. (2014). Influence of carbon nano-fiber on mechanical property of PALC. *Applied Mechanics and Materials*, *535*, 785–787.. doi:10.4028/www.scientific.net/AMM.535.785

Wang, L., Teles, M. P. R., Arabkoohsar, A., Yu, H., Ismail, K. A. R., Mahian, O., & Wongwises, S. (2022). A holistic and state-of-the-art review of nanotechnology in solar cells. In *Sustainable Energy Technologies and Assessments* (Vol. 54). Elsevier Ltd., doi:10.1016/j.seta.2022.102864

Wang, R., Xu, Y., Zhang, T., & Jiang, Y. (2015). Rapid and sensitive detection of Salmonella typhimurium using aptamer-conjugated carbon dots as fluorescence probe. *Analytical Methods*, *7*(5), 1701–1706. doi:10.1039/C4AY02880E

Wang, X., Xu, C., Yi, M., & Zhang, H. (2011). Effects of Nano-ZrO2 on the microstructure and mechanical properties of Ti(C, N)-based cermet die materials. *Advanced Materials Research*, *154–155*, 1319–1323.. doi:10.4028/www.scientific.net/AMR.403-408.1319

Wanger, G., Onstott, T. C., & Southam, G. (2008). Stars of the terrestrial deep subsurface: A novel "star-shaped" bacterial morphotype from a South African platinum mine. *Geobiology*, *6*(3), 325–330. doi:10.1111/j.1472-4669.2008.00163.x PMID:18498531

Wiechers, J. W., & Musee, N. (2010). Engineered inorganic nanoparticles and cosmetics: Facts, issues, knowledge gaps and challenges. In Journal of Biomedical Nanotechnology (Vol. 6, Issue 5, pp. 408–431). doi:10.1166/jbn.2010.1143

Williford, J. M., Santos, J. L., Shyam, R., & Mao, H. Q. (2015). Shape control in engineering of polymeric nanoparticles for therapeutic delivery. *Biomaterials Science*, *3*(7), 894–907. doi:10.1039/C5BM00006H PMID:26146550

Wu, Q., Liu, J., Wang, X., Feng, L., Wu, J., Zhu, X., Wen, W., & Gong, X. (2020). Organ-on-a-chip: Recent breakthroughs and future prospects. In BioMedical Engineering Online (Vol. 19, Issue 1). BioMed Central Ltd. doi:10.1186/s12938-020-0752-0

Wu, X., Xing, Y., Zeng, K., Huber, K., & Zhao, J. X. (2018). Study of Fluorescence Quenching Ability of Graphene Oxide with a Layer of Rigid and Tunable Silica Spacer. *Langmuir*, *34*(2), 603–611. doi:10.1021/acs.langmuir.7b03465 PMID:29275632

Wu, X., Yuan, X., Zhao, J., Ji, D., Guo, H., Yao, W., Li, X., & Zhang, L. (2023). Study on the effects of different pectinase/cellulase ratios and pretreatment times on the preparation of nanocellulose by ultrasound-assisted bio-enzyme heat treatment. *RSC Advances*, *13*(8), 5149–5157. doi:10.1039/D2RA08172E PMID:36777933

Xia, Y., Xiong, Y., Lim, B., & Skrabalak, S. E. (2009). Shape-controlled synthesis of metal nanocrystals: Simple chemistry meets complex physics? In Angewandte Chemie - International Edition (Vol. 48, Issue 1, pp. 60–103). doi:10.1002/anie.200802248

Xu, X., Ray, R., Gu, Y., Ploehn, H. J., Gearheart, L., Raker, K., & Scrivens, W. A. (2004). Electrophoretic analysis and purification of fluorescent single-walled carbon nanotube fragments. *Journal of the American Chemical Society*, *126*(40), 12736–12737. doi:10.1021/ja040082h PMID:15469243

Yakoubi, A., El, C., & Dhafer, B. (2023). Advanced Plasmonic Nanoparticle-Based Techniques for the Prevention, Detection, and Treatment of Current COVID-19. *Plasmonics*, *19*(18), 311–347. doi:10.1007/s11468-022-01754-0 PMID:36588744

Yamada, K., Kim, C. T., Kim, J. H., Chung, J. H., Lee, H. G., & Jun, S. (2014). Single walled carbon nanotube-based junction biosensor for detection of escherichia coli. *PLoS One*, *9*(9), e105767. Advance online publication. doi:10.1371/journal.pone.0105767 PMID:25233366

Yang, D. (2021). Application of Nanotechnology in the COVID-19 Pandemic This article was published in the following Dove Press journal. *International Journal of Nanomedicine*, *16*, 623–649. Advance online publication. doi:10.2147/IJN.S296383 PMID:33531805

Yang, X., & Wang, Z. (2022). Effect of Particle Size and Concentration of Nanocellulose on Thermal Properties of Nanocellulose Aerogel. *Journal of Physics: Conference Series*, *2437*(1), 012061. doi:10.1088/1742-6596/2437/1/012061

Zanoni, I., Blosi, M., Fiorini, V., Crosera, M., Ortelli, S., Stagni, S., Stefan, A., Psilodimitrakopoulos, S., Stratakis, E., Filon, F. L., & Luisa Costa, A. (2020). Use of Cotton Textiles Coated by Ir(III) Tetrazole Complexes within Ceramic Silica Nanophases for Photo-Induced Self-Marker and Antibacterial Application. *Nanomaterials (Basel, Switzerland)*, *10*(6), 1020. doi:10.3390/nano10061020 PMID:32471044

Zhang, H., Ma, X., Liu, Y., Duan, N., Wu, S., Wang, Z., & Xu, B. (2015). Gold nanoparticles enhanced SERS aptasensor for the simultaneous detection of Salmonella typhimurium and Staphylococcus aureus. *Biosensors & Bioelectronics*, *74*, 872–877. doi:10.1016/j.bios.2015.07.033 PMID:26241735

Zhang, L., Qian, M., Cui, H., Zeng, S., Wang, J., & Chen, Q. (2021). Spatiotemporal Concurrent Liberation of Cytotoxins from Dual-Prodrug Nanomedicine for Synergistic Antitumor Therapy. *ACS Applied Materials & Interfaces*, *13*(5), 6053–6068. doi:10.1021/acsami.0c21422 PMID:33525873

Zhang, Y., Zhang, K., & Ma, H. (2009). Electrochemical DNA Biosensors Based on Gold Nanoparticles / Cysteamine / Poly(glutamic acid). *American Journal of Biomedical Sciences*, *1*(2), 115–125. doi:10.5099/aj090200115

Zhao, Y., Ye, M., Chao, Q., Jia, N., Ge, Y., & Shen, H. (2009). Simultaneous detection of multifood-borne pathogenic bacteria based on functionalized quantum dots coupled with immunomagnetic separation in food samples. *Journal of Agricultural and Food Chemistry*, *57*(2), 517–524. doi:10.1021/jf802817y PMID:19154162

Zou, B., Huang, C. Z., Wang, J., & Liu, B. Q. (2006). Effect of Nano-Scale TiN on the Mechanical Properties and Microstructure of Si_3N_4 Based Ceramic Tool Materials. *Key Engineering Materials*, *315–316*, 154–158. . doi:10.4028/www.scientific.net/KEM.315-316.154

ADDITIONAL READING

Crosera, M., Mauro, M., Bovenzi, M., Adami, G., Baracchini, E., Maina, G., & Larese Filon, F. (2018). In vitro permeation of palladium powders through intact and damaged human skin. *Toxicology Letters*, *287*, 108–112. doi:10.1016/j.toxlet.2018.02.009 PMID:29421332

Psilodimitrakopoulos, S., Mouchliadis, L., Paradisanos, I., Lemonis, A., Kioseoglou, G., & Stratakis, E. (2018). Ultrahigh-resolution nonlinear optical imaging of the armchair orientation in 2D transition metal dichalcogenides. *Light, Science & Applications*, *7*(5), 18005–18009. doi:10.1038/lsa.2018.5 PMID:30839517

Zanoni, I., Crosera, M., Ortelli, S., Blosi, M., Adami, G., Filon, L., & Costa, A. L. (2019). CuO nanoparticle penetration through intact and damaged human skin. *New Journal of Chemistry*, *43*(43), 17033–17039. doi:10.1039/C9NJ03373D

Zhang, L., Li, Y., Che, W., Zhu, D., Li, G., Xie, Z., Song, N., Liu, S., Tang, B. Z., Liu, X., Su, Z., & Bryce, M. R. (2019). AIE Multinuclear Ir(III) Complexes for biocompatible organic nanoparticles with highly enhanced photodynamic performance. *Advancement of Science*, *6*(5), 1802050. doi:10.1002/advs.201802050 PMID:30886811

Zhang, W. Y., Wang, Y. J., Du, F., He, M., Gu, Y. Y., Bai, L., Yang, L. L., & Liu, Y. J. (2019). Evaluation of anticancer effect in vitro and in vivo of iridium(III) complexes on gastric carcinoma SGC-7901 cells. *European Journal of Medicinal Chemistry*, *178*, 401–416. doi:10.1016/j.ejmech.2019.06.003 PMID:31202128

Chapter 2
Bioinspired Synthesis of Nanocomposites

Semiha Ekrikaya
Nuh Naci Yazgan University, Turkey

Hikmetnur Danisman
https://orcid.org/0000-0002-4203-2575
Nuh Naci Yazgan University, Turkey

Seda Baktir
Nuh Naci Yazgan University, Turkey

Soley Arslan Arslan
Erciyes University, Turkey

Ismail Ocsoy
Erciyes University, Turkey

ABSTRACT

The design of nanocoatings and nanostructured based surfaces stands out in many fields and biomedical applications due to their superior advantages over traditional coatings. With current nanotechnological approaches, properties such as antimicrobial activity, biocompatibility, increased product life-time, quality interface bonding, enhanced chemical and mechanical properties, high hardness, heat insulation, dirt and water repellency, corrosion resistance, improved energy efficiency, and reduced moisture absorption have been significantly improved. In recent years, the use of NPs and nanocoatings in the design and development of many dental materials has become highly popular in dentistry due to their ability to provide unique combinations of properties. This section will primarily focus on nanocoating materials used in dentistry.

DOI: 10.4018/979-8-3693-1094-6.ch002

Copyright © 2024, IGI Global. Copying or distributing in print or electronic forms without written permission of IGI Global is prohibited.

INTRODUCTION

Up to now, nanotechnology and nanomaterials (NMs) have made significant advances in various industrial and scientific applications. Nanotechnological research involves producing small structures ranging in size from 1 to 100 nanometers (nm) by using different methods (chemical, biological, and physical) and making them available for a wide range of applications (Karaagac et al., 2021; Prasad et al., 2017). NMs can be produced in various shapes and forms, such as spherical, rods, nanofibers, and cages (Shankar et al., 2004; Skrabalak et al., 2008). NMs are widely used in various fields owing to their unique physical, chemical and biological applications (Dolai et al., 2021; Karaagac et al., 2021; Lohse et al., 2012). Among these synthesis methods, the green synthesis method offers advantages in terms of biocompatibility, cost-effectiveness, a simple process, fast synthesis, and large-scale synthesis (Shankar et al., 2004; Ocsoy et al., 2018; Akhtar et al., 2013).

The design of nanocoatings and nanostructured-based surfaces stands out in many fields and biomedical and dentistry applications due to their superior advantages over traditional coatings (Gajanan & Tijare, 2018; Damodharan, 2021; Ramos et al., 2017; Malik et al., 2023) . With current nanotechnological approaches, properties such as antimicrobial activity, biocompatibility, increased product life, quality interface bonding, improved chemical and mechanical properties, high hardness, thermal insulation, dirt and water repellency, corrosion resistance, improved energy efficiency and reduced moisture absorption have been significantly improved (Heo et al., 2016; Jandt & Watts, 2020; Yılmaz et al., 2019). However, taking advantage of the advantages of the green synthesis method in NPs synthesis contributes to the development of biomaterials (Janaki et al. 2015; Ildiz et al. 2017; Narayanan et al. 2021).

In recent years, nanotechnological approaches have also been frequently used in dentistry (Arslan et al., 2023; Ekrikaya et al., 2021). The use of NPs and nanocoatings in the design and development of many dental materials has become extremely popular due to their ability to provide unique combinations of properties (Heo et al., 2016; Hossain et al., 2022; Malik et al., 2023; Jandt & Watts, 2020). In this section, we will primarily focus on the use of nanotechnology in dentistry and the nanocoating materials used.

SYNTHESIS AND CHARACTERIZATION OF NANOCOMPOSITES AND NANOCOATINGS

NMs are synthesized using physical, chemical, and biological methods (Chavali et al., 2019; Demirbas et al., 2020; Karaagac et al., 2021). Metal and metal oxide

Figure 1. The bottom-up and the top-down approach approach

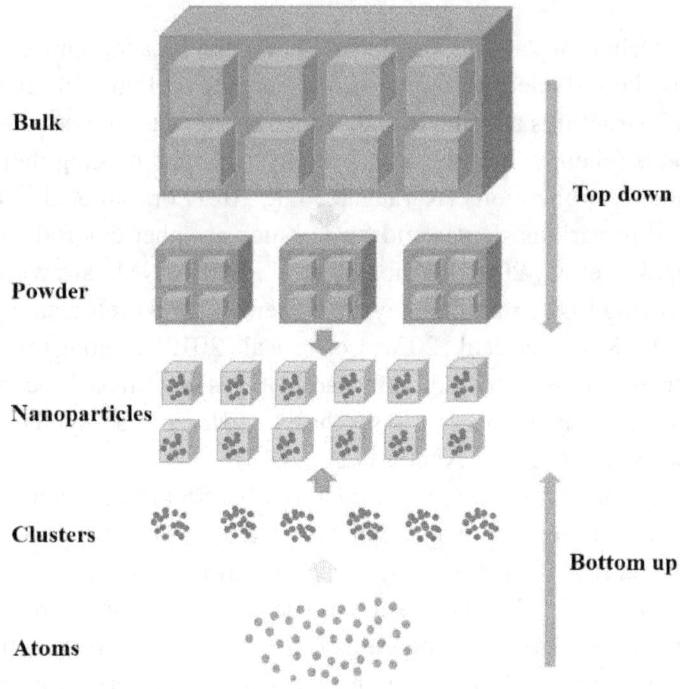

nanoparticles (NPs), magnetic NPs, quantum dots, fullerenes, dendrimers, carbon nanotubes, and other NMs can be obtained using these synthesis methods (Sharma et al., 2022; Ying et al., 2022; Shreyash et al., 2021, Karaagac et al., 202). The goal of the bottom-up approach is to obtain NPs by providing energy to materials at the atomic or molecular level through chemical reactions (see Figure 1). Among the synthesis methods, biological synthesis (green synthesis) has gained popularity in recent years because it is more cost-effective, biocompatible, and environmentally friendly than other methods (Ocsoy et al., 2017). Biological sources, including bacteria, fungi, yeast, algae, and various parts of plants such as roots, stems, leaves, and flowers, can be used for green synthesis (Arslan et al., 2023; Ekrikaya et al., 2021; Ocsoy et al., 2017).

Green synthesis of NPs has received significant attention in various industries, such as medicine, food, pharmaceuticals, and agriculture, because of their environmentally friendly, non-toxic, and eco-friendly nature (Karaagac et al., 2021; Demirbas et al., 2019; Ocsoy et al., 2017; Ocsoy et al., 2018; Lu et al., 2020; Celik et al., 2020; Chirayil et al., 2017; Dadi et al., 2021; Koca et al., 2020; Titus et al., 2019). Ultraviolet and visible light (UV-Vis) absorption spectroscopy uses electromagnetic radiation in the 190-800 nm wavelength range (Rajeshkumar et al., 2019; Jose Chirayil et al.,

2017). UV-Vis spectroscopy was used to determine the characteristic absorbance point, concentration, and stability of NPs. As particles become less stable, changes in peak shape and intensity values can be observed because of the depletion of stable the NPs, resulting in either a decrease in peak intensity, broadening of the peak, or the appearance of a second peak at longer wavelengths owing to changes in particle size (Jose Chirayil et al., 2017).

Scanning Electron Microscopy (SEM) is a nondestructive characterization tool that provides comprehensive information about the surface morphology, arrangement, and composition of NPs. (Rajeshkumar et al., 2019; Titus et al., 2019).Transmission Electron Microscopy (TEM) is used to obtain morphological (size, shape, surface features, and structure), compositional, and crystallographic information from the samples. Energy-Dispersive X-ray Spectroscopy (EDX) analyses the surface and near-surface elements, providing an element map of the sample and the percentage of different elements at various locations on the surface. EDX is often used in conjunction with SEM (Titus et al., 2019).

X-ray Diffraction (XRD) is a non-destructive analysis technique used to determine the crystallographic properties of samples and the phases they contain. When monochromatic X-rays are directed at a crystalline structure, they are diffracted differently, depending on the types and arrangement of atoms in the structure (Jose Chirayil et al., 2017). Fourier Transform Infrared Spectroscopy (FTIR) analysis is a spectroscopic technique in which organic, inorganic, and polymeric samples are scanned with infrared light (Titus et al., 2019). When infrared light passes through a sample, some of the light is absorbed by the sample, while some of it passes through. The resulting absorption and emission spectra have characteristic patterns, that can be used to determine the identity of a sample. Changes in the characteristic patterns of the absorption bands can help identify changes in sample compositions. FTIR is particularly useful for in detecting organic functional groups on the surface of metal NPs and for determining surface chemistry (Rajeshkumar et al., 2019).

Organic-Inorganic Nanocomposites and Nanocoatings

Green synthesis of NPs has been achieved using various biomaterials, such as bacteria (Hulkoti and Taranath, 2014), fungi, yeast (Moghaddam et al., 2015), viruses (Dujardin et al., 2003), microalgae (Schrofel et al., 2011), and plant biomass and extracts (Aslan et al., 2023; Mittal et al., 2013) (Figure 2). The use of different biological organisms in this field is rapidly developing owing to their increased success and ease of synthesis. NPs with various shapes, sizes, contents, and physicochemical properties can be synthesized using green methods. Molecules found in plants and microorganisms, enzymes, such as proteins, amines, alkaloids, and pigments, phenolic compounds, can reduce metal ions, resulting in NP production in a single step (Karnani and Chowdhary, 2013; Shah et al., 2015). Microorganisms can synthesize NPs under mild conditions without the need for toxic chemicals or harsh environments.

Compared with other natural sources, bacteria have a broad research field for green synthesis because they can be manipulated relatively easily. Many bacteria have the potential to produce NPs with various morphologies under mild culture conditions and room temperature (Thakkar et al., 2010). The locations where NPs are formed differ based on the reducing species of bacteria. Reducing enzymes or functional groups found within the cell or on the cell wall are involved in the intracellular formation. As a result, the generated NPs are limited to a certain area and can be challenging to extract from cells. Extracellular NPs are formed when bacteria release their enzymes into the surrounding environment, where reducing enzymes dissolve in the solvent medium (Nath and Banerjee, 2013). Similar to other green methods, algae-mediated NP synthesis is environmentally friendly. Algae are essential for applications because of their resistance to harsh environmental conditions and their ability to synthesize NPs in the presence of reducing agents, such as phenolic compounds and organic substances (Khan et al., 2019). The synthesis of NPs using algae involves incubating algal extract, metal precursor solutions, and algal extract-metal precursor solutions (Sharma et al., 2016). The reaction was initiated by mixing the liquid algal extract with a the metal precursor solution. A change in the color of the mixture is a typical indicator of the starting reaction, followed by the aggregation of particles and growth in NPs in different sizes and shapes, resulting in thermodynamically stable NPs (Fawcett et al., 2017; Prasad et al., 2016).

Figure 2. Schematic representation for plant-mediated biosynthesis of nanoparticles, characterization and application areas in dentistry

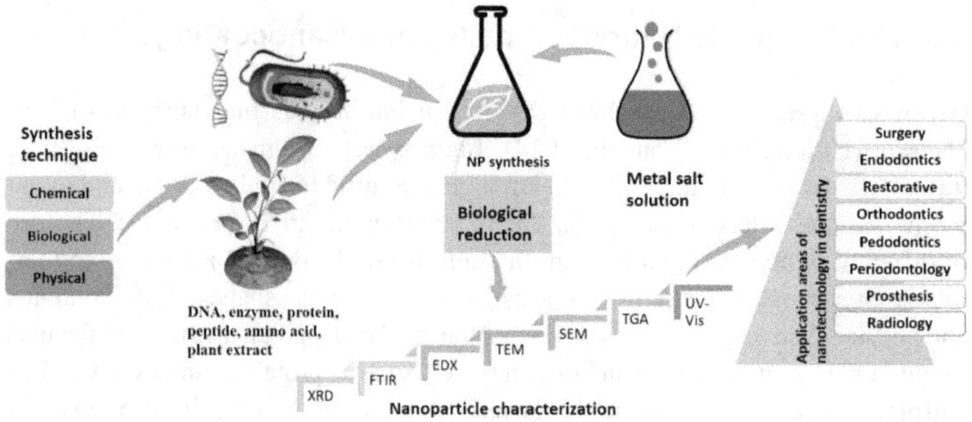

In traditional chemical and physical methods, the reducing agents involved in the reduction of metal ions and the stabilizing agents added to prevent unwanted aggregation of synthesized NPs pose environmental and toxicity risks to cells. However, when NP synthesis is achieved using the green method of biocompatible NPs, these agents naturally exist in biological organisms (Hussain et al., 2016).

Inorganic-Inorganic Nanocomposites and Nanocoatings

NPs can be classified into three main types based on their sources and mechanisms of action: inorganic NPs, organic NPs, and natural macromolecular compound NPs. Inorganic nanocomposites and nanocoatings include metals, metal oxides, their doped forms, and surface-modified NPs (Yun et al., 2022).

Metal -Metal Nanoparticles

Metal nanoparticles (MNPs) are one of the most common types of inorganic NPsand are considered as an effective solution against antibiotic resistance, preventingthe formation of resistant strains by attacking various biomolecules. MNPs are various nanometer-scale materials combined with other inorganic or organic elements that mainly carry metal ions. Metal compounds have been used as antimicrobial agents since ancient times (Allaker, 2010). Metal silver (Ag), gold (Au), copper (Cu, CuI, CuO, and Cu_2O), iron (Fe, Fe_2O_3), magnesium (Mg, MgO), titanium (Ti, TiO_2), and zinc (Zn, ZnO) wereused to produce MNPs (Slavin et al., 2017). Among these, the most widely used nanometals in dental materials are Ag, Zn, Ti, Cu, Au, and zirconium (Zr). While Au and Ag are used in their pure form, nano-oxides of Zr, Cu, Ti, and Zn are more commonly used. These MNPs have been incorporated into composite resins, endodontic materials, acrylic base materials, dental implants, and dental adhesives. They have been used as coating materials in orthodontic brackets and archwires and have been shown to improve the antimicrobial, mechanical and regenerative properties of these novel NP-containing materials (Agnihotri et al., 2020; Jandt & Watts, 2020; Wang et al., 2023). Plant extracts have been used as reducing and stabilizing agents that enable the production of MNPs without harming the environment or human health. In the past few years, there have been successful reports on the effective and rapid extracellular synthesis of Au, Ag, and Cu NPs using various plant extracts, such as *Aloe vera, lemongrass, geranium, Hibiscus sabdariffa, Arena sativa, alfalfa* (Yazdanian et al., 2022).

Ag NPs are one of the most widely used NPs in dentistry, primarily due to their exceptionally high antibacterial characteristics. Ag NPs exhibit significant antibacterial activity through oxidative stress and the release of Ag ions (Li et al.,

2019). Silver ions are highly reactive and can rapidly bind to negatively charged proteins, DNA and RNA, and this property constitutes a key component of their antibacterial mechanism. It has been observed that Ag NPs have the ability to bind to the bacterial cell membrane and can also penetrate the bacterial inner respiratory chain. This leads to disruption of the bacterial cell membrane and, consequently, cell death (Metin-Gursoy et al., 2017).Some plants, such as pomegranate, rice extract, basil, red cabbage, and green tea, are used to produce Ag NPs (Ocsoy et al., 2018; Yazdanian et al., 2022). It was reported that these Ag@SiO_2 NPs did not show cytotoxicity in human dental cells,were light-colored as desired, had antibacterial activity against *Streptococus Mutans* (*S. Mutans)* and showed very promising results for biofilm control (Rodrigues et al., 2020).

Cu NPs have been synthesized using the green synthesis method, employing extracts from various plants such as *Zingiber officinale, Celastrus paniculatus, Eclipta prostrata, Citrus medica Linn., Cardiospermum halicacabum, Eryngium caucasicum, Punica granatum, Plectranthus amboinicus, Azadirachta indica, and Madhuca longifolia* (Yazdanian et al., 2022, Fujimori et al., 2012). In dentistry, Cu NPs are used as surface coatings to enhance the bioactivity and antibacterial efficacy of titanium implants. Coating the healing caps of the implant with Cu NPs inhibits biofilm formation around the cap and reduces bacterial activity (Thukkaram et al., 2021). It has been shown that the addition of Cu NPs to nickel-titanium alloy reduces the load-bearing stress in orthodontic appliances obtained by adding Cu NPs and this increases tooth mobility (Xu et al., 2022; (Prabhu et al., 2010).

Currently, in the field of biomedicine, Au NPs find applications in various areas such as imaging, dental caries, periodontology, tissue engineering, bone regeneration, implantology, and cancer diagnosis (Bapat et al., 2020; Yougbare et al., 2019; Al-Nafori et al., 2017; Heo et al., 2016; M. Zhang et al., 2020). Au NPs were deposited onto Ti implant surfaces, the newly obtained implant surface exhibited the potential to reduce postoperative inflammatory responses and positively impact osteoblastic activity by aiding bone-implant integration and bone formation due to its scaffold release (Heo et al., 2016). Biocompatible Au NPs, synthesized through a green method using Salacia chinensis, have demonstrated their potential as environmentally friendly nanomolecules. These particles not only promote bone formation and diminish bone resorption but also serve as an effective material for inducing bone activity during implant placement (Jadhav et al., 2018). The application of Au NPs in cancer treatment has also gained attention because of their unique physical properties, known assurface plasmon resonance. This property allows them to respond to electromagnetic radiation by generating heat. This heat can impact cancer cells, causing them to thermal disintegration (Yazdanian et al., 2022).

Metal NP-Metal Oxide Nanoparticles

CuO NPs have been confirmed to show potent antibacterial activity against bacteria such as methicillin-resistant *Escherichia coli (E. coli)*, *Staphylococcus aureus (S. aureus)*, and *Pseudomonas aeruginosa (P. aeruginosa)* when applied as nanocoatings (Hasheminya et al., 2018; LewisOscar et al, 2015) LewisOscar et al. showed that CuO NPs used at a concentration of only 0.1 µg mL-1 inhibited *P. aeruginosa* biofilm formation by 94% and had a strong antibiofilm effect (LewisOscar et al., 2015). The effectiveness of CuO NPs at relatively low concentrations in exhibiting strong antibiofilm effects makes these NPs a suitable option for antimicrobial nanocoatings. There are studies in the literature that specifically use CuO, especially in surface nanocoating of orthodontic materials (Padmanabhan, 2023; S. Zhang et al., 2022). In recent years, green synthesis methods have been used to synthesize CuO NPs with antibacterial activities. CuO NPs preparedfrom aloe vera gel have excellent anticariogenic properties and can be used in various dental applications (Rajeshkumar et al., 2019). As mentioned previously, the cytotoxicity of Cu NPs limits their use.

Green synthesis can enhance the bioreactivity of ZnO NPs and increase their potential for biological applications. Plant extracts, especially those containing components, such as phenols and flavonoids, can be used for the production of ZnO NPs. Substances such as phenols and flavonoids found in these extracts can stabilize NPs through electrostatic, van der Waals, steric, and hydration, forces. This allowed the NPs to be stably held together. Extensive research has been conducted on plants, such as *Plectranthus amboinicus*, *Anisochilus carnosus*, and *Vitex negundo*, to form ZnO NPs in various sizes and shapes, encompassing spherical, hemispherical, hexagonal,rod-shaped, and agglomerate formations (Yazdanian et al., 2022).

The application of ZnO nanocoatings on orthopedic or dental implants has been shown to enhance osteoblast activity andsupport new bone formation. In addition, zinc functions as a cofactor for collagen synthesis and promates bone mineralization viaalkaline phosphatase. Therefore, ZnO NPs, which have a strong relationship with bone formation and mineralization, are ideal candidates for antimicrobial nanocoatings for materials to be used in close proximity to calcified tissues (Huo et al., 2013; Shen et al., 2014).

TiO_2 NPs, similar to other NPs, can be produced using green synthesis methods. Some plants used for the production of TiO_2 NPs include *Azadirachta indica branches, Syzygium aromaticum, Ficus benghalensis, Mentha arvensis, Echinacea purpurea, Citrus aurantifolia*, and *Acanthophyllum laxiusculum* extracts (Yazdanian et al., 2022). Ahmed and colleagues conducted a study in which they produced TiO_2 nanoparticles through green synthesis, using titanium tetraisopropoxide as a precursor and *Mentha arvensis* leaf extract as a reducing agent, and found that these nanoparticles exhibited highly promising antimicrobial effects (Ahmad et al., 2020).

ZrO_2, which enhances osteoconductivity in implants and reduces bacterial colonization, is of great interest in prosthetic dentistry. Applying cubic ZrO2 as a nanocoating on Ti surfaces can improve the surface properties of dental implants and enhance corrosion resistance, wear resistance, biocompatibility, and bone integration. Additionally, it can reduce bacterial colonization, thereby reducing the inflammatory response caused by prostheses in oral tissues (Das et al., 2016; Osman & Swain, 2015). ZrO_2 NPs were produced from clove and cardamom using a green synthesis method to coat orthodontic archwires and the antibacterial, antiinflammatory, and cytotoxic effects of this nanocoating were analyzed. Archwires coated with ZrO_2 NPs have been shown to exhibit anti-inflammatory properties and effective antibacterial activity against bacteria such as *S. mutans* and *S. aureus* (Selvaraj et al., 2021).

Metal Nanoparticle/Metal Oxide Nanoparticle Nanostructures

Combining metal or metal-oxide NPs with 2D NMs can enhance their surface properties and application potential. For instance, the deposition of metal oxide NPs on 2D NMs, such asgraphene, $MoS_2$and WSe_2can be utilized in various applications, (2023; Quintero et al., 2023; Nair et al., 2008; Quintero et al., 2023; Chen et al., 2021;). Ag@GO nanocomposite (Ag@GO NC) surrounds the bacteria, ensuring direct contact of Ag NPs with the bacterial surface and increasing antibacterial activity (Figure 3). In a study conducted in the field of dentistry, GO was used in dental adhesives in the form of Ag@GO NC to increase antibacterial activity, and successful results were obtained(Arslan et al., 2023).

Figure 3. Schematic representation of Ag@GO nanocomposite formation and interaction between Ag@GO nanocomposite and bacteria

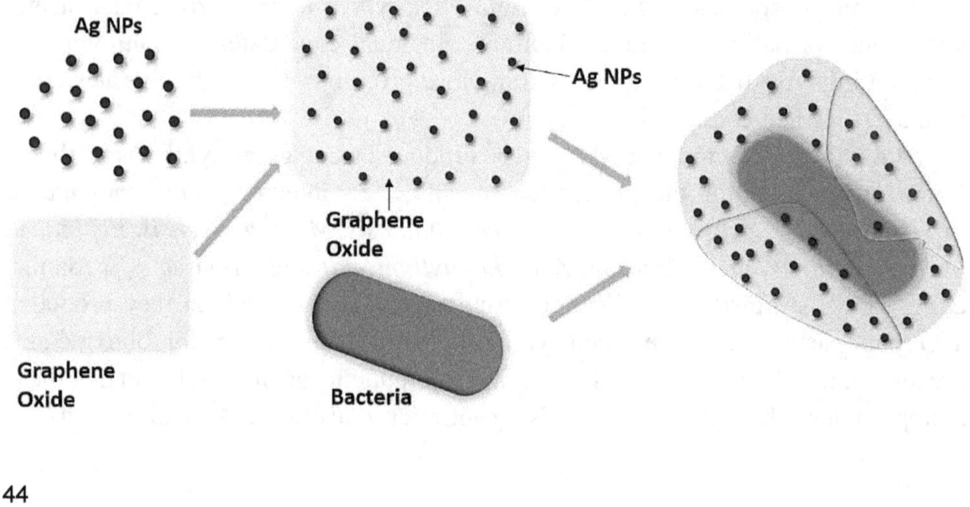

APPLICATIONS OF NANOCOMPOSITES AND NANOCOATINGSIN DENTISTRY

In the field of healthcare, nanomaterials are used for drug delivery, gene transfer, inhibition of pathogens (bacteria, fungi, and viruses), protein biosensing, DNA structure analysis, tissue engineering, cancer cell imaging, diagnosis and inhibition, purification of biological molecules, implants, prosthetics, optoelectronic devices, and sensors (Gajanan & Tijare, 2018; Damodharan, 2021; Ramos et al., 2017).

Nanocomposites and nanocoatings are actively employed in all disciplines of dentistry, including orthodontics, endodontics, periodontology, restorative dentistry, prosthodontics, and oral and maxillofacial surgeies. Implants, acrylic resins, restorative materials, cement, sealants, adhesives, orthodontic brackets, wires, and prosthetic base systems are among the dental materials in which NMs are used (Table 1). Furthermore, their applications are being explored in the field of tissue engineering and their potential impact on oral cancers (Agnihotri et al., 2020; Jandt & Watts, 2020; Slavin et al., 2017).

Dental caries, defined as a polymicrobial biofilm disease by the World Health Organization, is responsible for dental pain in almost 100% of adults worldwide (Balhaddad et al., 2019). Research in restorative dentistry is often aimedto preventing the formation of dental caries. Therefore, studies heve focused on common restorative materials used in dentistry, such as amalgams, composites, ceramics, and glass ionomer cements (Ferrando-Magraner et al., 2020). Currently, there is significant research into incorporating NPs into restorative materials to enhance their physical, mechanical, and antibacterial properties. Beyond restorative materials, the addition of NPs to toothpaste and mouth washes aims to prevent caries formation and promote remineralization (Ghafar et al., 2020). In restorative dentistry, NPs are widely used to reduce the activity of bacteria associated with dental caries. Both organic and inorganic NPs, such as TiO_2, Al_2O_3, SiO_2, AlN, ZnO, curcumin, chitosan, and graphene, have been used in conjunction with restorative filling materials to effectively address this issue (Ferrando-Magraner et al., 2020; Khurshid et al., 2015).

In previos studies reported in the literature, successful results were obtained by adding 0.05% Ag NPs to both two-stage and universal adhesives (Arslan et al., 2023; Ekrikaya et al., 2023). In a study utilizing rice extract for the synthesis of Ag NPs, it was demonstrated that rice extract can effectively be used as a reducing agent in the production of Ag NPs. The Ag NPs synthesized using this green synthesis method exhibited strong antibacterial effects against *S. mutans*, a causative agent of severe oral infections such as dental caries (Suwan et al., 2018).

Raorane and colleagues, through the incorporation of TiO_2 NPs synthesized using Citrus aurantifolia extract into dental composites, revealed that TiO_2 nanohybrids could serve as excellent fillers to enhance the physical properties of light-cured dental composite materials. These nanohybrids not only improved the antibacterial, hydrophilic, and self-cleaning characteristics of dental composites but also demonstrated their potential as fillers (Raorane et al., 2019).

Root canal infections are characterized by a polymicrobial structure caused by various bacteria and microorganisms, including *S. mutans, Enterococcus faecalis (E. faecalis), Streptococcus anginosus (S. anginosus), Staphylococcus aureus (S. aureus)* and *Fusobacterium nucleatum (F. nucleatum)* among approximately 150 species. Consequently, complete eradication of these infections is impossible (Poggio et al., 2011). During endodontic treatment, the biological compatibility and antimicrobial properties of instruments, sealers, filling materials, irrigation solutions, intracanal medications (e.g., calcium hydroxide [$Ca(OH)_2$]), periapical materials (e.g., Portland cement and mineral trioxide aggregate (MTA)), and filling materials (e.g., gutta-percha) play critical roles in achieving successful treatment. NMs have been reported to contributesignificantly to regenerative endodontics and enhance the antibacterial properies of endodontic materials (Raura et al., 2020).

Acrylic partial/full dentures or dental implants are often used for full mouth rehabilitation when all the teeth are lost. These dentures are made of polymethyl methacrylate (PMMA) based resins,which have a rough surface that, leads to increased biofilm accumulation (Verran & Maryan, 1997). To overcome these problems and improve the mechanical and biological properties of the materials, NPs such as ZnO, GO, Ag, $ZrO_{2,}$ and TiO_2 have been added to PMMA and used as coating materials on the surfaces of dental implants (Heo et al., 2016; Hossain et al., 2022; Jandt & Watts, 2020).

The inclusion of GO NPs in PMMA resin has been shown to enhance its mechanical properties, and exhibit an antibiofilm effect against microbial species found in artificial saliva; this effect lasts for up to 28 days. The advantage of this approach is that it does not require metal ions or antibiotics for its antimicrobial activity (Lee et al., 2018). G-Ag NPs embedded in PMMA demonstrated good safety profiles and antibacterial activity against all the tested microbial strains. Additionally, it enhanced the activation of antioxidant defense mechanisms and increased the flexural strength of PMMA (Bacali et al., 2020).

The surface structures of orthodontic appliances (brackets and archwires) tend to accumulate plaque and promote biofilm attachment because of their prolonged intraoral presence, which makes it challenging to maintain oral hygiene. Changes in the quantity, structure, and composition of dental plaque lead to increased microbial colonization and affect the oral ecosystem, potentially affecting both the hard and soft oral tissues. During orthodontic treatment, owing to prolonged contact with intraoral soft and hard tissues, these appliances can create a conducive environment for microbial growth and potentially lead to periodontal infections such as gingivitis and periodontitis. Furthermore, as a result of the long-term retention of dental plaque

Table 1. NPs produced by green synthesis methods and their application areas

Plants name(s)	NM(s)	Size (nm) \ Shape(s)	Field of application (in dentistry)	Reference(s)
Morinda citrifolia	Ag	3-11 nm (spherical)	Enhance the self- etching adhesive's conversion rate and antibacterial effectiveness	Morales-Lozoya et al. (2021)
Nymphae odorata	Ag	10-20 nm (spherical)	Antibacterial effectiveness	Gudimalla et al. (2021)
Capparis zeylanica	Ag	23 nm (spherical)	Antibacterial effectiveness	Nilavukkarasi et al. (2020)
Achillea bieberstein	Ag	12±2 nm	Antibacterial effectiveness	Some et al. (2018)
Rhizophora mucronata	Ag	60-95 nm	Antibacterial effectiveness	Asmathunisha and Kathiresan (2013)
Trifolium resupinatum	Ag	17 nm (spherical)	Antibacterial effectiveness	Some et al. (2018)
Zea mays	Ag	249.12 nm (spherical)	Antibacterial effectiveness	Patra and Baek (2016)
Zingiber officinale	Ag	10-20 nm	Antibacterial effectiveness	Some et al. (2018)
Ropaeolum majus	Ag	Cubic	Antibacterial effectiveness	Valsalam et al. (2019)
Leucasaspera extract	Ag	25-80 nm (clustere, irregular)	Antibacterial effectiveness	Kumaran and Vijayaraj (2017)
Dodonaeviscosa, Capparisdeciduas leaf	Ag	60-90 nm (Spherical)	Antibacterial effectiveness	Giridharan et al. (2014)
Viburnum opulus	Ag	10-50 nm (Spherical)	Antibacterial effectiveness	Moldovan et al. (2017)
Pteristriparita Sw leaf	Ag	32 nm Hexagonal, spherical abd Rod-shaped	Antibacterial effectiveness	Baskaran et al. (2016)
Ginkgo biloba	Ag	15-500 nm (cubic)	Antibacterial effectiveness	Song et al.(2009)
Terminalia catappa, Terminalia mellueri, Terminalia bentazoe, Terminalia bellerica	Ag	7-10-11 nm (spherical)	Antibacterial effectiveness	Mohamed El-Rafie and Hamed (2014)
Calophyllum tomentosum	Ag	Spherical	Antibacterial effectiveness	Govindappa et al. (2018)
Piper nigrum fruit	Ag	40-100 nm Spherical, Cuboidal	Antibacterial effectiveness	Seethalakshmi (2015)
Salvia officinalis leaf	Ag	16 nm Spherical	Antibacterial effectiveness	Baharara et al. (2017)
Litchi chinensis	Au, Ag	-	Antibacterial effectiveness	Murad et al. (2018)
Prunus domestica gum	Au, Ag	7-30 nm Spherical	Antibacterial effectiveness	Islam et al. (2017)
Caesalpinia pulcherrima	Ag	9 nm (spherical)	Antibacterial effectiveness	Moteriya et al. (2020)
Mentha piperita	Ag	35 nm (spherical)	Antibacterial effectiveness	Khatoon et al. (2018)

continued on following page

Table 1. Continued

Plants name(s)	NM(s)	Size (nm) \ Shape(s)	Field of application (in dentistry)	Reference(s)
Hibiscus rosa sinensis	Ag	14 nm (spherical/prism)	Antibacterial effectiveness	Philip (2010)
Cinnamomum camphora	Ag	3.2-20 nm (Cubic hexagonal crystalline)	Antibacterial effectiveness	Huang et al. (2008)
Rhododedendron dauricam	Ag	25-40 nm (spherical)	Antibacterial effectiveness	Mittal et al. (2012)
Tribulus terrestris	Ag	16-28 nm (spherical)	Antibacterial effectiveness	Gopinath et al. (2012)
Sesuvium portulacastrum	Ag	5-20 nm (spherical)	Antibacterial effectiveness	Nabikhan et al. (2010)
Camelia Sinensis	Ag,Ag@SiO_2	10,64 nm (spherical) 11,13 nm (spherical)	Antimicrobial effectiveness and obtaining aesthetic color in dental applications	Rodrigues et al. (2020)
Oryza sativa	Ag	346.4 ± 36.8 nm	Antimicrobial effectiveness on microorganisms that cause dental caries	Suwan et al. (2018)
Aloe vera	Ag	-	Endodontic root-canal disinfection	Haripriya and Ajitha (2017)
Curcuma aromatica	Ag	10–30 nm (irregular shape)	S. mutans biofilm inhibition, PMMA/AgNP nanocoating	Thomas et al. (2018)
Rubus fruticosus, Rubus idaeus	Ag	-	Antibacterial effectiveness	Ekrikaya at al. (2021)
Camelia sinensis	Ag(GO)	-	Antibacterial effectiveness in dental adhesives	Arslan at al. (2023)
Ledebouria revoluta	Ti	47 nm (tetragonal)	Antimicrobial properties, improve biocompatibility and used in dental implant	Aswini et al. (2021)
Pouteria campechiana	Ti	73-140 nm (spherical)	Antimicrobial properties, improve biocompatibility and used in dental implant	Narayanan et al. (2021)
Syzygium cumini	Ti	22 nm (spherical round)	Used in dental implant	Sethy et al. (2020)
Nyctanthes arbor-tristis	Ti	100-150, 100-200 nm	Offer antimicrobial properties and used in dental implant	Sundrarajan et al. (2011)
Citrus sinensis	Ti	19 nm (tetragonal)	Offer antimicrobial properties and used in dental implant	Rao et al. (2015)
Hibiscus-rosa-sinensis	Ti	7-24 nm (spherical)	Offer antimicrobial properties and used in dental implant	Kumar et al. (2019)
Citrus aurantifolia	TiO_2	30–40 nm (crystalline)	İncreased mechanical strength and decreased percentage polymerization shrinkage of dental composite, antimicrobial effectiveness	Raorane et al. (2019)
Mentha arvensis	TiO_2	20–70 nm (crystalline)	Antimicrobial effectiveness	Ahmad et al. (2020)
Nephelium lappaceum (peel)	ZnO	20 nm (spherical)	Boost antibacterial effectivenees	Yuvakkumar et al. (2014)
Ginger	ZnO	23-26 nm (crystalline)	Boost antibacterial effectivenees Used in sealing material	Janaki et al. (2015)

continued on following page

Table 1. Continued

Plants name(s)	NM(s)	Size (nm) \ Shape(s)	Field of application (in dentistry)	Reference(s)
Polygala tenuifolia	ZnO	33-73.5 nm Spherical	Boost antibacterial effectivenees Used in sealing material	Zikalala et al. (2018)
Coptidis rhizome	ZnO	2.9-25.2 nm (Spherical and rod shaped)	Offer antimicrobial activity and used in dental implant	Nagajyothi et al. (2014)
Aloe vera	ZnO	22.18 nm (hexagonal)	Antibacterial effectivenees	Varghese et al. (2015)
Curcuma pseudomontana	Au	20 nm (spherical)	Uses in gene expression, catalysis, nanoelectronics, and disease diagnostics	Muniyappan et al. (2021)
Sorbus aucuparia	Au	16-18 nm (Spherical, triangular, hexagonal)	Uses in gene expression, catalysis, nanoelectronics, and disease diagnostics	Dubey et al. (2010)
Cinnamon	Au	35 nm (spherical)	Uses in gene expression, catalysis, nanoelectronics, and disease diagnostics	El Mitwalli et al. (2020)
Croton caudatus	Au	20 nm Spherical	Uses in gene expression, catalysis, nanoelectronics, and disease diagnostics	Kumar et al. (2019)
Canna indica, Quisqualis indica	Au	30-130 nm (Polymorphic/ stable)	Uses in gene expression, catalysis, nanoelectronics, and disease diagnostics	Shukla et al. (2012)
Sansevieria raxburghiana	Au	-	Uses in gene expression, catalysis, nanoelectronics, and disease diagnostics	Kumar et al. (2019)
Salacia chinensis	Au	1.5 ± 0.8 nm (distorted spherical shape)	Osteoinductive effectiveness during dental implant therapy	Jadhav et al. (2018)
Viburnum opulus L.	Cu	Hybrid nanoflower	Antimicrobial effectiveness	Ildiz et al. (2017)
Eucalyptus globulus	Cu	44-145 nm Nanowires	Antimicrobial effectiveness	Pinto et al. (2019)
Azadirachta indica and Aloe vera	CuO	-	Anticariogenic activity	Rajeshkumar et al. (2019)
Pisonia grandis	Zn, Mn	-	Improved microleakage and antibacterial capabilities. Used in rejenerative endodontics	Joghee et al. (2019)
Cardamom and clove	ZrO_2	5 -20 nm (spherical)	Coating orthodontic archwires, Antimicrobial and antienflamatuar effectiveness	Selvaraj et al. (2021)
Peganum harmala	Pd	22.5 ± 5.7 nm (spherical)	Antimicrobial effectiveness	Fahmy et al. (2021)
Coleus amboinicus	Pd	40-50 nm (spherical)	Antimicrobial effectiveness	Bathula et al. (2020)
Anogeissus latifolia	Pd	4.8 ± 1.6 nm (spherical)	Antimicrobial effectiveness	Kora et al. (2018)
Filicium decipiens	Pd	2-22 nm (spherical)	Antimicrobial effectiveness	Sharmila et al. (2017)
Musa paradisica	Pd	50 nm (crystalline)	Antimicrobial effectiveness	Bankar et al. (2010)

and biofilm on the surfaces of appliances, they may contribute to the development of demineralized lesions on the enamel, known as white spot lesions, disrupting the balance between demineralization and remineralization (Boke et al., 2014; 1999; Diamanti-Kipioti et al., 1987).

NPs commonly used in orthodontic applications are integrated into orthodontic materials or applied as a special coating on the surface of the material (Padmanabhan, 2023). The surface structure of implants, the gold standard treatment option for tooth loss, plays a critical role in ensuring biocompatibility and osteointegration. Coating implants with NPs such as TiO_2, HA, calcium phosphate ($Ca_3(PO_4)_2$), and calcium silicate (Ca_2O_4Si) can promote cell differentiation and proliferation around the implant, providing a suitable environment for early and long-term bone formation, particularly in cortical bone, where increased vascularity is necessary (Mendonça et al., 2008). Notably, a complication called peri-implantitis can lead to bone loss around the implant and ultimately to implant failure. TiO_2 nanocoatings significantly reduce microbial adhesion, surface roughness, and chemical wear on the implant surface, thus reducing the number of bacteria attached to the implant surface and the risk of peri-implantitis (Priyadarsini et al., 2018; Hossain et al., 2022).

CONCLUSION

The applications of NMs and the innovations they bring to the medical field are remarkable. NMs, especially those obtained through green synthesis methods, are more sustainable and environmentally friendly compared to other approaches. NMs with superior functional properties can be used in various fields by modifying their characteristics such as size, shape, surface coating type, and surface charge. This current chapter highlights the presence of existing innovative dental nanomaterials that can be used in these areas and the significant increase in research activities with the aim of establishing a new paradigm based on nanotechnology.

Particularly, NMs obtained through the green synthesis method, which are biocompatible, have been concluded to have significant potential applications in various technological domains, including the field of biotechnology, and especially within all branches of dentistry. In summary, the applications of nanocomposites and nanocoatings in various disciplines of dentistry are becoming widespread and contribute to the production of advanced materials with innovative functions. In the event of the development and market introduction of such innovative nano materials, the benefits for the patient and the quality of dental treatment will clearly emerge. Green synthesis biogenic nanoparticles, representing a revolutionary technology with significant innovative potential, will overcome the limitations of traditional treatments in the field of dentistry and pave the way for the development of green synthesis nanoparticle formulations.

REFERENCES

Agnihotri, R., Gaur, S., & Albin, S. (2020). Nanometals in dentistry: Applications and toxicological implications-A systematic review. *Biological Trace Element Research*, *197*(1), 70–88. doi:10.1007/s12011-019-01986-y PMID:31782063

Ahmad, W., Jaiswal, K. K., & Soni, S. (2020). Green synthesis of titanium dioxide (TiO2) nanoparticles by using Mentha arvensis leaves extract and its antimicrobialproperties. *Inorganic and Nano-Metal Chemistry*, *50*(10), 1032–1038. doi:10.1080/24701556.2020.1732419

Akhtar, M. S., Panwar, J., & Yun, Y. S. (2013). Biogenic synthesis of metallic nanoparticles by plant extracts. *ACS Sustainable Chemistry & Engineering*, *1*(6), 591–602. doi:10.1021/sc300118u

Al-Nafori, M. K., Elshal, M. G., & Refai, W. M. (2017). The effect of incorporating gold and silver nanoparticles in orthodontic adhesive system on bond strength of orthodontic bracket. *EC Dental Science*, *11*(4), 119–131.

Allaker, R. P. (2010). The use of nanoparticles to control oral biofilm formation. *Journal of Dental Research*, *89*(11), 1175–1186. doi:10.1177/0022034510377794 PMID:20739694

Ariga, K. (2023). Molecular nanoarchitectonics: Unification of nanotechnology and molecular/materials science. *Beilstein Journal of Nanotechnology*, *14*(1), 434–453. Advance online publication. doi:10.3762/bjnano.14.35 PMID:37091285

Arslan, S., Ekrikaya, S., Ildiz, N., Yusufbeyoglu, S., & Ocsoy, I. (2023). Evaluation of the antibacterial activity of dental adhesive containing biogenic silver nanoparticles decorated nanographene oxide nanocomposites (Ag@ nGO NCs) and effect on bond strength to dentine. *Odontology*, 1–14. doi:10.1007/s10266-023-00836-7 PMID:37436660

Aslan, T., Dadi, Ş., Kafdag, O., Temur, N., Ildiz, N., Ocsoy, I., & Ustun, Y. (2023). Rational design of EDTA-incorporated nanoflowers as novel and effective endodontic disinfection against biofilms. *Odontology*, 1–9. doi:10.1007/s10266-023-00857-2 PMID:37787827

Asmathunisha, N., & Kathiresan, K. (2013). A review on biosynthesis of nanoparticles by marine organisms. *Colloids and Surfaces. B, Biointerfaces*, *103*, 283–287. doi:10.1016/j.colsurfb.2012.10.030 PMID:23202242

Aswini, R., Murugesan, S., & Kannan, K. (2021). Bio-engineered TiO_2 nanoparticles using Ledebouria revoluta extract: Larvicidal, histopathological, antibacterial and anticancer activity. *International Journal of Environmental Analytical Chemistry*, *101*(15), 2926–2936. doi:10.1080/03067319.2020.1718668

Bacali, C., Baldea, I., Moldovan, M., Carpa, R., Olteanu, D. E., Filip, G. A., & Badea, F. (2020). Flexural strength, biocompatibility, and antimicrobial activity of a polymethyl methacrylate denture resin enhanced with graphene and silver nanoparticles. *Clinical Oral Investigations*, *24*(8), 2713–2725. doi:10.1007/s00784-019-03133-2 PMID:31734793

Bahara, J., Ramezani, T., Mousavi, M., & Asadi-Samani, M. (2017). Antioxidant and anti- inflammatory activity of green synthesized silver nanoparticles using Salvia officinalis extract. *Annals of Tropical Medicine and Public Health*, *10*(5), 1265–1270. doi:10.4103/ATMPH.ATMPH_174_17

Balhaddad, A. A., Kansara, A. A., Hidan, D., Weir, M. D., Xu, H. H., & Melo, M. A. S. (2019). Toward dental caries: Exploring nanoparticle-based platforms and calcium phosphate compounds for dental restorative materials. *Bioactive Materials*, *4*, 43–55. doi:10.1016/j.bioactmat.2018.12.002 PMID:30582079

Bankar, A., Joshi, B., Kumar, A. R., & Zinjarde, S. (2010). Banana peel extract mediated novel route for the synthesis of silver nanoparticles. *Colloids and Surfaces. A, Physicochemical and Engineering Aspects*, *368*(1-3), 58–63. doi:10.1016/j.colsurfa.2010.07.024

Bapat, R. A., Chaubal, T. V., Dharmadhikari, S., Abdulla, A. M., Bapat, P., Alexander, A., & Kesharwani, P. (2020). Recent advances of gold nanoparticles as biomaterial indentistry. *International Journal of Pharmaceutics*, *586*, 119596. doi:10.1016/j.ijpharm.2020.119596 PMID:32622805

Baskaran, X., Geo Vigila, A. V., Parimelazhagan, T., Muralidhara-Rao, D., & Zhang, S. (2016). Biosynthesis, characterization, and evaluation of bioactivities of leaf extract-mediated biocompatible silver nanoparticles from an early tracheophyte, Pteris tripartita Sw. *International Journal of Nanomedicine*, *11*, 5789–5806. doi:10.2147/IJN.S108208 PMID:27895478

Bathula, C., Subalakshmi, K., Kumar, A., Yadav, H., Ramesh, S., Shinde, S., & Kim, H. (2020). Ultrasonically driven green synthesis of palladium nanoparticles by Coleusamboinicus for catalytic reduction and Suzuki-Miyaura reaction. *Colloids and Surfaces. B, Biointerfaces*, *192*, 111026. doi:10.1016/j.colsurfb.2020.111026 PMID:32344163

Bauerlein, E. (2003). Biomineralization of unicellular organisms: An unusual membrane biochemistry for the production of ınorganic nano-and microstructures. *Angewandte Chemie International Edition*, *42*(6), 614–641. doi:10.1002/anie.200390176 PMID:12574993

Boke, F., Gazioglu, C., Akkaya, S., & Akkaya, M. (2014). Relationship between orthodontic treatment and gingival health: A retrospective study. *European Journal of Dentistry*, *8*(03), 373–380. doi:10.4103/1305-7456.137651 PMID:25202219

Celik, C., Ildiz, N., & Ocsoy, I. (2020). Building block and rapid synthesis of catecholamines-inorganic nanoflowers with their peroxidase-mimicking and antimicrobial activities. *Scientific Reports*, *10*(1), 2903. doi:10.1038/s41598-020-59699-5 PMID:32075999

Chang, H. S., Walsh, L. J., & Freer, T. J. (1999). The effect of orthodontic treatment onsalivary flow, pH, buffer capacity, and levels of mutans streptococci and lacto bacilli. *Australian Orthodontic Journal*, *15*(4), 229–234. https://search.informit.org/doi/10.3316/informit.986899452640633 PMID:11933357

Chavali, M. S., & Nikolova, M. P. (2019). Metal oxide nanoparticles and their applications in nanotechnology. *SN Applied Sciences*, *1*(6), 607. doi:10.1007/s42452-019-0592-3

Chen, L. X., Chen, Z. W., Jiang, M., Lu, Z., Gao, C., Cai, G., & Singh, C. V. (2021). Insights on the Dual Role of Two-Dimensional Materials as Catalysts and Supports for Energy and Environmental Catalysis. *Journal of Materials Chemistry. A, Materials for Energy and Sustainability*, *9*(4), 2018–2042. doi:10.1039/D0TA08649E

Chirayil, C. J., Abraham, J., Mishra, R. K., George, S. C., & Thomas, S. (2017). Instrumental techniques for the characterization of nanoparticles. In S. Thomas, R. Thomas, A. K. Zachariah, & R. Kumar Mishra (Eds.), *Thermal and rheological measurement techniques for nanomaterials characterization* (pp. 1–36). Elsevier. doi:10.1016/B978-0-323-46139-9.00001-3

Dadi, S., Celik, C., Mandal, A. K., & Ocsoy, I. (2021). Dopamine and norepinephrine assistant-synthesized nanoflowers immobilized membrane with peroxidase mimic activity for efficient detection of model substrates. *Applied Nanoscience*, *11*(1), 117–125. doi:10.1007/s13204-020-01577-7

Damodharan, J. (2021). Nanomaterials in medicine–An overview. *Materials Today: Proceedings*, *37*, 383–385. doi:10.1016/j.matpr.2020.05.380

Danisman, H., Çelebi, F., Danışman, Ş., & Bıçakçı, A. (2022). Effects of diamond-like carbon coating on frictional and mechanical properties of orthodontic brackets: An invitro study. *Apos Trends In Orthodontıcs*, *12*(1), 13–19. doi:10.25259/APOS_79_2021

Das, I., Chattopadhyay, S., Mahato, A., Kundu, B., & De, G. (2016). Fabrication of a cubic zirconia nanocoating on a titanium dental implant with excellent adhesion, hardness andbiocompatibility. *RSC Advances*, *6*(64), 59030–59038. doi:10.1039/C6RA10661G

Demirbas, A., Büyükbezirci, K., Celik, C., Kislakci, E., Karaagac, Z., Gokturk, E., Kati, A., Cimen, B., Yilmaz, V., & Ocsoy, I. (2019). Synthesis of long-term stable gold nanoparticles benefiting from red raspberry (Rubus idaeus), strawberry (Fragaria ananassa), and blackberry (Rubus fruticosus) extracts–gold ion complexation and investigation of reaction conditions. *ACS Omega*, *4*(20), 18637–18644. doi:10.1021/acsomega.9b02469 PMID:31737823

Demirbas, A., Karaytuğ, T., Arabaci, N., Yilmaz, E. S., & Ocsoy, I. (2020). Synthesis of metallic and metal oxide nanomaterials. In N. Srivastava, M. Srivastava, P. K. Mishra, & V. K. Gupta (Eds.), *Green Synthesis of Nanomaterials for Bioenergy Applications* (pp. 99–123). Wiley., doi:10.1002/9781119576785.ch4

Diamanti-Kipioti, A., Gusberti, F. A., & Lang, N. P. (1987). Clinical and microbiological effects of fixed orthodontic appliances. *Journal of Clinical Periodontology*, *14*(6), 326–333. doi:10.1111/j.1600-051X.1987.tb00979.x PMID:3509967

Dolai, J., Mandal, K., & Jana, N. R. (2021). Nanoparticle size effects in biomedical applications. *ACS Applied Nano Materials*, *4*(7), 6471–6496. doi:10.1021/acsanm.1c00987

Dubey, S. P., Lahtinen, M., Särkkä, H., & Sillanpää, M. (2010). Bioprospective of Sorbus aucuparia leaf extract in development of silver and gold nanocolloids. *Colloids and Surfaces. B, Biointerfaces*, *80*(1), 26–33. doi:10.1016/j.colsurfb.2010.05.024 PMID:20620889

Dujardin, E., Peet, C., Stubbs, G., Culver, J. N., & Mann, S. (2003). Organization of metallic nanoparticles using tobacco mosaic virus templates. *Nano Letters*, *3*(3), 413–417. doi:10.1021/nl034004o

Ekrikaya, S., Yilmaz, E., Arslan, S., Karaaslan, R., Ildiz, N., Celik, C., & Ocsoy, I. (2023). Dentin bond strength and antimicrobial activities of universal adhesives containing silver nanoparticles synthesized with Rosa canina extract. *Clinical Oral Investigations*, *27*(11), 1–12. doi:10.1007/s00784-023-05306-6 PMID:37821654

Ekrikaya, S., Yilmaz, E., Celik, C., Demirbuga, S., Ildiz, N., Demirbas, A., & Ocsoy, I. (2021). Investigation of ellagic acid rich-berry extracts directed silver nanoparticles synthesis and their antimicrobial properties with potential mechanisms towards Enterococcus faecalis and Candida albicans. *Journal of Biotechnology*, *341*, 155–162. doi:10.1016/j.jbiotec.2021.09.020 PMID:34601019

Fahmy, S. A., Fawzy, I. M., Saleh, B. M., Issa, M. Y., Bakowsky, U., & Azzazy, H. M. E. S. (2021). Green synthesis of platinum and palladium nanoparticles using Peganum harmala L. seed alkaloids: Biological and computational studies. *Nanomaterials (Basel, Switzerland)*, *11*(4), 965. doi:10.3390/nano11040965 PMID:33918743

Fawcett, D., Verduin, J. J., Shah, M., Sharma, S. B., & Poinern, G. E. J. (2017). A review of current research into the biogenic synthesis of metal and metal oxide nanoparticles via marine algae and seagrasses. *Journal of Nanoscience*, *2017*(4), 1–15. doi:10.1155/2017/8013850

Ferrando-Magraner, E., Bellot-Arcís, C., Paredes-Gallardo, V., Almerich-Silla, J. M., García-Sanz, V., Fernández-Alonso, M., & Montiel-Company, J. M. (2020). Antibacterial propertiesof nanoparticles in dental restorative materials. A systematic review and meta-analysis. *Medicina*, *56*(2), 55. doi:10.3390/medicina56020055 PMID:32013103

Fujimori, Y., Sato, T., Hayata, T., Nagao, T., Nakayama, M., Nakayama, T., & Suzuki, K. (2012). Novel antiviral characteristics of nanosized copper (I) iodide particles showing inactivation activity against 2009 pandemic H1N1 influenza virus. *Applied and Environmental Microbiology*, *78*(4), 951–955. doi:10.1128/AEM.06284-11 PMID:22156433

Gajanan, K., & Tijare, S. N. (2018). Applications of nanomaterials. *Materials Today: Proceedings*, *5*(1), 1093–1096. doi:10.1016/j.matpr.2017.11.187

Geyter, N. (2021). Biological activity and antimicrobial property of Cu/aC: Hnanocomposites and nanolayered coatings on titanium substrates. *Materials Science and Engineering C*, *119*, 111513. doi:10.1016/j.msec.2020.111513 PMID:33321606

Ghafar, H., Khan, M. I., Sarwar, H. S., Yaqoob, S., Hussain, S. Z., Tariq, I., & Sohail, M. F. (2020). Development and characterization of bioadhesive film embedded with lignocaine and calcium fluoride nanoparticles. *AAPS PharmSciTech*, *21*(2), 1–12. doi:10.1208/s12249-019-1615-5 PMID:31912272

Giridharan, T., Masi, C., Sindhu, S., & Arumugam, P. (2014). Studies on green synthesis, characterization and anti-proliferative potential of silver nano particle using Dodonaea viscosa and Capparis decidua. *Biosciences Biotechnology Research Asia*, *11*(2), 665–673. doi:10.13005/bbra/1320

Gopinath, V., MubarakAli, D., Priyadarshini, S., Priyadharsshini, N. M., Thajuddin, N., & Velusamy, P. (2012). Biosynthesis of silver nanoparticles from Tribulus terrestris and itsantimicrobial activity: A novel biological approach. *Colloids and Surfaces. B, Biointerfaces*, *96*, 69–74. doi:10.1016/j.colsurfb.2012.03.023 PMID:22521683

Govindappa, M., Hemashekhar, B., Arthikala, M. K., Rai, V. R., & Ramachandra, Y. L. (2018). Characterization, antibacterial, antioxidant, antidiabetic, anti-inflammatory and antityrosinase activity of green synthesized silver nanoparticles using Calophyllum tomentosum leaves extract. *Results in Physics*, *9*, 400–408. doi:10.1016/j.rinp.2018.02.049

Grass, G., Rensing, C., & Solioz, M. (2011). Metallic copper as an antimicrobial surface. *Applied and Environmental Microbiology*, *77*(5), 1541–1547. doi:10.1128/AEM.02766-10 PMID:21193661

Gudimalla, A., Jose, J., Varghese, R. J., & Thomas, S. (2021). Green synthesis of silver nanoparticles using Nymphae odorata extract incorporated films and antimicrobial activity. *Journal of Polymers and the Environment*, *29*(5), 1412–1423. doi:10.1007/s10924-020-01959-6

Haripriya, S., & Ajitha, P. (2017). Antimicrobial efficacy of silver nanoparticles of Aloe vera. *Journal of Advanced Pharmacy Education and Research*, *7*(2), 2249–3379.

Hasheminya, S. M., Mokarram, R. R., Ghanbarzadeh, B., Hamishekar, H., & Kafil, H. S. (2018). Physicochemical, mechanical, optical, microstructural and antimicrobial properties ofnovel kefiran-carboxymethyl cellulose biocomposite films as influenced by copper oxide nanoparticles (CuONPs). *Food Packaging and Shelf Life*, *17*, 196–204. doi:10.1016/j.fpsl.2018.07.003

Heo, D. N., Ko, W. K., Lee, H. R., Lee, S. J., Lee, D., Um, S. H., & Kwon, I. K. (2016). Titanium dental implants surface-immobilized with gold nanoparticles as osteoinductive agents for rapid osseointegration. *Journal of Colloid and Interface Science*, *469*, 129–137. doi:10.1016/j.jcis.2016.02.022 PMID:26874978

Hossain, N., Islam, M. A., Chowdhury, M. A., & Alam, A. (2022). Advances of nanoparticles employment in dental implant applications. *Applied Surface Science Advances*, *12*, 100341. doi:10.1016/j.apsadv.2022.100341

Huang, J., Li, Q., Sun, D., Lu, Y., Su, Y., Yang, X., Wang, H., Wang, Y., Shao, W., He, N., Hong, J., & Chen, C. (2007). Biosynthesis of silver and gold nanoparticles by novel sundried Cinnamomum camphora leaf. *Nanotechnology*, *18*(10), 105104. doi:10.1088/0957-4484/18/10/105104

Hulkoti, N. I., & Taranath, T. C. (2014). Biosynthesis of nanoparticles using microbes—A review. *Colloids and Surfaces. B, Biointerfaces*, *121*, 474–483. doi:10.1016/j.colsurfb.2014.05.027 PMID:25001188

Huo, K., Zhang, X., Wang, H., Zhao, L., Liu, X., & Chu, P. K. (2013). Osteogenic activity and antibacterial effects on titanium surfaces modified with Zn-incorporated nanotube arrays. *Biomaterials*, *34*(13), 3467–3478. doi:10.1016/j.biomaterials.2013.01.071 PMID:23439134

Ildiz, N., Baldemir, A., Altinkaynak, C., Özdemir, N., Yilmaz, V., & Ocsoy, I. (2017). Self assembled snowball-like hybrid nanostructures comprising Viburnum opulus L. extract and metal ions for antimicrobial and catalytic applications. *Enzyme and Microbial Technology*, *102*, 60–66. doi:10.1016/j.enzmictec.2017.04.003 PMID:28465062

Islam, N. U., Amin, R., Shahid, M., Amin, M., Zaib, S., & Iqbal, J. (2017). A multi-target therapeutic potential of Prunus domestica gum stabilized nanoparticles exhibited prospective anticancer, antibacterial, urease-inhibition, anti-inflammatory and analgesic properties. *BMC Complementary and Alternative Medicine*, *17*(1), 1–17. doi:10.1186/s12906-017-1791-3 PMID:28535789

Jadhav, K., Deore, S., Dhamecha, D., Hr, R., Jagwani, S., Jalalpure, S., & Bohara, R. (2018). Phytosynthesis of silver nanoparticles: Characterization, biocompatibility studies, and anticancer activity. *ACS Biomaterials Science & Engineering*, *4*(3), 892–899. doi:10.1021/acsbiomaterials.7b00707 PMID:33418773

Jadhav, K., Rajeshwari, H. R., Deshpande, S., Jagwani, S., Dhamecha, D., Jalalpure, S., & Baheti, D. (2018). Phytosynthesis of gold nanoparticles: Characterization, biocompatibility, and evaluation of its osteoinductive potential for application in implant dentistry. *Materials Science and Engineering C*, *93*, 664–670. doi:10.1016/j.msec.2018.08.028 PMID:30274099

Janaki, A. C., Sailatha, E., & Gunasekaran, S. (2015). Synthesis, characteristics and antimicrobial activity of ZnO nanoparticles. *Spectrochimica Acta. Part A: Molecular and Biomolecular Spectroscopy*, *144*, 17–22. doi:10.1016/j.saa.2015.02.041 PMID:25748589

Jandt, K. D., & Watts, D. C. (2020). Nanotechnology in dentistry: Present and future perspectives on dental nanomaterials. *Dental Materials*, *36*(11), 1365–1378. doi:10.1016/j.dental.2020.08.006 PMID:32981749

Joghee, S., Ganeshan, P., Vincent, A., & Hong, S. I.Leaf Extract and Their Antimicrobial Activity. (2019). Ecofriendly biosynthesis of zinc oxide and magnesium oxide particles from medicinal plant Pisonia grandis R. Br. leaf extract and their antimicrobial activity. *BioNanoScience*, *9*(1), 141–154. doi:10.1007/s12668-018-0573-9

Karaagac, Z., Dadi, S., Ekrikaya, S., & Ocsoy, I. (2021). Plant Extract Guided Nanomaterials and Plant Disease Management. In M. A. Mallick, M. K. Solanki, B. Kumari, & S. K. Verma (Eds.), *In Nanotechnology in Sustainable Agriculture* (pp. 271–290). CRC Press.

Karnani, R. L., & Chowdhary, A. (2013). Biosynthesis of silver nanoparticle by eco-friendly method. *Indian journal of Nanoscience*, 1(1), 25-31.

Khan, I., Saeed, K., & Khan, I. (2019). Nanoparticles: Properties, applications and toxicities. *Arabian Journal of Chemistry*, *12*(7), 908–931. doi:10.1016/j.arabjc.2017.05.011

Khatoon, A., Khan, F., Ahmad, N., Shaikh, S., Rizvi, S. M. D., Shakil, S., Al-Qahtani, M. H., Abuzenadah, A. M., Tabrez, S., Ahmed, A. B. F., Alafnan, A., Islam, H., Iqbal, D., & Dutta, R. (2018). Silver nanoparticles from leaf extract of Mentha piperita: Eco-friendly synthesis andeffect on acetylcholinesterase activity. *Life Sciences*, *209*, 430–434. doi:10.1016/j.lfs.2018.08.046 PMID:30138593

Khurshid, Z., Zafar, M., Qasim, S., Shahab, S., Naseem, M., & AbuReqaiba, A. (2015). Advances in nanotechnology for restorative dentistry. *Materials (Basel)*, *8*(2), 717–731. doi:10.3390/ma8020717 PMID:28787967

Koca, F. D., Yilmaz, D. D., Onmaz, N. E., & Ocsoy, I. (2020). Peroxidase-like activity and antimicrobial properties of curcumin-inorganic hybrid nanostructure. *Saudi Journal of Biological Sciences*, *27*(10), 2574–2579. doi:10.1016/j.sjbs.2020.05.025 PMID:32994713

Kora, A. J., & Rastogi, L. (2018). Green synthesis of palladium nanoparticles using gum ghatti (Anogeissus latifolia) and its application as an antioxidant and catalyst. *Arabian Journal of Chemistry*, *11*(7), 1097–1106. doi:10.1016/j.arabjc.2015.06.024

Kumar, P. S. M., Francis, A. P., & Devasena, T. (2014). Biosynthesized and chemically synthesized titania nanoparticles: Comparative analysis of antibacterial activity. *Journal of Environmental Nanotechnology*, *3*(3), 73–81. doi:10.13074/jent.2014.09.143098

Kumar, P. V., Kala, S. M. J., & Prakash, K. S. (2019). Green synthesis of gold nanoparticles using Croton Caudatus Geisel leaf extract and their biological studies. *Materials Letters*, *236*, 19–22. doi:10.1016/j.matlet.2018.10.025

Kumaran, N. S., & Vijayaraj, R. (2017). Biosynthesis of silver nano particles from Leucas aspera (Willd.) link and its anti-inflammatory potential against carrageen induced paw edema in rats. *International Journal of Pharmaceutical Sciences and Research*, *8*(6), 2588–2593. doi:10.13040/IJPSR.0975-8232.8(6).2588-93

Lee, J. H., Jo, J. K., Kim, D. A., Patel, K. D., Kim, H. W., & Lee, H. H. (2018). Nano-graphene oxide incorporated into PMMA resin to prevent microbial adhesion. *Dental Materials*, *34*(4), e63–e72. doi:10.1016/j.dental.2018.01.019 PMID:29402540

LewisOscar, F., MubarakAli, D., Nithya, C., Priyanka, R., Gopinath, V., Alharbi, N. S., & Thajuddin, N. (2015). One pot synthesis and anti-biofilm potential of copper nanoparticles(CuNPs) against clinical strains of Pseudomonas aeruginosa. *Biofouling*, *31*(4), 379–391. doi:10.1080/08927014.2015.1048686 PMID:26057498

Li, Y., Yang, C., Yin, X., Sun, Y., Weng, J., Zhou, J., & Feng, B. (2019). Inflammatory responses to micro/nano-structured titanium surfaces with silver nanoparticles in vitro. *Journal of Materials Chemistry. B*, *7*(22), 3546–3559. doi:10.1039/C8TB03245A

Lohse, S. E., & Murphy, C. J. (2012). Applications of colloidal inorganic nanoparticles: From medicine to energy. *Journal of the American Chemical Society*, *134*(38), 15607–15620. doi:10.1021/ja307589n PMID:22934680

Lu, J., Guo, J., Song, S., Yu, G., Liu, H., Yang, X., & Lu, Z. (2020). Preparation of Ag nanoparticles by spark ablation in gas as catalysts for electrocatalytic hydrogen production. *RSC Advances*, *10*(63), 38583–38587. doi:10.1039/D0RA06682F PMID:35517560

Maliael, M. T., Jain, R. K., & Srirengalakshmi, M. (2022). Effect of nanoparticle coatings on frictional resistance of orthodontic archwires: A systematic review and meta-analysis. *World Journal of Dentistry*, *13*(4), 417–424. doi:10.5005/jp-journals-10015-2066

Malik, S., Muhammad, K., & Waheed, Y. (2023). Nanotechnology: A revolution in modern industry. *Molecules (Basel, Switzerland)*, *28*(2), 661. doi:10.3390/molecules28020661 PMID:36677717

Mendonça, G., Mendonça, D. B., Aragao, F. J., & Cooper, L. F. (2008). Advancing dental implant surface technology–from micron-to nanotopography. *Biomaterials*, *29*(28), 3822–3835. doi:10.1016/j.biomaterials.2008.05.012 PMID:18617258

Mittal, A. K., Chisti, Y., & Banerjee, U. C. (2013). Synthesis of metallic nanoparticles usingplant extracts. *Biotechnology Advances*, *31*(2), 346–356. doi:10.1016/j.biotechadv.2013.01.003 PMID:23318667

Mittal, A. K., Kaler, A., & Banerjee, U. C. (2012). Free Radical Scavenging and AntioxidantActivity of Silver Nanoparticles Synthesized from Flower Extract of Rhododendron dauricum. *Nano Biomedicine and Engineering*, *4*(3). doi:10.5101/nbe.v4i3.p118-124

Moghaddam, A., Namvar, F., Moniri, M., Md. Tahir, P., Azizi, S., & Mohamad, R. (2015). Nanoparticles biosynthesized by fungi and yeast: A review of their preparation, properties, and medical applications. *Molecules (Basel, Switzerland)*, *20*(9), 16540–16565. doi:10.3390/molecules200916540 PMID:26378513

Mohamed El-Rafie, H., & Hamed, M. A. A. (2014). Antioxidant and anti-inflammatory activities of silver nanoparticles biosynthesized from aqueous leaves extracts of four Terminalia species. *Adv. Nat. Sci. Nanosci. Nanotechnol*, *5*(3), 1–11. doi:10.1088/2043-6262/5/3/035008

Moldovan, B., David, L., Vulcu, A., Olenic, L., Perde-Schrepler, M., Fischer-Fodor, E., Baldea, I., Clichici, S., & Filip, G. A. (2017). In vitro and in vivo anti-inflammatory properties of green synthesized silvernanoparticles using Viburnum opulus L. fruits extract. *Materials Science and Engineering C*, *79*, 720–727. doi:10.1016/j.msec.2017.05.122 PMID:28629073

Morales-Lozoya, V., Espinoza-Gómez, H., Flores-López, L. Z., Sotelo-Barrera, E. L., Núñez-Rivera, A., Cadena-Nava, R. D., & Rivero, I. A. (2021). Study of the effect of the differentparts of Morinda citrifolia L.(noni) on the green synthesis of silver nanoparticles and theirantibacterial activity. *Applied Surface Science*, *537*, 147855. doi:10.1016/j.apsusc.2020.147855

Moteriya, P., & Chanda, S. (2020). Green Synthesis of Silver Nanoparticles fromCaesalpinia pulcherrima Leaf Extract and Evaluation of Their Antimicrobial, Cytotoxic and Genotoxic Potential (3-in-1 System). *Journal of Inorganic and Organometallic Polymers and Materials*, *30*(10), 3920–3932. doi:10.1007/s10904-020-01532-7

Muniyappan, N., Pandeeswaran, M., & Amalraj, A. (2021). Green synthesis of gold nanoparticles using Curcuma pseudomontana isolated curcumin: Its characterization, antimicrobial, antioxidant and anti-inflammatory activities. *Environmental Chemistry and Ecotoxicology*, *3*, 117–124. doi:10.1016/j.enceco.2021.01.002

Murad, U., Khan, S. A., Ibrar, M., Ullah, S., & Khattak, U. (2018). Synthesis of silver and gold nanoparticles from leaf of Litchi chinensis and its biological activities. *Asian Pacific Journal of Tropical Biomedicine*, *8*(3), 142. doi:10.4103/2221-1691.227995

Nabikhan, A., Kandasamy, K., Raj, A., & Alikunhi, N. M. (2010). Synthesis of antimicrobialsilver nanoparticles by callus and leaf extracts from saltmarsh plant, Sesuviumportulacastrum L. *Colloids and Surfaces. B, Biointerfaces*, *79*(2), 488–493. doi:10.1016/j.colsurfb.2010.05.018 PMID:20627485

Nagajyothi, P. C., Sreekanth, T. V. M., Tettey, C. O., Jun, Y. I., & Mook, S. H. (2014). Characterization, antibacterial, antioxidant, and cytotoxic activities of ZnO nanoparticles using Coptidis Rhizoma. *Bioorganic & Medicinal Chemistry Letters*, *24*(17), 4298–4303. doi:10.1016/j.bmcl.2014.07.023 PMID:25088397

Nair, R. R., Blake, P., Grigorenko, A. N., Novoselov, K. S., Booth, T. J., Stauber, T., & Geim, A. K. (2008). Fine structure constant defines visual transparency of graphene. *science, 320*(5881), 1308-1308. https://doi.org/ doi:10.1126/science.115696

Narayanan, M., Devi, P. G., Natarajan, D., Kandasamy, S., Devarayan, K., Alsehli, M., & Pugazhendhi, A. (2021). Green synthesis and characterization of titanium dioxide nanoparticles using leaf extract of Pouteria campechiana and larvicidal and pupicidal activity on Aedes aegypti. *Environmental Research*, *200*, 111333. doi:10.1016/j.envres.2021.111333 PMID:34051198

Nath, D., & Banerjee, P. (2013). Green nanotechnology–a new hope for medicalbiology. *Environmental Toxicology and Pharmacology*, *36*(3), 997–1014. doi:10.1016/j.etap.2013.09.002 PMID:24095717

Nilavukkarasi, M., Vijayakumar, S., & Kumar, S. P. (2020). Biological synthesis and characterization of silver nanoparticles with Capparis zeylanica L. leaf extract for potentantimicrobial and anti proliferation efficiency. *Materials Science for Energy Technologies*, *3*, 371–376. doi:10.1016/j.mset.2020.02.008

Ocsoy, I., Tasdemir, D., Mazicioglu, S., & Tan, W. (2018). Nanotechnology in plants. In R. Varshney, M. Pandey, & A. Chitikineni (Eds.), *Plant Genetics and Molecular Biology* (pp. 263–275). Springer., doi:10.1007/10_2017_53

Ocsoy, I., Temiz, M., Celik, C., Altinsoy, B., Yilmaz, V., & Duman, F. (2017). A green approach for formation of silver nanoparticles on magnetic graphene oxide and highlyeffective antimicrobial activity and reusability. *Journal of Molecular Liquids*, *227*, 147–152. doi:10.1016/j.molliq.2016.12.015

Osman, R. B., & Swain, M. V. (2015). A critical review of dental implant materials with anemphasis on titanium versus zirconia. *Materials (Basel)*, *8*(3), 932–958. doi:10.3390/ma8030932 PMID:28787980

Padmanabhan, S. (2023). *No. 1* (Vol. 29). Nanotechnology in Orthodontics. In Seminars in Orthodontics. WB Saunders., doi:10.1053/j.sodo.2023.01.005

Patra, J. K., & Baek, K. H. (2016). Biosynthesis of silver nanoparticles usingaqueous extract of silky hairs of corn and investigation of its antibacterial and anticandidal synergistic activity and antioxidant potential. *IET Nanobiotechnology / IET*, *10*(5), 326–333. doi:10.1049/iet-nbt.2015.0102 PMID:27676382

Philip, D. (2010). Green synthesis of gold and silver nanoparticles using Hibiscus rosa sinensis. *Phys. E Low-Dimens. Physica E, Low-Dimensional Systems and Nanostructures*, *42*(5), 1417–1424. doi:10.1016/j.physe.2009.11.081

Pinto, R. J., Lucas, J. M., Silva, F. M., Girão, A. V., Oliveira, F. J., Marques, P. A., & Freire, C. S. (2019). Bio-based synthesis of oxidation resistant copper nanowires using an aqueous plant extract. *Journal of Cleaner Production*, *221*, 122–131. doi:10.1016/j.jclepro.2019.02.189

Poggio, C., Lombardini, M., Colombo, M., Dagna, A., Saino, E., Arciola, C. R., & Visai, L. (2011). Antibacterial effects of six endodontic sealers. *The International Journal of Artificial Organs*, *34*(9), 908–913. doi:10.5301/ijao.5000055 PMID:22094573

Prabhu, B. M., Ali, S. F., Murdock, R. C., Hussain, S. M., & Srivatsan, M. (2010). Copper nanoparticles exert size and concentration dependent toxicity on somatosensory neurons of rat. *Nanotoxicology*, *4*(2), 150–160. doi:10.3109/17435390903337693 PMID:20543894

Prasad, R., Bhattacharyya, A., & Nguyen, Q. D. (2017). Nanotechnology in sustainable agriculture: Recent developments, challenges, and perspectives. *Frontiers in Microbiology*, *8*, 1014. doi:10.3389/fmicb.2017.01014 PMID:28676790

Prasad, R., Pandey, R., & Barman, I. (2016). Engineering tailored nanoparticles withmicrobes: Quo vadis? *Wiley Interdisciplinary Reviews. Nanomedicine and Nanobiotechnology*, *8*(2), 316–330. doi:10.1002/wnan.1363 PMID:26271947

Rajeshkumar, S., Bharath, L. V., & Geetha, R. (2019). Broad spectrum antibacterial silver nanoparticle green synthesis: Characterization, and mechanism of action. In A. K. Shukla & S. Iravani (Eds.), *In Micro and Nano Technologies,Green synthesis, characterization and applications of nanoparticles* (pp. 429–444). Elsevier. doi:10.1016/B978-0-08-102579-6.00018-6

Ramos, A. P., Cruz, M. A., Tovani, C. B., & Ciancaglini, P. (2017). Biomedical applications of nanotechnology. *Biophysical Reviews*, *9*(2), 79–89. doi:10.1007/s12551-016-0246-2 PMID:28510082

Raorane, D. V., Chaughule, R. S., Pednekar, S. R., & Lokur, A. (2019). Experimentalsynthesis of size-controlled TiO2 nanofillers and their possible use as composites in restorative dentistry. *The Saudi Dental Journal*, *31*(2), 194–203. doi:10.1016/j.sdentj.2019.01.008 PMID:30983829

Raura, N., Garg, A., Arora, A., & Roma, M. (2020). Nanoparticle technology and its implications in endodontics: A review. *Biomaterials Research*, *24*(1), 1–8. doi:10.1186/s40824-020-00198-z PMID:33292702

Rodrigues, M. C., Rolim, W. R., Viana, M. M., Souza, T. R., Gonçalves, F., Tanaka, C. J., Bueno-Silva, B., & Seabra, A. B. (2020). Biogenic synthesis and antimicrobial activity of silica-coated silvernanoparticles for esthetic dental applications. *Journal of Dentistry*, *96*, 103327. doi:10.1016/j.jdent.2020.103327 PMID:32229160

Selvaraj, A., George, A. M., & Rajeshkumar, S. (2021). Efficacy of zirconium oxidenanoparticles coated on stainless steel and nickel titanium wires in orthodontictreatment. *Bioinformation*, *17*(8), 760. doi:10.6026/97320630017760 PMID:35540697

Sethy, N. K., Arif, Z., Mishra, P. K., & Kumar, P. (2020). Green synthesis of TiO2 nanoparticles from Syzygium cumini extract for photo-catalytic removal of lead (Pb) in explosive industrial wastewater. *Green Processing and Synthesis*, *9*(1), 171–181. doi:10.1515/gps-2020-0018

Shah, M., Fawcett, D., Sharma, S., Tripathy, S. K., & Poinern, G. E. J. (2015). Greensynthesis of metallic nanoparticles via biological entities. *Materials (Basel)*, *8*(11), 7278–7308. doi:10.3390/ma8115377 PMID:28793638

Shankar, S. S., Rai, A., Ahmad, A., & Sastry, M. (2004). Rapid synthesis of Au, Ag, and bimetallic Au core–Ag shell nanoparticles using Neem (Azadirachta indica) leaf broth. *Journal of Colloid and Interface Science*, *275*(2), 496–502. doi:10.1016/j.jcis.2004.03.003 PMID:15178278

Sharma, N. K., Vishwakarma, J., Rai, S., Alomar, T. S., AlMasoud, N., & Bhattarai, A. (2022). Green route synthesis and characterization techniques of silver nanoparticles and their biological adeptness. *ACS Omega*, *7*(31), 27004–27020. doi:10.1021/acsomega.2c01400 PMID:35967040

Sharma, V. K., Yngard, R. A., & Lin, Y. (2009). Silver nanoparticles: Green synthesis and theirantimicrobial activities. *Advances in Colloid and Interface Science*, *145*(1-2), 83–96. doi:10.1016/j.cis.2008.09.002 PMID:18945421

Sharmila, G., Fathima, M. F., Haries, S., Geetha, S., Kumar, N. M., & Muthukumaran, C. (2017). Green synthesis, characterization and antibacterial efficacy of palladium nanoparticlessynthesized using Filicium decipiens leaf extract. *Journal of Molecular Structure*, *1138*, 35–40. doi:10.1016/j.molstruc.2017.02.097

Shen, X., Hu, Y., Xu, G., Chen, W., Xu, K., Ran, Q., Ma, P., Zhang, Y., Li, J., & Cai, K. (2014). Regulation of the biological functions of osteoblasts and bone formation by Zn-incorporated coating on microrough titanium. *ACS Applied Materials & Interfaces*, *6*(18), 16426–16440. doi:10.1021/am5049338 PMID:25148131

Shreyash, N., Bajpai, S., Khan, M. A., Vijay, Y., Tiwary, S. K., & Sonker, M. (2021). Green synthesis of nanoparticles and their biomedical applications: A review. *ACS Applied Nano Materials*, *4*(11), 11428–11457. doi:10.1021/acsanm.1c02946

Shukla, D., & Vankar, P. S. (2012). Synthesis of plant parts mediated gold nanoparticles. *International Journal of Green Nanotechnology*, *4*(3), 277–288. do i:10.1080/19430892.2012.706175

Skrabalak, S. E., Chen, J., Sun, Y., Lu, X., Au, L., Cobley, C. M., & Xia, Y. (2008). Gold nanocages: Synthesis, properties, and applications. *Accounts of Chemical Research*, *41*(12), 1587–1595. doi:10.1021/ar800018v PMID:18570442

Slavin, Y. N., Asnis, J., Hńfeli, U. O., & Bach, H. (2017). Metal nanoparticles: Understanding the mechanisms behind antibacterial activity. *Journal of Nanobiotechnology*, *15*(1), 1–20. doi:10.1186/s12951-017-0308-z PMID:28974225

Some, S., Sen, I. K., Mandal, A., Aslan, T., Ustun, Y., Yilmaz, E. Ş., & Ocsoy, I. (2018).Biosynthesis of silver nanoparticles and their versatile antimicrobialproperties. *Materials Research Express*, *6*(1), 01200. https://doi.org/ doi:10.1088/2053-1591/aae23e

Song, J. Y., & Kim, B. S. (2009). Rapid biological synthesis of silver nanoparticles using plant leaf extracts. *Bioprocess and Biosystems Engineering*, *32*(1), 79–84. doi:10.1007/s00449-008-0224-6 PMID:18438688

Sundrarajan, M., & Gowri, S. (2011). Green synthesis of titanium dioxide nanoparticles by Nyctanthes arbor-tristis leaves extract. *Chalcogenide Letters*, *8*(8), 447–451.

Suwan, T., Khongkhunthian, S., & Okonogi, S. (2018). Green synthesis and inhibitory effects against oral pathogens of silver nanoparticles mediated by rice extracts. *Drug Discoveries & Therapeutics*, *12*(4), 189–196. doi:10.5582/ddt.2018.01034 PMID:30224591

Thakkar, K. N., Mhatre, S. S., & Parikh, R. Y. (2010). Biological synthesis of metallicnanoparticles. *Nanomedicine; Nanotechnology, Biology, and Medicine*, *6*(2), 257–262. doi:10.1016/j.nano.2009.07.002 PMID:19616126

Thomas, R., Snigdha, S., Bhavitha, K. B., Babu, S., Ajith, A., & Radhakrishnan, E. K. (2018). Biofabricated silver nanoparticles incorporated polymethyl methacrylate as a dental adhesive material with antibacterial and antibiofilm activity against Streptococcus mutans. *Biotech*, *8*(9), 1–10. doi:10.1007/s13205-018-1420-y PMID:30221117

ThukkaramM.VaidulychM.KyliánO.RigoleP.AliakbarshiraziS.AsadianM., & De

Titus, D., Samuel, E. J. J., & Roopan, S. M. (2019). Nanoparticle characterization techniques. In A. K. Shukla & S. Iravani (Eds.), *In Micro and Nano Technologies, Green synthesis, characterization and applications of nanoparticles* (pp. 303–319). Elsevier., doi:10.1016/B978-0-08-102579-6.00012-5

Valsalam, S., Agastian, P., Arasu, M. V., Al-Dhabi, N. A., Ghilan, A. K. M., Kaviyarasu, K., & Arokiyaraj, S. (2019). Rapid biosynthesis and characterization of silver nanoparticles fromthe leaf extract of Tropaeolum majus L. and its enhanced in-vitro antibacterial, antifungal,antioxidant and anticancer properties. *Journal of Photochemistry and Photobiology. B, Biology*, *191*, 65–74. doi:10.1016/j.jphotobiol.2018.12.010 PMID:30594044

Varghese, E., & George, M. (2015) Green synthesis of zincoxide nanoparticles. *International Journal of Advance Research In Science And Engineering.*

Verran, J., & Maryan, C. J. (1997). Retention of Candida albicans on acrylic resin and silicone of different surface topography. *The Journal of Prosthetic Dentistry*, *77*(5), 535–539. doi:10.1016/S0022-3913(97)70148-3 PMID:9151275

Wang, N., Yu, J., Yan, J., & Hua, F. (2023). Recent advances in antibacterial coatings for orthodontic appliances. *Frontiers in Bioengineering and Biotechnology*, *11*, 1093926. doi:10.3389/fbioe.2023.1093926 PMID:36815889

Yazdanian, M., Rostamzadeh, P., Rahbar, M., Alam, M., Abbasi, K., Tahmasebi, E., & Yazdanian, A. (2022). The potential application of green-synthesized metal nanoparticles indentistry: A comprehensive review. *Bioinorganic Chemistry and Applications, 2311910*, 1–27. Advance online publication. doi:10.1155/2022/2311910 PMID:35281331

Yilmaz, O., & Yorgancioglu, A. (2019). Nanocoatings: preparation, properties, and biomedical applications. In C. Vasile (Ed.), *In Micro and Nano Technologies, Polymeric Nanomaterials In Nanotherapeutics* (pp. 299–331). Elsevier. doi:10.1016/B978-0-12-813932-5.00008-X

Ying, S., Guan, Z., Ofoegbu, P. C., Clubb, P., Rico, C., He, F., & Hong, J. (2022). Green synthesis of nanoparticles: Current developments and limitations. *Environmental Technology & Innovation, 26*, 102336. doi:10.1016/j.eti.2022.102336

Yougbare, S., Chang, T. K., Tan, S. H., Kuo, J. C., Hsu, P. H., Su, C. Y., & Kuo, T. R. (2019). Antimicrobial gold nanoclusters: Recent developments and future perspectives. *International journal of molecular sciences, 20*(12), 2924.n doi:10.3390/ijms20122924

Yuvakkumar, R., Suresh, J., & Hong, S. I. (2014). Green synthesis of zincoxide nanoparticles. *Advanced Materials Research, 952*, 137–140. . doi:10.4028/www.scientific.net/AMR.952.137

Zhang, M., Liu, X., Xie, Y., Zhang, Q., Zhang, W., Jiang, X., & Lin, J. (2020). Biological safe gold nanoparticle-modified dental aligner prevents the Porphyromonas gingivalisbiofilm formation. *ACS Omega, 5*(30), 18685–18692. doi:10.1021/acsomega.0c01532 PMID:32775870

Zikalala, N., Matshetshe, K., Parani, S., & Oluwafemi, O. S. (2018). Biosynthesis protocols for colloidal metal oxide nanoparticles. *Nano-Structures & Nano-Objects, 16*, 288–299. doi:10.1016/j.nanoso.2018.07.010

Chapter 3
Multifaceted and Diverse Applications of Nanocomposites

Sunil Jayant Kulkarni
https://orcid.org/0000-0002-5988-3448
Gharda Institute of Technology, India

ABSTRACT

The materials produced on nanoscale have a new set of properties. Nanotechnology and nanomaterials have positively affected the human lifestyle. It has an impact on agricultural, energy, petroleum, sports, chemical and catalysis, environmental, paints and coating, medical, beauty, and textile sectors. Feeding the increasing needs of the population has given rise to different innovations in various sectors. The agricultural sector also has benefitted from such innovations. Nanotechnology and nanomaterials have come to the rescue of the farmers. Micronutrients are supplied to the crops through fertilizers. Nanomaterials are explored for fuel borne catalysis for soot prevention. Nanoparticle reinforced tires have greater abrasive resistance and gas permeability. Nanoparticles are used in coating for better gloss, scratch and corrosion resistance, and anti-fog coatings. Nanoparticles and nanomaterials like cubosomes, nanodots, liposomes, dendrimers, nano-emulsions have become common ingredients in cosmetics.

INTRODUCTION

Advancements in technologies are touching the human life across the world. The technology has enabled human beings to modify plants, animals and materials,

DOI: 10.4018/979-8-3693-1094-6.ch003

Copyright © 2024, IGI Global. Copying or distributing in print or electronic forms without written permission of IGI Global is prohibited.

according to their needs. The research and investigations in biotechnology, nanotechnology and environmental technology have become very important areas of research. Bionanotechnology has potential to manipulate the plants and animal for better productivity and yield. Nanotechnology is being explored in various applications ranging from chemical engineering to farming. Environmental technology has gained importance due to increasing need to develop the materials and processes with sustainable approach. If we look carefully at these three fields, they can be integrated to provide sustainable solutions to many material synthesis and waste treatment problems. Nanotechnology is being explored due to its huge potential and endless possibilities.

"The technology in which the materials and systems are designed, synthesized, characterized and utilized by facilitating control over their size in nanoscale(1 to 100 nm)", can be termed as nanotechnology(Kreyling et al., 2010). For fabrication of nanomaterials, two approaches, namely top down and bottom up, are employed(Chaudhary et al., 2016). Carbon nanotubes are replacing silicon chip due to their semiconducting properties with better performance.

The material when disintegrated to nanosize, exhibits different properties than the original material like gold and its nano-counterpart. Gold is shiny yellow metal that absorbs green light when disintegrated to 10 nm and appears red(Roduner, 2006). The nanomaterial can be defined as the one that has minimum one dimension less than 100 nm. (Vollath, 2008). Green nanotechnology focuses on synthesis and application of nanotechnology in sustainable manner (Mahajan et al., 20022). Green synthesis of nanomaterial is widely explored. Surface reactivity and properties of nanomaterials can be enhanced by functionalization. Carbon nanotubes are functionalized by bromine (Mazov et al., 2009). Direct or indirect functionalization methods can be used for nanomaterials. Direct functionalizations include covalent, non-covalent, inorganic, doping of heteroatoms and mediated by immobilization (Mahajan et al., 2022). Grafting is indirect method of functionalization. Nanotechnology and nanomaterials have positively affected the lifestyle of human. It has impact on agricultural, energy, petroleum, sports, chemical and catalysis, environmental, paints and coating, medical, beauty and textile sectors. Over next 30 years global energy demand is predicted to increase by 60%(Esmaeili, 2011). One way to partially counter this issue is, use of effective transmission of energy. Nanotechnology can provide solutions to the issuesconcerningaccess to stranded natural-gas resources. Development of nanocatalysts and nanoscale membranes for gas to liquid production can serve this purpose(Esmaeili, 2011). Nanostructured material can be developed for transport of compressed-natural-gas or electricity transmission between long distance points. In oil industry, for offshore drilling, light weight, stronger, and more corrosion-resistant structural materials can be used in platforms. Investigators have reported 50% reduction in urea requirement by using Hydroxypatite Nanoparticle with urea (Kottegoda et al.,2017). However, some experts have expressed concerns about acute health effects of Hydroxypatite Nanoparticles (Suppan, 2017; Scientific Committee on Consumer Safety, 2016).

Nanoparticles are nanosize particles with three external dimensions of nanosize, whereas nanomaterials have any internal or external structure with nanosize(Dobson et al., 2019). In nanomaterials at least one dimension is of 1 to 100 nm size(Armarego, 2022). Nanoparticles are isolated solid phase objects of size 1 to 100 nm. Carbon nanotubes, graphene, nanowire and fullerenes are examples of nanomaterials. Zinc sulphide, cadmium sulphide, gold, silver and aluminium are examples of nanoparticles. Nanomaterials are made from nanoparticles. In nanocomposites, one of the phases have at least one dimension in nano range of 1 to 100 nm. The nanocomposites are classified metal matrix, ceramic and polymer matrix (Ducheyne,2015; Pacheco and Buzea, 2021; Malhotra, et al., 2018).

Recent investigations on 2D nanomaterials indicate that 2 D nanomaterials derived from graphene have applications in energy storage, sensors, capacitors and semiconductors(Yadav A. et al.,2023). As properties of nanomaterials depend on size to volume ratio, in 2D nanomaterial processing, one can have better control. Nanomaterials have become important tool for treating soil and water for removal of contaminants because of their reactivity(Mukhopadhyay et al., 2021). Natural polysaccharides can be used for preparation of nanomaterials for their application in water treatment(Nasrollahzadeh et al., 2021; Raza et al., 2023). Combination of green chemistry and engineering principles are explored for sustainable synthesis of nanoparticles(Noah et al. and Ndangili, 2022). In analytical chemistry, the nanomaterials are being explored due to their unique physicochemical properties(Kori et al., 2023). The nanomaterials have potential to solve the problem of drug overdose and its side effects. Targeted drug delivery can be used with the application of nanomaterials for selectively injecting the drug to the patient. Mesoporous materials can be used for better results with better predictability(Ansari et al., 2022). Plants, microorganisms and biological derivatives can be employed for nanomaterial synthesis(Irshad et al., 2021). Nanomaterials also find application in modern day agricultural revolution. Nanofertilizers, nanopesticides and nanosensors are used to enhance sustainability in agricultural sector through calculated and targeted use of these materials(Yasmine et al., 2023; An et al., 2022; Badgar et al, 2022; Singh et al., 2021). Electrochemical sensors employing nanomaterials are being explored for sensor application in detecting pesticides, biomarkers etc.(Wang et al., 2023). Nanomaterials are being explored for novel sustainable applications like hydrogen storage, asymmetric photocatalysis, enhancing sustainability of cement concrete and pollutant sensing(Zheng et al., 2021; Qiu et al.,2021; Onaizi et al., 2021; Rasheed, 2023). This article discusses applications of nanoparticles and nanomaterials in

Figure 1. Fields of application of nanomaterials

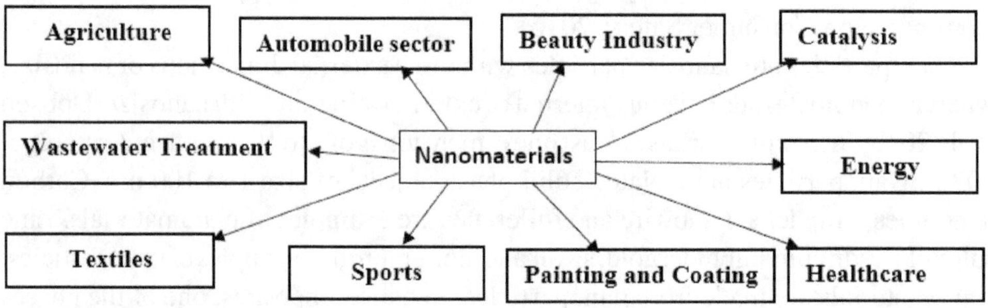

various sectors and related research. Figure 1 depicts applications of nanomaterials in various sectors.

INVESTIGATIONS ON APPLICATIONS OF NANOPARTICLES AND NANOMATERIALS

Nanomaterials are used in various sectors as indicated in fig.1 based on their properties. The specific properties of nanomaterials relevant to the applications are tabulated in Table 1. Nanoparticles can be of different types based on their origin namely Carbon, organic, inorganic and hybrid. Hydrogel is example of hybrid nanomaterial whereas graphene and fullerenes are carbon-based nanoparticles. Dentrimers and liposomes are organic whereas metallic nanomaterials are inorganic. Figure 2 depicts various nanoparticles used in industrial applications.

Agriculture

Increase in the populations at fast speed and its increase in the living standards has pose a challenging task for the scientists. Feeding the increasing needs of the population has given rise to different innovations in various sectors. The agricultural sector also has benefitted from such innovations. Nanotechnology and nanomaterials have come to the rescue of the farmers. Micronutrients are supplied to the crops through fertilizers. The drawback of the conventional method of fertilizer supply is bulk quantity and waste of nutrients. Targeted dosage of these components through the nano-fertilizers improves intake and utilization efficiency. This also reduces environmental footprints if pesticides and other plant protection products are fed by targeted delivery system (Khandelwal et al., 2022). Precision farming is frequently used term in the modern-day agriculture. Bioactivity, bioavailability and bioreactivity

Table 1. Properties and related applications of nanomaterials

Property type	Property	Field of application
Mechanical and electrical properties	Reliability and fatigue resistance, semiconducting or insulating properties	Electrical components and micro-electromechanical system
Physical properties	Melting and sintering temperature	Better material processing
Physicochemical property	Better atomic transportation	Hydrogen storage, energy storage batteries
Electrical and structural properties	Resistivity and grain size	Electronic components sensors
Mechanical properties	Hardness, wear resistance and grain size	Coatings
Physicochemical property	Surface volume ratio	Catalysis
Physicochemical and optical properties	Optical and antimicrobial activity	Agriculture
Physicochemical, mechanical properties	Conductivity, UV stability, Scratch resistance, atomic transport, catalytic activity, mechanical strength	Automobile sector
Physical and chemical properties	Colour, transparency, solubility	Beauty industry
Biological and physical properties	High surface volume ratio, biocompatibility and biodegradability	Healthcare industry
Physical properties	Resilience, strength, weight, scratch resistance	Sports
Physical, chemical and biological properties	Conductive, optical, hydrophobic, electrical, antimicrobial, flame retardant properties	Textile industry

Figure 2. Types of nanoparticles

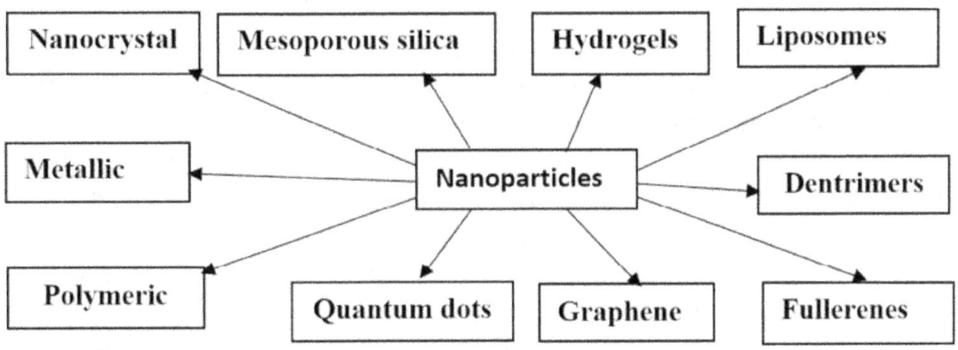

are the attributes of nanomaterials that make them most sought-after alternative in precision farming. Nano fertilizers refer to the macro and micro nutrient fertilizers. These have size less than 100 nanometer. Microorganisms such as bacteria, fungi and algae are metal resistant. They can be used for preparing metallic nanoparticles due to their ability to tolerate metal concentrations. Microorganisms can be considered as nano-factories for metallic nanoparticles (Ndaba et al., 2022). Packaging food material is spoiled due to less shelf life. Oxidation and lack of antimicrobial properties is the reason for the same. Use of oxygen scavenger material can help increase shelf life of packaged food. Silver nanoparticles can be used to avoid ripening of the fruits by reducing ethylene gas.

Nanomaterials in agricultural sector help in meeting the needs of the sector considering climate change and limited availability of the nutrients (Pramila, 2008). Increase in the yield, minimizing loss of nutrient and minimizing use of plant protection products are some of the key applications of the nanomaterials in agricultural sector. Enhancement in nutrient uptake, treatment of diseases and efficient use of the essential components can be facilitated through nanomaterials. Target specific delivery can reduce the side effects and damages to the crops. Nanotechnology also plays critical role in plant breeding and genetic modifications. The nanomaterials such as fullerenes, carbon nanotubes and nanofibers are used as sensor for mechanism in detecting soil quality, humidity, waste management and delivery, nutrient and pesticide management and delivery (Pramila, 2008). Also, the waste materials like wheat straw and soy hulls can be used for value products such as plastic composites. Large number of nano fertilizers are being synthesized from waste materials for effective and sustainable applications in fertilizers (Soltani et al., 2023; Kalaivani et al., 2023; Singh et al., 2023; Rojas-Velázquez et al., 2023; Yadav et al.,2023). Agroformulations from bio nanocomposites can work many purposes in farming sector such as intensification via energy saving, low environmental footprints and increase in the yield. There is the need to be careful while adopting the nanobiocomposites due to some concern's such nanotoxicity, government regulation, public knowledge, and production costs(Menossi et al., 2022). Novel materials are synthesized from various raw material with nanoparticle activities. Antibacterial nanomaterials are derived from graphene oxide by (Zankana et al., 2023). Bio-nanocomposites are considered as the next generation packaging material due to their antimicrobial properties, physicochemical and mechanical properties. These properties make them durable solution for food packaging (Dubey et al., 2023).

New age materials such as microcrystalline cellulose with different moieties can be used as reducing agent and capping agent for synthesis of nanocrystalline cellulose-silver nanocomposites (Al-Shemy et al., 2023). Wide range of applications of nanomaterials in agriculture includesmanure, aquaculture, water supply and distribution, water filtration, feed and vaccines for animals, food processing and packaging (Kumari et al., 2023). There needs to be regulatory framework to cater for safety and environmental aspects. Nano molecules are injected to plant and mammalian cells by using nanodrug delivery systems. Different countries are already taking measures to make the nanocomposite application in various sectors safe and sustainable (Kumari et al., 2023).

Automobile Sector

Automobile industry also is beneficiary of the revolution in nanotechnology. Lighter and harder nanomaterials can improve performance and cost effectiveness of the nanomaterials. Fullerene-C60, Carbon Nano tubes (CNTs) and grapheme sheets are promising materials due to excellent mechanical, thermal and electrical properties(Goyal et al., 2014, Al-Mutairi, 2022). Nanocomposites are mixture of conventional material with nanomaterials. Bulk properties of the conventional materials and size dependent properties of the nanomaterial produces composite with novel properties. Nanomaterials are finding variety of applications in automotive industry. Nanomaterials are explored for fuel borne catalysis for soot prevention. Nanoparticle reinforced tires with greater abrasive resistance and gas permeability, nanoparticles in coating for better gloss, scratch and corrosion resistance, anti-fog coatings, plastic body parts with better properties, are examples of application of nanocomposites and nanomaterials. There are some limitations in the production of nanoparticles for automotive sector. The compatibility of nanoparticles with limited bulk materials restricts its processing to limited sources. Commercial grade polymer matrix depends on crude oil prices, also carbon nanotubes are costly. Stringer norms and acceptance by the stakeholder takes time and give rise to longer lead time. Nanotechnology also provides option of improved cooling with better thermal performance. The solder material can be modified to provide better joint at low temperature. Nano sensing elements, Nano based coatings, batteries and fuel cells with better capacities, displays with lower costs are some of the applications of nanomaterials and nanocomposites in automobile industry (Goyal et al., 2014). Energy storage systems in the Electrical vehicles (EVs) pose a challenge to the scientists. High energy density, specific power, recyclability, low self-discharge rate and longer life make Lithium-ion batteries, a better choice(Al-Mutairi, 2022). A battery pack is series of electrochemical cells. Heating is a major issue which needs to be addressed. It is difficult to remove heat by air and liquid cooling. Organic and salt hydrates as phase change materials can be used for heat absorption. These materials have low mechanical resistance and thermal conductivity. Also, they have limited stability. Literature is reported on the IC engine lining using aluminium based nanocomposite (Tiruvenkadam et al., 2015). Disk brake and drum making from aluminium and CNT based nanocomposite is also explored with success(Sundaram and Mahamani,

2015). Automotive market has largest share in the plastic nanocomposites(Gutowski et al., 2017).Bearing, ratchet, and electrical contact applications can be boosted by silver based matrix metals. Ceramic reinforced nanocomposites with superior tribomechanical propertiesare being explored (Venkatesh et al.,2022). Investigations on nanocomposites also include research carried out towards betterment of application of nanocomposites. Polypropylene used in housing of headlight was combined with a graphene/montmorillonite hybrid (MMT-G) to reduce moisture absorption(Lee et al., 2023). Toughness of graphene epoxy materials can be improved by addition of grapheme (Khalid et al., 2023).

Beauty Industry

Nanoparticles like cubosomes, nanodots, liposomes, dendrimers, nano-emulsions have become common ingredients in cosmetics (Effiong et al., 2020; Katz, 2015). The Nanoparticles with unique properties have increased their role in beauty industry due to their unique properties. Wrinkling, dehydrated and inelastic skin associated with aging and dispersed hyperpigmentation can be handled better with the applications of nanoparticle-based materials. Nanosized particles are also used in sunscreens, anti-ageing products, razors and curling tongs(Raj et al., 2012)). Cosmetics can be broadly classified as skin, hair, nail and face care products and fragrance products (Alka et al., 2014).

Micellar nanoparticles are latest field of application of nanotechnology (Aziz et al., 2019). Bioactive component transport is facilitated by these nanosized micellar nanoparticles due to high surface area availability. For applications like make-up remover, facial cleanser, anti-aging lotion, sun-screens, oil in water nano-emulsions are playing very crucial role. Skin has multi-layered structure. Micellar nanotechnology can enable high active compound dosage concentration to penetrate through the skin. Nano-emulsions have superior properties as cosmetic basis resulting in high stability, larger surface area, pleasant aesthetic character, and skin feel (Yukuyama et al.2016; Sonneville-Aubrun et al., 2004). Sunscreen contains Titanium dioxide and Zinc Oxide(Iyer et al., 2017). Nano silver act as preservative in the cosmetic products.

Liposomes are made of hydrophobic phospholipid bilayers enclosing an aqueous medium (Effiong et al., 2020). Liposomes can trap the cosmetic ingredients that are released at target point. They can also be used for encapsulation of vitamins delivered to refresh skin epidermis. Niosomes are made from non-ionic surfactants and cholesterol in aqueous medium(Effiong et al., 2020). They can be used for entrapping hydrophilic drugs and lipophilic drugs within the aqueous core and membrane bilayer of niosomes respectively. Polymeric nanocapsules can be used for delivery of the drug deep into the skin. Nano emulsions are finding applications due to desirable properties. Due to their fine structure, they can be sprayed. They are used for controlled delivery of lotions, shampoo, nail enamels, and hair conditioners. For enhanced hydration, bioavailability, stability and controlled occlusion, solid lipid Nano lipids can be used(Effiong et al., 2020).Concerns about nanoparticle application is due to possibility of their entering into blood stream. It is necessary to share the information on cosmetic ingredients and their impact on health across the countries. In appropriate and safe use of nanoparticles in the cosmetic industry; researchers, manufacturers, customers and regulators have important role to play. Risk arising from nanomaterial use in cosmetics was similar as that seen in conventional material (Karlz, 2007). There is need to investigate the finished products for their effects on human health and skin.

Catalysis

Graphene oxide, reduced graphene oxide and their metal oxide composites have wide range of applications in catalysis and super capacitor technology (Tamang et al., 2023). Graphene and its derivatives have excellent catalytic properties. Especially the presence of carbon nanotubes in its derivatives makes them promising candidate for catalyst applications (Sachdeva, 2020).Unsupported nanoparticles can be used as a catalyst in organic reactions. Nanoparticles may agglomerate and experience reduction in catalytic activity. Graphene oxide and reduced graphene oxide can work as support and loading of various metal nanoparticles on them can improve the conversion efficiency. Investigation has revealed that the reduced graphene based CuOx exhibited far better catalytic activity than only CuOx nanoparticles (Choi et al., 2017). Also, a research indicates that CuO-rGO has better catalyst properties than CuOx -rGO (Choi, 2017).CuO NPs/rGO can increase the potential of the reaction mixtures to absorb ultra-wave radiations(Dandiaet al., 2018). Catalyst activities of platinum nanoparticles supported on graphene are also well versed. A highly crystalline, colloidally stable and highly monodisperse reduced graphene supported platinum nanoparticles are used to obtain acridine Dione derivatives(Aday, 2016). Acridine dione derivatives are used in applications related to anticancer, antimicrobial, myorelaxant, free radical scavenging (Ulus et al., 2015;Gündüz et. al., 2014; Moallem et al., 2015). RGO-ZnO, Zinc oxide supported on graphene oxide can be used for synthesis of biologically active 3 substituted indole (Rajesh et al., 2015). Nanoparticle can act as amphiphilic heterogeneous catalyst in water (Choi et al., 2017). Investigations are also reported on Pyrene-tagged metal complexes supported onto GO as catalyst (Choi et al., 2017). Graphitised solids are used as support for pyrene-tagged metal complexes (Rodriguez-Perez et al., 2013). Investigations arealso reported on cyclization of acetylenic carboxylic acid and the coupling of diphenylcyclopropenone with substituted phenylacetylenes, hydrogenation of alkenes and alcohol oxidation using pyrene-tagged metal complexes supported on rGO(Gutiérrez-Blanco et al., 2018). Nanostructured artificial enzyme synthesis has been reported with better catalytic properties than natural enzymes (Villalba-Rodríguez et al., 2023). Silver nanoparticle composites have high catalytic activity and antimicrobial ability (Zhang et al., 2023). Silver nanoparticles can be imbedded with many other metals to yield the catalyst with customized activities for the specific reactions (Rajendiran et al, 2023;Biehler et al., 2023; Noreen et al., 2023). The solid base nanocatalysts KF/CaO cab be utilized for biodiesel production(Wen et al., 2010). Catalytic properties of nanomaterials are established in photocatalysis, drug delivery, water treatment, dye degradation, energy applications and hydrogen fuel cells (Chaturvedi et al., 2012). Synthesis of graphene and graphene based nanocomposite by artificial intelligence is also being explored and utilized(Tariq., 2022).

Energy

Nanotechnology is contributing to the energy savings through technology and through the nanomaterials that are used to minimize energy consumption. Semiconductors used in light emitting displays are made up of nanoscale materials. Advances in nanotechnology-based lighting systems can reduce the energy consumption by 10%(Elcock, 2007). Quantum dots, which are nanocrystals, are capable of producing distinct colours of light as the size of the individual crystals is varied. Nanofluids can also increase heat transfer coefficient in heat exchanger. Also, they can prevent scaling inside the pipes. (Elcock, 2007).

Nanotechnology is being looked at as a technology that can intensify the energy storage and generation by using novel methods. Solar energy is considered as sustainable source of energy. Nanomaterials find application in photovoltaics, wind energy, geothermal and biomass energy (TERI ENVIS Resource Partner on Renewable Energy and Climate Change, 2021).The solar cells, in the initial years were having crystalline silicone semiconductors. These are expensive and have only 33% efficiency. In the second generation solar cells, cadmium telluride (CdTe) and copper indium gallium selenide (CIGS) were used. It was improvement over first generation solar cells in terms of cost and efficiency. The third generation solar cells are exploring new materials and methods namely quantum dots, nanostructured semiconductors, and amorphous silicon. Novel technologies like nanowires, quantum dots and radial junctions have the potential to redefine the solar cells. A thin film on glass, plastic or metal of cadmium telluride (CdTe), copper indium gallium diselenide (CIGS), and amorphous thin-film silicon (a-Si, TF-Si) offer more absorption of photons. Thin film boosts absorption efficiency by 60% (TERI ENVIS Resource Partner on Renewable Energy and Climate Change, 2021). Metal oxide based nanomaterials are utilized for transparent electrode in solar cells.

Enhancement in heat transfer by using nanofluids can save considerable amount of energy required for heating or cooling the working and utility fluids. Low thermal conductivities of fluids like water, mineral oil, and ethylene glycol affect the energy conservation. The nanofluid utilization can considerably reduce the energy requirement. An investigation suggests that 0.5% addition of nanomaterials can result in 11% energy savings (Krishna, 2016). Al_2O_3/water nanofluid is explored for heat exchange applications. Shekarian et.al. (2016) were able to reduce energy consumption by addition of nanomaterial (Al_2O_3-water). Number of investigators have explored water based nanofluid with Cu, Ag, CuO, Al_2O_3, and TiO_2 for heat exchanger applications(Senthilkumar et al, 2023;Cieslinski&Kaczmarczyk 2013) Significant rise in heat transfer is reported in literature with 0.2 to 0.5% volume / volume concentration of nanofluids(Kulkarni, 2017)). Nanomaterials can be used in insulation to reduce the heat loss in buildings (Elbony,2020;Elcock, 2007). Advance insulating material like aerogel can reduce energy loss considerably.

Healthcare

Nano silver powder can be absorbed on surface of bacterial cells. This powder has the ability to influence the metabolism and reproduction of bacteria. Compared to silver, Silver nanoparticles are many folds effective in antibacterial applications. Zinc oxide nanoparticles also have unique properties in terms of antibacterial nature and cytocompatibility. They can absorb wide range of radiations. These nanoparticles are safer for human skin (Josef et al., 2023).Stability of nanomaterial is very important aspect. The properties, namely,size of particles and particle distribution, morphology, surface chemistry, pH, kinetics and surface charge are tested for stability check. Many treatments involving high toxicity like chemotherapeutic cancer drugs, can be carried out by using nanotechnology. Nanoparticles in medicinal applications can be organic, inorganic and carbon based. Organic molecules like proteins, carbohydrates, lipids can be converted into nanomaterials. Inorganic nanoparticles include metal oxides and salts. Carbon nanotubes, graphene and fullerenes are examples of carbon based nanomaterials. In surgical textiles and implants, tissue technology, artificial organ and wound dressing, nanofibers are used (Haleem et al., 2023)).

Nanoparticles can be used to decorate different molecules so that they can't be recognized by immune system, and hence they can reach target efficiently (Kawadkar et al., 2011). Nanoparticles can help to overcome physiological barriers like brain barrier. Targeted drug delivery can be done with nanoparticles. Nanoparticles which have diagnostic and threptic properties are being utilized in applications and are evolving continuously. Quantum dots, DNA-protein and nanoparticle conjugates are used in diagnostic applications. Gold and magnetic nanoparticles also find application. Gold nanoparticles can be used for labelling, delivery, heating and sensing. Gold nanoparticles provide high spatial resolution and specificity in many labelling applications. Optical properties like strong absorption, scattering and especially Plasmon resonance, make them useful in light dependent applications namely photo thermal and photo acoustic imaging. Also, nanoparticles can serve as carriers for drug delivery. Optical properties of nanomaterial change after binding with other molecules. This can be used for DNA detection. Magnetic nanoparticles contain elements such as iron, nickel and cobalt and their chemical compounds. Magnetic resonance imaging (MRI) utilizes iron oxide nanoparticles for in vivo detection of cells. In some medical applications, nanoparticles can act as medicines also (Shrivastava & Dash, 2009). Metals containing nanoparticles can be injected to tumour cells. These nanoparticles can be heated by near-infrared radiation to kill the tumour cells. In cancer therapy by hyperthermia, a magnetic nanoparticles (MNP) system under the influence of an external high frequency magnetic field can be used for the temperature enhancement (Hergt et al., 2006; Hilger et al., 2005). Titanium dioxide nanoparticles can be used for killing bacteria. Titanium dioxide can also be used to prepare the antiseptic surfaces that can only work under ultraviolet light. The disinfectant properties of silver nanoparticles are well known. Nano based medicines have contributed to the fight against malaria. Biocompatible systems are developed that are capable of incorporating drugs, lowering the resistance progress, contributing for diagnosis, control and treatment of malaria by target delivery(Pujar and Bagawan, 2016). Nanotechnology based materials have revolutionized the tissue engineering. Gold nanoparticles are used for protein-protein interaction in immunohistochemistry (Zaib and Iqbal, 2019).

There are some concerns and difficulties with regard to synthesis and applications of nanotechnology-based drugs. Clinical trials are required for the new medicine before using those in hospitals. Animal experimentation has its own limitations. There may be some effects and side effects of nanomedicine which are not significant in animals but significant in humans. Also, animal studies do not characterize long term effects of nanomedicines(Keskinbora& Jameel, 2018). Based on circumstances and benefits of the drugs, the risk is evaluated arising out of clinical trials. Nanobiotechnology offers very important inputs to development of detection devices (Bazwa et al., 2017).While using any nanoparticle in medical application, different considerations of toxicology should be used(Anderson et al., 2016). In diagnostic applications, pharmacokinetic properties should be assessed for nano particles.

Painting and Coating

Corrosion resistant paints and coatings are becoming popular due to their ability to safeguard materials. Nanomaterial based coating are capable exhibiting enhanced corrosion inhibition. A nanocomposite coating can be defined as material made of minimum two immiscible phases, separated from one another by interface region(Nguyen-Tri et al., 2018). Based on the type of nanostructured fillers, the nanocomposite coatings can be classified as 0 D where fillers are nanoparticles, 1 D, where fillers are nanotubes, 2D where fillers are nanolayers. Based on types of matrix, the nanocomposite coatings are classified as organic-inorganic, organic-organic, inorganic-inorganic and inorganic-organic. For organic matrix or polymer-based nanocomposites, polymers like epoxy, polyvinylidene fluoride, polyethylene glycol, polystyrene, polyvinyl alcohol, polyacrylic etc. can be used(Nguyen et al., 2015;Lowe et al., 2015; He et al., 2011; Hou et al., 2009;Golgoon, 2015). For inorganic nanofilling; oxides, clay, carbides, borides etc. can be used. For coating; the methods namely, cold spray, sol gel, thermal spray, physical vapour deposition and chemical vapour deposition methods can be used. The most used type of nanocomposite in coating is organic-inorganic nanocomposite.

Solvent-based polymer manufacturing processes have disadvantage regard to health and safety aspects. Emulsion-based coatings and adhesives are replacing solvent based coating and adhesives (Dastjerdiet al., 2019). Cellulose nanocrystal (CNC) surfaces can be functionalized with well-known carbohydrate chemistries. CNC compatibility and dispersion increases due to functionalization(Dastjerdi et al., 2019).CNCs can be used as filling materials for more sustainable polymer composite production(Eichhorn et al., 2010; Roman, 2015). In CNC-poly latex matrix, it was observed that 6% CNC can enhance composite modulus by 300 times(Favier, 1995).Abrasion and scratch resistance for coating enhances by addition of CNCs prepared by blending technic(Trovatti, 2010). Shear strength, tack and peel strength enhances for adhesives with CNCs and synthesized by in situ emulsion polymerization(Dastjerdi, 2018).

Polypyrrole- zinc oxide nanoparticles can be used in anticorrosion coatings as additives(Valençaet al., 2015). Addition of 0.2% nanocomposite to epoxy coatings increased the anticorrosive efficiency considerably. Excess addition of the nanocomposite may decrease the protection efficiency due to increased porosity of the coating due to particles agglomeration(Valença, 2015).Silver and zinc oxide nanoparticle-based coatings can be used for splash zone applications(Alam et al., 2019). Thermal and electrochemical properties of polyaniline can be improved significantly by using different nanomaterials. Oxides of titanium, zinc and silica can be used for this purpose(Hermas et al., 2013; Gomez et al., 2010;Alam et al., 2017). Strong bonding between inorganic and organic phases along with interaction between nanomaterials with polyaniline is responsible for enhanced properties(Ates&Topkaya). Field emission scanning electron microscope investigation of silver and zinc oxide nanoparticles with sodium chloride test solution, indicated that fabrication coating were uniform, crack free and homogeneous. Energy dispersive x-ray analysis indicated the presence of carbon and chlorine ions. This indicates that coating layer covered the panel completely. Electrochemical impedance spectroscopy for the coated surfaces indicated that the presence of Ag and ZnO nanoparticles improves the mechanical and the anticorrosion properties for the manufactured coatings(Alam et al., 2019).

Sports

The sports are modernizing and expectations from the players are ever increasing. Competitions are at very high levels. The skillsets acquired by the sportspersons need to be complemented with high quality sports gears. Light weight, water proof, flexible sport gears with customized properties for the specific sports are becoming need of the hour. Nanotechnology plays very significant role in the modernization of sports gears. In Tennis, silver nanoparticle infuses 22% more ball hitting ability than normal racquets(Kumar, 2019). Particles like Fullerenes, nanoclay, and nano-titanium increases bounce of the balls. In Formula 1 racing, the weight of the cars is decreased by using nanomaterials for various parts of the car(Kumar, 2019). Different properties of nanomaterials are exploited in sports materials for better performance and specific applications. Water resistance, bacterial protection, UV protection, heat and cold protection are some of the aspects of the nanomaterials in sports application(Rahman et al., 2023). Nano-silicate can be used for coating a polyester fabric that can stay dry even after two months in the water. Shoes with water resistant properties also contains titanium dioxide-zinc oxide nanomaterials. Sweating during sports activities can give rise to microbial infection. Chitosan based fibres with moisture-controlling, anti-bacterial, and anti-fungal properties can be used for sportswear and socks(Heine et al., 2007). Nanomaterial is incorporated to these fabrics during spinning or finishing stages. Silver nanoparticles are used in the sports gear, sports shoes, and insoles. Metal oxide nanomaterials can be used for UV protection during outdoor sports. Some nanoparticles have ability to produce hydrophilic surfaces that clean themselves. Different types of nanomaterials are used for these sports activities. Nanofibers, metal matrix composites, ceramic matrix composites and polymer matrix composite are used for different applications. In weightlifting, nanocomposites can be used to increase wear resistance of the equipments (Heine et al., 2007). Titanium based nanoparticles can be used for increasing elasticity in tennis (Hiremath& Bhat, 2017).Nano clays are used to maintain pressure in the cricket ball and football(Bissessur, 2020). Carbon nanotubes, nano-silicates, fullerenes, carbon fibers, zinc oxide nanoparticles are used in sports equipments for customized requirements (Rahman et al., 2023; Yang et al., 2004;Pan et al., 2015;Rasheed, & Khalid,2014). New development in the nanotechnology applications in sports is the mechanical strain sensors that are flexible (Shalaby& Saad, 2020).Out of the total carbon fiber consumption across the world, 25% is used in sports(Yang, 2023). Carbon fiber demand in the field of stationery and sports products is increasing by

48% annually(Yang, 2023). Ni-TiN and SiC nanocomposite was used by Bai et al. (2022) on Dumbbell to reduce sports injuries(Bai et al., 2022). They used pulse electrodeposition method for this purpose with encouraging results.

Textiles

Wrinkle resistant, water repellent and antibacterial fabrics with odor control can be manufactured by incorporating nanomaterial. Also, optical displays and nanosensors can be used for different applications (Asif & Hasan, 2018).Invisible coatings and Fabric touch pads have turned the basic textile into nano textile.Artificial muscles are developed from inherently conductive polymers. Multiple functions like deformation detection, light transmission; sensing and data transmission device can be facilitated in woven optical fibers. Light emitting textiles are now developed that have graphic and colourful display. The stress and failure in human body can be monitored through smart textile (Asif & Hasan, 2018). Different types of nano-finishes of semi-conductor ceramics and metallic oxides can infuse properties like water repellency, oil repellency, thermal resistivity, antimicrobial properties and flame retardancy. Silver, platinum and iron based nanoparticles are capable of infusing mechanical, electrical and magnetic properties. Polyesters treated with silver nanoparticles are bacteriostatic. In socks, silver based nanoparticles can be used for antibacterial properties.

Wastewater Treatment

Nanoparticles, nanomembrane and nanopowder find wide range of applications in wastewater treatment. Removal of chemical and biological pollutants can be facilitated by using these nanomaterials. Heavy metals, nutrients and oxygen demanding pollutants can be eliminated from wastewater by using nanomaterials. The classes of materials used in wastewater treatment are metal-containing nanoparticles, carbonaceous nanomaterials, zeolites and dendrimers (Tiwari et al., 2008). Generally reverse osmosis and nanofiltration is used for removal of pollutants with molar mass of 1000 Da. These methods and even ultrafiltration is ineffective for removal of dissolved inorganic and organic materials with molar mass below 3000 Da. Dendritic polymers can be used for removal of these compounds. Studies have revealed that magnesium oxide nanoparticles are effective biocides against gram positive and gram negative bacteria (Stoimenov et al., 2002). Coliform can be neutralized effectively by using silver nanoparticles(Jain and Pradeep, 2005). For removal of arsenic from waste water zinc oxide nanoparticles are found to be effective (Brittany et al., 2006). Zeolite based nanomaterials can be used for removal of heavy metals from wastewater (Alvarez-Ayuso et al., 2003). High surface area availability is main advantage of nanomaterials in sorption processes. Nanomaterials find application in adsorption, membrane separation, photocatalysis, disinfection and sensing-monitoring applications. Carbon nanotubes and oxide nanoparticles are good adsorbents due to high specific area, short intraparticle diffusion distance and assessable adsorbent sites (Qu et al., 2013). Nano-zeolite, nano-silver, carbon nanotubes, nano-magetite can be used in membrane separation for their antimicrobial activities. Mechanical and chemical stability is important factor in their application in membrane separation. In photocatalysts, fullerene and titanium oxide nanoparticles are used due to their photocatalytic ability. Silver and titanium oxide nanoparticles, carbon nanotubes are used for disinfection owing to strong and wide-spectrum of antimicrobial activities. Characteristics of quantum dots like broad absorption spectrum, narrow, bright and stable emission that scales with the particle size and chemical component, make them useful in the applications related to sensing and monitoring in optical detection. Harmful pollutants like chlorophenols can be completely neutralized by nano-TiO_2 photocatalytic oxidation(Tinucci et al., 1993).

CASE STUDIES AND LATEST DEVELOPMENTS

Studies carried out by Globe Newswire indicated that the global market for nanotechnology fabric will experience a growth of 24 to 25% and increase from 5.75 billion $ in 2022 to 7.16 billion$ in 2023(globenewswire.2023). Nanomaterials find wide application in wastewater treatment for removal of various contaminant from effluent. Industrial effluent treatment plants need to remove heavy metals from the wastewater. An investigation on application of calcium-based nanomaterials for heavy metal removal was reported by Roy and Bhattacharya(2015). They used calcium hydroxide, oxide and sulfide-based nanoparticles for removal of heavy metals from wastewater. Their research indicated that the Ca-based nanoparticles dissolved readily in water. The solubility was dependent on solubility. It was observed that heavy metal removal by supplied nanoparticles was high at high pH values. The dose of nanoparticles affected the heavy metal removal up to certain optimum vale of dosage. Case study of wastewater treatment in Bangalore, India indicated that nanotechnology-based treatments can increase the effeteness of treatment considerably(Sahu et al., 2019). Annual growth for nanotechnology based emerging products in Bangalore was estimated at 30%. Research and development expenditure for nanotechnology in Bangalore increased by 24.3% between 2007 to 2015(Sahu et al., 2019). Investigators have demonstrated successful use of nanomaterials for construction materials with enhanced properties(Shilar et al., 2022).

Smart nanomaterials are being explored for various engineering applications. Intelligent and smart nanomaterials are responsive to the stimuli which can be

Table 2. Smart nanomaterials

Stimuli type	Applications	Reference
Thermoresponsive	Targeted drug delivery	Kayal, 2021
Piezoelectric	Sensors, actuators, implantables	Roy et al., 2022
Responsive to electrochemical stimuli	Electrochemical sensors	Luo, 2006
Responsive to magnetic stimuli	Magnetic resonance imaging	Nikazar, 2020
Responsive to photo stimuli	Drug delivery	Nikazar, 2020
Responsive to pH	Tumor therapy	Chu et al., 2022
Responsive to enzymes	Drug delivery, biosensing, cancer diagnosis	Hu et al., 2012; Hu et al., 2014;Yousefiasl, 2023

indicator of certain properties or interaction(Alam, 2023). This phenomenon can be used in various applications as indicated in Table 2.

Bionanotechnology and nanobiotechnology are two emerging areas related to nanotechnology. Nanotechnology that is inspired and guided by biological systems and processes is termed as bionanotechnology whereas nanobiotechnology uses biological nano-technological tools for studying biological systems. Biotechnology in nanotechnology inspires green synthesis of nanomaterials. This bionanotechnology approach in synthesis is eco-friendly. Nanotechnology in biotechnology has enabled advanced biological applications such as tissue engineering and nano array diagnostics. The combined approach of biotechnology and nanotechnology utilizes advantages of both for better and novel applications. Applications of bionanotechnology are depicted in fig.3. Bionanomaterials are replacing the nanomaterials because of sustainable synthesis methodologies and low environmental footprints. Nanomaterial synthesis and applications has disadvantages like environmental concerns and health effects(Jeevanandam, 2022). Bionanomaterials are synthesized from biomolecules and its synthesis and application involves negligible environmental and health effects. These materials are modern day nanomaterials that are used in tissue engineering, cancer therapy, fabric materials, agricultural applications and food packaging(Jalilinejad,2022; Yadav et al., 2022; Nesa, 2022; Srivastava et al., 2023).

CONCLUSION

Rapid increase in the populations and its living standards has posed a challenging task for the scientists. Nanotechnology is being explored in various applications ranging from chemical engineering to farming. Environmental technology has

Figure 3. Applications of bionanotechnology

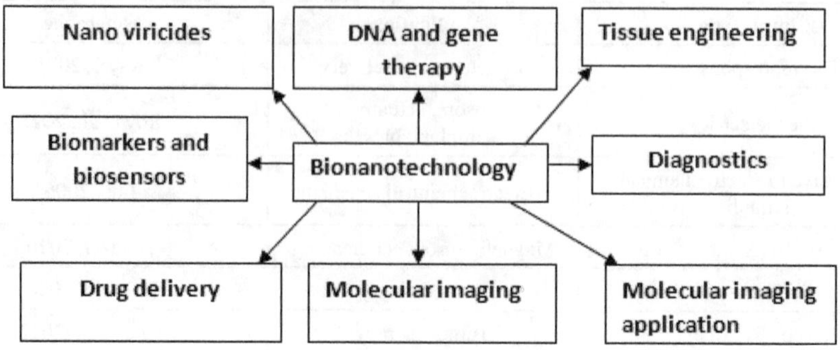

gained importance due to increasing need to develop the materials and processes with sustainable approach. If we look carefully at these three fields, they can be integrated to provide sustainable solutions to many material synthesis and waste treatment problems. Nanotechnology is being explored due to its huge potential and endless possibilities. Feeding the increasing needs of the population has given rise to different innovations in various sectors.

The agricultural sector also has benefitted from such innovations. Nanotechnology and nanomaterials have come to the rescue of the farmers. Micronutrients are supplied to the crops through fertilizers. Automobile industry also is beneficiary of the revolution in nanotechnology. Lighter and harder nanomaterials can improve performance and cost effectiveness of the nanomaterials. Fullerene-C60, Carbon Nano tubes (CNTs) and grapheme sheets are promising materials due to excellent mechanical, thermal and electrical properties.

Nanoparticles like cubosomes, nanodots, liposomes, dendrimers, Nano-emulsions have become common ingredients in cosmetics (Effiong et al., 2020; Katz, 2015). The Nanoparticles with unique properties have increased their role in beauty industry. Graphene oxide, reduced graphene oxide and their metal oxide composites have wide range of applications in catalysis and super capacitor technology. Nanotechnology is contributing to the energy savings through the nanomaterials that are used to minimize energy consumption. Semiconductors used in light emitting displays are made up of nanoscale materials. Nano-silver powder can be absorbed on surface of bacterial cells. This powder has the ability to influence the metabolism and reproduction of bacteria.

Biotechnology in nanotechnology inspires green synthesis of nonmaterial. This bio nanotechnology approach in synthesis is eco-friendly. Nanotechnology in biotechnology has enabled advanced biological applications such as tissue engineering and nano array diagnostics. The combined approach of biotechnology and nanotechnology utilizes advantages of both for better and novel applications. Bionanomaterials are replacing the nanomaterials because of sustainable synthesis methodologies and low environmental footprints. Nanomaterial synthesis and applications has disadvantages like environmental concerns and health effects. Smart nanomaterials are being explored for various engineering applications. Intelligent and smart nanomaterials are responsive to the stimuli which can be indicator of certain properties or interaction.

ACKNOWLEDGEMENT

Author is thankful to the affiliating institute, Gharda Institute of Technology.

REFERENCES

Aday, B., Yıldız, Y., Ulus, R., Eris, S., Sen, F., Kaya, M.(2016). One-pot, efficient and green synthesis of acridinedione derivatives using highly monodisperse platinum nanoparticles supported with reduced graphene oxide. *N J Chem., 40*(1), 748–54. doi: 02 098 K. doi:10.1039/C5NJ

Al-Mutairi, N. H., Mehdi, A. H., & Kadhim, B. J. (2022). Nanocomposites Materials Definitions, Types and Some of Their Applications: A Review. *European Journal of Research Development and Sustainability*, *3*(2), 102–108.

Al-Shemy, M. T., El-Shafei, A. M., Al-Sayed, A., & Adel, A. M. (2023). Facile in-situ synthesis of nanocrystalline celluloses-silver bio-nanocomposite for chitosan based active packaging. *Egyptian Journal of Chemistry*, *66*(4), 335–352.

Alam, M. A., Samad, U. A., Khan, R., Alam, M., & Al-Zahrani, S. M. (2017). Anti-corrosive performance of epoxy coatings containing various nano-particles for splash zone applications. *Korean Journal of Chemical Engineering*, *34*(8), 2301–2310. doi:10.1007/s11814-017-0114-1

Alam, M. A., Samad, U. A., Sherif, E. S. M., Seikh, A., Al-Zahrani, S. M., Alharthi, N. H., & Alam, M. (2019). Synergistic effect of Ag and ZnO nanoparticles on polyaniline incorporated epoxy/2pack coatings for splash zone applications. *Journal of Coatings Technology and Research*, *16*(3), 835–845. doi:10.1007/s11998-018-00156-4

Alam, M. K., Karim, R., & Ahammad, A. S. (2023). Smart Nanomaterials and Their Applications. *Emerging Nanomaterials and Their Impact on Society in the 21st Century, 135*, 100-124. doi:10.21741/9781644902172-5

Alka, L., Anurag, V., Himanshi, J., Niti, Y., & Neha, K. (2014). NanotechnologyBased Cosmeceuticals. *ISRN Dermatology, 2014*, 843687. doi:10.1155/2014/843687 PMID:24963412

Alvarez-Ayuso, E., Garcıa-Sánchez, A., & Querol, X. (2003). Purification of metal electroplating waste waters using zeolites. *Water Research, 37*(20), 4855–4862. doi:10.1016/j.watres.2003.08.009 PMID:14604631

An, C., Sun, C., Li, N., Huang, B., Jiang, J., Shen, Y., Wang, C., Zhao, X., Cui, B., Wang, C., Li, X., Zhan, S., Gao, F., Zeng, Z., Cui, H., & Wang, Y. (2022). Nanomaterials and nanotechnology for the delivery of agrochemicals: Strategies towards sustainable agriculture. *Journal of Nanobiotechnology, 20*(1), 1–19. doi:10.1186/s12951-021-01214-7 PMID:34983545

Anderson, D. S., Sydor, M. J., Fletcher, P., & Holian, A. (2016). Nanotechnology: The risks and benefits for medical diagnosis and treatment. *Journal of Nanomedicine & Nanotechnology, 7*(4), e143.

Ansari, M. J., Unnisa, A., Singh, A., Verma, D., Kanaoujiya, R., & Gonzales, J. L. A. (2022). Application of porous nanomaterials for sustained and targeted drug release. *Advances in Science and Technology (Owerri, Nigeria), 117*, 105–110. https://www.scientific.net/AST.117.105. doi:10.4028/p-b2484c

Armarego, W. L. F. (2022). Nanomaterials. In editor(s): Wilfred L.F. Armarego, Purification of Laboratory Chemicals (Ninth Edition). Butterworth-Heinemann, 586-630. doi:10.1016/B978-0-323-90968-6.50005-9

Asif, A., & Hasan, M. Z. (2018). Application of nanotechnology in modern textiles: A review. *International Journal of Current Engineering and Technology, 8*(2), 227–231.

Ates, M., & Topkaya, E. (2015). Nanocomposite film formations of polyaniline via TiO2, Ag, and Zn, and their corrosion protection properties. *Progress in Organic Coatings, 82*, 33–40. doi:10.1016/j.porgcoat.2015.01.014

Aziz, Z. A. A., Mohd-Nasir, H., Ahmad, A., Mohd. Setapar, S. H., Peng, W. L., Chuo, S. C., Khatoon, A., Umar, K., Yaqoob, A. A., & Mohamad Ibrahim, M. N. (2019). Role of Nanotechnology for Design and Development of Cosmeceutical: Application in Makeup and Skin Care. *Frontiers in Chemistry, 7*, 739. doi:10.3389/fchem.2019.00739 PMID:31799232

Badgar, K., Abdalla, N., El-Ramady, H., & Prokisch, J. (2022). Sustainable applications of nanofibers in agriculture and water treatment: A review. *Sustainability (Basel)*, *14*(1), 464. doi:10.3390/su14010464

Bai, H., & Li, Q. (2022). Electrodeposited Ni/TiN-SiC Nanocomposites on the Dumbbell: Reducing Sport Injuries. *Coatings*, *12*(2), 177. doi:10.3390/coatings12020177

Bajwa, S. Z., Munawar, A., & Khan, W. S. (2017). Nanotechnology in medicine: Innovation to market. *Pharmaceutical Bioprocessing*, *5*(2), 11–15.

Barreneche, C., Martín, M., Calvo-de la Rosa, J., Majó, M., & Fernández, A. I. (2019). Own-synthetize nanoparticles to develop nano-enhanced phase change materials (NEPCM) to improve the energy efficiency in buildings. *Molecules (Basel, Switzerland)*, *24*(7), 1232. doi:10.3390/molecules24071232 PMID:30934832

Biehler, E., Quach, Q., & Abdel-Fattah, T. M. (2023). Silver-Nanoparticle-Decorated Fused Carbon Sphere Composite as a Catalyst for Hydrogen Generation. *Energies*, *16*(13), 5053. doi:10.3390/en16135053

Bissessur, R. (2020). Nanomaterials applications. In *Polymer Science and Nanotechnology* (pp. 435–453). Elsevier. doi:10.1016/B978-0-12-816806-6.00018-2

Brittany, L., Carino, V., Kuo, J., Leong, L., & Ganesh, R. (2006). Adsorption of organic Compounds to metal oxide nanoparticles. In *Conference presentation is part of.* General Environmental.

Chaturvedi, S., Dave, P. N., & Shah, N. K. (2012). Applications of nano-catalyst in new era. *Journal of Saudi Chemical Society*, *16*(3), 307–325. doi:10.1016/j.jscs.2011.01.015

Chaudhary, L. S., Ghatmale, P. R., & Chavan, S. S. (2016). Review on: Application of nanotechnology in computer science. [IJSR]. *International Journal of Scientific Research*, *5*, 1542–1545.

Choi, J., Oh, H., Han, S. W., Ahn, S., Noh, J., & Park, J. B. (2017). Preparation and characterization of graphene oxide supported Cu, Cu2O, and CuO nanocomposites and their high photocatalytic activity for organic dye molecule. *Current Applied Physics*, *17*(2), 137–145. doi:10.1016/j.cap.2016.11.020

Chu, S., Shi, X., Tian, Y., & Gao, F. (2022). pH-responsive polymer nanomaterials for tumor therapy. *Frontiers in Oncology*, *12*, 855019. doi:10.3389/fonc.2022.855019 PMID:35392227

Cieslinski, J. T., & Kaczmarczyk, T. Z. (2011). Pool boiling of water-Al_2O_3 and water-Cu nanofluids on horizontal smooth tubes. *Nanoscale Research Letters*, *6*(1), 1–9. doi:10.1186/1556-276X-6-220 PMID:21711741

Dandia, A., Bansal, S., Sharma, R., Rathore, K. S., & Parewa, V. (2018). Microwave-assisted nanocatalysis: A CuO NPs/rGO composite as an efficient and recyclable catalyst for the Petasisborono–Mannich reaction. *RSC Advances*, *8*(53), 30280–30288. doi:10.1039/C8RA05203D PMID:35546842

Dastjerdi, Z., Cranston, E. D., Berry, R., Fraschini, C., & Dubé, M. A. (2019). Polymer Nanocomposites for Emulsion-Based Coatings and Adhesives. *Macromolecular Reaction Engineering*, *13*(2), 1800050. doi:10.1002/mren.201800050

Dastjerdi, Z., Cranston, E. D., & Dubé, M. A. (2018). Pressure sensitive adhesive property modification using cellulose nanocrystals. *International Journal of Adhesion and Adhesives*, *81*, 36–42. doi:10.1016/j.ijadhadh.2017.11.009

Dobson, P. King. Stephen and Jarvie. Helen (2019, May 14). Nanoparticle. *Encyclopedia Britannica.* https://www.britannica.com/science/nanoparticle

Dubey, A., Irudhayaraj, S., & Jaiswal, A. (2023). Bio-nanocomposites: A Next Generation Food Packaging Materials. In *Synthesis and Applications of Nanomaterials and Nanocomposites* (pp. 197–231). Springer Nature Singapore. doi:10.1007/978-981-99-1350-3_8

Ducheyne, P. (2015). 1.109. Bioactive Ceramics. *Comprehensive Biomaterials*, *1*, 157–179. doi:10.1016/B978-0-08-055294-1.00021-0

Effiong, D. E., Uwah, T. O., Jumbo, E. U., & Akpabio, A. E. (2020). Nanotechnology in Cosmetics: Basics, Current Trends and Safety Concerns—A Review. *Advances in Nanoparticles*, *9*, 1–22. doi:10.4236/anp.2020.91001

Eichhorn, S. J., Dufresne, A., Aranguren, M., Marcovich, N. E., Capadona, J. R., Rowan, S. J., Weder, C., Thielemans, W., Roman, M., Renneckar, S., Gindl, W., Veigel, S., Keckes, J., Yano, H., Abe, K., Nogi, M., Nakagaito, A. N., Mangalam, A., Simonsen, J., & Peijs, T. (2010). Current international research into cellulose nanofibres and nanocomposites. *Journal of Materials Science*, *45*(1), 1–33. doi:10.1007/s10853-009-3874-0 PMID:32836382

Elbony, F. A. (2020). Nanotechnology and Energy Efficiency In Buildings. JL. *EGYPTIAN SOCIETY OF ENGNEERS*, *59*(1), 29–35.

Elcock, D. (2007). *Potential impacts of nanotechnology on energy transmission applications and needs (No. ANL/EVS/TM/08-3). Argonne National Lab*. ANL.

Esmaeili, A. (2011, December). Applications of nanotechnology in oil and gas industry. In AIP conference proceedings (Vol. 1414, No. 1, pp. 133-136). American Institute of Physics. doi:10.1063/1.3669944

Favier, V., Canova, G. R., Cavaillé, J. Y., Chanzy, H., Dufresne, A., & Gauthier, C. (1995). Nanocomposite materials from latex and cellulose whiskers. *Polymers for Advanced Technologies*, *6*(5), 351–355. doi:10.1002/pat.1995.220060514

GLOBE NEWSWIRE. (2023). Research and markets. Global Nanotechnology Clothing Market Report 2023: Rising Use of Nanotechnology Clothing in the Medical Industry Owing to its Anti-Microbial Properties Drives Growth. *Nanotechnology Clothing Global Market Report 2023*. https://www.researchandmarkets.com/

Golgoon, A., Aliofkhazraei, M., Toorani, M., Moradi, M. H., & Rouhaghdam, A. S. (2015). Corrosion and wear properties of nanoclay-polyester nanocomposite coatings fabricated by electrostatic method. *Procedia Materials Science*, *11*, 536–541. doi:10.1016/j.mspro.2015.11.042

Gomez, H., Ram, M. K., Alvi, F., Stefanakos, E., & Kumar, A. (2010). Novel synthesis, characterization, and corrosion inhibition properties of nanodiamond– polyaniline films. *The Journal of Physical Chemistry. C, Nanomaterials and Interfaces*, *114*(44), 18797–18804. doi:10.1021/jp106379e

Goyal, R., Sharma, M., & Amberiya, U. K. (2014). Innovative nano composite materials and applications in automobiles. *International Journal of Engineering Research & Technology (Ahmedabad)*, *3*(1), 3001–3009.

Gündüz, M. G., İşli, F., El-Khouly, A., Yıldırım, Ş., Fincan, G. S., & Şimşek, R. (2014). Microwave-assisted synthesis and myorelaxant activity of 9-indolyl-1, 8-acridinedione derivatives. *European Journal of Medicinal Chemistry*, *75*, 258–266. doi:10.1016/j.ejmech.2014.01.059 PMID:24534541

Gutiérrez-Blanco, A., Peris, E., & Poyatos, M. (2018). Pyrene-connected tetraimidazolylidene complexes of iridium and rhodium. Structural features and catalytic applications. *Organometallics*, *37*(21), 4070–4076. doi:10.1021/acs.organomet.8b00633

Gutowski, V., Yang, W., Li, S., Dean, K., & Zhang, X. (2017). Lightweight Nanocomposite Materials, In book: Lightweight and Sustainable Materials for Automotive Applications, 277-302, Haleem, A., Javaid, M., Singh, R. P., Rab, S., & Suman, R. (2023). Applications of Nanotechnology in Medical field. *Global Health Journal (Amsterdam, Netherlands)*, *7*(2), 70–77.

He, S., Lu, C., & Zhang, S. (2011). Facile and efficient route to polyimide-TiO2 nanocomposite coating onto carbon fiber. *ACS Applied Materials & Interfaces, 3*(12), 4744–4750. doi:10.1021/am201239h PMID:22081943

Heine, E., Knops, H. G., Schaefer, K., Vangeyte, P., & Moeller, M. (2007). Antimicrobial functionalisation of textile materials. *Multifunctional Barriers for Flexible Structure: Textile, Leather and Paper,* 23-38.

Hergt, R., Dutz, S., Müller, R., & Zeisberger, M. (2006). Magnetic particle hyperthermia: Nanoparticle magnetism and materials development for cancer therapy. *Journal of Physics Condensed Matter*, *18*(38), S2919–S2934. doi:10.1088/0953-8984/18/38/S26

Hermas, A. E. A., Salam, M. A., & Al-Juaid, S. S. (2013). In situ electrochemical preparation of multi-walled carbon nanotubes/polyaniline composite on the stainless steel. *Progress in Organic Coatings*, *76*(12), 1810–1813. doi:10.1016/j.porgcoat.2013.05.021

Hilger, I., Hergt, R., & Kaiser, W. A. (2005, February). Use of magnetic nanoparticle heating in the treatment of breast cancer. [). IET Digital Library.]. *IEE Proceedings. Nanobiotechnology*, *152*(1), 33–39. doi:10.1049/ip-nbt:20055018 PMID:16441156

Hiremath, N., & Bhat, G. (2017). High-performance carbon nanofibers and nanotubes. In *Structure and properties of high-performance fibers* (pp. 79–109). Woodhead Publishing. doi:10.1016/B978-0-08-100550-7.00004-8

Hou, W., & Wang, Q. (2009). UV-driven reversible switching of a polystyrene/titania nanocomposite coating between superhydrophobicity and superhydrophilicity. *Langmuir*, *25*(12), 6875–6879. doi:10.1021/la900151y PMID:19388630

Hu, J., Zhang, G., & Liu, S. (2012). Enzyme-responsive polymeric assemblies, nanoparticles and hydrogels. *Chemical Society Reviews*, *41*(18), 5933–5949. doi:10.1039/c2cs35103j PMID:22695880

Hu, Q., Katti, P. S., & Gu, Z. (2014). Enzyme-responsive nanomaterials for controlled drug delivery. *Nanoscale*, *6*(21), 12273–12286. doi:10.1039/C4NR04249B PMID:25251024

Irshad, M. A., Nawaz, R., Rehman, M. Z., Adrees, M., Rizwan, M., Ali, S., Ahmad, S., & Tasleem, S. (2021). Synthesis, characterization and advanced sustainable applications of titanium dioxide nanoparticles: A review. *Ecotoxicology and Environmental Safety*, *212*, 111978. doi:10.1016/j.ecoenv.2021.111978 PMID:33561774

Iyer, K. S., & Chavan, S. S. (2017). Nanotechnology and Its Application in Cosmetic Industry A Concise Review. *International Journal of Innovative Science and Research Technology*, *2*(9), 263–267.

Jain, P., & Pradeep, T. (2005). Potential of silver nanoparticle-coated polyurethane foam as an antibacterial water filter. *Biotechnology and Bioengineering*, *90*(1), 59–63. doi:10.1002/bit.20368 PMID:15723325

Jalilinejad, N., Rabiee, M., Baheiraei, N., Ghahremanzadeh, R., Salarian, R., Rabiee, N., Akhavan, O., Zarrintaj, P., Hejna, A., Saeb, M. R., Zarrabi, A., Sharifi, E., Yousefiasl, S., & Zare, E. N. (2023). Electrically conductive carbon-based (bio)-nanomaterials for cardiac tissue engineering. *Bioengineering & Translational Medicine*, *8*(1), e10347. doi:10.1002/btm2.10347 PMID:36684103

Jeevanandam, J., Ling, J. K. U., Barhoum, A., San Chan, Y., & Danquah, M. K. (2022). Bionanomaterials: Definitions, sources, types, properties, toxicity, and regulations. In Fundamentals of Bionanomaterials (pp. 1-29). Elsevier.

Joseph, T. M., Kar Mahapatra, D., Esmaeili, A., Piszczyk, Ł., Hasanin, M. S., Kattali, M., Haponiuk, J., & Thomas, S. (2023). Nanoparticles: Taking a unique position in medicine. *Nanomaterials (Basel, Switzerland)*, *13*(3), 574. doi:10.3390/nano13030574 PMID:36770535

Kalaivani, M., & Ravi, S. (2023). Green Synthesis of ZnO NPs and CdO-ZnO Nanocomposites using aqueous Extract of Water Hyacinth (Eichhorniacrassipes) Characterization, Structural and Nano-fertilizer using Application. *Indian Journal of Science and Technology*, *16*(25), 1918–1926. doi:10.17485/IJST/v16i25.180

Katz, L. M., Dewan, K., & Bronaugh, R. L. (2015). Nanotechnology in cosmetics. *Food and Chemical Toxicology*, *85*, 127–137. doi:10.1016/j.fct.2015.06.020 PMID:26159063

Kawadkar, J., Chauhan, M. K., & Maharana, M. (2011). Nanobiotechnology: Application of nanotechnology in diagnosis, drug discovery and drug development. *Asian Journal of Pharmaceutical and Clinical Research*, *4*(1), 23–25.

Kayal, S. (2021). Thermoresponsive magnetic/polymer composite nanoparticles for biomedical applications. *Materials Today: Proceedings*, *41*, 1116–1119. doi:10.1016/j.matpr.2020.08.605

Keskinbora, K. H., & Jameel, M. A. (2018). Nanotechnology applications and approaches in medicine: A review. *J NanosciNanotechnol Res*, *2*(2), 6.

Khalid, M. Y., Kamal, A., Otabil, A., Mamoun, O., & Liao, K. (2023). Graphene/epoxy nanocomposites for improved fracture toughness: A focused review on toughening mechanism. *Chemical Engineering Journal Advances*, *16*, 100537. doi:10.1016/j.ceja.2023.100537

Khandelwal, A., Joshi, R., & Shrivastava, M. (2022). Application of Nanomaterials in Agriculture and their Impact on Environment. *International Journal of Agriculture Environment and Biotechnology*, *15*(2), 195–204.

Kori, A. H., Jagirani, M. S., & Soylak, M. (2023). Graphene-Based Nanomaterials: A Sustainable Material for Solid-Phase Microextraction (SPME) for Environmental Applications. *Analytical Letters*, *56*(15), 1–16. doi:10.1080/00032719.2023.2173221

Kottegoda, N., Sandaruwan, C., Priyadarshana, G., Siriwardhana, A., Rathnayake, U. A., Berugoda Arachchige, D. M., Kumarasinghe, A. R., Dahanayake, D., Karunaratne, V., & Amaratunga, G. A. J. (2017). Urea-hydroxyapatite nanohybrids for slow release of nitrogen. *ACS Nano*, *11*(2), 1214–1221. doi:10.1021/acsnano.6b07781 PMID:28121129

Kreyling, W. G., Semmler-Behnke, M., & Chaudhry, Q. (2010). A complementary definition of nanomaterial. *Nano Today*, *5*(3), 165–168. doi:10.1016/j.nantod.2010.03.004

Krishna, V. M. (2016). Heat Transfer Enhancement by using ZnO Water Nanofluid in a Concentric Tube Heat Exchanger under Forced Convection Conditions. *International Journal of Innovations in Engineering and Technology*, *7*(4), 177–184.

Kulkarni, S. J. (2017). Summary of Research and Studies on Heat Transfer Enhancements by Nanofluids. *International Journal of Engineering Research and General Science*, *5*(2), 175–181.

Kumar, M. (2019). Applications of Nano science in sports: An analytical study. International Journal of Physiology. *Nutrition and Physical Education*, *4*(1), 2642–2646.

Kumar, N., Sangma, S. N., Ray, D. P., Ghosh, R. K., Ammayappan, L., & Chattopadhyay, S. N. (2020). Nanomaterial regulation and its applications in textile sector: A review. *International Journal of Bioresource Science*, *7*(2), 75–81. doi:10.30954/2347-9655.02.2020.5

Kumari, R., Suman, K., Karmakar, S., Lakra, S. G., Saurav, G. K., & Mahto, B. K. (2023). Regulation and safety measures for nanotechnology-based agri-products. *Frontiers in Genome Editing*, *5*, 1200987. doi:10.3389/fgeed.2023.1200987 PMID:37415849

Lee, P. C., Kang, D., Oh, J. T., Seo, J. Y., Shin, D., Jung, J. U., Ko, Y. K., Ha, J. U., & Kim, M. G. (2023). Reducing Moisture Absorption in Polypropylene Nanocomposites for Automotive Headlamps Using Hydrophobicity-Modified Graphene/Montmorillonite. *Nanomaterials (Basel, Switzerland)*, *13*(9), 1439. doi:10.3390/nano13091439 PMID:37176984

Lowe, S., O'Brien-Simpson, N. M., & Connal, L. A. (2015). Antibiofouling polymer interfaces: Poly (ethylene glycol) and other promising candidates. *Polymer Chemistry*, *6*(2), 198–212. doi:10.1039/C4PY01356E

Luo, X., Morrin, A., Killard, A. J., & Smyth, M. R. (2006). Application of nanoparticles in electrochemical sensors and biosensors. Electroanalysis. *An International Journal Devoted to Fundamental and Practical Aspects of Electroanalysis*, *18*(4), 319–326.

Ma, B. (2021). *Aluminum matrix nanocomposites for applications in automobile industry* [Doctoral dissertation, Sorbonne Université].

Mahajan, G., Kaur, M., & Gupta, R. (2022). Green functionalized nanomaterials: Fundamentals and future opportunities. In *Green Functionalized Nanomaterials for Environmental Applications* (pp. 21–41). Elsevier. doi:10.1016/B978-0-12-823137-1.00003-8

Maheshwari, R., Joshi, G., Mishra, D. K., & Tekade, R. K. (2019). Bionanotechnology in pharmaceutical research. In *Basic fundamentals of drug delivery* (pp. 449–471). Academic Press. doi:10.1016/B978-0-12-817909-3.00012-1

Malhotra, B. D., Ali, M. A., Malhotra, B., & Ali, M. (2018). In B. D. Malhotra & M. A. Ali (Eds.), *Nanocomposite Materials: Biomolecular Devices. Nanomaterials for Biosensors* (pp. 145–159). William Andrew Publishing., doi:10.1016/B978-0-323-44923-6.00005-4

Mazov, I., Krasnikov, D., Stadnichenko, A., Kuznetsov, V., Romanenko, A., Anikeeva, O., & Tkachev, E. (2012). Direct vapor-phase bromination of multiwall carbon nanotubes. *Journal of Nanotechnology*, *2012*, 2012. doi:10.1155/2012/954084

Menossi, M., Casalongué, C., & Alvarez, V. A. (2021). Bio-nanocomposites for Modern Agricultural Applications. In *Handbook of Consumer Nanoproducts*. Springer., doi:10.1007/978-981-15-6453-6_68-1

Moallem, S. A., Dehghani, N., Mehri, S., Shahsavand, S., Alibolandi, M., & Hadizadeh, F. (2015). Synthesis of novel 1, 8-acridinediones derivatives: Investigation of MDR r eversibility on breast cancer cell lines T47D and tamoxifenresistant T47D. *Research in Pharmaceutical Sciences*, *10*(3), 214. PMID:26600848

Mukhopadhyay, R., Sarkar, B., Khan, E., Alessi, D. S., Biswas, J. K., Manjaiah, K. M., Eguchi, M., Wu, K. C. W., Yamauchi, Y., & Ok, Y. S. (2022). Nanomaterials for sustainable remediation of chemical contaminants in water and soil. *Critical Reviews in Environmental Science and Technology*, *52*(15), 2611–2660. doi:10.1080/10643389.2021.1886891

Nasrollahzadeh, M., Sajjadi, M., Iravani, S., & Varma, R. S. (2021). Starch, cellulose, pectin, gum, alginate, chitin and chitosan derived (nano) materials for sustainable water treatment: A review. *Carbohydrate Polymers*, *251*, 116986. doi:10.1016/j.carbpol.2020.116986 PMID:33142558

Ndaba, B., Roopnarain, A., Haripriya, R. A. M. A., & Maaza, M. (2022). Biosynthesized metallic nanoparticles as fertilizers: An emerging precision agriculture strategy. *Journal of Integrative Agriculture*, *21*(5), 1225–1242. doi:10.1016/S2095-3119(21)63751-6

Nesa, S. H. S., & Tarangini, K. (2023). A review on augmentation of natural fabric materials with novel bio/nanomaterials and their multifunctional perspectives. *Hybrid Advances, 100020.*

Nguyen, T. A., Nguyen, T. V., Thai, H., & Shi, X. (2016). Effect of nanoparticles on the thermal and mechanical properties of epoxy coatings. *Journal of Nanoscience and Nanotechnology*, *16*(9), 9874–9881. doi:10.1166/jnn.2016.12162

Nguyen-Tri, P., Nguyen, T. A., Carriere, P., & Ngo Xuan, C. (2018). Nanocomposite coatings: Preparation, characterization, properties, and applications. *International Journal of Corrosion*, *2018*, 2018. doi:10.1155/2018/4749501

Nikazar, S., Barani, M., Rahdar, A., Zoghi, M., & Kyzas, G. Z. (2020). Photo-and magnetothermally responsive nanomaterials for therapy, controlled drug delivery and imaging applications. *ChemistrySelect*, *5*(40), 12590–12609. doi:10.1002/slct.202002978

Noah, N. M., & Ndangili, P. M. (2022). Green synthesis of nanomaterials from sustainable materials for biosensors and drug delivery. *Sensors International*, *3*, 100166. doi:10.1016/j.sintl.2022.100166

Noreen, I., Hashmi, A., Iqbal, Y., Malik, F. P., Iqbal, M., & Amin, F. (2023). In-situ formation of Azo dye capped-silver nanoparticles and their nanocomposite with reduced graphene oxide for dye degradation. *Materials Chemistry and Physics*, *308*, 128299. doi:10.1016/j.matchemphys.2023.128299

Onaizi, A. M., Huseien, G. F., Lim, N. H. A. S., Amran, M., & Samadi, M. (2021). Effect of nanomaterials inclusion on sustainability of cement-based concretes: A comprehensive review. *Construction & Building Materials*, *306*, 124850. doi:10.1016/j.conbuildmat.2021.124850

Pacheco, I., & Buzea, C. (2021). Nanomaterials and Nanocomposites: Classification and Toxicity. In O. V. Kharissova, L. M. Torres-Martínez, & B. I. Kharisov (Eds.), *Handbook of Nanomaterials and Nanocomposites for Energy and Environmental Applications*. Springer. doi:10.1007/978-3-030-36268-3_

Pan, S., Lin, H., Deng, J., Chen, P., Chen, X., Yang, Z., & Peng, H. (2015). Novel wearable energy devices based on aligned carbon nanotube fiber textiles. *Advanced Energy Materials*, *5*(4), 1401438. doi:10.1002/aenm.201401438

Prameela, K. L. (2017). Nanomaterial's applications in agriculture. *J Chem Pharm Sci*, *10*(1), 593–596.

Pujar, A. S., & Bagawan, A. M. (2016). Applications of Nanotechnology: In Medicine. *International Journal of Scientific Research*, *7*(6), 1211–1215.

Qiu, X., Zhang, Y., Zhu, Y., Long, C., Su, L., Liu, S., & Tang, Z. (2021). Applications of nanomaterials in asymmetric photocatalysis: Recent progress, challenges, and opportunities. *Advanced Materials*, *33*(6), 2001731. doi:10.1002/adma.202001731 PMID:32672886

Qu, X., Alvarez, P. J., & Li, Q. (2013). Applications of nanotechnology in water and wastewater treatment. *Water Research*, *47*(12), 3931–3946. doi:10.1016/j.watres.2012.09.058 PMID:23571110

Rahman, K. T., Moin, T. S., Chowdhury, M. F. M., & Khan, M. N. (2023). Emerging Nano-Enable Materials in the Sports Industry. *Emerging Applications of Nanomaterials*, *141*, 75–100. doi:10.21741/9781644902288-4

Rahman, K. T., Moin, T. S., Chowdhury, M. F. M., & Khan, M. N. (2023). Emerging Nano-Enable Materials in the Sports Industry. *Emerging Applications of Nanomaterials*, *141*, 75–100. doi:10.21741/9781644902288-4

Raj, S., Jose, S., Sumod, U. S., & Sabitha, M. (2012). Nanotechnology in Cosmetics: Opportunities and Challenges. *Journal of Pharmacy & Bioallied Sciences*, *4*(3), 186–193. doi:10.4103/0975-7406.99016 PMID:22923959

Rajendiran, R., Seelam, P. K., Patchaiyappan, A., Balla, P., Shankar, H., Ravi, B., Perupogu, V., & Lassi, U. (2023). Morphologically tailored facet dependent silver nanoparticles supported α-Al2O3 catalysts for chemoselective reduction of aromatic nitro compounds. *Chemical Engineering Journal*, *451*, 138507. doi:10.1016/j.cej.2022.138507

Rajesh, U. C., Wang, J., Prescott, S., Tsuzuki, T., & Rawat, D. S. (2015). RGO/ ZnO nanocomposite: An efficient, sustainable, heterogeneous, amphiphilic catalyst for synthesis of 3-substituted indoles in water. *ACS Sustainable Chemistry & Engineering*, *3*(1), 9–18. doi:10.1021/sc500594w

Ramsden, J. J. (1990). What is nanotechnology? *Nature*, *344*, 524–526.

Rasheed, A., & Khalid, F. A. (2014, June). Fabrication and properties of CNTs reinforced polymeric matrix nanocomposites for sports applications. []. IOP Publishing.]. *IOP Conference Series. Materials Science and Engineering*, *60*(1), 012009. doi:10.1088/1757-899X/60/1/012009

Rasheed, T. (2023). Carbon dots as potential greener and sustainable fluorescent nanomaterials in service of pollutants sensing. *Trends in Analytical Chemistry*, *158*, 116841. doi:10.1016/j.trac.2022.116841

Raza, A., Altaf, S., Ali, S., Ikram, M., & Li, G. (2022). Recent advances in carbonaceous sustainable nanomaterials for wastewater treatments. *Sustainable Materials and Technologies*, *32*, e00406. doi:10.1016/j.susmat.2022.e00406

Rodriguez-Perez, L., Herranz, M. Á., & Martín, N. (2013). The chemistry of pristine graphene. *Chemical Communications (Cambridge)*, *49*(36), 3721–3735. doi:10.1039/c3cc38950b PMID:23539514

Roduner, E. (2006). Size matters: Why nanomaterials are different. *Chemical Society Reviews*, *35*(7), 583–592. doi:10.1039/b502142c PMID:16791330

Rojas-Velázquez, Á. N., Guillén-Castillo, O. I., Alcalá-Jauregui, J. A., Loredo-Osti, C., Ramírez-Tobías, H. M., Romero-Méndez, M. J., Méndez-Cortés, H., & Hernández-Montoya, A. (2023). Effect of a nitrogenous nanocomposite on leaching and N content in lettuce in soil columns. *Discover Nano*, *18*(1), 98. doi:10.1186/s11671-023-03874-w PMID:37523022

Roman, M. (2015). Toxicity of cellulose nanocrystals: A review. *Industrial Biotechnology (New Rochelle, N.Y.)*, *11*(1), 25–33. doi:10.1089/ind.2014.0024

Roy, A., & Bhattacharya, J. (2015). *Nanotechnology in industrial wastewater treatment*. IWA Publishing. https://library.oapen.org/handle/20.500.12657/25811

Roy, A., Dwari, D., Ram, M. K., & Datta, P. (2022). Piezoelectric nanomaterials for biomedical applications. In Food, Medical, and Environmental Applications of Nanomaterials (pp. 355-377). Elsevier. doi:10.1016/B978-0-12-822858-6.00012-1

Sachdeva, H. (2020). Recent advances in the catalytic applications of GO/rGO for green organic synthesis. *Green Process Synth*, *9*(1), 515–537. doi:10.1515/gps-2020-0055

Sahu, A. K., Acharya, P. P., Kar, B. R., Nemani, R., & Gowda, M. (2019). Nanotechnology In Water Treatment, Case Study In Bangalore. *International Journal of Advanced Research*, *7*(2), 142–149. doi:10.21474/IJAR01/8470

Sandra, V., Stojanovic, B., Ivanović, L., Miladinovic, S., & Milojević, S. (2018). *Application of nanocomposites in the automotive industry. Mobility and Vehicle Mechanics*. MVM.

Scientific Committee on Consumer Safety. (2016). *Hydroxypatite (nano)*. European Commission. https://ec.europa.eu/ health/scientific_committees /consumer_safety /docs /sccs_ o_191.pdf

Senthilkumar, R., Vaidyanathan, S., & Sivaraman, B. (2013). Effect of copper nanofluid concentration on thermal performance of heat pipes. *Frontiers in Heat Pipes*, *4*, 1–5. doi:10.5098/fhp.v4.1.3004

Shalaby, M. N., & Saad, M. M. (2020). Advanced material engineering and nanotechnology for improving sports performance and equipment. *International Journal of Psychosocial Rehabilitation*, *24*(10), 2314–2322.

Shekarian, E., Jafari Nasr, M. R., Tarighaleslami, A. H., Walmsley, T. G., Atkins, M. J., Sahebjamee, N., & Alaghehbandan, M. (2016). Impact of hybrid heat transfer enhancement techniques in shell and tube heat exchanger design. *Chemical Engineering Transactions*, *52*, 1159–1164. doi:10.3303/CET1652194

Shilar, F. A., Ganachari, S. V., & Patil, V. B. (2022). Advancement of nano-based construction materials-A review. *Construction & Building Materials*, *359*, 129535. doi:10.1016/j.conbuildmat.2022.129535

Shrivastava, S., & Dash, D. (2009). Applying nanotechnology to human health: Revolution in biomedical sciences. *Journal of Nanotechnology*, *2009*, 2009. doi:10.1155/2009/184702

Singh, A., Dasauni, K., Kumar Nailwal, T., & Nenavathu, B. P. (2023). Formulation of dual functional gCN/TeO2-ZnO nanocomposites as a controlled release nanofertilizer and antibacterial agent. *Nanotechnology*, *34*(15), 155602. doi:10.1088/1361-6528/acb2d1 PMID:36715368

Singh, H., Sharma, A., Bhardwaj, S. K., Arya, S. K., Bhardwaj, N., & Khatri, M. (2021). Recent advances in the applications of nano-agrochemicals for sustainable agricultural development. *Environmental Science. Processes & Impacts*, *23*(2), 213–239. https://pubs.rsc.org/en/content/articlelanding/2021/em/d0em00404a/unauth. doi:10.1039/D0EM00404A PMID:33447834

Soltani, S., Gacem, A., Choudhary, N., Yadav, V. K., Alsaeedi, H., Modi, S., Patel, A., Khan, S. H., Cabral-Pinto, M. M. S., Yadav, K. K., & Patel, A. (2023). Scallion Peel Mediated Synthesis of Zinc Oxide Nanoparticles and Their Applications as Nano fertilizer and Photocatalyst for Removal of Organic Pollutants from Wastewater. *Water (Basel)*, *15*(9), 1672. doi:10.3390/w15091672

Sonneville-Aubrun, O., Simonnet, J. T., & L'alloret, F. (2004). Nanoemulsions: A new vehicle for skincare products. *Advances in Colloid and Interface Science*, *108*, 145–149. doi:10.1016/j.cis.2003.10.026 PMID:15072937

Srivastava, R., Varghese, R., & Abraham, A. R. (2023). Bio-nanomaterials: applications and utility in agricultural and medical field. In Applications of Multifunctional Nanomaterials (pp. 693-707). Elsevier.

Stoimenov, P. K., Klinger, R. L., Marchin, G. L., & Klabunde, K. J. (2002). Metal oxide nanoparticles as bactericidal agents. *Langmuir*, *18*(17), 6679–6686. doi:10.1021/la0202374

Sundaram, M. U., & Mahamani, A. (2015). Development of Carbon Nanotube Reinforced Aluminium Matrix Composite Brake Drum for Automotive Applications. Research and Innovation in Carbon Nanotube-Based Composites, Dr. BrahimAttaf (Ed.), Chapter, 2015.

Suppan, S. (2017). Applying nanotechnology to fertilizer: rationales, research, risks and regulatory challenges. The Institute for Agriculture and Trade Policy works locally and globally, Brazil.

Tamang, S., Rai, S., Bhujel, R., Bhattacharyya, N. K., Swain, B. P., & Biswas, J. (2023). A concise review on GO, rGO and metal oxide/rGO composites: Fabrication and their supercapacitor and catalytic applications. *Journal of Alloys and Compounds*, *947*, 169588. doi:10.1016/j.jallcom.2023.169588

Tariq, W., Ali, F., Arslan, C., Nasir, A., Gillani, S. H., & Rehman, A. (2022). Synthesis and applications of graphene and graphenebased nanocomposites: Conventional to artificial intelligence approaches. *Frontiers in Environmental Chemistry, 3*, 890408. doi:10.3389/fenvc.2022.890408

TERI ENVIS Resource Partner on Renewable Energy and Climate Change. (2021). *Application of Nanotechnology in Solar Energy*. 1-32. http://www.envis.nic.in/WriteReadData /userfile s/file/n ewsletters/19- TERI.pdf

Tinucci, L., Borgarello, E., Minero, C., & Pelizzetti, E. (1993). Treatment of industrial wastewaters by photocatalytic oxidation on TiO2. In *Proceedings of the 1st International Conference on TiO2 Photocatalytic Purification and treatment of Water and Air* (pp. 585-594). Ontario, Canada.

Tiruvenkadam, N., Thyla, P. R., Senthilkumar, M., Bharathiraja, M., & Murugesan, A. (2015). Synthesis of new aluminum nano hybrid composite liner for energy saving in diesel engines. *Energy Conversion and Management, 98*, 440–448. doi:10.1016/j.enconman.2015.04.017

Tiwari, D. K., Behari, J., & Sen, P. (2008). Application of nanoparticles in waste water treatment. *World Applied Sciences Journal, 3*(3), 417–433.

Trovatti, E., Oliveira, L., Freire, C. S., Silvestre, A. J., Neto, C. P., Pinto, J. J. C., & Gandini, A. (2010). Novel bacterial cellulose–acrylic resin nanocomposites. *Composites Science and Technology, 70*(7), 1148–1153. doi:10.1016/j.compscitech.2010.02.031

Ulus, R., Yeşildağ, İ., Elmastaş, M., & Kaya, M. (2015). Rapid synthesis of novel 1,8-dioxoacridine carboxylic acid derivatives by microwave irradiation and their free radical scavenging activity. *Medicinal Chemistry Research, 24*(10), 3752–3759. doi:10.1007/s00044-015-1417-6

Valença, D. P., Alves, K. G. B., Melo, C. P. D., & Bouchonneau, N. (2015). Study of the efficiency of polypyrrole/ZnO nanocomposites as additives in anticorrosion coatings. *Materials Research, 18*(suppl 2), 273–278. doi:10.1590/1516-1439.371614

Venkatesh, R., Sakthivel, P., Vivekanandan, M., Kannan, C. R., Krishna, J. P., Dhanabalan, S., Thirugnanasambandham, T., & Majora, M. (2023). Synthesis and Thermal Adsorption Characteristics of Silver-Based Hybrid Nanocomposites for Automotive Friction Material Application. *Adsorption Science and Technology, 2023*, 2023. doi:10.1155/2023/1003492

Villalba-Rodríguez, A. M., Martínez-Zamudio, L. Y., Martínez, S. A. H., Rodríguez-Hernández, J. A., Melchor-Martínez, E. M., Flores-Contreras, E. A., González-González, R. B., & Parra-Saldívar, R. (2023). Nanomaterial Constructs for Catalytic Applications in Biomedicine: Nanobiocatalysts and Nanozymes. *Topics in Catalysis*, *66*(9-12), 707–722. doi:10.1007/s11244-022-01766-4 PMID:36597435

Vollath, D. (2008). Nanomaterials an introduction to synthesis, properties and application. *Environmental Engineering and Management Journal*, *7*(6), 865–870.

Wang, Q., Han, N., Shen, Z., Li, X., Chen, Z., Cao, Y., Si, W., Wang, F., Ni, B.-J., & Thakur, V. K. (2023). MXene-based electrochemical (bio) sensors for sustainable applications: Roadmap for future advanced materials. *Nano Materials Science*, *5*(1), 39–52. doi:10.1016/j.nanoms.2022.07.003

Wen, L., Wang, Y., Lu, D., Shengyang, H., & Heyou, H. (2010). Preparation of KF/CaO nanocatalyst and its application in biodiesel production from Chinese tallow seed oil. *Fuel*, *89*(9), 2267–2271. doi:10.1016/j.fuel.2010.01.028

Yadav, A., Kumar, H., Sharma, R., & Kumari, R. (2023). Synthesis, processing, and applications of 2D (nano) materials: A sustainable approach. *Surfaces and Interfaces*, *39*, 102925. doi:10.1016/j.surfin.2023.102925

Yadav, A., Yadav, K., & Abd-Elsalam, K. A. (2023). Nanofertilizers: Types, Delivery and Advantages in Agricultural Sustainability. *Agrochemicals*, *2*(2), 296–336. doi:10.3390/agrochemicals2020019

Yadav, D., Kwak, M., Chauhan, P. S., Puranik, N., Lee, P. C., & Jin, J. O. (2022, November). Cancer immunotherapy by immune checkpoint blockade and its advanced application using bio-nanomaterials. [). Academic Press.]. *Seminars in Cancer Biology*, *86*, 909–922. doi:10.1016/j.semcancer.2022.02.016 PMID:35181474

Yang, H., Zhu, S., & Pan, N. (2004). Studying the mechanisms of titanium dioxide as ultraviolet-blocking additive for films and fabrics by an improved scheme. *Journal of Applied Polymer Science*, *92*(5), 3201–3210. doi:10.1002/app.20327

Yang, L. (2023). Challenges and recent progress in carbon-based nanocomposites for sportswear and sensing applications. *Materials Research Express*, *10*(5), 052001. doi:10.1088/2053-1591/accf64

Yasmine, R., Ahmad, J., Qamar, S., & Qureshi, M. I. (2023). Engineered nanomaterials for sustainable agricultural production, soil improvement, and stress management: An overview. *Engineered Nanomaterials for Sustainable Agricultural Production, Soil Improvement and Stress Management*, 1-23. doi:10.1016/B978-0-323-91933-3.00022-2

Yousefiasl, S., Zare, I., Ghovvati, M., & Ghomi, M. (2023). Enzyme-Responsive Materials: Properties, Design, and Applications. In Stimuli-Responsive Materials for Biomedical Applications (pp. 203-229). American Chemical Society.

Yukuyama, M. N., Ghisleni, D. D., Pinto, T. J., & Bou-Chacra, N. A. (2016). Nanoemulsion: Process selection and application in cosmetics–a review. *International Journal of Cosmetic Science*, *38*(1), 13–24. doi:10.1111/ics.12260 PMID:26171789

Zaib, S., & Iqbal, J. (2019). Nanotechnology: Applications, techniques, approaches, & the advancement in toxicology and environmental impact of engineered nanomaterials. *Importance & Applications of Nanotechnology, 8*.

Zankana, M. M., Al-dalawy, S. M., & Barzinjy, A. A. (2023). Synthesis and characterization of bio-nanocomposites: Functionalization of graphene oxide with a biocompatible amino acid. *Hybrid Advances*, *3*, 100070. doi:10.1016/j.hybadv.2023.100070

Zhang, S., Jiang, W., Liu, G., Liu, S., Chen, H., Lyu, G., Yang, G., Liu, Y., & Ni, Y. (2023). Preparation of ultrafine and highly loaded silver nanoparticle composites and their highly efficient applications as reductive catalysts and antibacterial agents. *Journal of Colloid and Interface Science*, *629*, 766–777. doi:10.1016/j.jcis.2022.09.018 PMID:36099844

Zheng, J., Wang, C. G., Zhou, H., Ye, E., Xu, J., Li, Z., & Loh, X. J. (2021). *Current research trends and perspectives on solid-state nanomaterials in hydrogen storage*. Research. doi:10.34133/2021/3750689

Chapter 4
Applications of Nanocomposites in Water Remediation:
A Mechanistic Overview

Rajni Garg
https://orcid.org/0000-0002-9629-8141
Galgotias College of Engineering and Technology, India

Rishav Garg
https://orcid.org/0000-0001-6565-9234
Galgotias College of Engineering and Technology, India

Nnabuk Okon Eddy
https://orcid.org/0000-0001-7704-3082
University of Nigeria, Nigeria

ABSTRACT

Water is a crucial natural resource for human and ecosystem health. However, rapid industrialization, urban expansion, and exploitation of natural resources have negatively impacted water quality. Effluents, containing various contaminants, negatively impact water quality. Various techniques have been used to treat effluents, but these methods have drawbacks such as incomplete removal, energy demands, and hazardous sludge generation. T Adsorption is an efficient method for wastewater treatment due to its abundant adsorbents, exceptional performance in pollutant removal, straightforward operational procedures, and cost-effectiveness. The reusability of adsorbents is crucial for financial feasibility. The sorption approach also offers advantages such as the proliferation of sorbents within the ecosystem, their capacity for sorption, remarkable recyclability, regeneration, cost-effectiveness, and the facile desorption of contaminants. This chapter aims to assess the efficacy of nanocomposites in remediating organic pollutants from aquatic ecosystems.

DOI: 10.4018/979-8-3693-1094-6.ch004

Copyright © 2024, IGI Global. Copying or distributing in print or electronic forms without written permission of IGI Global is prohibited.

INTRODUCTION

Water is an indispensable natural resource that plays a vital role in ensuring the survival of the living organisms. The availability of uncontaminated water is of utmost importance, not only for human well-being but also for the overall health and balance of the ecosystem (Nicola et al., 2020). However, over the past few decades, the quality of water has been adversely impacted by the steady growth of the population, rapid industrialization, expanding urban areas, and irresponsible exploitation of natural resources. Freshwater resources across the globe are inherently constrained, posing significant challenges for sustaining human existence (Odoemelam, Oji, Okon, et al., 2023). The impact of industrialization and the rapid advancement of technology has led to a concerning decline in water sources and a simultaneous rise in the quantity of contaminated water (Garg, Garg, Sillanpää, et al., 2023). The annual occurrence of diseases in humans and animals is attributed to the presence of contaminants in water. Polluted water not only jeopardises human well-being, but also endangers the integrity of the entire ecosystem (Rani et al., 2022).

Effluents contain a diverse array of contaminants, encompassing toxic ions, heavy metals, pharmaceuticals, organic matter, nutrients, plastics, nanoparticles, organic compounds, pharmaceutical and personal care products, dyes, pesticides, radionuclides, and pathogens (Figure 1). The existence of these pollutants, particularly inorganic and organic substances, exerts a negative impact on the overall quality of water (Kalantari et al., 2015). To date, numerous techniques have been employed for the treatment of effluents, encompassing sedimentation, flotation, electrolysis, electrodialysis, reverse osmosis, distillation, chemical precipitation, flocculation, ion exchange, solvent extraction, microfiltration, oxidation, catalytic procedures, and adsorption (Garg, Garg, Khan, et al., 2023). The treatment methods, such as electrochemical removal, chemical precipitation, and ion exchange, exhibit several drawbacks in the context of environmental conservation (Garg et al., 2022). These drawbacks encompass incomplete removal of pollutants, substantial energy demands, and the generation of hazardous sludge (Sherlala et al., 2019). The challenges that may arise are closely linked to the financial and technical resources needed for the establishment, functioning, and upkeep of these technologies leading to insufficient implementation (Andelescu et al., 2018). Hence, it is imperative to formulate efficacious and ecologically conscious remedies, encompassing techniques such as adsorption utilising biopolymers, biological interventions, and nanotechnology strategies.

Adsorption process emerges as a highly efficient technique for the treatment of wastewater. This is primarily due to its numerous benefits, including the availability of abundant adsorbents, exceptional performance in pollutant removal, straightforward operational procedures, and a cost-effective nature (Sharma et al., 2023a). Through

Figure 1. Various types of contaminants present in wastewater

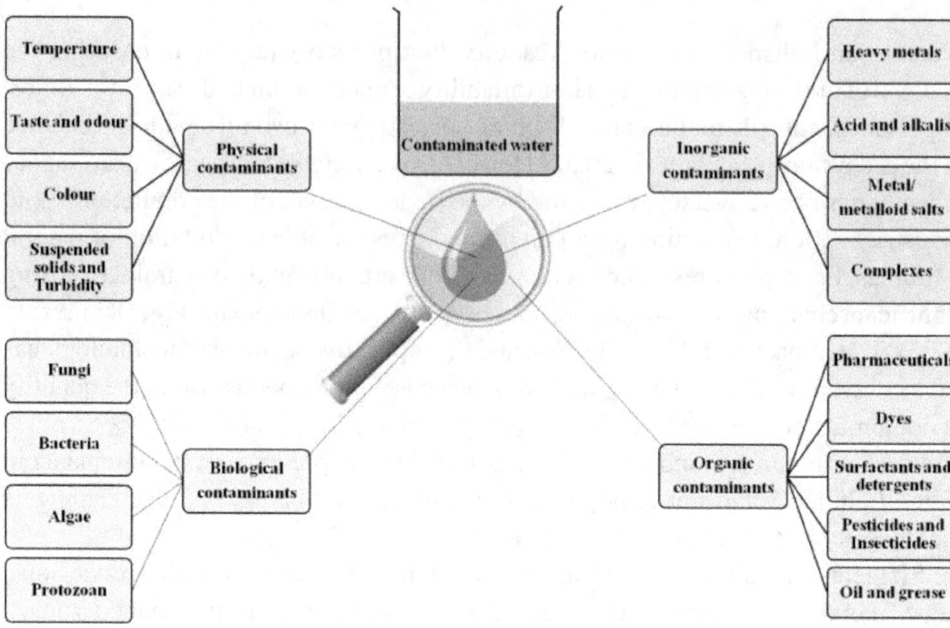

the implementation of this particular procedure, it is possible to effectively reduce the levels of various organic pollutants in a manner that is both economically viable and environmentally beneficial, thereby enhancing the overall quality of the water resource (Suwannahong et al., 2021). The reusability of adsorbents is a crucial aspect that warrants careful consideration, as it plays a pivotal role in determining its financial feasibility. The retrieval of the absorbed contaminant from the surface of the sorbent is necessary in this process (Eddy, Garg, et al., 2023). Additional advantages of the adsorption approach that warrant consideration encompass the profusion of sorbents within the ecosystem, their notable capacity for adsorption, remarkable recyclability, regeneration, and the facile desorption of contaminants (Marbawi et al., 2019). The primary objective of this chapter is to assess and contrast the efficacy of nanocomposites in the remediation of organic pollutants from aquatic ecosystems. In order to accomplish this, a comprehensive investigation was conducted on diverse categories of pollutants, a wide range of adsorbents, and the significant impact of various variables in the process of contaminant removal. The investigation also encompassed the examination of the adsorption mechanism pertaining to organic contaminants. A thorough investigation was conducted to assess the efficacy of nanocomposites and biocomposites in removing various types of pollutants.

HEAVY METAL IONS

Heavy metal ions, known for their high release rates, are a cause for concern due to their status as prominent contaminants. Heavy metals and metalloids are a group of elements characterised by their high atomic density. This group encompasses various elements such as arsenic (As), silver (Ag), chromium (Cr), copper (Cu), cadmium (Cd), lead (Pb), zinc (Zn), mercury (Hg), iron (Fe), and platinum (Pt) etc. (Amari et al., 2021). The water is continuously exposed to a variety of natural and human-induced activities, resulting in the regular release of metallic toxic elements. In various locations across the globe, the mean levels of Fe, Co, Ni, Cr, Mn, As, and Cd detected in surface water ecosystems exceed the permissible thresholds for potable water (Tizro & Baseri, 2017). Heavy metals exhibit a lack of biodegradability, leading to their propensity for bioaccumulation, wherein their concentration gradually increases within living organisms over time. Additionally, these organisms exhibit a remarkable level of persistence, which can have both direct and indirect impacts on a wide range of other organisms as a result of biomagnifications (Sharma et al., 2023b). Numerous heavy metal ions exhibit toxicity or carcinogenic properties. Even when present in extremely low quantities, these substances have the potential to cause harm to various organs, including the lungs, kidneys, liver, prostate, oesophagus, stomach, and skin (Jayakaran et al., 2019; Rani et al., 2022). Additionally, they have been linked to the development of neurodegenerative disorders such as Parkinson's and Alzheimer's diseases. Aquatic organisms, such as phytoplankton, zooplankton, and fish, can be impacted by metals as well (Nikalje, 2015). These metals have the ability to accumulate in various organs of these organisms, leading to oxidative damage, disruption of the endocrine system, and suppression of the immune system. These effects can have consequences on their survival and growth. These impacts necessitate the advancement of technologies aimed at effectively extracting them from water sources (Zanna et al., 2018).

A significant proportion of the Earth's lakes and rivers have got contamination due to the presence of heavy metals. The natural sources of metal contamination in the environment arise from the intricate interplay between metal-containing rocks and volcanic eruptions . The ecological impact of volcanoes can manifest through infrequent yet substantial emissions, such as explosive volcanic events, as well as ongoing but less intense emissions, which encompass geothermal activity and degassing (Da'ana et al., 2021). Anthropogenic sources encompass a range of activities that contribute to environmental degradation. These activities stem from industrial processes such as the combustion of fossil fuels and metal processing, as well as agricultural practises involving the use of pesticides. Additionally, domestic activities such as the disposal of garbage and the use of detergents also contribute to anthropogenic sources of pollution, a pivotal sector that warrants careful consideration, as it holds a prominent position in the sustainable development of both developed and developing nations (Aragaw et al., 2021). The extraction of minerals releases significant amount of heavy metals that are subsequently carried via flowing rivers and streams, where they have the potential to dissolve in water or become part of sediments. The presence of these metallic species has been observed to infiltrate the groundwater, potentially leading to water scarcity (Khatoon et al., 2018). Additionally, their presence can contribute to soil erosion, impeding the growth of crops and posing significant health risks to local animal populations as well as human communities. The rise in concentration of these pollutants, especially in recent decades, can be attributed to anthropogenic activities, urbanisation, and industrialization (Garg, Garg, Sharma, et al., 2023).

ORGANIC POLLUTANTS

Dyes, hydrocarbons, pesticides, and drugs belong to the category of organic components. Organic compounds play a crucial role in diverse industries and can be produced through a wide range of industrial processes. The introduction of these contaminants has the potential to infiltrate sewage systems, thereby leading to their pollution. Various types of organic compounds, including pesticides, exhibit notable toxicological effects on organisms within ecosystems (Modi et al., 2023).

Dyes encompass a vast and varied array of organic pollutants that are present in wastewater, gaining entry into effluent via multiple industrial sectors, including printing, paper production, painting, and leather manufacturing. These organisms occur in their natural habitats and can also be artificially synthesized (Shah, 2023). Annually, a substantial quantity of approximately 100 metric tonnes of diverse dyes is released into effluent through both direct and indirect means. These entities are categorised into three distinct categories, namely anionic, cationic, and non-ionic. The potential impacts of dyes on organisms can vary as per their concentration in aqueous environments and the extent of contact (Ali, 2018). For example, certain ecological consequences encompass genetic mutations, dermatological conditions, skin irritations, allergic reactions, carcinogenic effects, and so forth. The structure of cationic dyes is characterised by the presence of positive charges, which classifies them as basic dyes. The water-soluble properties of these dyes make them extensively utilised in various applications, including silk dyeing, acrylic, and nylon (Eddy, Odiongenyi, Garg, Ukpe, Garg, ChimaOgoko, et al., 2023).

Cationic dyes exhibit notable characteristics such as elevated colour intensity and luminosity. The group comprises several dyes that are of utmost importance, rhodamine 6G (RG), methylene blue (MB), including basic yellow (BY), malachite green (MG), and methyl violet (MV). One additional detrimental consequence of high concentrations of cationic dyes in water is their inhibitory effect on light penetration, thereby impacting the photosynthesis process (Karthikeyan et al., 2021). In contrast to cationic dyes, anionic dyes possess negative charges within their molecular structure. Anionic dyes, also referred to as direct dyes, encompass a diverse array of reactive dyes within their classification. Several crucial anionic dyes, such as direct red (DR), direct green (DG), methyl orange (MO), congo red (CR), and reactive black (RB), are known to have significant ecological implications (Pandey et al., 2023). Dispersing dyes, also known as non-ionic dyes, represent the final category of dyes that find extensive application in the production of polyamide, coloured plastics, polyester, and other related materials. One additional adverse consequence of these dyes is their potential to induce allergic reactions within the human body. Their existence exert a detrimental impact on the quality of water, as well as the well-being of human beings. As a result, numerous industries make concerted efforts to mitigate the presence of these harmful dyes within sewage prior to their release into aquatic ecosystems. The pivotal role of scientific communities in the advancement of effective methodologies cannot be overstated (Ali, 2018).

Pharmaceutical substances possess diverse applications and can be harnessed for the purpose of mitigating ailments. Nevertheless, the release of pharmaceutical substances in elevated extent into wastewater exerts a deleterious influence on the environment and poses a grave peril to the survival of various organisms within ecosystems (Mandal et al., 2022). In addition, certain antibiotics Pesticides, as an organic compound, are widely employed in safeguarding plants from the detrimental effects of pests, fungi, and weeds. The presence of these hazardous substances has significant ecological impacts on aquatic ecosystems (Eddy, Odiongenyi, Garg, Ukpe, Garg, Nemr, et al., 2023). Groundwaters can be disseminated through diverse mechanisms, including the application of sprays and the dispersion of dust particles. Atrazine, a pesticide, is known to cause hormonal disruption in living organisms. The existence of certain pesticides within aquatic environments, even when present in minimal concentrations, can lead to mortality. The removal of pesticides, thus, holds significant ecological significance (Odoemelam, Oji, Eddy, et al., 2023).

Volatile organic compounds (VOCs), including benzene, toluene, phenol, and other aromatic compounds, are a prevalent type of organic pollutants found in industrial effluent, particularly in refinery wastewater systems and have been observed to have adverse effects on various biological systems (Teiri et al., 2020). These effects include but are not limited to vomiting, liver and kidney dysfunction, impairment of the nervous system, feelings of nausea, development of cataracts, irritation of the eyes, and respiratory issues. Hydrocarbon compounds, including oil and gas condensate, represent an additional category of organic contaminants found in wastewater (D'Ugo et al., 2021).

APPROACHES FOR WATER REMEDIATION

A multitude of techniques exist for the removal of organic contaminants from water-based environments. These methods encompass electrodialysis, distillation, demulsification, electrochemical procedures, chemical precipitation, coagulation, sedimentation, membrane processes, ion exchange, filtration, and adsorption (Garg, Garg, Khan, et al., 2023). Figure 2 provides a comprehensive overview of various processes employed for the removal of organic and inorganic contaminants, along with their respective advantages and disadvantages. Understanding these characteristics is of utmost importance for the effective implementation of appropriate methods in the mitigation of pollutants. Electrical treatment processes encompass an extensive range of technologies such as electrochemical processes, electro-coagulation, electro-precipitation, electro-flotation, electrophoresis, and other related methodologies (Yadav et al., 2019). The electrochemical process represents a remarkably efficient methodology for the elimination of diverse categories of pollutants. It encompasses several chemical processes, including electro-coagulation, electrodeposition, and electro-flotation. Here, the redox reactions involve the exchange of electrons between organic contaminants and electrodes. The expenditure associated with the apparatus employed in this technology is relatively elevated, and the execution of this procedure demands a significant fiscal dedication. One additional limitation pertains to the elevated levels of sludge disposal (Stefaniuk et al., 2016). Nevertheless, this process does not involve the use of any harmful chemicals and exhibits a remarkable ability to specifically target and remove metallic pollutants.

Flocculation, a widely observed phenomenon in wastewater treatment systems, involves the utilisation of chemicals to induce the coagulation of pollutants, effectively facilitating their separation from the contaminated water. Flocculation is a process that disrupts the stability of colloids by counteracting the attractive forces that keep them apart. Aluminium, $FeSO_4$, and $FeCl_3$ are essential and extensively employed flocculants in the field of water decontamination. The primary limitations associated with this process are its considerable financial implications and its reliance on substantial quantities of chemical substances (Amari et al., 2021). Coagulants are commonly employed in the initial stage of wastewater treatment. Additionally, it is worth noting that ion-exchange represents a highly efficient mechanism for the removal of both cationic and anionic pollutants. This particular technology plays a crucial role in the reduction of heavy metal contamination due to its exceptional removal capabilities, remarkable regenerative properties, and rapid reaction kinetics. A diverse array of synthetic and natural resins can be harnessed, with synthetic resins often being the favoured choice. Ion-exchange entails the exchange of undesirable ions with their more desirable counterparts. Nevertheless, it is important to acknowledge that this particular process exhibits certain significant limitations, including its

Figure 2. Methods for removal of pollutants

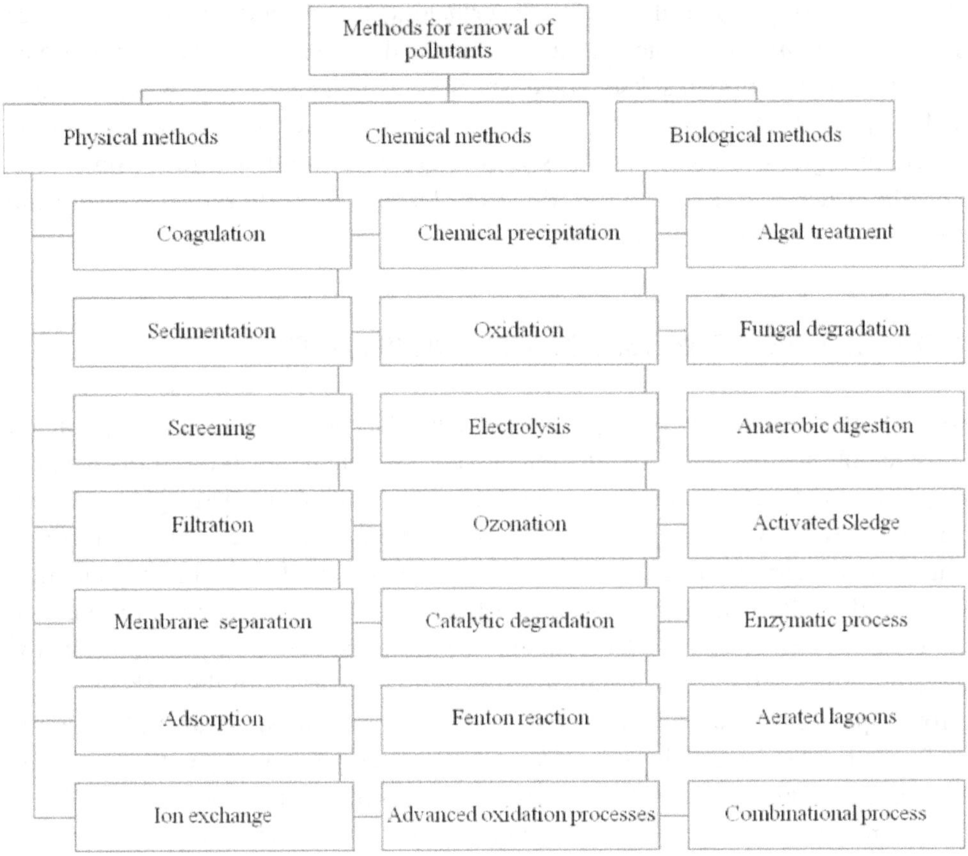

substantial financial implications and its limited effectiveness in eliminating metal ions (Ghaedi, 2021).

In wastewater treatment systems, the application of filtration methodologies, including membrane separation, is extensively observed, particularly in the case of seawater treatment. The utilisation of reverse osmosis, ultra and nanofiltration techniques has elicited considerable attention in modern times. These sequestrative processes operate under elevated pressure conditions to effectively remove pollutants from wastewater. Furthermore, they are employed to mitigate turbidity and microbial contaminants (Cheriyamundath & Vavilala, 2021). Membrane technology exhibits certain drawbacks, namely a substantial operational expenditure and a diminished flow rate, thereby constraining its overall utilisation. Nevertheless, this particular technological advancement exhibits remarkable efficacy in eliminating various types of ions, rendering it a distinctive method for purifying water, particularly seawater. Additionally, chemical precipitation presents itself as a cost-effective and straightforward method that has the potential to effectively remove a wide range of pollutants (Oyewo et al., 2018). Nevertheless, it is important to acknowledge that this particular technological advancement does possess certain limitations, notably the generation of substantial quantities of sludge and challenges associated with its proper disposal. These factors can restrict its widespread implementation in certain regions or contexts. The sediments present in this particular procedure have the potential to undergo filtration in order to effectively separate them from the wastewater (Kapoor et al., 2018).

ADSORPTIVE REMOVAL OF CONTAMINANTS

Among the various removal processes, the adsorptive removal has garnered considerable interest due to its numerous advantages, including its cost-effectiveness and ability to efficiently remove organic contaminants. Adsorbents play a pivotal role in the economic discourse within this methodology. The crucial features that warrant consideration include its affordability, eco-friendliness, and recyclability (Uko et al., 2022). Indeed, the process of adsorption takes place at the interface between the surface of the adsorbent and the adsorbed contaminant. Additionally, the process of contaminants being adsorbed onto the surface of the sorbent occurs through physical or chemical means. In this context, physical adsorption is the preferred mechanism, as it allows for easy recovery and subsequent reuse of the sorbent (Egbosiuba et al., 2022). During the process of physisorption, van der Waals forces are established between the contaminant and the surface of the adsorbent. Hence, the system exhibits a propensity for facile disruption, resulting in the removal of the contaminant from the surface of the sorbent. Conversely, the chemisorption phenomenon, being irreversible in nature, necessitates a substantial amount of energy. This technology has the potential to effectively eliminate pollutants, even when present in low concentrations (Kulal & Badalamoole, 2020).

NANOCOMPOSITES AS NANOADSORBENT

The identification of an appropriate sorbent poses a significant obstacle in the realm of industrial applications. In order to achieve this objective, the development of an effective sorbent for the purpose of eliminating organic contaminants assumes a crucial role within this undertaking. In the scientific literature, one can find a diverse array of adsorbents. These classifications can be understood as encompassing two distinct categories: biosorbents, which pertain to natural materials with inherent adsorption capabilities, and conventional sorbents, referring to synthetic substances commonly employed for adsorption purposes (Naseem & Durrani, 2021). A diverse array of nanoadsorbents have been reported in the literature including metallic nanomaterials, carbonaceous nanomaterials, silicon nanomaterials, metal oxide nanomaterials, metal chalcogenide nanomaterials, and magnetic nanomaterials (Garg, Garg, Sharma, et al., 2023). Silver, gold, platinum and zerovalent iron are the most popular metallic nanoadsorbents. Graphene, graphene oxide, Carbon nanotubes (CNTs), and their composites belong to the category of carbon-based nanosorbents. The silicon nanotubes and sheets are classified within the category of silicon nanomaterials. Aluminium oxide, iron oxide, magnesium oxide, nickel oxide, zirconium oxide, and titanium oxide are among the various metal-based oxide nanosorbents (Liu et al., 2022). FeS, CuSe, MoS_2, WSe_2, MnS, and NiS are notable examples of metal chalcogenide nanoadsorbents, which have demonstrated significant adsorption potential. Chitosan, cellulose, chitin, dendrimers, electrospun fabric, alginate, polysiloxane/graphene oxide, and polyacrylonitrile, are among the key polymeric nanosorbents that play a crucial role in sequestration of pollutants. Composites of nano iron oxides hold significant importance as magnetic nanoadsorbents (Sharma et al., 2023b).

Nanocomposites have emerged as a noteworthy sorbent variant that has garnered considerable interest in recent times. The system is comprised of two distinct phases, namely the matrix and reinforcement, both of which exhibit dimensions at the nanoscale. The key characteristics that make nanosorbents highly significant include their remarkable ability to undergo reversibility, possession of a large specific surface area, exceptional capacity for adsorption, compatibility with biological systems, and abundance of active sites (Garg, Garg, Khan, et al., 2023). Additionally, the pore size of these systems can be adjusted. A range of analytical techniques can be utilised to evaluate nanocomposites, such as dynamic light scattering (DLS), transmission electron microscopy (TEM), scanning electron microscopy (SEM), and X-ray diffraction (XRD) to determine the particle sizes in nanocomposites. Additionally, the utilisation of SEM and XRD analyses allows for the determination of pore sizes and crystal sizes, respectively (Sharma et al., 2023a).

The magnetic property of the composite structure is attributed to the presence of iron oxide, allowing for effortless separation using a magnet. The consideration of polydispersity and size control is of utmost importance in the pursuit of an optimal sorbent with high efficiency (Sarwar et al., 2021). Polymeric nanocomposites, due to their affordability and ease of use, have experienced a surge in usage and can be produced through diverse methods like the intercalation polymerization, sol-gel process, and more. Metal-based composites that receive adequate support possess numerous benefits (Sherlala et al., 2019). These include the presence of substantial active areas, exceptional stability in both chemical and thermal environments, and noteworthy catalytic activity. These distinctive characteristics render them as truly exceptional materials. Carbon-based materials possess remarkable characteristics, including well-structured pores, remarkable specific surface areas, and a significant capacity for the removal of organic contaminants (S. Wang et al., 2017).

Applications of Magnetic Nanocomposites in Water Remediation

Magnetic nanoparticles possess several benefits, such as exceptional stability, effortless separation from water-based solutions, straightforward synthesis, a substantial specific surface area, and a remarkable ability to efficiently eliminate contaminants.

Tizro & Baseri (2017) reported the mitigation of environmentally hazardous Co(II) ions through the utilisation of magnetic nanoadsorbents. The nanocomposites, characterised by an average particle size of 60±10 nm, exhibited a peak removal efficiency of 92% when subjected to optimised conditions. The Langmuir isotherm model elucidated a discernible maximum adsorption capacity of 43.292 mg/g at a temperature of 313 K. The nanoadsorbent demonstrated commendable recyclability, as evidenced by its ability to achieve a desorption efficiency of 95% when subjected to a nitric acid solution. In the study conducted by Kalantari et al. (2015), a nanocomposite consisting of Fe_3O_4/montmorillonite (Fe_3O_4/MMT NC) was successfully synthesised. This nanocomposite featured magnetic nanoparticles with an average diameter of approximately 8.24 nm. The primary objective of this synthesis was to develop a highly effective solution for the removal of Pb^{2+}, Cu^{2+}, and Ni^{2+} ions from aqueous solutions. The adequacy of the selected models was confirmed through variance analysis. The nanocomposite material exhibited removal efficiencies of 89.72% (Pb^{2+}), 94.89% (Cu^{2+}), and 76.15% (Ni^{2+}), correspondingly. The empirical findings exhibited a high degree of concordance with the theoretical projections, thereby affirming the efficacy of Fe_3O_4/MMT nanocomposites in the sequestration of heavy metal contaminants from aqueous matrices.

The synthesis of a Chitosan-magnetic-graphene oxide (CMGO) nanocomposite was successfully conducted by Sherlala et al. (2019). The efficacy of the nanocomposite in the removal of arsenic from aqueous solutions was demonstrated through a series of batch adsorption experiments. The process exhibited a positive correlation with augmented adsorbent dosage and prolonged contact time. Notably, an optimal pH of 7.3 was identified as the most conducive condition, resulting in the attainment of the maximum adsorption capacity of 45 mg/g and an efficiency of 61%. Langmuir isotherm model fitting suggested a uniform distribution of adsorption sites on the adsorbent surface. The thermodynamic analysis conducted in this study has elucidated that the process of adsorption of As(III) exhibits characteristics of being exothermic and spontaneous in nature. The facile separation and recovery of the nanoparticles through the utilisation of an external magnetic field, owing to their superparamagnetic attributes, serves to augment the potential of the nanocomposite in the realm of arsenic remediation.

Wang et al. (2017) undertook a comprehensive investigation into the potential of pine-derived biochar (PB) as a stabilising support medium for nanoscale zero-valent iron (nZVI), with the ultimate goal of augmenting its efficacy in removing arsenate (As(V)). In the conducted batch experiments, it was observed that the biochar-supported nanoscale zero-valent iron (nZVI/BC) exhibited a remarkable capacity for the removal of arsenic in its pentavalent form (As(V)) over a broad pH spectrum spanning from 3 to 8. Equilibration was successfully attained within a time frame of 60 minutes, yielding a Langmuir maximum adsorption capacity of 124.5 grammes per kilogramme at a pH level of 4.1. The experimental results demonstrate that the presence of anoxic conditions led to a notable enhancement of 8% in the removal of As(V) in comparison to oxic conditions. This observation strongly suggests the occurrence of As(V) reduction within anoxic environments. In the realm of continuous flow, the utilisation of completely mixed reactors (CMRs) has proven to be highly effective in the removal of As(V) concentrations from influent solutions. Notably, the application of nZVI/BC in CMRs has yielded remarkable results, achieving a complete removal of 100% for an influent As(V) concentration of 2.1 mg/L, and an impressive 90% removal for an influent As(V) concentration of 5.5 mg/L. The facile retrieval of the nZVI/BC composite from aqueous solutions, owing to its magnetic characteristics, highlights its promising capability for effective arsenate remediation.

Bhaumik et al. (2011) reported the synthesis of a magnetic nanocomposite comprised of polypyrrole (PPy) coated with Fe_3O_4. The nanocomposite was prepared using an in situ polymerization method, wherein pyrrole monomer was utilised as the precursor. The primary objective of this synthesis was to develop a material capable of effectively removing Cr(VI), a highly toxic substance. The structural and morphological properties of the nanocomposite were investigated through the utilisation of advanced analytical techniques that provided valuable insights into the composition and arrangement of the nanocomposite's constituent The nanocomposite material exhibited complete adsorption efficacy, reaching a maximum removal efficiency of 100% for a concentration of 200 mg/L of Cr(VI) in an aqueous medium with a pH value of 2. The surface of the nanocomposite was confirmed to undergo Cr(VI) adsorption through the utilisation of ATR-FTIR and XPS techniques. The XPS analysis further indicated the potential involvement of ion exchange and reduction mechanisms in the removal process. The observed phenomenon of decreased adsorption efficiency as a function of increasing pH was documented in this study, characterised by endothermicity and spontaneity. Notwithstanding the suboptimal recovery of Cr(VI) during the desorption process, the regenerated adsorbent exhibited commendable efficacy in terms of reusability, as it successfully maintained its adsorption capacity over the course of two cycles without any noteworthy diminishment.

Nicola et al. (2020) successfully synthesised and characterised magnetic iron oxide-silica shell nanocomposites with different iron oxide/silica ratios. The composite material, which demonstrates superior characteristics such as an ideal contact surface area and maximum saturation magnetization, has been chosen for the purpose of adsorbing Pb^{2+} ions from aqueous solutions. The investigated material exhibited a remarkable capacity for absorption, as evidenced by its maximum adsorption capacity of 14.9 mg/g. This value is in line with the reported capacities of analogous materials documented in existing literature. The nanocomposite's commendable chemical and physical stability, coupled with its remarkable adsorption capacity, establish its standing as a financially viable adsorbent material for the purpose of lead removal.

Andelescu et al.(2018) used magnetite/carbon nanocomposites for exploration of diverse adsorption parameters. The solution pH was systematically varied within the range of 2 to 9. Four different quantities of sorbent, specifically 10, 20, 40, and 60 mg, were employed. The initial metal ion concentration was varied across five levels, namely 10, 30, 50, 100, and 150 mg/L. Furthermore, three distinct temperatures, namely 25, 45, and 65°C, were considered in the investigation. The efficiency of metal ion removal exhibited a positive correlation with the pH of the solution, displaying an upward trend as the pH increased. Moreover, the removal efficiency demonstrated an upward trajectory with an increase in carbon content, magnetite/carbon nanocomposite dose, and temperature. Conversely, a decrease in the initial metal ion concentration resulted in a reduction in the removal efficiency. The maximum adsorption capacities for copper, cadmium, and zinc were quantified as 41.11 mg/g, 76.67 mg/g, and 48.45 mg/g, respectively. The thermodynamic analysis conducted in this study revealed a notable decrease in the Gibbs free energy, which is indicative of a spontaneous adsorption phenomenon. The nanocomposites exhibited remarkable efficacy in the removal of metal ions from authentic wastewaters, thereby showcasing their promising utility in real-world systems, given the ideal circumstances of 1 g/L adsorbent concentration, a temperature of 25°C, and a pH value of 6.

Garg et al. (2023) reported the eradication of heavy metal ions from simulated water by means of biosynthesized silica-supported iron oxide nanocomposites (nano-IOS). These nanocomposites are fabricated via an environmentally friendly synthesis procedure employing agricultural and garden wastes. The efficacy of the nanocomposites in the removal of Pb^{2+}, Cd^{2+}, Ni^{2+}, Cu^{2+}, and Zn^{2+} has been demonstrated through a series of batch mode experiments. The experimental parameters that yielded the most favourable outcomes encompassed a nano-IOS concentration of 0.75 g/L, a contact time of 70 minutes, and pH levels of 5.0 for Pb^{2+}, Cd^{2+}, and Cu^{2+}, while a pH of 6.0 was deemed optimal for Ni^{2+} and Zn^{2+}. The adsorption behaviour of nano-IOS was effectively characterised that provided a comprehensive description of the adsorption process, suggesting a chemisorption phenomenon occurring on the surface of the nano-IOS material. The adsorption process exhibited spontaneous characteristics, as evidenced by the observed thermodynamic spontaneity. Additionally, the process was found to be endothermic, indicating a net absorption of heat during the adsorption phenomenon. The bio-synthesized nano-IOS has been demonstrated to possess both environmental compatibility and remarkable efficacy as a nano-adsorbent, specifically designed for the rapid sequestration of heavy metal ions from aqueous solutions, including water and wastewater.

Nassar et al. (2023) reported the successful synthesis of a novel photocatalyst, namely UV-light-active MCC/S-VO_2, through a hydrothermal route to assess the photocatalytic efficiencies of V_2O_5, doped S-VO_2, and MCC/S-VO_2 nanostructures through the observation of methylene blue (MB) concentration degradation under UV-light exposure. The experimental findings revealed that the MCC/S-VO_2 composite exhibited enhanced photocatalytic efficiency in the degradation of MB, surpassing the performance of both pure V_2O_5 and S-VO_2 with doping. Significantly, the degradation of 72.3% of Methylene Blue (MB) was observed subsequent to a 6-hour exposure to MCC/S-VO_2. Furthermore, the introduction of Hydrogen Peroxide (H2O2) exhibited a pronounced effect on the photodegradation process, resulting in a notable enhancement and achieving a degradation efficiency of 92.5% within a time span of 55 minutes. The investigation into the degradation of MB through kinetic studies has provided valuable insights. It has been determined that the degradation process can be accurately described by a pseudo-first-order model. The rate constant (kobs) associated with this model has been determined to be 3.9×10^{-2} min^{-1}. The experimental investigations pertaining to the scavenging of active species have indicated that hydroxyl radicals and positive holes play a pivotal role in the degradation mechanism. The demonstrated stability and reusability of the nanostructures substantiated their potential applicability in the effective removal of MB dye from aqueous media.

The synthesis of a magnetic photocatalytic material, denoted as nZVI@TiO_2, was accomplished by Petala et al. (2016) through a meticulously designed multistage chemical procedure. The structural and chemical characterization of the samples was conducted using X-ray powder diffraction and Raman spectroscopy techniques. These methods allowed for the determination of the crystal structure and chemical composition of the materials under investigation. In addition, the physical and textural properties of the samples were assessed through bulk magnetization measurements and electron microscopy analysis. These techniques provided valuable insights into the magnetic behaviour and microstructural features of the materials. The material exhibited remarkable efficacy in the removal of hexavalent chromium (Cr(VI)) from aqueous solutions. The exceptional efficacy observed in this study can be primarily ascribed to the collaborative efforts of both nano zero-valent iron (nZVI) and titanium dioxide (TiO_2), as substantiated by comparative investigations conducted on pure TiO_2 under diverse experimental conditions. The present study elucidates novel perspectives on the synergistic degradation mechanism of chromium(VI) and the concomitant inhibition of oxidation of nanoscale zero-valent iron (nZVI) particles within the composite material.

Mechanisms for the Removal of Contaminants

The investigation of the complex processes involved in the removal of heavy metal and metalloid ions from solutions is still an unresolved issue. Most current remediation methods mainly focus on the processes of adsorption, adsorption-reduction, and/or photocatalytic degradation. In this specific context, it is crucial to determine the physicochemical characteristics of these pollutants (Tizro & Baseri, 2017). Furthermore, it is essential to incorporate sophisticated characterization methodologies, such as spectroscopy measurements, alongside theoretical calculations at the molecular scale. It is essential to adopt this approach in order to understand the potential ramifications of these pollutants on the natural environment and to develop alternative techniques for their elimination (Kalantari et al., 2015).

Biogenic composites are more advantageous than conventional sorbents like activated carbon for adsorbing pollutants because they have a larger number of functional groups on their surfaces. The adsorption mechanism arises from the complex interaction and intricate electrostatic forces between the adsorbent and the adsorbate (S. Wang et al., 2017). The adsorption process is enhanced by the complex interaction between the solute (contaminant) and the surface of the adsorbent material, resulting in their subsequent physical adsorption into the irregular pores on the adsorbent surface. The contaminant is captured on the adsorbent surface by intentionally introducing the adsorbent into the solution, leading to the merging, clumping, and accumulation of the sludge (Vahdat et al., 2020).

Figure 3 illustrates the process by which pollutant molecules form chemical bonds with adsorbent particles. Based on empirical observations, it is clear that there are two distinct categories of bonds: chemical bonds and physical bonds (Nicola et al., 2020). Physical adsorption is defined by the feeble van der Waals attraction that occurs between the adsorbent and the adsorbate. In contrast, chemical adsorption occurs through the establishment of a chemical bond between these entities. The uptake process typically consists of three sequential phases: (1) the movement of the pollutant from the liquid solution to the immediate surroundings of the sorbent surface, (2) the spreading of the pollutant from the immediate surroundings of the sorbent surface to the interconnected pores of the sorbent, and (3) the attachment

Figure 3. Illustrative mechanism for adsorptive removal of contaminants

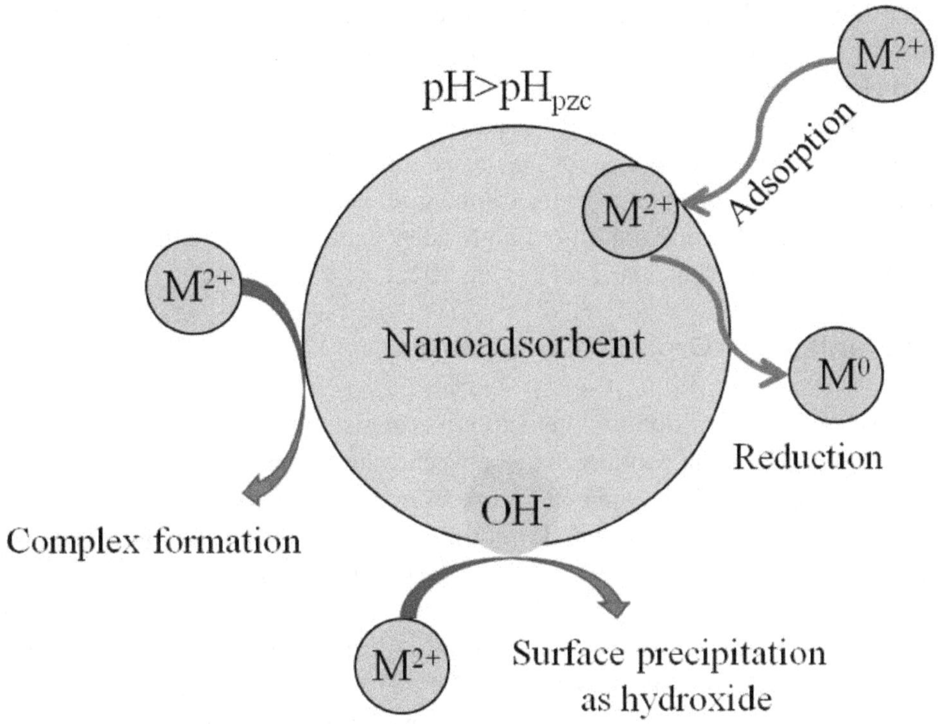

of the pollutant to the active sites on the sorbent surface (Kataria & Garg, 2018). Literature offers further elucidation on the diverse adsorption mechanisms exhibited by organic pollutants when they come into contact with the surface of the sorbent. These mechanisms entail the interaction of pore filling, π-π interactions, hydrogen bonding, electrostatic interactions, hydrophobic partitioning interactions, and van der Waals forces (Manna et al., 2018).

Mechanism for Inorganic Compounds

The absorption of pollutants from solutions can be enhanced through an adsorption process that is propelled by chemical interactions between functional groups. Functional groups, such as those that include oxygen, hydroxyl, or carbonyl groups, are essential for binding with metal ions (Gurgur et al., 2020). The operation of this mechanism is controlled by the formation of robust surface complexes through π-π donor-acceptor interactions, hydrogen bond interactions, hydrophobic interactions, and/or electrostatic attraction (Liu et al., 2022). Similarly, the adsorption-reduction process acts as an additional mechanism that aids in immobilising and subsequently reducing metal ions with high valency to ones with lower valency. In this process, metal ions with higher valence undergo reduction, resulting in the formation of denser particles or clusters that quickly precipitate (Rahman et al., 2018). Typical examples of adsorption-reduction mechanisms include the conversion of Cr^{6+} to Cr^{3+} and As^{5+} to As^{3+}, among other instances (C. Wang et al., 2014). Therefore, the occurrence of low valent metal ions in the natural environment is significantly more advantageous. The current consensus among scientists is that this phenomenon is affected by the differing mobility of high valence metal ions in comparison to low valence metal ions (Ghanbarizadeh et al., 2022).

Mechanism for Organic Compounds

The hydrophobic partitioning interaction mechanism is thought to arise from the partitioning of organic contaminants between water molecules and the surface of the nanocomposite. Hydrogen bonds arise from the electrostatic attraction between positively charged hydrogen ions (H^+) and other molecules. These bonds are relatively weak in nature (Ahmadpour et al., 2015). The existence of functional groups on the composite's surface has the capability to affect the hydrogen bonding interactions. Furthermore, the interactions among various species can be considered a crucial factor that influences the behaviour and dynamics of aromatic compounds (P. Wang et al., 2021). An electrostatic interaction occurs when organic pollutants and composite particles carry electric charges. Moreover, it is crucial to acknowledge that the van der Waals force, a relatively weak force, arises from the interactions between the pollutant and the composite surface during the process of adsorption. These mechanisms have been consistently recorded in numerous previous studies (Santhosh et al., 2018). Eddy et al. observed that the adsorption of glyphosate from wastewater follows the π–π interaction mechanism (Odoemelam, Oji, Eddy, et al., 2023). Researchers have conducted a distinct study and found that the adsorption of carbamazepine (Pérez-Lucas et al., 2023) and glyphosate (Hnana et al., 2019) is controlled by hydrophobic interactions. They have also provided evidence for the existence of hydrophobic and electrostatic interactions between the adsorbate and the surface of the sorbent. The application of the photocatalytic degradation mechanism using visible light is commonly used to eliminate persistent organic pollutants (POPs) and other organic contaminants (Figure 4). Moreover, it has garnered substantial attention for its ability to efficiently remove small amounts of heavy metal and metalloid ions from solutions by utilising a complex combination of photocatalytic reactions, catalyst structures, mass transport processes, absorption of visible light, and the arrangement of active sites on the surface (Ahmadpour et al., 2015). A commonly used method involves developing tailored nanoparticles or composites, where adjusting the energy gap leads to improved photocatalytic

Figure 4. Illustrative photocatalytic degradation mechanism for nanocomposites

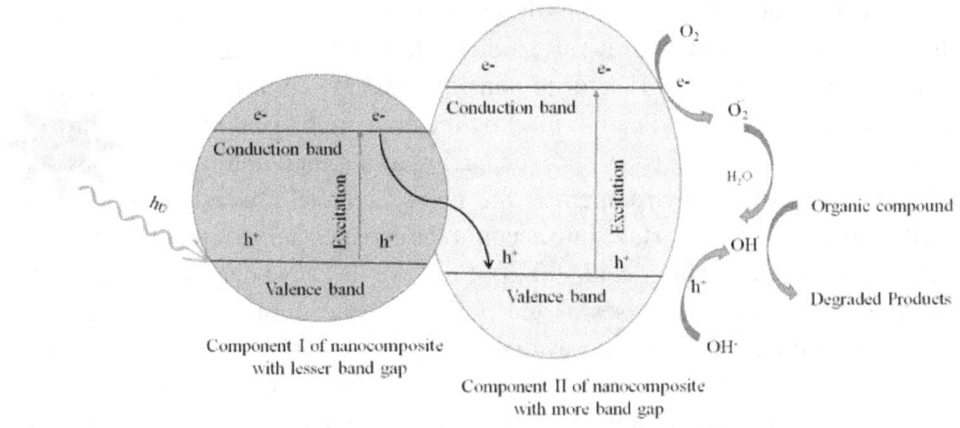

effectiveness. Organic compounds can be effectively removed from water by using adsorption or adsorption-reduction techniques (Riyanti et al., 2023).

FUTURE PROSPECTIVE

The performance of an adsorbent in the sequestration of contaminants is influenced by various environmental factors. These factors include pH levels, duration and intensity of stirring, temperature variations, concentrations of pollutants, and the amount of sorbent used. Among all the factors involved, it is crucial to emphasise that pH plays a vital role in the process of distinguishing and isolating contaminants in water-based environments (Iftekhar et al., 2018). By adjusting the pH of a solution, changes are induced in the levels of positive and negative charges present in the solution, specifically related to hydronium and hydroxide ions. These alterations subsequently have an impact on the rate of adsorption. In an acidic ecosystem, a significant number of active sites are covered by hydrogen ions (H^+), which reduces the ability of the sorbent surface to adsorb pollutants (Garg, Garg, Khan, et al., 2023). The efficacy of removing cationic pollutants typically diminishes in an acidic setting. Within an alkaline setting, the presence of OH^- ions in the solution increases, impeding the assimilation of anionic pollutants at these specific pH levels. This phenomenon arises from intense competition between OH^- ions and anionic contaminants for adsorption (Sharma et al., 2023a). Consequently, a notable phenomenon arises in the natural environment where H^+ ions and cationic contaminants experience an electrostatic repulsion force under conditions of low pH. Conversely, under high pH conditions, OH^- ions and anionic ions generate electrostatic repulsion forces. In environments with a high pH, there is no antagonism observed between cationic contaminants and OH^- ions. This lack of competition ultimately results in an increased efficiency of adsorption, as indicated by previous studies (Ocheje Ameh et al., 2020).

Temperature is a crucial factor to consider when examining the adsorption of pollutants, as it has a significant impact on the economic feasibility of the process. Therefore, the process of adsorption occurring at a temperature that is consistent with the surrounding environment is the more financially advantageous option. Additionally, it is crucial to take into account the concentration of pollutants present in the solution (Eddy, Garg, et al., 2023). Prior studies have demonstrated that the adsorption efficiency decreases as the pollutant concentration increases, due to the limited number of active sites available in relation to the low concentrations of pollutants. The dosage of sorbents plays a pivotal role in the effective removal of contaminants. Through the augmentation of the sorbent dosage, there is an increase in the quantity of active sites available for the entrapment of the adsorbate (Pi et al., 2018). By optimising the sorbent dosage, the maximum adsorption capacity can be achieved. Based on previous research findings, it has been observed that the efficiency of uptake increases as the concentration of the sorbent is increased. The temporal duration of the adsorption process exerts a substantial influence, thereby resulting in the most pronounced alteration in adsorption efficacy during initial periods (Hassan et al., 2020). During the state of equilibrium, it can be observed that the rate of adsorption remains constant, indicating that the sorbent sites have reached saturation with pollutants. Consequently, no further ions have the capacity to be adsorbed. Additionally, the velocity of agitation serves to reduce the hindrance of material exchange (Rao et al., 2013). Additionally, it delineates the energy expenditure necessary for agitating the mixture. The rate of the adsorption process tends to exhibit an upward trend as the stirring rate is intensified. An elevated stirring velocity induces the formation of vortices within the solution, resulting in a reduction in the uniformity of the solution. Furthermore, the introduction of additional agitation results in a decrease in the duration of interaction between the particles of the sorbent and the contaminants, thereby causing a decline in the rate at which adsorption occurs (Odoemelam, Oji, Eddy, et al., 2023).

The reusability of the adsorbent plays a crucial role in determining the economic viability of the uptake process, as it is closely linked to the chemical and mechanical stability of the adsorbent. In order to facilitate the regeneration process of the nanosorbent, various reagents have been employed thus far. These reagents encompass EDTA, NaOH, acetone, ethanol, as well as certain acids like HCl and acetic acid (Rani et al., 2022). The challenge of isolating nanocomposites from the solution arises due to their diminutive particle size. However, it is worth noting that numerous nanosorbents have exhibited noteworthy recyclability following multiple cycles of reuse that marks their beneficial applicability (Babapour et al., 2022).

CONCLUSION

The removal of various contaminants from effluents can be achieved through a range of processes, with the adsorption process being widely recognised as an effective method. This approach offers numerous advantages, including a notable capability to remove contaminants of any kind, a substantial adsorption capacity, and a plentiful supply of adsorbents. The careful selection of an appropriate sorbent, which encompasses both cost-effectiveness and high reusability, holds significance in this particular process. The attention garnered by the utilisation of nanocomposites stems from their notable recyclability, substantial uptake capacity, and minimal contact time. This study aimed to investigate the removal performance of nanocomposites in the context of pollutant removal from aqueous media. Various sorbents, including magnetic nanocomposites, have exhibited impressive specific surface areas and notable adsorption efficiencies in the removal of diverse contaminants, particularly organic compounds. Based on prior research, it has been observed that several nanocomposites exhibit exceptional performance characteristics and remarkable efficacy in the elimination of water pollutants aiding in sustainable water management.

REFERENCES

Ahmadpour, A., Zare, M., Behjoomanesh, M., & Avazpour, M. (2015). Photocatalytic decolorization of methyl orange dye using nano-photocatalysts. *Advances in Environmental Technology*, *1*(3), 121–127. doi:10.22104/aet.2015.371

Ali, S. M. (2018). Fabrication of a nanocomposite from an agricultural waste and its application as a biosorbent for organic pollutants. *International Journal of Environmental Science and Technology*, *15*(6), 1169–1178. doi:10.1007/s13762-017-1477-x

Amari, A., Alzahrani, F. M., Katubi, K. M., Alsaiari, N. S., Tahoon, M. A., & Ben Rebah, F. (2021). Clay-polymer nanocomposites: Preparations and utilization for pollutants removal. *Materials (Basel)*, *14*(6), 1365. doi:10.3390/ma14061365 PMID:33799810

Andelescu, A., Nistor, M. A., Muntean, S. G., & Rădulescu-Grad, M. E. (2018). Adsorption studies on copper, cadmium, and zinc ion removal from aqueous solution using magnetite/carbon nanocomposites. *Separation Science and Technology*, *53*(15), 2352–2364. doi:10.1080/01496395.2018.1457696

Aragaw, T. A., Bogale, F. M., & Aragaw, B. A. (2021). Iron-based nanoparticles in wastewater treatment: A review on synthesis methods, applications, and removal mechanisms. *Journal of Saudi Chemical Society*, *25*(8), 101280. doi:10.1016/j.jscs.2021.101280

Babapour, M., Hadi Dehghani, M., Alimohammadi, M., Moghadam Arjmand, M., Salari, M., Rasuli, L., Mubarak, N. M., & Ahmad Khan, N. (2022). Adsorption of Cr(VI) from aqueous solution using mesoporous metal-organic framework-5 functionalized with the amino acids: Characterization, optimization, linear and nonlinear kinetic models. *Journal of Molecular Liquids*, *345*, 117835. doi:10.1016/j.molliq.2021.117835

Bhaumik, M., Maity, A., Srinivasu, V. V., & Onyango, M. S. (2011). Enhanced removal of Cr(VI) from aqueous solution using polypyrrole/Fe3O4 magnetic nanocomposite. *Journal of Hazardous Materials*, *190*(1–3), 381–390. doi:10.1016/j.jhazmat.2011.03.062 PMID:21497438

Cheriyamundath, S., & Vavilala, S. L. (2021). Nanotechnology-based wastewater treatment. *Water and Environment Journal : the Journal / the Chartered Institution of Water and Environmental Management*, *35*(1), 123–132. doi:10.1111/wej.12610

D'Ugo, E., Bruno, M., Mukherjee, A., Chattopadhyay, D., Giuseppetti, R., De Pace, R., & Magurano, F. (2021). Characterization of microbial response to petroleum hydrocarbon contamination in a lacustrine ecosystem. *Environmental Science and Pollution Research International*, *28*(20), 26187–26196. doi:10.1007/s11356-021-13885-8 PMID:33871774

Da'ana, D. A., Zouari, N., Ashfaq, M. Y., Abu-Dieyeh, M., Khraisheh, M., Hijji, Y. M., & Al-Ghouti, M. A. (2021). Removal of Toxic Elements and Microbial Contaminants from Groundwater Using Low-Cost Treatment Options. *Current Pollution Reports*, *7*(3), 300–324. doi:10.1007/s40726-021-00187-3

Eddy, N. O., Garg, R., Garg, R., Eze, S. I., Ogoko, E. C., Kelle, H. I., Ukpe, R. A., Ogbodo, R., & Chijoke, F. (2023). Sol-gel synthesis, computational chemistry, and applications of Cao nanoparticles for the remediation of methyl orange contaminated water. *Advances in Nano Research*, *15*(1), 35–48. doi:10.12989/anr.2023.15.1.035

Eddy, N. O., Odiongenyi, A., Garg, R., Ukpe, R. A., & Garg, R. ChimaOgoko, E., Kelle, H. I., & Ogbodo, R. (2023). Ecotoxicological aspects of Nanotechnology. In Nanotechnology for Sustainable Agriculture, Food and Environment (pp. 211–223). CRC Press. doi:10.1201/9781003397861-16

Eddy, N. O., Odiongenyi, A. O., Garg, R., Ukpe, R. A., Garg, R., El Nemr, A., Ngwu, C. M., & Okop, I. J. (2023). Quantum and experimental investigation of the application of Crassostrea gasar (mangrove oyster) shell–based CaO nanoparticles as adsorbent and photocatalyst for the removal of procaine penicillin from aqueous solution. *Environmental Science and Pollution Research International, 30*(23), 64036–64057. doi:10.1007/s11356-023-26868-8 PMID:37059957

Egbosiuba, T. C., Egwunyenga, M. C., Tijani, J. O., Mustapha, S., Abdulkareem, A. S., Kovo, A. S., Krikstolaityte, V., Veksha, A., Wagner, M., & Lisak, G. (2022). Activated multi-walled carbon nanotubes decorated with zero valent nickel nanoparticles for arsenic, cadmium and lead adsorption from wastewater in a batch and continuous flow modes. *Journal of Hazardous Materials, 423*(PB), 126993. doi:10.1016/j.jhazmat.2021.126993

Garg, R., Garg, R., Khan, M. A., Bansal, M., & Garg, V. K. (2023). Utilization of biosynthesized silica-supported iron oxide nanocomposites for the adsorptive removal of heavy metal ions from aqueous solutions. *Environmental Science and Pollution Research International, 30*(34), 81319–81332. doi:10.1007/s11356-022-21111-2 PMID:35672639

Garg, R., Garg, R., Okon Eddy, N., Ibrahim Almohana, A., Fahad Almojil, S., Amir Khan, M., & Ho Hong, S. (2022). Biosynthesized silica-based zinc oxide nanocomposites for the sequestration of heavy metal ions from aqueous solutions. *Journal of King Saud University. Science, 34*(4), 101996. doi:10.1016/j.jksus.2022.101996

Garg, R., Garg, R., Sharma, R., Anjum, A., & Eddy, N. O. (2023). Introduction to Nanotechnology and Nanomaterials. In Nanotechnology for Sustainable Agriculture, Food and Environment (pp. 1–15). CRC Press. doi:10.1201/9781003397861-1

Garg, R., Garg, R., Sillanpää, M., Alimuddin, Khan, M. A., Mubarak, N. M., & Tan, Y. H. (2023). Rapid adsorptive removal of chromium from wastewater using walnut-derived biosorbents. *Scientific Reports, 13*(1), 6859. doi:10.1038/s41598-023-33843-3 PMID:37100812

Ghaedi, M. (2021). Adsorption: Fundamental Processes and Applications. In *Adsorption: Fundamental Processes and Applications, 33*(1). https://linkinghub.elsevier.com/retrieve/pii/B9780128188057000011

Ghanbarizadeh, P., Parivazh, M. M., Abbasi, M., Osfouri, S., Dianat, M. J., Rostami, A., Dibaj, M., & Akrami, M. (2022). Performance Enhancement of Specific Adsorbents for Hardness Reduction of Drinking Water and Groundwater. *Water (Basel), 14*(17), 2749. doi:10.3390/w14172749

Gurgur, E., Oluyamo, S. S., Adetuyi, A. O., Omotunde, O. I., & Okoronkwo, A. E. (2020). Green synthesis of zinc oxide nanoparticles and zinc oxide–silver, zinc oxide–copper nanocomposites using Bridelia ferruginea as biotemplate. *SN Applied Sciences, 2*(5), 1–12. doi:10.1007/s42452-020-2269-3

Hassan, S. S., Al-Ghouti, M. A., Abu-Dieyeh, M., & McKay, G. (2020). Novel bioadsorbents based on date pits for organophosphorus pesticide remediation from water. *Journal of Environmental Chemical Engineering, 8*(1), 103593. doi:10.1016/j.jece.2019.103593

Hnana, K., Barhoun, A., & Draoui, K. (2019). Improvement of glyphosate adsorption using new composites based on ghassoul and chitosan: Kinetics and equilibrium study. *Mediterranean Journal of Chemistry, 9*(4), 266–279. doi:10.13171/mjc9419111121087kd

Iftekhar, S., Ramasamy, D. L., Srivastava, V., Asif, M. B., & Sillanpää, M. (2018). Understanding the factors affecting the adsorption of Lanthanum using different adsorbents: A critical review. *Chemosphere, 204*, 413–430. doi:10.1016/j.chemosphere.2018.04.053 PMID:29677649

Jayakaran, P., Nirmala, G. S., & Govindarajan, L. (2019). Qualitative and Quantitative Analysis of Graphene-Based Adsorbents in Wastewater Treatment. *International Journal of Chemical Engineering, 2019*, 1–17. doi:10.1155/2019/9872502

Kalantari, K., Ahmad, M. B., Fard Masoumi, H. R., Shameli, K., Basri, M., & Khandanlou, R. (2015). Rapid and high capacity adsorption of heavy metals by Fe3O4/montmorillonite nanocomposite using response surface methodology: Preparation, characterization, optimization, equilibrium isotherms, and adsorption kinetics study. *Journal of the Taiwan Institute of Chemical Engineers, 49*, 192–198. doi:10.1016/j.jtice.2014.10.025

Kapoor, V., Phan, D., & Pasha, A. B. M. T. (2018). Effects of metal oxide nanoparticles on nitrification in wastewater treatment systems: A systematic review. *Journal of Environmental Science and Health. Part A, Toxic/Hazardous Substances & Environmental Engineering, 53*(7), 659–668. doi:10.1080/10934529.2018.1438825 PMID:29469639

Karthikeyan, P., Nikitha, M., Pandi, K., Meenakshi, S., & Park, C. M. (2021). Effective and selective removal of organic pollutants from aqueous solutions using 1D hydroxyapatite-decorated 2D reduced graphene oxide nanocomposite. *Journal of Molecular Liquids*, *331*, 115795. doi:10.1016/j.molliq.2021.115795

Kataria, N., & Garg, V. K. (2018). Green synthesis of Fe3O4 nanoparticles loaded sawdust carbon for cadmium (II) removal from water: Regeneration and mechanism. *Chemosphere*, *208*, 818–828. doi:10.1016/j.chemosphere.2018.06.022 PMID:29906756

Khatoon, N., Alam, H., Manzoor, N., & Sardar, M. (2018). Removal of toxic contaminants from water by sustainable green synthesised non-toxic silver nanoparticles. *IET Nanobiotechnology / IET*, *12*(8), 1090–1096. doi:10.1049/iet-nbt.2018.5075 PMID:30964019

Kulal, P., & Badalamoole, V. (2020). Efficient removal of dyes and heavy metal ions from waste water using Gum ghatti - graft - poly(4-acryloylmorpholine) hydrogel incorporated with magnetite nanoparticles. *Journal of Environmental Chemical Engineering*, *8*(5), 104207. doi:10.1016/j.jece.2020.104207

Liu, W. X., Song, S., Ye, M. L., Zhu, Y., Zhao, Y. G., & Lu, Y. (2022). Nanomaterials with Excellent Adsorption Characteristics for Sample Pretreatment: A Review. *Nanomaterials (Basel, Switzerland)*, *12*(11), 1845. doi:10.3390/nano12111845 PMID:35683700

Mandal, A. K., Katuwal, S., Tettey, F., Gupta, A., Bhattarai, S., Jaisi, S., Bhandari, D. P., Shah, A. K., Bhattarai, N., & Parajuli, N. (2022). Current Research on Zinc Oxide Nanoparticles: Synthesis, Characterization, and Biomedical Applications. *Nanomaterials (Basel, Switzerland)*, *12*(17), 3066. doi:10.3390/nano12173066 PMID:36080103

Manna, S., Gopakumar, D. A., Roy, D., Saha, P., & Thomas, S. (2018). Nanobiomaterials for removal of fluoride and chlorophenols from water. In *New Polymer Nanocomposites for Environmental Remediation*. Elsevier Inc. doi:10.1016/B978-0-12-811033-1.00020-2

Marbawi, H., Arif, M., Ros, M., Khudri, S., Othman, A. R., Effendi, M. I., Gansau, J. A., & Yasid, N. A. (2019).. Kinetic Analysis of the Adsorption of Glyphosate onto Palm Oil Fronds Activated Carbon. *BIOREMEDIATION SCIENCE AND TECHNOLOGY, 7*(1), 29–33.

Modi, S., Yadav, V. K., Amari, A., Alyami, A. Y., Gacem, A., Harharah, H. N., & Fulekar, M. H. (2023). Photocatalytic Degradation of Methylene Blue Dye from Wastewater by Using Doped Zinc Oxide Nanoparticles. *Water (Basel)*, *15*(12), 2275. doi:10.3390/w15122275

Naseem, T., & Durrani, T. (2021). The role of some important metal oxide nanoparticles for wastewater and antibacterial applications: A review. *Environmental Chemistry and Ecotoxicology*, *3*, 59–75. doi:10.1016/j.enceco.2020.12.001

Nassar, M. Y., NourEldien, M. S., Ibrahim, I. M., & Aly, H. M. (2023). A Facile Hydrothermal Synthesis of S-VO2-Cellulose Nanocomposite for Photocatalytic Degradation of Methylene Blue Dye. *Processes (Basel, Switzerland)*, *11*(5), 1322. doi:10.3390/pr11051322

Nicola, R., Costişor, O., Ciopec, M., Negrea, A., Lazău, R., Ianăşi, C., Picioruş, E. M., Len, A., Almásy, L., Szerb, E. I., & Putz, A. M. (2020). Silica-coated magnetic nanocomposites for Pb2+ removal from aqueous solution. *Applied Sciences (Basel, Switzerland)*, *10*(8), 2726. Advance online publication. doi:10.3390/app10082726

Nikalje, A. P. (2015). Nanotechnology and its Applications in Medicine. *Medicinal Chemistry (Sunnyvale, Calif.)*, *5*(2), 81–89. doi:10.4172/2161-0444.1000247

Ocheje Ameh, P., Okon Eddy, N., Ameh, P. O., & Eddy, N. O. (2020). Theoretical and Experimental Investigations of the Corrosion Inhibition Action of Piliostigma Thonningii Extract on Mild Steel in Acidic Medium. *Communication in Physical Science, 2018*(1), 27–42. https://journalcps.com/

Odoemelam, S. A., Oji, E. O., Eddy, N. O., Garg, R., Garg, R., Islam, S., Khan, M. A., Khan, N. A., & Zahmatkesh, S. (2023). Zinc oxide nanoparticles adsorb emerging pollutants (glyphosate pesticide) from aqueous solutions. *Environmental Monitoring and Assessment*, *195*(6), 78195. doi:10.1007/s10661-023-11255-0 PMID:37166547

Odoemelam, S. A., Oji, E. O., Okon, N., & Rajni, E. (2023). Zinc oxide nanoparticles adsorb emerging pollutants (glyphosate pesticide) from aqueous solutions. *Environmental Monitoring and Assessment*, *195*(6), 658. doi:10.1007/s10661-023-11255-0 PMID:37166547

Oyewo, O. A., Agboola, O., Onyango, M. S., Popoola, P., & Bobape, M. F. (2018). Current Methods for the Remediation of Acid Mine Drainage Including Continuous Removal of Metals From Wastewater and Mine Dump. In M. N. V. Prasad, P. J. de C. Favas, & S. K. B. T.-B.-G. for M. S. R. Maiti (Eds.), Bio-Geotechnologies for Mine Site Rehabilitation (pp. 103–114). Elsevier. doi:10.1016/B978-0-12-812986-9.00006-3

Pandey, S., Makhado, E., Kim, S., & Kang, M. (2023). Recent developments of polysaccharide based superabsorbent nanocomposite for organic dye contamination removal from wastewater — A review. *Environmental Research*, *217*, 114909. doi:10.1016/j.envres.2022.114909 PMID:36455632

Pérez-Lucas, G., El Aatik, A., Aliste, M., Navarro, G., Fenoll, J., & Navarro, S. (2023). Removal of Contaminants of Emerging Concern from a Wastewater Effluent by Solar-Driven Heterogeneous Photocatalysis: A Case Study of Pharmaceuticals. *Water, Air, and Soil Pollution*, *234*(1), 55. Advance online publication. doi:10.1007/s11270-023-06075-4

Petala, E., Baikousi, M., Karakassides, M. A., Zoppellaro, G., Filip, J., Tuček, J., Vasilopoulos, K. C., Pechoušek, J., & Zbořil, R. (2016). Synthesis, physical properties and application of the zero-valent iron/titanium dioxide heterocomposite having high activity for the sustainable photocatalytic removal of hexavalent chromium in water. *Physical Chemistry Chemical Physics*, *18*(15), 10637–10646. doi:10.1039/C6CP01013J PMID:27035846

Pi, Y., Li, X., Xia, Q., Wu, J., Li, Y., Xiao, J., & Li, Z. (2018). Adsorptive and photocatalytic removal of Persistent Organic Pollutants (POPs) in water by metal-organic frameworks (MOFs). *Chemical Engineering Journal*, *337*, 351–371. doi:10.1016/j.cej.2017.12.092

Rahman, M., Zahin, F., Saadi, M. A. S. R., Sharif, A., & Hoque, M. E. (2018). *Surface Modification of Advanced and Polymer Nanocomposites*., doi:10.1007/978-3-319-76090-2_6

Rani, S., Sharma, S., Bansal, M., Garg, R., & Garg, R. (2022). *Enhanced Zn(II) adsorption by chemically modified sawdust based biosorbents*. Environmental Science and Pollution Research, Ii., doi:10.1007/s11356-022-22963-4

Rao, A., Bankar, A., Kumar, A. R., Gosavi, S., & Zinjarde, S. (2013). Removal of hexavalent chromium ions by Yarrowia lipolytica cells modified with phyto-inspired Fe0/Fe3O4 nanoparticles. *Journal of Contaminant Hydrology*, *146*, 63–73. doi:10.1016/j.jconhyd.2012.12.008 PMID:23422514

Riyanti, F., Hasanudin, H., Rachmat, A., Purwaningrum, W., & Hariani, P. L. (2023). Photocatalytic degradation of methylene blue and Congo red dyes from aqueous solutions by bentonite-Fe3O4 magnetic. *Communications in Science and Technology*, *8*(1), 1–9. doi:10.21924/cst.8.1.2023.1007

Santhosh, C., Malathi, A., Daneshvar, E., Kollu, P., & Bhatnagar, A. (2018). Photocatalytic degradation of toxic aquatic pollutants by novel magnetic 3D-TiO2@ HPGA nanocomposite. *Scientific Reports*, *8*(1), 1–15. doi:10.1038/s41598-018-33818-9 PMID:29311619

Sarwar, A., Wang, J., Khan, M. S., Farooq, U., Riaz, N., Nazir, A., Mahmood, Q., Hashem, A., Al-Arjani, A. B. F., & Alqarawi, A. A., & Abd_allah, E. F. (. (2021). Iron oxide (Fe3o4)-supported sio2 magnetic nanocomposites for efficient adsorption of fluoride from drinking water: Synthesis, characterization, and adsorption isotherm analysis. *Water (Switzerland)*, *13*(11), 1514. doi:10.3390/w13111514

Shah, R. K. (2023). Efficient photocatalytic degradation of methyl orange dye using facilely synthesized α-Fe2O3 nanoparticles. *Arabian Journal of Chemistry*, *16*(2), 104444. doi:10.1016/j.arabjc.2022.104444

Sharma, R., Garg, R., Bali, M., & Eddy, N. O. (2023a). Biogenic synthesis of iron oxide nanoparticles using leaf extract of Spilanthes acmella: Antioxidation potential and adsorptive removal of heavy metal ions. *Environmental Monitoring and Assessment*, *195*(11), 1–17. doi:10.1007/s10661-023-11860-z PMID:37857875

Sharma, R., Garg, R., Bali, M., & Eddy, N. O. (2023b). Potential applications of green-synthesized iron oxide NPs for environmental remediation. *Environmental Monitoring and Assessment*, *195*(11), 1397. Advance online publication. doi:10.1007/s10661-023-12035-6 PMID:37910248

Sherlala, A. I. A., Raman, A. A. A., Bello, M. M., & Buthiyappan, A. (2019). Adsorption of arsenic using chitosan magnetic graphene oxide nanocomposite. *Journal of Environmental Management*, *246*, 547–556. doi:10.1016/j.jenvman.2019.05.117 PMID:31202019

Stefaniuk, M., Oleszczuk, P., & Ok, Y. S. (2016). Review on nano zerovalent iron (nZVI): From synthesis to environmental applications. *Chemical Engineering Journal*, *287*, 618–632. doi:10.1016/j.cej.2015.11.046

Suwannahong, K., Wongcharee, S., Kreetachart, T., Sirilamduan, C., Rioyo, J., & Wongphat, A. (2021). Evaluation of the microsoft excel solver spreadsheet-based program for nonlinear expressions of adsorption isotherm models onto magnetic nanosorbent. *Applied Sciences (Basel, Switzerland)*, *11*(16), 7432. doi:10.3390/app11167432

Teiri, H., Hajizadeh, Y., Samaei, M. R., Pourzamani, H., & Mohammadi, F. (2020). Modelling the phytoremediation of formaldehyde from indoor air by Chamaedorea Elegans using artificial intelligence, genetic algorithm and response surface methodology. *Journal of Environmental Chemical Engineering*, *8*(4), 103985. Advance online publication. doi:10.1016/j.jece.2020.103985

Tizro, S., & Baseri, H. (2017). Removal of Cobalt Ions from Contaminated Water Using Magnetite Based Nanocomposites: Effects of Various Parameters on the Removal Efficiency. *J. Water Environ. Nanotechnol.*, *2*(23), 174–185. 10.22090

Uko, C. A., Tijani, J. O., Abdulkareem, S. A., Mustapha, S., Egbosiuba, T. C., & Muzenda, E. (2022). Adsorptive properties of MgO/WO3 nanoadsorbent for selected heavy metals removal from indigenous dyeing wastewater. *Process Safety and Environmental Protection*, *162*, 775–794. doi:10.1016/j.psep.2022.04.057

Vahdat, A., Ghasemi, B., & Yousefpour, M. (2020). Mechanical properties of the hydroxyapatite and magnetic nanocomposite of hydroxyapatite adsorbents. *South African Journal of Chemical Engineering*, *33*, 90–94. doi:10.1016/j.sajce.2020.05.007

Wang, C., Luo, H., Zhang, Z., Wu, Y., Zhang, J., & Chen, S. (2014). Removal of As(III) and As(V) from aqueous solutions using nanoscale zero valent iron-reduced graphite oxide modified composites. *Journal of Hazardous Materials*, *268*, 124–131. doi:10.1016/j.jhazmat.2014.01.009 PMID:24480523

Wang, P., Fu, F., & Liu, T. (2021). A review of the new multifunctional nano zero-valent iron composites for wastewater treatment: Emergence, preparation, optimization and mechanism. *Chemosphere*, *285*(July), 131435. doi:10.1016/j.chemosphere.2021.131435 PMID:34256206

Wang, S., Gao, B., Li, Y., Creamer, A. E., & He, F. (2017). Adsorptive removal of arsenate from aqueous solutions by biochar supported zero-valent iron nanocomposite: Batch and continuous flow tests. *Journal of Hazardous Materials*, *322*, 172–181. doi:10.1016/j.jhazmat.2016.01.052 PMID:26852252

Yadav, K. K., Kumar, S., Pham, Q. B., Gupta, N., Rezania, S., Kamyab, H., Yadav, S., Vymazal, J., Kumar, V., Tri, D. Q., Talaiekhozani, A., Prasad, S., Reece, L. M., Singh, N., Maurya, P. K., & Cho, J. (2019). Fluoride contamination, health problems and remediation methods in Asian groundwater: A comprehensive review. *Ecotoxicology and Environmental Safety*, *182*(January), 109362. doi:10.1016/j.ecoenv.2019.06.045 PMID:31254856

Zanna, H. K., Gunda, H. J., Tikau, F. W., & Gero, S. M. (2018).. . *Assessment of Heavy Metals in Trona from Yobe State, Nigeria.*, *6*(1), 10–15.

Chapter 5
Nanophotocatalyst-Infused Textile Composites for Environmental Remediation

Vanshika Sharma
Department of Textile Engineering, Jawaharlal Nehru Government Engineering College, Mandi, India

Amit Kumar
Department of Textile Engineering, Jawaharlal Nehru Government Engineering College, Mandi, India

Priya Jaswal
Department of Textile Engineering, Jawaharlal Nehru Government Engineering College, Mandi, India

Chirag Porwal
School of Engineering, Indian Institute of Technology, Mandi, India

Akshay Gaur
School of Engineering, Indian Institute of Technology, Mandi, India

Rahul Vaish
School of Engineering, Indian Institute of Technology, Mandi, India

ABSTRACT

The development of innovative materials for environmental cleaning applications has been the focus of substantial research efforts due to the growing global concern over environmental pollution and the need for sustainable solutions. In light of this, textile composites based on nano photocatalysts have become an exciting new way to tackle a range of environmental issues. Nanophotocatalyst-based textile composites have a wide range of environmental uses, including materials for oil spill cleanup, water purification filters, and self-cleaning textiles and masks for air purification. These composites are appropriate for a variety of real-world situations due to their benefits of scalability, durability, and adaptability. The research status on textile composites based on nano photocatalysts and their uses in environmental remediation is described in this chapter.

DOI: 10.4018/979-8-3693-1094-6.ch005

Copyright © 2024, IGI Global. Copying or distributing in print or electronic forms without written permission of IGI Global is prohibited.

INTRODUCTION

In today's world, environmental sanitation has become a concerning issue. People all across the globe are actively looking for sustainable solutions, inspiring a widespread pursuit of sustainability in nearly every aspect of life. The rising environmental pollution and growing energy demands have spurred the global development of new techniques that prioritize both non-toxicity and environmental friendliness (Varadarajan & Venkatachalam, 2016). However, in light of recent advancements, it is crucial that the cleaning techniques we embrace not only reflect long-term sustainability but also guarantee a lasting, beneficial influence on our environment both now and in the future (Sivaram et al., 2019). By combining polymer and textile science with new energy and environmental technologies, traditional polymers and fabrics may now perform a wide range of new activities. As a result, certain unique technical fabrics and polymers have started to appear, including textiles and polymers that are photovoltaic and photocatalytic. Aside from textiles, nanotechnology has grown significantly in the last few decades and has shown enormous promise for use in design, research, recreation, and environmental cleanup (Yetisen et al., 2016). During the processing of textiles, large amounts of harmful chemicals are used for various activities. These chemicals cause immense pollution in the water bodies. This chapter introduces different photoactive textile materials and delves deeply into their characteristics and photo-reactivity, which help to find sustainable ways of breaking down the chemicals and getting rid of malicious microbes. We highlight various methods to improve their activities when exposed to visible light, emphasizing recent breakthroughs in semiconductor photocatalysts combined with soft materials. We also investigate other methods for producing soft materials for various uses.

BACKGROUND

The fact that wastewater contains both organic and inorganic contaminants highlights the need for innovative and useful solutions to be developed globally to address water quality issues in a way that is beneficial to both society and the environment. Therefore, it is essential to develop strong photocatalysts and degradation techniques that can effectively remove pollutants from wastewater. Although several mechanisms have been found for the degradation of environmental pollutants, one that stands out as being very effective is photocatalysis (Tahir et al., 2022). Catalysis is a technique for accelerating a process by including an additional component, which is frequently referred to as a catalyst, which decomposes but then regenerates after the catalytic cycle (Herrmann, 2017). Photocatalysts are agents that absorb light and cause a system to undergo a photochemical change, as opposed to chemicals, materials, or processes that are produced or enabled by light. When we talk about photocatalysis it is the catalysis that occurs in the presence of light energy. The name "photocatalysis" is the result of combining the words "photo" and "catalysis," and it generally refers to the process where a substance changes the rate of a reaction when exposed to light. Photocatalysis is a specialized field within chemistry that revolves around chemical reactions occurring in the presence of both light and a key catalyst known as a photocatalyst. A photocatalyst essentially performs the role of a semiconductor material, greatly accelerating the reaction rate when exposed to light. When a substance comes into contact with photons with a high enough energy level, free radical processes begin to operate, activating the chemical reaction known as photocatalysis (Ameta et al., 2018). In recent years, photocatalysis has become one of the most efficient processes for dissolving extremely complicated chemical compounds. Photocatalytic procedures offer a possible substitute for conventional water filtration techniques by efficiently eliminating bacterial pollutants and dissolved organic compounds from wastewater. Photocatalysis is considered an environmentally friendly and green way of dealing with challenges and problems such as environmental pollution and reducing energy consumption (Kumar et al., 2023; Kumar, Sharma, & Vaish, 2022a, 2022b; Liu et al., 2017).

The textile industry encompasses a multifaceted spectrum of processes. In this journey from raw materials to finished textile clothing, a substantial amount of waste is also generated. Among the myriad challenges that this waste presents, there is an insidious number of synthetic dyes, heavy metals, and hazardous chemicals. Nanotechnology has also spread its roots during the past few decades (Yetisen et al., 2016). Electron mobility is drastically reduced as a result of the substantial scattering that occurs during this phase. Soft materials, in particular textiles and polymers, make good hosts for semiconductors, increasing efficiency and opening up a wide range of useful uses. The visible light absorption range of polymers is expanded by the incorporation of semiconductors, making them suitable for photovoltaic applications. Additionally, it enables the sensitization or immobilization of nanocrystals for photocatalytic use in polymers. Our daily lives are greatly influenced by light-activated materials, which power solar-powered calculators and watches. They are used for diagnostic purposes in medical laboratories utilizing fluorescent probes, making use of their capacity to absorb ultraviolet (UV) radiation (Mather & Wilson, 2017). Similar to this, the use of light-activated materials in healthcare is growing quickly, with the same approach in mind lots of textile composites have paved their way towards functional use instead of only aesthetic pleasing. Photocatalysis is providing a major change in sustainability. Due to this, the chances of environmentally friendly methods have risen which can be taken into consideration soon.

TEXTILES AND NANOTECHNOLOGY

Nanotechnology, which includes both nanoscience and nanoparticles, is one of the most promising fields of the 21st century. Its scope extends across diverse scientific domains, including physics, chemistry, biology, materials science, and engineering, as it involves the study and application of tiny entities. In the upcoming years, nanotechnology is surely going to become integrated with every aspect of the textile industry. Its incorporation into modern textiles focuses on bringing novel ideas into robust, multipurpose textile structures without sacrificing the natural qualities of textiles, like flexibility, breathability, and aesthetic appeal. An extensive analysis of the state of nanotechnology applications in the textile industry today highlights the improvement of different textile attributes (Hosne Asif & Hasan, 2018; Kumar, Sharma, & Vaish, 2022a, 2022c; Kumar, Sharma, Vaish, et al., 2022).

Why Textiles Textile-Based Composites for Photocatalysis?

Photocatalysts are often utilized in their powdery solid form when they are immobilized on the surfaces of the porous material. So, the question arises why don't we use the photocatalysts as bulk materials in their pure, solid form? The photocatalyst may be less useful or effective for a variety of reasons:

a) Limited Surface Area: Photocatalysts have a small amount of surface area accessible for catalytic reactions in their solid state. A bigger surface area can increase catalytic activity since photocatalysis mostly takes place at the surface where photons are absorbed. To improve the surface area, photocatalysts are utilized in powdered or nanoparticulate forms.

b) Mass Transport Limitations: Solid photocatalysts may experience mass transport restrictions, especially when reactants must diffuse to the material's surface where the catalytic sites are located. Better interaction between reactants and catalytic sites can be facilitated by photocatalysts in solutions that are powdered or distributed.

c) Recovery and Reuse: It might be difficult to effectively recover and repurpose solid photocatalysts from reaction solutions in various applications. Cost increases and material waste may follow from this.

d) Light Absorption: For some reactions, some solid photocatalysts may only be able to absorb a small amount of light in the desired spectral range (for example, visible light). The photocatalyst can be changed, or sensitizers can be used, to get around this restriction.

e) Uniform Distribution: Uneven catalytic activity can result from the inability to achieve a uniform distribution of solid photocatalysts in a reaction mixture.

Despite these difficulties, solid photocatalysts continue to be utilized in a variety of applications, particularly when they are included in coatings, films, or immobilized on surfaces where their solid nature offers stability and longevity. However, utilizing photocatalysts in more dispersed or powdered forms can have benefits in terms of catalytic performance and usability for particular reactions and circumstances. The option to use will rely on the precise conditions of the photocatalytic reaction and the desired results.

A variety of semiconductor materials are utilized in photocatalysis, in both solid and powder form. These substances hold the key to utilizing light energy's transforming potential for a variety of applications. However, the choice between solid and powdered photocatalysts has different benefits and disadvantages that influence the field of practical applications. Just as photocatalysts exhibit certain drawbacks in their solid state, these limitations persist when they are utilized in their powdery form. The practical utilization of powdered semiconductor photocatalysts has several notable limitations when used practically:

a) Laborious Retrieval Processes: Using powdered semiconductor photocatalysts frequently requires lengthy procedures to remove them from the reaction fluid. In addition to taking up important time, this procedure inevitably results in the loss of photocatalyst components.
b) Health Risks: The fine dust produced by powdered photocatalysts can be harmful to people's health, especially if it is inhaled or comes into contact with the skin. It is crucial to take precautions against these health risks.
c) Aggregation-Induced Surface Reduction: During photocatalytic reactions, photocatalyst particle aggregation can reduce the amount of accessible surface area. Given that the surface is where light absorption and catalytic reactions predominantly take place, this lower surface area directly correlates with decreased photocatalytic activity.
d) Charge Carrier Recombination: Semiconductor nanocrystals, which are frequently utilized as photocatalysts, are prone to rapid charge carrier recombination. This inherent property can greatly reduce the total photocatalytic efficiency, which reduces the efficiency with which light energy is captured for catalytic reactions.
e) Limited Spectral Range: One drawback of many currently available semiconductor photocatalysts is that they primarily act in the UV or near-UV region of the electromagnetic spectrum. Examples of these materials include TiO_2 and ZnO. This reduces their versatility by limiting their ability to use solar or visible light for catalytic reactions.

The ongoing research efforts are focused on the creation of novel photocatalytic materials and technologies to address these issues and improve the practical usability of photocatalysts. Realizing the full potential of photocatalysis across numerous applications, such as environmental remediation, renewable energy production, and advanced materials synthesis, requires addressing issues related to material retrieval, health and safety, surface area preservation, charge carrier dynamics, and spectral responsiveness.

One of the strategies to solve the above-listed problem is to effectively immobilize the photocatalyst on a polymer or textile substrate to address these (Saleem & Zaidi, 2020). Organic polymers typically constitute an appealing substrate, for example, to immobilize a semiconductor for heterogeneous photocatalysis applications since they are inexpensive, lightweight, and easy to shape. Charge separation can be efficiently aided by nanocomposite photocatalysts made of conducting polymers and semiconductor nanocrystals. Conducting polymers can sensitize semiconductor nanocrystals, which results in increased carrier utilization and light harvesting efficiency. These two benefits are advantageous for photocatalytic applications. The polymers also operate as a host material for immobilizing semiconductor nanocrystals, increasing the stability of the photocatalysts and maintaining the photocatalytic activity. The disadvantages of filtering and suspension of tiny photocatalyst particles are also avoided.

Numerous polymers exhibit excellent light transmittance within the TiO_2 activation range, which corresponds to the UV-A region, particularly in the context of the solar spectrum. This characteristic, which enables the irradiation of numerous photocatalytic surfaces in one direction with a single radiation source and hence increases the effectiveness of the process, is highly desirable for the photocatalytic treatment of water and air. The integration or coating of tiny powder nanoparticles on textiles has remained a difficulty, despite the current surge in interest in using nanotechnology to modify textile and polymeric surfaces.

The use of textile materials for photocatalysis is an innovative approach to address environmental, health, and sustainable challenges. These substances can make common items and surfaces become self-cleaning, antibacterial, and pollutant-removing solutions, enhancing living conditions and resource preservation. Beyond traditional clothing and furnishings, textiles, the adaptable materials that surround us every day, are finding creative uses. One such frontier is their incorporation into photocatalysis, a game-changing technology that uses light to catalyze chemical reactions. Photocatalytic textiles are a possible solution for a variety of problems because they combine environmental engineering with materials science. These textiles have made it possible to create improved medical devices, self-cleaning surfaces, and air and water purification systems. These photocatalytic elements have been selectively incorporated into these fabrics. This investigation digs into the varied world of photocatalytic fabrics, highlighting their various applications, and environmental advantages.

Numerous compelling factors promote the use of textiles in photocatalysis, highlighting their crucial significance in a variety of cutting-edge applications, which are stated as follows:

a) Surface Area: Textiles have a lot of fibers, a lot of surface area, and a porous structure. This offers plenty of room for applying and immobilizing photocatalytic materials, enabling effective exposure to light and the ensuing catalytic processes.
b) Integration: During the manufacturing process, photocatalytic elements can be smoothly included in textiles. The photocatalyst is equally dispersed and firmly fixed to the textile substrate due to this integration.
c) Versatility: Self-cleaning materials, personal protection equipment, and the purification of air and water are just a few uses for photocatalytic textiles' versatility. They provide a flexible framework for addressing issues relating to the environment and public health.
d) Sustainability: By promoting self-cleaning surfaces and lowering the need for chemical cleaning agents, photocatalytic textiles can aid in sustainability initiatives. Savings in energy and resources may result from this.
e) Purification of the Air and Water: Photocatalytic textiles can assist in the removal of contaminants from the air and water, including bacteria and volatile organic compounds (VOCs). This is especially helpful in applications involving interior conditions and water treatment.
f) Anti-Microbial properties: Textiles with photocatalytic capabilities can also have antimicrobial and antifungal qualities, which makes them acceptable for usage in healthcare facilities and personal hygiene products.
g) Longevity: Because photocatalytic materials may maintain their catalytic activity for long periods when integrated into textiles, they are a strong and long-lasting solution.
h) Clean Energy Generation: Photocatalysis can use solar energy to power chemical reactions in some applications, such as self-cleaning textiles, which eliminates the need for external power sources.

A cutting-edge field of scientific research has been launched by the interaction between photocatalysis and textiles. This ground-breaking collaboration not only improves the aesthetic aspects of daily life but also gives us the tools to take on today's most pressing problems. The development of photocatalytic fabrics is a monument to human ingenuity and our unwavering dedication to finding solutions that promote sustainability, hygienic practices, and the general welfare of our planet

Figure 1. Applications of photocatalyst-NPs embedded textile composites

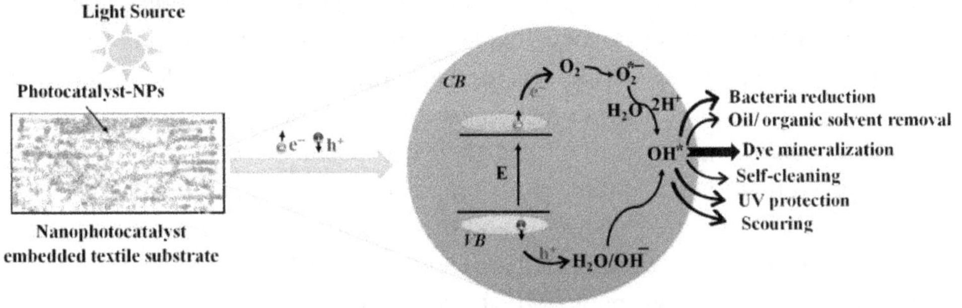

and its people. These compelling reasons support the concept that textiles are well suited to integrate with photocatalysis and anticipate a promising future. Figure 1 illustrates various applications of photocatalyst nanoparticle-embedded textile-based composites.

Preparation of Nano Photocatalysts-Based Textile Composites

Creating techniques for applying or embedding nonstructural coating onto textile substrates is a difficult task for R&D centers. The perfect polymeric support, nevertheless, should meet the following criteria: excellent adherence to the photocatalyst with a high capacity for adsorption and a large specific surface area good chemical inertness, great sensitivity to the reaction species, and mechanical sturdiness. Nanoparticles, which are extremely small materials, can change the surface and can be utilized as fillers in synthetic fibers at the textile-interface level, fiber surface manufacturing, or synthetic synthesis(Komeily-Nia et al., 2019). Numerous research have examined the impact of ZnO and TiO_2 nanoparticles on various aspects of fabric characteristics. Focused on Montazer et al.'s study was the photocatalyst functionalization of fibers. A more comprehensive understanding of the uses for nanoparticle-finishing textile fabrics. Furthermore, Maja Radeti showcased some recent developments in the finishing of various TiO_2 nanoparticle-infused textiles. Semiconductors with metallic coatings have many advantageous improvements in the mechanical or electrical properties of textiles fire resistance, or self-cleaning UV catalytic surface chemistry while keeping the textiles' fundamental structure.

To give textiles a variety of functionality, Kiwi and Pulgarin have recently worked diligently on the deposition and testing of semiconductor and metallic deposits using cutting-edge techniques. Inorganic active-based nanostructured materials can be roughly divided into two groups: inorganic nanostructured loaded organic carriers and inorganic nanostructured materials and their nanocomposites.

Inorganic nanostructured materials that have been loaded into a variety of organic carriers, such as liposomes, cyclodextrins, nanocapsules, microcapsules, and dendrimers, fall under this category. Readers can consult Dastjerdi and Montazer's work for comprehensive discussions on the use of these structures in functionalizing textile products and modification techniques. We shall pay particular attention to inorganic nanostructured materials and their composites in this work. Inorganic nanostructured materials and their nanocomposites include nanocomposites made of inorganic nanostructured materials and their composites. There are three main ways to incorporate inorganic nanoparticles into textiles: functional finishing, coating, and mixing. Various photocatalytic nanoparticles have been applied to both natural and manufactured polymers, including textile materials, in numerous trials. These works have incorporated nanoparticles into polymers and fabrics using the aforementioned techniques.

Numerous in-situ solution-based procedures can be used to incorporate nanoparticles into textile goods during the finishing process. Typically, these procedures include coating the textiles with the nanoparticle solution using simple techniques like pad or dip coating. Chemical methods provide a variety of nanoparticle dispersions but are frequently favored because they are straightforward and inexpensive. With such flexibility, it is possible to alter the size, morphology, and stability of particles by varying important experimental variables like concentrations, reducing agents, temperature, pH, surfactants, capping agents, and solvents. In contrast, high-quality metal nanoparticles can be produced physically, but these procedures are typically more expensive and call for specialized equipment. As a result, chemical methods are more efficient and feasible for lab studies, allowing the creation of a variety of nanoparticle dispersions.

Various standard techniques frequently used in the textile industry can effectively apply nanoparticles to fabrics. Another technique called ex-situ involves assembling nanoparticles on textiles using binders as a medium. The creation of a coating composition, application of the coating compound to the textile surface, and subsequent drying of the textile are all processes in this process. Both organic and inorganic binder materials can be utilized for this. Organic binders are frequently selected, in particular, those that include latex, polyacrylate, vinyl ester, polyurethane, polyethylene-vinyl acetate, polyolefin, polyester, polyamide, polyether, poly(styrene-co-butadiene), polyisoprene, and polychloroprene, as well as their mixtures. The assembly of nanoparticles on textiles is made possible by these organic binders. Ozone gas treatment, supercritical carbon dioxide, enzymatic alteration, and alkaline or acidic etching are examples of additional techniques that cannot be categorized as direct nano-coating. They can be utilized for both the initial nanocoating process and the pretreatment. Other physical techniques for creating nano topography on textile surfaces include corona discharge and laser treatments. All physical techniques can be used as a pretreatment before covering substrates with nanoparticles later on. The second category is blending, which comprises electrospinning, melt mixing, and bicomponent. This group could be divided further into blends. Either during in situ polymerization or melt spinning. the transformation surfaces of nanoparticles are frequently necessary to prevent the aggregation of nanoparticles in melt during polymerization because nanoparticles have a high surface energy, they mix. For wastewater, the electrospinning process is highly advised. Its capabilities for treatment and water purification should be further improved. either by injecting semiconductors directly, like TiO_2, or these semiconductor types on nanofibers.

Semiconductor-Based Photocatalysts

In general, organic and inorganic substances can be employed as semiconductor materials, with crystalline inorganic solids being the most popular. The periodic table's groupings of atoms, to which their constituents belong, are commonly used to categorize these inorganic semiconductors. However, the field of semiconductor technology has developed to include organic electronics which provides a fascinating frontier in high-performance electronics. The ability of some plastics to exhibit metallic conductivity after being modified was uncovered by researchers in a crucial discovery made in the late 1970s. Normally, plastics act as insulators of electrical current, but the development of specific organic polymers with this property has opened the door for revolutionary advancements in the field of organic semiconductors (Marszewski et al., 2015).

Significant progress has been made in the field of organic/inorganic devices due to the combination of organic materials and nanocrystalline semiconductors. The development of numerous technologies, including dye-sensitized solar cells (DSCs), organic solar cells, hybrid and organic light-emitting diodes (OLEDs), organic field-effect transistors, as well as electrochemical sensors and displays, has been significantly accelerated by this partnership. Transitions within a single energy band can be used to precisely adjust the electrical characteristics of semiconductors. It is possible to introduce charge carriers into semiconductors by a variety of techniques, including doping with atoms from various periodic system groups, electronic injection, or optical excitation).

The two types of semiconductors, organic and inorganic, also have their differences.

The charge transport mechanism between these two classes of semiconductors accounts for the majority of the differences. Atoms are covalently bound together in inorganic materials, and electrons move in broad bands due to delocalized plane waves' very high mobility and very little dispersion. Atoms are kept together by comparatively modest van der Waals forces in organic semiconductors. This stops charge transport in CBs and delocalized valence. Charge carriers localized at molecular entities, such as oligomers or segments of a conjugated polymer chain, travel between neighboring sites with statistically varying energies when they are in organic semiconductors.

a) Inorganic semiconductors: Homogeneous and heterogeneous catalysis are the two most common forms. While the catalyst and reactants are in the same phase during homogeneous catalysis, they are in distinct phases during heterogeneous catalysis. There are benefits and drawbacks to each of these groups. However, scientists have been working hard to develop new catalyst systems that combine the advantageous traits of both homogeneous and heterogeneous catalysis. In this endeavor, the development of nanocatalysts and nanocatalysis has been crucial. The characteristics of nanoparticle catalyst's selectivity, efficacy, and recyclability align well with the contemporary desire for environmentally benign or "green" catalysts. As a result, their applications have increased, and in the upcoming years, it is anticipated that their usage will increase significantly across a variety of industries. Metal oxides can function as efficient photocatalytic materials because they make electron transmission easier. WO_3, CdO_3, In_2O_3, Ag_2O, MnO_2, Cu_2O_3, Fe_2O_3, V_2O_5, TiO_2, ZrO_2, RuO_2, Cr_2O_3, Co_3, Ni, SnO_2, CeO_2, Nb_2O_3, $KTaO_3$, $SrTiO_3$, K_4NbO_{17}, and a host of other metal oxides (Komeily-Nia et al., 2019) can be chosen from a large variety. The cation-oxygen bonding has a significant impact on the electrical characteristics of compounds with an oxide basis. Solid-state band theory, ionic bonding ideas from solid-state chemistry, or a mix of the two can all be used to explain this phenomenon. A wide range of electrical characteristics in metal oxides is the outcome of this interaction between localized and itinerant characters. Extensive research has been done on many types of metal oxides in a variety of forms, including nanoparticles, clusters, encapsulated particles, thin films, aerogels, and nanofibers. These many forms have been used in the creation of self-cleaning materials, solar energy conversion, and extremely effective photocatalysis.

The photocatalytic degradation of solid and liquid organic compounds using inorganic metal oxide nanocrystal photocatalysts is influenced by a range of factors. The surface and solid-state characteristics of the photocatalyst powder, the type of organic compound being degraded, oxygen concentrations, gas-phase humidity, and temperature are important determinants of gas-solid photocatalytic degradation events. The formation of high hydrophilicity and photoinduced redox reactions of adsorbed compounds are two of the main outcomes of these interactions. There are oxidative and reductive pathways in the semiconductor mechanism, which involve the valence band (VB) hole and the conduction band (CB) electron, respectively. The reaction's outcome is inherently linked to the chemical potential of the carriers, leading to the production of oxidizing and reducing species with a range of abilities. Redox reactions at the oxide particle's surface, which are frequently started when organic pollutants are adsorbed onto it, are another element that influences the reactions. The amount of pollutant adsorption is substantially influenced by the oxide's surface area and chemical functionality. In addition, several factors, such as the recombination and trapping of charged carriers in inaccessible sites, the reaction temperature, the media used in the reaction, the thermal and temporal stability of the semiconducting oxides, the morphology, and the accessible surface area of the oxides, are crucial in determining the effectiveness of synthesized oxide photocatalysts.

b) Organic Semiconductors: Semiconductors with extended π-conjugated organic backbones exhibit unique optoelectrical properties. The electrical properties of metals and the mechanical qualities of polymers are compellingly combined in inherently conductive polymers (ICPs). Exploring the characteristics of organic semiconductors has garnered a lot of attention lately, particularly for use in electronic devices. LEDs, EMI shielding, conductive layers for organic light-emitting diodes (OLEDs) and organic field-effect transistors (OFETs), optically active layers for OLEDs, and anticorrosion coatings for iron and steel are some of these applications. Based on molecular weight, organic semiconductors can be generically categorized. They can be divided into two categories: heterocyclic polymers, which have molecular weights above 1000, and conjugated polycyclic compounds, which have molecular weights below 1000. Pyrylium salts, aromatics, heteroaromatics, chlorins, porphyrins, and phthalocyanines are further classified systematically. Due to their simplicity in producing thin films with substantial surface areas, polymers are preferred materials for semiconductors. Similar to their inorganic counterparts, organic molecules/polymers with a conjugated (hetero) aromatic backbone are ideal as semiconductors in optoelectronic devices due to their capacity for charge transport and effective interaction with light. Pentacene, anthracene, and rubrene are examples of aromatic hydrocarbons, which are tiny molecules that conduct electricity. Poly(3-hexylthiophene), poly(p-phenylenevinylene), polyacetylene, and its variants are examples of polymeric organic semiconductors. The processes of photochemical electron transfer and energy transfer involve a variety of organic substances, including dyes, porphyrins, phthalocyanines, and transition metal complexes. Similar to inorganic semiconductors, organic semiconductors can be doped. Organic metals are another name for dopant-susceptible organic semiconductors like polyaniline (Ormecon) and PEDOT: PSS. As a result, they have been useful as photocatalysts for a variety of processes, including photopolymerization, the creation of solar cells, and photodynamic treatment. Organic photocatalysts have been used for water and air purification for the past ten years.

Contrary to inorganic semiconductors, the solid-state structure of organic materials is mostly based on weak interactions between nearby molecules or polymer chains, particularly van der Waals and dipole-dipole interactions. These interactions endow objects with qualities ranging from insulation to conductivity. However, it's still unclear to what degree organic materials can mimic the properties of inorganic semiconductors. The comparatively lower chemical stability of organic materials in comparison to their inorganic counterparts, as well as their decreased charge mobility as a result of higher electron-phonon coupling and increased structural disorder, are two potential obstacles. Organic semiconductors are subject to a variety of degradation mechanisms, which can be divided into groups based on the internal and external variables causing the degradation. For instance, Scholz et al. clarified the molecular pathways of material degradation in several organic light-emitting diode (OLED) materials(Scholz et al., 2015) Major pathways for deterioration within different layer configurations of polymeric organic solar cells and modules were investigated in separate study. Inorganic catalysts are frequently used in applications for the textile industry because of their variety, inexpensive preparation techniques, and decreased toxicity. Researchers' interest in these catalysts' nanotechnology developments has increased, leading to the creation of more sophisticated catalysts. Additionally, these metal-oxide catalysts' properties, such as their composition, size, shape, surface area, and interrelationships, can be altered to fine-tune their performance. As a result, considering their benefits and adaptability, our focus will switch to the creation of photoactive polymers and fabrics employing inorganic semiconductors.

FUTURE RESEARCH DIRECTION-ENVIRONMENTAL CLEANING

Semiconductors' photocatalytic characteristics have a wide range of uses in water treatment, air purification, and the elimination of bacteria and contaminants from diverse surfaces. In addition to their UV absorbance properties, metal oxide nanoparticles including TiO_2, Al_2O_3, ZnO, and MgO also have photocatalytic and antibacterial properties. Especially for healthcare items, these nanoparticles have been studied for treating textile materials to impart antibacterial qualities, self-decontamination abilities, and UV-blocking activities. The removal of contaminants and pathogens from various surfaces, air purification, and water treatment are just a few of the many uses for semiconductors' photocatalytic characteristics. TiO_2, Al_2O_3, ZnO, and MgO are examples of metal oxide nanoparticles that display photocatalytic and antimicrobial properties in addition to UV absorption. Investigations on the use of these nanoparticles for treating textile materials to provide antibacterial, self-decontaminating, and UV-blocking qualities have focused on healthcare items. The compatibility with aqueous solutions, solar activation, and reliance on ambient oxygen as the main oxidizing agent make organic photocatalysts promising for environmental applications. When exposed to light, electrons move from the valence band (VB) to the conduction band (CB), leaving the semiconductor with an equal number of empty sites. The energized electrons and holes then go to the semiconductor's surface. Third, whereas the excited holes interact with electron acceptors (A), the excited electrons interact with absorbed electron donors (D) at the surface. However, a sizable number of electron-hole pairs recombine in the second stage, wasting the input energy as heat or radiated light. Organic semiconductors can also use the same set of procedures. However, it is important to remember that -bonding orbitals are crucial for charge transfer in organic semiconductors. As a result, electron excitation mainly takes place between the highest occupied molecular orbitals. Materials that display spontaneous electric polarization are referred to as ferroelectric materials. In his studies on the dielectric characteristics of a particular salt rock, Joseph Valasek was the first to describe ferroelectricity (Valasek, 1921). Following are the points on how photocatalytic textile composite work and help in environmental cleaning:

a) Dye degradation:

To address environmental issues and ensure sustainable industrial practices, dye degradation is crucial. The capacity of dye degradation to lessen the negative environmental effects of dye pollution released into water bodies makes it important. These synthetic dyes can stay in the environment for long periods and represent serious dangers to aquatic ecosystems and human health. They are frequently used in the textile, paper, and dyeing industries. By reducing dyes to less hazardous byproducts, we can both remove their color and complicated chemical structures. Some of the most important methods used to achieve dye degradation include advanced oxidation processes, photocatalysis, and enzymatic degradation. In line with international efforts to combat water pollution and encourage sustainable practices in many industries, this procedure helps to clean up water resources, reduce pollution, and create a healthier environment. Additionally, the creation of effective eco-friendly dye degradation techniques is essential for adhering to strict environmental standards and reducing the environmental impact of industrial processes. As shown in Figure 2, organic dyes are broken down and removed from wastewater or other contaminated sources by a method called photocatalytic dye degradation, which makes use of photocatalytic materials.

Often employed in water treatment and wastewater cleanup procedures, this technology is environmentally beneficial. It has been proven that heterogeneous photocatalysis is an effective method for removing organic pollutants from both the atmosphere and water. The following factor influences the dye degradation process:

i) Photocatalytic material: The main object in photocatalytic dye degradation is choosing a photocatalytic material. Photocatalytic materials are those that can use light energy to carry out a chemical reaction, typically TiO_2 and other

Figure 2. Dye breakdown by photocatalyst NPs loaded textiles

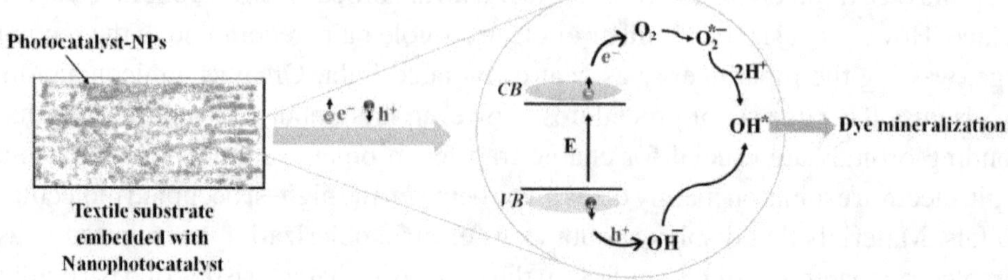

semiconductor materials are used as photocatalytic materials (Chen et al., 2020a).TiO_2 is used due to its high photocatalytic activity when exposed to UV or visible light.

ii) Adsorption of dye: The surface of the photocatalytic material is adsorbed with the organic dye molecules found in the contaminated water or solution. This happens as a result of physical interactions between the dye molecules and the surface of the photocatalyst, such as Van der Waals forces or electrostatic interactions.

iii) Photocatalytic activation: The photocatalytic substance becomes activated when it is exposed to UV or visible light (Feizpoor & Habibi-Yangjeh, 2018). The photocatalyst absorbs the light's energy, which encourages the transfer of electrons from the valence band to the conduction band. e^-/h^+ couples are produced as a result (Chen et al., 2020b).

iv) Generation of reactive species: The holes (h^+) and separated electrons (e^-) are extremely reactive entities. While the holes can oxidize water molecules (H_2O) to make hydroxyl radicals (<<ch005.m1.eps>>), the electrons can decrease oxygen molecules (O_2) to form superoxide radicals (<<ch005.m2.eps>>). The degradation process depends heavily on these reactive species (Singh & Dutta, 2018).

v) Dye degradation: The reactive species that are produced, in particular hydroxyl radicals, attack and oxidize the dye molecules that have been adsorbed. The color molecules are broken down into smaller, less complicated chemicals during this oxidative reaction. The process is repeated until all color molecules have been mineralized into more basic and benign compounds like carbon dioxide (CO_2) and water (H_2O).

vi) Photocatalyst regeneration: The photocatalytic material is unaltered and can be recycled numerous times when the dye is broken down. As long as there is a sufficient amount of light, the photocatalyst can continue to adsorb and destroy dye molecules.

vii) Water clarification: The photocatalytic process can result in the elimination of additional contaminants and the purification of water in addition to the destruction of dyes. This is true because other organic and inorganic pollutants in the water can also be broken down by the reactive species produced during photocatalysis.

The in situ synthesis of Cu_2O on cotton fibers with antibacterial properties and reusable photocatalytic degradation of dyes has been reported. A flexible and recyclable composite material (COCO) made of cotton textiles and cuprous oxide-nanocellulose (Cu_2O-NC) with highly effective photocatalytic dye degradation characteristics was created. Cu_2O was produced in situ using flexible cotton fabrics as substrates, growing uniformly on the fibers, and then being wrapped in nano cellulose (NC). Methylene blue (MB), whose degradation rate reached as high as 98.32%, was used to demonstrate COCO-5's photocatalytic degradation capabilities. Using electron paramagnetic resonance (EPR) spectroscopy, transient photocurrent response (TPR) spectroscopy, Fourier transform infrared (FTIR) spectroscopy, and ion chromatography (IC), the mechanism of COCO-5 photocatalysis and the process of dye degradation were examined. This study examined every step of the process, from dye degradation to electron excitation to harmless tiny molecules. TiO_2 is considered a wonderful photocatalytic agent due to high photocatalytic activity and because TiO_2 has a broad bandgap energy (3.2 eV for the anatase phase), it may absorb a sizable amount of ultraviolet (UV) light. In TiO_2, electron-hole pairs are known to be produced very efficiently by UV light and are necessary for photocatalysis. TiO_2 has been used in a lot of studies for photosynthesis where the textile material has been treated with TiO_2 to check its dye degradation property. there are also others where TiO_2 is used in combination with some other semiconductor material.

Figure 3 illustrates the dye mineralization performance of $CaCu_3Ti_4O_{12}$-NPs screen-printed cotton fabric. The 82% MB dye breakdown was shown by CCTO-NPs loaded fabric during the photocatalysis process in 4 h, while pure cotton fabric presented nearly 11% dye degradation in 4 h (Kumar et al., 2023).

b) Nano-photocatalyst-based antibacterial textiles:

In the field of environmental sanitation and cleaning, antibacterial photocatalysis is incredibly important. Antibacterial compounds can successfully attack and neutralize dangerous microorganisms, such as bacteria and viruses, found on various surfaces and in the air by harnessing the power of photocatalytic processes. This method considerably lowers the danger of the emergence of antibiotic resistance in addition to providing a green and sustainable substitute for chemical disinfectants. Reactive oxygen species (ROS), which have strong antibacterial capabilities, are produced

Table 1. Textile-based composites for dye degradation application

Material	Pollutants	Textile Substrate	Degradation Rate	Reference
CeO_2/Ag/CdS	MB dye	Cotton	89.4%, 180 min	(Y. Wang et al., 2022)
C_3N_4/Cu-MOF	EBT	Cotton	95%, 15 min	(H. Wang et al., 2023)
TiO_2-AC/HDTMS	MB and RhB dye	Cotton	90% and 76.5%; 4 h	(Sun et al., 2022)
konjac glucomannan	MB dye	Fabric	>80%	(Zeng et al., 2018)
ZIF-8/Ag_2CO_3	RhB dye	Cotton	91%, 30 min	(L. Wang et al., 2023)
Ag/AgCl/ZIF-8/TiO_2	MB dye	Cotton	98.5%,105 min	(Guan et al., 2019)
GO/TiO_2	MB dye	Cotton	90%, 4 days	(Karimi et al., 2016)
TiO_2	MB dye	Cotton	90%, 14 h	(Cheng et al., 2018)
Ag/ZnO	MB dye	Cotton	90%, 120 min	(Ibănescu et al., 2014)
rGO/$BiVO_4$	MB dye	Cotton	96.9%, 190 min	(Guo et al., 2021)
g-C_3N_4	Methyl Orange dye	Cotton	90%, 1 h	(Han et al., 2021)
TiO_2-bamboo charcoal and polydimethylsiloxane	RhB dye	Cotton	85%, 4 h	(Wu et al., 2021)

when minerals like titanium dioxide (TiO_2) are exposed to light in antibacterial photocatalysis. This technique is very adaptable and has uses in decontaminating surfaces, air filtration, and water purification, all of which promote a safer and healthier environment. The significance of antibacterial photocatalysis in environmental cleaning is becoming more and more clear as concerns about infectious diseases and public health increase. This technology provides a potent and effective tool for stopping the spread of dangerous pathogens and maintaining a clean and hygienic living and working environment. When exposed to light, particularly ultraviolet (UV) or visible light, photocatalytic materials, kill or impede the growth of bacteria and other microbes. Applications for this technology include self-cleaning surfaces, medical equipment, water treatment, and air purification. Table 2 shows the various photocatalysts embedded in textile substrates for bacteria reduction. Using TiO_2, ZnO, WO_3, ZnS, Fe_2O_3, and graphene-based nanomaterials, the advanced oxidation

Figure 3. Absorbance spectrum illustrating MB dye mineralization by (a) pure cotton, (b) CCTO-NPs coated cotton, and (c) MB dye degradation percentage
(Kumar et al., 2023)

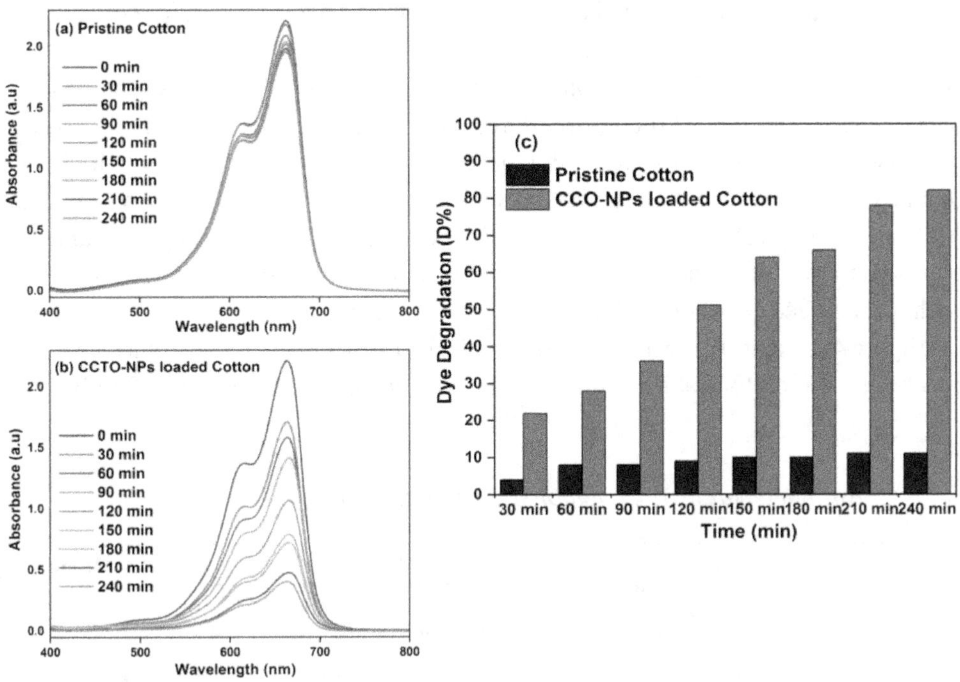

Figure 4. E. Coli reduction by: (a.1-a.4) Pure cotton, and (b.1-b.4) SrTiO3-NPs embedded cotton
(Kumar et al., 2024)

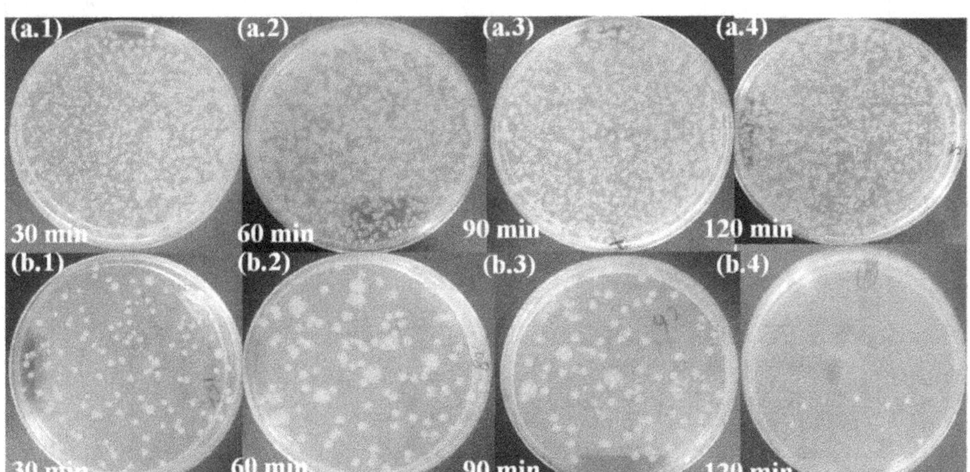

process (AOP) is advantageous for removing environmental pollutants from water sources ZnO, one of these photocatalysts has received a lot of interest for its use in the photo-reduction of microorganisms, the creation of hydrogen, and the treatment of synthetic dyes. The various factors that affect the bacteria reduction efficacy of nano photocatalysts are the type of photocatalytic material, bacterial adsorption, activation by light, and generation of reactive species. The bacteria on the surface or in the fluid are either killed or rendered inactive as a result of the photocatalytic processes. Bacterial cell structure and metabolic functions are interfered with by the oxidative and photocatalytic processes, which inhibit bacterial growth and survival. As long as the photocatalytic material is exposed to light, it is still active and can continue to produce reactive species. This enables continuous antibacterial action without the need for chemical or disinfectant replenishment.

Figure 4 presents bacterial reduction by $SrTiO_3$-NPs loaded cotton surface. After two hours, the cotton fabric with SrTiO3 incorporated in it inhibited 99.3% of E.

Table 2. Antibacterial textiles demonstrating photocatalytic effect

Sr. No	**Photocatalyst**	**Coating Application Method**	**Antibacterial Activity**			**Reference**
			Pathogens	**R% before washing**	**Substrate**	
1	$g\text{-}C_3N_4/ZnO$ (CNZ)	Coating	S. *aureus*, E. *coli*	>98	Cotton	(Hosseini-Hosseinabad et al., 2022)
2	$rGO\text{-}TiO_2/QAS\text{-}SiO_2$	Coating	S. *aureus*, E. *coli*	~99.9	Cotton	(Zhu et al., 2022)
3	MoS_2	Spray coating	S. *aureus*, E. *coli*	>98	Cotton	(Kumar, Sharma, & Vaish, 2022a)
4	Cu_2O-nanocellulose	Pad-dry-cure	S. *aureus*, E. *coli*	93.25, 89.73	Cotton	(Su et al., 2021)
5	polysiloxane–Ag@ZnO	Grafting	S. *aureus*, E. *coli*	Very good	Cotton	(M. Wang et al., 2019)
6	Copper oxide	Coating	S. *aureus*, E. *coli*	~100	Cotton	(Bashiri Rezaie et al., 2018)
7	ZnO/BiOBr	Coating	S. *aureus*, P. *aeruginosa*	96,76	Cotton	(Xu et al., 2017)
8	Fe_3O_4	Dip-coating/ Soaking	S. *aureus*, Candida *albicans*	95,99	Cotton/ polyester	(Rastgoo et al., 2016)
9.	$CaCu_3Ti_4O_{12}$	Screen printing	S. *aureus*, E. *coli*	>94	Cotton	(Kumar et al., 2023)

Coli and 96.09% of S. aureus, respectively. Even after fifteen washing cycles, the coated fabric was still able to eliminate microorganisms by more than 92% (Kumar et al., 2024).

c. Nano- photocatalyst based Self-cleaning textiles:

Self-cleaning, made possible by photocatalysis, is a revolutionary technology that is crucial for sustainability and environmental cleanup. Utilizing the force of sunshine, photocatalytic self-cleaning materials, frequently coated with photocatalysts like titanium dioxide (TiO_2), degrade and remove organic pollutants, grime, and contaminants from surfaces. This not only adds considerably to environmental cleaning but also aids in maintaining the attractiveness of buildings and infrastructure. Self-cleaning surfaces minimize the release of hazardous chemicals and pollutants into the environment, encouraging cleaner air and water, by decreasing the need for frequent cleaning and chemical treatments. Additionally, they are essential in the fight against air pollution because they break down dangerous airborne chemicals. The use of self-cleaning materials in a variety of sectors, including architecture and transportation, improves the durability and aesthetics of buildings while also supporting sustainable practices by lowering maintenance costs and the environmental impact of cleaning procedures. Self-cleaning using photocatalysis is a useful tool in the pursuit of cleaner and more sustainable living spaces in a time of rising environmental awareness.

Figure 5 illustrates the working of self-cleaning fabrics. When exposed to light, photocatalytic materials are used in the technique of photocatalysis to degrade organic and inorganic contaminants on surfaces. It is frequently utilized for self-cleaning applications in several different industries, including automotive (self-cleaning car coatings), architecture (self-cleaning glass), and more. To achieve self-cleaning qualities, the lotus effect is frequently imitated. The use of fluorochemicals for oil repellency and the requirement for micro- or nanostructures are potential downsides of this strategy. Structured surfaces can also be somewhat soft and abrasion-prone even after being properly treated. It is critical to understand that passive protection methods, such as barriers, filters, permeable membranes, or adsorbents, may preserve surface toxicity and demand post-use sterilization in addition to other constricting limitations unique to each technique. Active protection, which incorporates concurrent agent degradation, is a different strategy as a result. The produced radicals target and degrade the contaminants that have been adsorbed on the surface at the molecular level. The contaminants undergo oxidation and are converted into less dangerous, simpler compounds like carbon dioxide (CO_2) and water (H_2O). A common name for this process is photocatalytic oxidation. Pollutants are carried away by rains or simply fall off the surface due to diminished adhesion as they are broken down and

Figure 5. Working of nano photocatalyst-induced self-cleaning fabrics

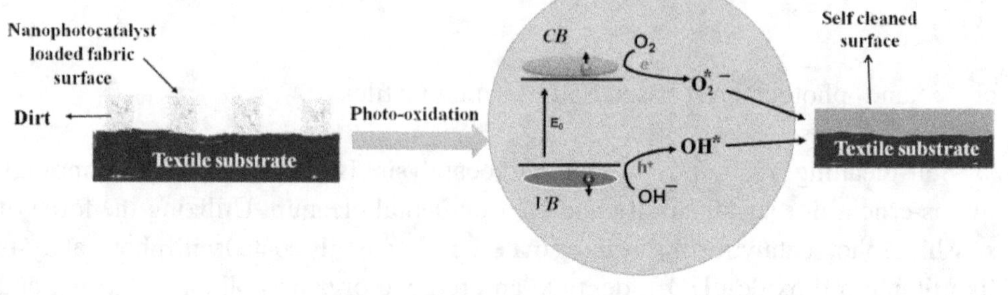

converted into harmless chemicals. As a result, the surface maintains its cleanliness and contaminant-free status (Banerjee et al., 2015). As long as there is light to activate the photocatalytic substance, the photocatalytic activity continues. This makes it an environmentally benign and low-maintenance method for keeping surfaces clean and free of pollutants because it enables the surface to continuously self-clean.

d) Nanophocatalyst application in de-sizing processes

Using photocatalytic compounds and light exposure, the textile industry uses photocatalytic desizing to remove sizing chemicals from fabrics. During the manufacturing process, sizing agents are added to textiles to enhance the effectiveness of the weaving process and the fabric's quality. An efficient and environmentally beneficial way to get rid of these substances is through photocatalytic desizing. Sizing agents, which are primarily chemicals based on starch, are frequently applied to fabrics created in the textile industry. The smoothness, stiffness, and manageability of the cloth are improved by these sizing agents. However, they must be taken out before any more processing or dyeing. The fabric's surface is coated with a photocatalytic substance. Photocatalytic nanoparticles can be applied to the fabric by spraying, padding, or coating it. Light is shone upon the fabric that has the photocatalytic substance applied to it. Depending on the particular photocatalyst and the available light source, ultraviolet (UV) or visible light is typically used. The photocatalytic substance becomes active when it is exposed to light. Electrons in the material are excited by the light's energy and travel from the valence band to the conduction band. Electron-hole pairs (e^-/h^+), a highly reactive species, are produced as a result. On the surface of the fabric, the separated electrons (e^-) and holes (h^+) participate in a variety of chemical processes: Holes (h^+) can immediately oxidize and degrade the sizing agents. When starch-based sizing agents are used, the holes may cause the starch molecules to break down into more straightforward, water-soluble chemicals. Superoxide radicals ($^-O2$) and other reactive oxygen species (ROS) are created when electrons (e^-) reduce oxygen molecules (O2). These ROS are also capable of attacking and weakening the sizing agents. The photocatalytic material's reactive species efficiently fragment the sizing agents on the fabric's surface into smaller, more soluble parts. The fabric can then be easily cleaned of these particles. The fabric is normally rinsed or washed to get rid of the degraded sizing agents and any leftover photocatalytic material after the photocatalytic desizing process. This process makes sure the fabric is free of pollutants and clean.

After the sizing agents are eliminated, the cloth can be dried and processed further as necessary, including dyeing, finishing, or further treatments. Wet processing of textiles has a significant negative environmental impact since it uses so much water and energy. High water use also contributes to the production of a large amount of effluents. To reduce the energy used in textile pretreatment, a method for desizing cotton cloth that has been starch-sized using ultraviolet C (UVC) assistance has been created. The whiteness index, water absorption, tensile strength, Fourier transform infrared (FTIR), and wide-angle X-ray diffraction of the UVC-assisted desized cloth are evaluated and contrasted with the control. The UVC-assisted de-sizing procedure has the potential to save more than 70% of the time, 80% of the energy, and 60% of the water. Additionally, a life cycle analysis was performed. Compared to the current method, the photocatalytic desizing process can lessen its negative effects on human health by over 85% and conserve 69% of mineral resources. An innovative strategy for sustainable downsizing can be quickly adopted by the textile sector (Panda et al., 2023).

e) Nano-photocatalysis application in Oil / organic solvent removal:

Oil and organic solvents can be broken down and eliminated from air or water by a process called photocatalytic degradation, which happens when photocatalytic materials are exposed to light. Applications for this technology include wastewater treatment, air purification, and environmental cleanup. Depending on the application, the first stage entails depositing a layer of photocatalytic material onto a surface or into a water treatment system. The photocatalytic substance may be a catalyst, a coating, or immobilized nanoparticles. The photocatalytic substance in photocatalytic degradation is activated by light. Depending on the particular photocatalyst and application, ultraviolet (UV) or visible light is frequently utilized. Sunlight from the sun or artificial UV lamps are both acceptable light sources. The photocatalytic substance becomes active when it is exposed to light. Electrons in the material are excited by the light's energy and travel from the valence band to the conduction band. Electron-hole pairs (e^-/h^+), a highly reactive species, are produced as a result. In the degrading process, the separated electrons (e^-) and holes (h^+) are essential. Oil and organic solvent molecules can be directly oxidized by holes (h^+). These chemicals are broken down by the oxidation process into simpler, less dangerous molecules like water (H_2O) and carbon dioxide (CO_2). Superoxide radicals ($^{-}O_2$) and other reactive oxygen species (ROS) are created when oxygen molecules are reduced by electrons (e^-). These ROS can further damage and degrade the oil and organic solvents due to their potent oxidative characteristics. Oil and organic solvents are efficiently broken down into simpler and less hazardous components by the reactive species produced by the photocatalytic material. This process is often termed photocatalytic oxidation. The degraded products can be removed from the treated medium in a variety of ways, depending on the particular application, including filtration, settling, or just letting them volatilize into the air. To remove the impurities that have been photocatalytically cleaned, separation methods may be used in water treatment applications.

f) Nano-photocatalysis application in scouring:

When exposed to light, photocatalytic materials can be used in the cleaning and purifying process known as photocatalytic scouring to remove organic and inorganic pollutants from water and surfaces. This technique is frequently employed in the purification of air, water, and self-cleaning surfaces. Applying a layer of photocatalytic material on the surface that has to be cleaned or into a water treatment system is the initial step. The substance may take the shape of a paint, or coating, or it may be incorporated into a substrate. The photocatalytic substance in photocatalytic scouring is activated by light. Typically, depending on the particular photocatalyst and application, either visible light or ultraviolet (UV) light is employed. The photocatalytic substance becomes active when it is exposed to light. Electrons in the material are excited by the light's energy and travel from the valence band to the conduction band. Electron-hole pairs (e^-/h^+), a highly reactive species, are produced as a result. On the water's surface or within it, the separated electrons (e^-) and holes (h^+) participate in a variety of chemical processes: Organic and inorganic pollutants that are present on the surface or in the water can be immediately oxidized by holes (h^+). These pollutants are broken down into simpler, less dangerous chemicals during the oxidation process. Superoxide radicals (<<ch005.m3.eps>>) and other reactive oxygen species (ROS) are created when electrons (e^-) reduce oxygen molecules (O_2). These ROS can further target and destroy pollutants due to their potent oxidative characteristics. Organic and inorganic pollutants on the surface of the water are efficiently destroyed by the reactive species produced by the photocatalytic material. A common name for this process is photocatalytic oxidation. The pollutants become simpler and less stick to surfaces as they get degraded. This enables them to be effortlessly removed through filtration in water treatment applications or washed away by water in the case of surface cleaning. As long as there is adequate light exposure, photocatalytic scouring can continue. By continuing to be active, the photocatalytic substance ensures continuous cleaning and purification.

g) Nano-photocatalysis application in UV-protection:

A system known as photocatalytic UV protection shields surface and textiles from UV (ultraviolet) radiation using photocatalytic compounds. Utilizing specific materials' photocatalytic characteristics to absorb and neutralize harmful UV radiation is how this technique operates. A photocatalytic substance, usually titanium dioxide (TiO_2), zinc oxide (ZnO), or other semiconductor materials, is an essential part of photocatalytic UV protection. UV rays can be absorbed and interacted with by these materials. The photocatalytic substance is either integrated into a material during manufacture or coated onto a surface (such as glass, textiles, or skin). For instance, it could be integrated into the fibers of a textile or applied as a surface coating.UV rays are absorbed by the photocatalytic substance when it is exposed to them. UV radiation includes energy that can damage surfaces and be detrimental to human skin, resulting in fading, deterioration, or discoloration. Electron-hole pairs (e^-/h^+) are created when the photocatalytic material is activated by the UV radiation that has been absorbed. These couples of electrons and holes are very flammable. In chemical processes that aid in UV radiation protection, the separated electrons (e^-) and holes (h^+) can take part. Holes (h^+) can combine with the water molecules or hydroxide ions in the area to create hydroxyl radicals (<<ch005.m4.eps>>). Strong oxidizing agents known as hydroxyl radicals can disarm and degrade UV light. Electrons (e^-) can take part in reducing oxygen (O_2) molecules to superoxide radicals (<<ch005.m5.eps>>). Additionally, these radicals may help to render damaging UV rays inactive. By breaking UV light into less hazardous components, the reactive species produced, especially hydroxyl and superoxide radicals, mitigate the deleterious effects of UV radiation. This procedure lessens the UV radiation's capacity to injure human skin or corrode surfaces.

CONCLUSION

The worldwide textile industry is a significant polluter and faces sustainability challenges. These days, conversations about chemicals, human rights, water shortages, and climate change revolve around textiles and apparel. These problems bring up important concerns regarding people's political, social, and economic lives. Common practices and opposing ideals related to production and consumption are at the heart of many problems(Boström & Micheletti, 2016). This chapter has analyzed and provided a quick summary of the most significant recent developments in the field of textiles modified with photocatalysts and how these studies can be used to pave the way for making the textile industry more sustainable and environmentally friendly in the near future (Kumar et al., 2023, 2024; Kumar, Sharma, & Vaish, 2022a). A new window of opportunity for the multifunctional transformation of textiles may be opened by the use of photocatalysts and their nanocomposites on textiles. Here most of the study has been done on how to reduce the harmful effect of Methylene Blue dye and RhB dye as shown in above Table 1 and Table 2. To increase the viability of photocatalytic textiles, two significant technological obstacles must be removed: the development of photocatalysts and efficient loading techniques (Kuspanov et al., 2023). Enhancing photocatalysts to better use the full solar spectrum for high photo-efficiency, improving photocatalyst immobilization on fabrics, durability, and the impact of photocatalysts on fibers during continuous UV irradiation are a few of these. Under UV light, the different metal oxides can be photo-catalytically active, however, the effectiveness of these processes is rather modest. It would be ideal to develop a photocatalytic material that is both commercially viable and has a catalytic activity that goes beyond this 4% range because UV light only makes up 4% of the solar radiation that the earth's surface receives. The use and upkeep of textiles that include photocatalysts is the main focus of the second issue. The low surface energy of flexible substrates and their irregular shape are two of the main challenges to achieving sufficient adhesion of photocatalysts on textiles. There is a need for a careful evaluation of the fate of textiles embedded with photocatalysts exposed to continuous UV radiation, including any potential degradation brought on by contact with photocatalysts. To make sure that these processes are safe for human consumption, the byproducts of catalysis should be examined.

REFERENCES

Ameta, R., Solanki, M. S., Benjamin, S., & Ameta, S. C. (2018). Photocatalysis. In *Advanced Oxidation Processes for Waste Water Treatment* (pp. 135–175). Elsevier. doi:10.1016/B978-0-12-810499-6.00006-1

Banerjee, S., Dionysiou, D. D., & Pillai, S. C. (2015). Self-cleaning applications of TiO2 by photo-induced hydrophilicity and photocatalysis. *Applied Catalysis B: Environmental, 176–177*, 396–428. doi:10.1016/j.apcatb.2015.03.058

Bashiri Rezaie, A., Montazer, M., & Mahmoudi Rad, M. (2018). Environmentally friendly low cost approach for nano copper oxide functionalization of cotton designed for antibacterial and photocatalytic applications. *Journal of Cleaner Production, 204*, 425–436. doi:10.1016/j.jclepro.2018.08.337

Boström, M., & Micheletti, M. (2016). Introducing the Sustainability Challenge of Textiles and Clothing. *Journal of Consumer Policy*, *39*(4), 367–375. doi:10.1007/s10603-016-9336-6

Chen, D., Cheng, Y., Zhou, N., Chen, P., Wang, Y., Li, K., Huo, S., Cheng, P., Peng, P., Zhang, R., Wang, L., Liu, H., Liu, Y., & Ruan, R. (2020a). Photocatalytic degradation of organic pollutants using TiO2-based photocatalysts: A review. *Journal of Cleaner Production*, *268*, 121725. doi:10.1016/j.jclepro.2020.121725

Chen, D., Cheng, Y., Zhou, N., Chen, P., Wang, Y., Li, K., Huo, S., Cheng, P., Peng, P., Zhang, R., Wang, L., Liu, H., Liu, Y., & Ruan, R. (2020b). Photocatalytic degradation of organic pollutants using TiO2-based photocatalysts: A review. *Journal of Cleaner Production*, *268*, 121725. doi:10.1016/j.jclepro.2020.121725

Cheng, D., He, M., Ran, J., Cai, G., Wu, J., & Wang, X. (2018). In situ reduction of TiO2 nanoparticles on cotton fabrics through polydopamine templates for photocatalysis and UV protection. *Cellulose (London, England)*, *25*(2), 1413–1424. doi:10.1007/s10570-017-1606-1

Feizpoor, S., & Habibi-Yangjeh, A. (2018). Ternary TiO2/Fe3O4/CoWO4 nanocomposites: Novel magnetic visible-light-driven photocatalysts with substantially enhanced activity through p-n heterojunction. *Journal of Colloid and Interface Science*, *524*, 325–336. doi:10.1016/j.jcis.2018.03.069 PMID:29656067

Guan, X., Lin, S., Lan, J., Shang, J., Li, W., Zhan, Y., Xiao, H., & Song, Q. (2019). Fabrication of Ag/AgCl/ZIF-8/TiO2 decorated cotton fabric as a highly efficient photocatalyst for degradation of organic dyes under visible light. *Cellulose (London, England)*, *26*(12), 7437–7450. doi:10.1007/s10570-019-02621-8

Guo, Q., Sun, H., Zhang, L., & Li, D. (2021). Cotton fabric-based rGO/BiVO4 recyclable photocatalytic nanocomposites for dye degradation under visible light. *Composites Communications*, *27*, 100846. doi:10.1016/j.coco.2021.100846

Han, J., Liu, Y., Tian, Z., Yu, F., Sun, X., & Wang, W. (2021). Microtubular carbonized cotton fiber modified g-C3N4 for the enhancement of visible-light-driven photocatalytic activity. *Materials Today. Communications*, *29*, 102926. doi:10.1016/j.mtcomm.2021.102926

Haug, H., & Koch, S. W. (2009). *Quantum Theory of the Optical and Electronic Properties of Semiconductors*. WORLD SCIENTIFIC., doi:10.1142/7184

Herrmann, J. (2017). Photocatalysis. In *Kirk-Othmer Encyclopedia of Chemical Technology* (pp. 1–44). Wiley. doi:10.1002/0471238961.1608152019051816.a01.pub3

Hosne Asif, A. K. M. A., & Hasan, M. Z. (2018). Application of Nanotechnology in Modern Textiles: A Review. *International Journal of Current Engineering and Technology*, *8*(02). doi:10.14741/ijcet/v.8.2.5

Hosseini-Hosseinabad, S. M., Minaeian, S., Tavakoli, A., Sabaei, M., Yousefi Zoshk, M., Laripour, R., Ramezani, S., Hoseini, M., & Chamanara, M. (2022). Development of g-C3N4/ZnO nanocomposite as a novel, highly effective and durable photocatalytic antibacterial coating for cotton fabric. *Ceramics International*. doi:10.1016/j.ceramint.2022.12.080

Ibănescu, M., Muşat, V., Textor, T., Badilita, V., & Mahltig, B. (2014). Photocatalytic and antimicrobial Ag/ZnO nanocomposites for functionalization of textile fabrics. *Journal of Alloys and Compounds*, *610*, 244–249. doi:10.1016/j.jallcom.2014.04.138

Karimi, L., Yazdanshenas, M. E., Khajavi, R., Rashidi, A., & Mirjalili, M. (2016). Functional finishing of cotton fabrics using graphene oxide nanosheets decorated with titanium dioxide nanoparticles. *Journal of the Textile Institute*, *107*(9), 1122–1134. doi:10.1080/00405000.2015.1093311

Komeily-Nia, Z., Montazer, M., Heidarian, P., & Nasri-Nasrabadi, B. (2019). Smart photoactive soft materials for environmental cleaning and energy production through incorporation of nanophotocatalyst on polymers and textiles. *Polymers for Advanced Technologies*, *30*(2), 235–253. doi:10.1002/pat.4480

Komeily-Nia, Z., Montazer, M., Heidarian, P., & Nasri-Nasrabadi, B. (2019). Smart photoactive soft materials for environmental cleaning and energy production through incorporation of nanophotocatalyst on polymers and textiles. *Polymers for Advanced Technologies*, *30*(2), 235–253. doi:10.1002/pat.4480

Kumar, A., Sharma, M., Amari, A., & Vaish, R. (2024). Sonocatalytic induced dye degradation and antibacterial performance of SrTiO3 nanoparticles embedded cotton fabric. *Environmental Research*, *240*, 117541. doi:10.1016/j.envres.2023.117541 PMID:37914006

Kumar, A., Sharma, M., & Vaish, R., & Ahmed, S. (2022). Poling effect on piezocatalytic antibacterial and dye degradation activities of BaTiO3 nanoparticles embedded cotton fabric. *Journal of Alloys and Compounds*, *168530*. doi:10.1016/j.jallcom.2022.168530

Kumar, A., Sharma, M., & Vaish, R. (2022a). Durable antibacterial cotton fabric via spray-coating of photocatalytic MoS2. *Materials Chemistry and Physics*, *126658*, 126658. doi:10.1016/j.matchemphys.2022.126658

Kumar, A., Sharma, M., & Vaish, R. (2022b). Screen printed calcium fluoride nanoparticles embedded antibacterial cotton fabric. *Materials Chemistry and Physics*, *288*, 126449. doi:10.1016/j.matchemphys.2022.126449

Kumar, A., Sharma, M., & Vaish, R. (2022c). BaTiO 3 Nanoparticles Embedded Antibacterial Cotton Fabric with UV Protection Characteristics. *Journal of Natural Fibers*, 1–14. doi:10.1080/15440478.2022.2139325

Kumar, A., Sharma, M., & Vaish, R. (2023). CaCu3Ti4O12 nanoparticle-loaded cotton fabric for dual photocatalytic antibacterial and dye degradation applications. *Environmental Science and Pollution Research International*, *30*(55), 117011–117021. Advance online publication. doi:10.1007/s11356-023-26835-3 PMID:37046162

Kuspanov, Z., Bakbolat, B., Baimenov, A., Issadykov, A., Yeleuov, M., & Daulbayev, C. (2023). Photocatalysts for a sustainable future: Innovations in large-scale environmental and energy applications. *The Science of the Total Environment*, *885*, 163914. doi:10.1016/j.scitotenv.2023.163914 PMID:37149164

Liu, S., Hu, Q., Qiu, J., Wang, F., Lin, W., Zhu, F., Wei, C., Zhou, N., & Ouyang, G. (2017). Enhanced Photocatalytic Degradation of Environmental Pollutants under Visible Irradiation by a Composite Coating. *Environmental Science & Technology*, *51*(9), 5137–5145. doi:10.1021/acs.est.7b00350 PMID:28379014

Marszewski, M., Cao, S., Yu, J., & Jaroniec, M. (2015). Semiconductor-based photocatalytic CO 2 conversion. *Materials Horizons*, *2*(3), 261–278. doi:10.1039/C4MH00176A

Mather, R., & Wilson, J. (2017). Fabrication of Photovoltaic Textiles. *Coatings*, *7*(5), 63. doi:10.3390/coatings7050063

Panda, S. K. B. C., Sen, K., & Mukhopadhyay, S. (2023). Sustainable Photocatalytic Desizing Process for the Starch-Based Size. *ACS Omega*, *8*(21), 18726–18734. doi:10.1021/acsomega.3c00713 PMID:37273639

Rastgoo, M., Montazer, M., Malek, R. M. A., Harifi, T., & Mahmoudi Rad, M. (2016). Ultrasound mediation for one-pot sonosynthesis and deposition of magnetite nanoparticles on cotton/polyester fabric as a novel magnetic, photocatalytic, sonocatalytic, antibacterial and antifungal textile. *Ultrasonics Sonochemistry*, *31*, 257–266. doi:10.1016/j.ultsonch.2016.01.008 PMID:26964948

Saleem, H., & Zaidi, S. (2020). Sustainable Use of Nanomaterials in Textiles and Their Environmental Impact. *Materials (Basel)*, *13*(22), 5134. doi:10.3390/ma13225134 PMID:33203051

Scholz, S., Kondakov, D., Lüssem, B., & Leo, K. (2015). Degradation Mechanisms and Reactions in Organic Light-Emitting Devices. *Chemical Reviews*, *115*(16), 8449–8503. doi:10.1021/cr400704v PMID:26230864

Singh, R., & Dutta, S. (2018). A review on H2 production through photocatalytic reactions using TiO2/TiO2-assisted catalysts. *Fuel*, *220*, 607–620. doi:10.1016/j.fuel.2018.02.068

Sivaram, N. M., Gopal, P. M., & Barik, D. (2019). Toxic Waste From Textile Industries. In *Energy from Toxic Organic Waste for Heat and Power Generation* (pp. 43–54). Elsevier. doi:10.1016/B978-0-08-102528-4.00004-3

Su, X., Chen, W., Han, Y., Wang, D., & Yao, J. (2021). In-situ synthesis of Cu2O on cotton fibers with antibacterial properties and reusable photocatalytic degradation of dyes. *Applied Surface Science*, *536*, 147945. doi:10.1016/j.apsusc.2020.147945 PMID:33012933

Sun, L., Fang, K., Chen, W., Liu, K., Zhu, J., & Zhang, C. (2022). Fabrication of a novel superhydrophobic cotton by HDTMS with TiO2 supported activated carbon nanocomposites for photocatalysis and oil/water separation. *Industrial Crops and Products*, *189*, 115836. doi:10.1016/j.indcrop.2022.115836

Tahir, M. B., Nawaz, T., Nabi, G., Sagir, M., Khan, M. I., & Malik, N. (2022). Role of nanophotocatalysts for the treatment of hazardous organic and inorganic pollutants in wastewater. *International Journal of Environmental Analytical Chemistry*, *102*(2), 491–515. doi:10.1080/03067319.2020.1723570

Valasek, J. (1921). Piezo-Electric and Allied Phenomena in Rochelle Salt. *Physical Review*, *17*(4), 475–481. doi:10.1103/PhysRev.17.475

Varadarajan, G., & Venkatachalam, P. (2016). Sustainable textile dyeing processes. *Environmental Chemistry Letters*, *14*(1), 113–122. doi:10.1007/s10311-015-0533-3

Wang, H., Meng, J., Li, F., & Li, T. (2023). Graphitic carbon nitride/ metal-organic framework composite functionalized cotton for efficient oil-water separation and dye degradation. *Journal of Cleaner Production*, *385*, 135758. doi:10.1016/j.jclepro.2022.135758

Wang, L., Wang, J., Tang, M., Wang, C., Gao, D., & Zhou, Y. (2023). Developing a Z-scheme Ag2CO3/ZIF-8 heterojunction for the surface decoration of cotton fabric toward repeatable photocatalytic dye degradation. *Applied Surface Science*, *610*, 155605. doi:10.1016/j.apsusc.2022.155605

Wang, M., Zhang, M., Zhang, M., Aizezi, M., Zhang, Y., Hu, J., & Wu, G. (2019). In-situ mineralized robust polysiloxane–Ag@ZnO on cotton for enhanced photocatalytic and antibacterial activities. *Carbohydrate Polymers*, *217*, 15–25. doi:10.1016/j.carbpol.2019.04.042 PMID:31079671

Wang, Y., Yang, L., Zou, R., Lan, J., Yao, A., Xiao, H., & Lin, S. (2022). Z-scheme CeO2/Ag/CdS heterojunctions functionalized cotton fibers as highly recyclable and efficient visible light-driven photocatalysts for the degradation of dyes. *Journal of Cleaner Production*, *380*, 135012. doi:10.1016/j.jclepro.2022.135012

Wu, Z., Fang, K., Chen, W., Zhao, Y., Xu, Y., & Zhang, C. (2021). Durable superhydrophobic and photocatalytic cotton modified by PDMS with TiO2 supported bamboo charcoal nanocomposites. *Industrial Crops and Products*, *171*, 113896. doi:10.1016/j.indcrop.2021.113896

Xu, Q., Xie, L., Diao, H., Li, F., Zhang, Y., Fu, F., & Liu, X. (2017). Antibacterial cotton fabric with enhanced durability prepared using silver nanoparticles and carboxymethyl chitosan. *Carbohydrate Polymers*, *177*, 187–193. doi:10.1016/j.carbpol.2017.08.129 PMID:28962757

Yetisen, A. K., Qu, H., Manbachi, A., Butt, H., Dokmeci, M. R., Hinestroza, J. P., Skorobogatiy, M., Khademhosseini, A., & Yun, S. H. (2016). Nanotechnology in Textiles. *ACS Nano*, *10*(3), 3042–3068. doi:10.1021/acsnano.5b08176 PMID:26918485

Zeng, M., Echols, I., Wang, P., Lei, S., Luo, J., Peng, B., He, L., Zhang, L., Huang, D., Mejia, C., Wang, L., Mannan, M. S., & Cheng, Z. (2018). Highly Biocompatible, Underwater Superhydrophilic and Multifunctional Biopolymer Membrane for Efficient Oil–Water Separation and Aqueous Pollutant Removal. *ACS Sustainable Chemistry & Engineering*, *6*(3), 3879–3887. doi:10.1021/acssuschemeng.7b04219

Zhu, J., Fang, K., Chen, W., Liu, K., Sun, L., & Zhang, C. (2022). Preparation of superhydrophobic, antibacterial and photocatalytic cotton by the synergistic effect of dual nanoparticles of rGO-TiO2/QAS-SiO2. *Industrial Crops and Products*, *189*, 115801. doi:10.1016/j.indcrop.2022.115801

KEY TERMS AND DEFINITIONS

In-situ: It is a technique in which several processes are carried out in sequence without exposing the sample to air.

Nanoparticles: These are the substances whose size is in between 10-500 nm.

Nano-photocatalyst: These are small semiconductor particles which are a few nanometers in at least one dimension.

Photocatalysis: The phenomenon of changing in rate of reaction due to presence of a substance in the presence of light is termned as photocatalysis.

Photocatalyst: These are the material which change the rate of chemical reaction when exposed to light.

Sustainable: This means using energy or resources in such a way that they are not harmful towards environment and can be used for a longer period of time.

Chapter 6
Characteristics, Applications, and Limitations of Nanocomposites in Biosensing

Jyoti Bhattacharjee
https://orcid.org/0009-0006-6835-2899
University of Calcutta, India

Debjani Bhakta
University of Calcutta, India

Subhasis Roy
https://orcid.org/0000-0003-4197-535X
University of Calcutta, India

ABSTRACT

To combat the exponentially increasing environmental challenges, there is a greater demand for eco-friendly and sustainable materials in various industries, including sensor technology. Green nanocomposites, which combine biodegradable polymers with nanoparticles obtained from renewable resources, offer a possible solution to the demand for effective biosensors and environmental sustainability. This chapter assesses current advances in developing green nanocomposites for skin cancer biosensors and energy storage devices such as lithium-ion batteries, supercapacitors, and fuel cells, emphasizing their prospective applications and performance enhancements. This chapter provides an in-depth examination of green nanocomposites as sustainable, efficient, and cost-effective biosensors, highlighting their manufacturing, unique features, and numerous sensing applications, and also explores the inclusion of nanomaterials such as cellulose nanofibers, chitosan, and graphene oxide into biopolymer matrices such as starch, polylactic acid, cellulose, proteins, and chitosan to build biosensors.

DOI: 10.4018/979-8-3693-1094-6.ch006

Copyright © 2024, IGI Global. Copying or distributing in print or electronic forms without written permission of IGI Global is prohibited.

INTRODUCTION

Biosensors are analytical instruments that combine a biological sensing element with a physical transducer to detect and quantify specific analytes. They have various uses in environmental monitoring, food safety, medical diagnostics, and other sectors. Nanocomposites are materials that mix two or more phases, one of which is nanoscale in size (Thakur et al., 2022; Mandal et al., 2023). They have distinct features that make them excellent for use in biosensor applications. Nanocomposites, for instance, can be engineered to have large surface areas, improving the sensitivity and selectivity of biosensors (Malhotra BD et al., 2018)

Furthermore, nanocomposites can be employed to construct biosensors that are more stable and robust than typical biosensors (Song et al.,2021). In recent years, there has been an increase in interest in producing sustainable nanocomposites for biosensor applications. In one review (Naresh et al., 2021), essential insights into the practical uses of these sustainable technologies by developing nanocomposite-based biosensors for detecting emerging environmental pollutants have been highlighted. Wearable biosensors have also been found to be used for sustainable nanocomposites. Graphene-based nanocomposites have been used to develop flexible, biodegradable sensors that measure physiological parameters such as glucose levels and pH (Bai et al., 2020). These sensors are not only environmentally beneficial, but they also provide real-time health monitoring.

This chapter delves into the fascinating world of sustainable nanocomposites as biosensors, providing light on their potential to transform environmental science. The search for sustainable biosensors has recently gained traction as researchers and scientists attempt to produce efficient, environmentally friendly, cost-effective alternatives (Navid Rabiee et al.,2021). In this effort, nanocomposites have emerged as a ray of hope, promising to revolutionize biosensor technology. The word "nanocomposite" refers to materials made up of nanoscale components, with one dimension typically less than 100 nanometers. These components, which are frequently nanoparticles, nanofibers, or nanosheets, are combined with a matrix material to form a composite structure that uses the unique features of both ingredients (Haleem et al.,2020). Using toxic chemicals and energy-intensive processes should be minimized in manufacturing nanocomposites. Green synthesis approaches, such as biological synthesis with microorganisms or plant extracts, can lower the environmental footprint (Dey et al., 2022, Arunadevi et al., 2018). Researchers created a biodegradable sensor for detecting heavy metal ions in water by combining cellulose nanofibers with quantum dots (D'souza et al., 2020). This biosensor has high sensitivity and degrades spontaneously, reducing its environmental impact. Biogenic synthesis uses microorganisms such as bacteria and fungi to create nanocomposites. These bacteria can bioaccumulate metal ions and intracellularly synthesize nanoparticles, providing a sustainable and biocompatible method for nanocomposite manufacture (Ba Yao et al., 2022). Waste materials can also be recycled or upcycled to create sustainable nanocomposites. For instance, incorporating waste-derived nanoparticles into composite materials can reduce environmental impact while adding value to discarded materials (Wang et al., 2020). Several challenges must be addressed before sustainable nanocomposites can be widely adopted for biosensor applications. For instance, it is important to develop methods for mass-producing sustainable approach nanocomposites with consistent quality (Araújo R et al., 2022). This chapter highlights the importance of developing methods for integrating different nanocomposites with other biosensor components, such as electrodes and signal amplifiers (Miao et al., 2022).

DIFFERENT NANOCOMPOSITES AS BIOSENSING MATERIALS

Graphene and Graphene Oxide (GO)

Graphene and graphene oxide (GO) are two-dimensional nanomaterials with distinct characteristics that make them promising biosensor materials. Graphene is a single layer of hexagonally organized carbon atoms. It is the most powerful and thin material known to science, with exceptional electrical and thermal conductivity. GO is a graphene derivative produced by oxidizing graphene. Due to the hydrophilic nature of GO, it is more compatible with biological systems than graphene (Al-Arjan et al., 2020). Graphene and GO can be employed in various methods to create biosensors. They can, for instance, be used to build scaffolds for immobilizing biological components, improve biological components' performance, or generate novel biosensors (Gupta et al., 2022, Dey et al., 2017).

Graphene and GO have extremely high surface areas, allowing for the immobilization of biomolecules such as antibodies, enzymes, and DNA probes. This improves their sensitivity and specificity in biosensing. The strong electrical conductivity of graphene enables the construction of electrochemical biosensors. Changes in electrical characteristics caused by biomolecule attachment can be translated into observable signals. Graphene and GO are both biocompatible materials, which reduces the potential of interference with biological samples in biosensing applications. The oxygen-containing functional groups in graphene oxide allow for simple chemical functionalization and customized surface changes for specific biosensing demands. The mechanical strength of graphene enables the development of robust biosensors that can resist a wide range of environmental conditions (Rajaman et al., 2018).

Applications:

Graphene-based glucose biosensors have been produced by immobilizing glucose oxidase (GOx) onto graphene surfaces. These biosensors provide rapid, sensitive, and selective glucose measurement, making them excellent tools for diabetes control (Vilian et al., 2017). Graphene and GO have been used in DNA sensors to identify certain DNA sequences. Functionalized graphene may adsorb single-stranded DNA, and the adsorption of complementary DNA sequences causes changes in electrical conductivity, allowing for sequence-specific detection. Graphene and GO have been functionalized with antibodies or aptamers to make protein biosensors. Biomarkers associated with diseases such as cancer, heart problems, and infectious infections can be detected by these biosensors. They are developing graphene-based biosensors for environmental monitoring to address pollution and resource management challenges (Wang et al., 2017). Graphene-based biosensors have been used to detect environmental pollutants and toxins. For instance, a graphene oxide-based sensor has been created to quickly detect heavy metals in water sources. Graphene-peptide-based sensors have been used to monitor neurotransmitters such as dopamine and serotonin in real time. These sensors could be used in neuroscience research and the development of neuromodulation devices.

Figure 1. Pictorial presentation of biosensors and components on a graphene platform (Peña et al., 2018)

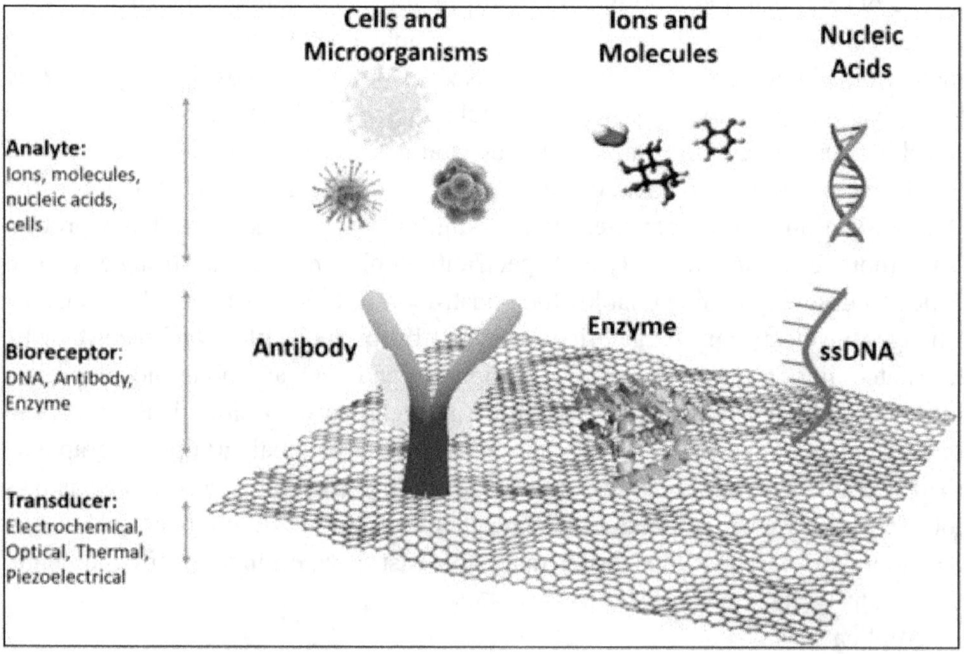

COVID-19 Diagnostics: Graphene oxide-based biosensors garnered interest during the COVID-19 pandemic for their promise in rapid and sensitive detection of the SARS-CoV-2 virus. These biosensors can provide immediate results (Beatriz et al., 2020).

As research continues, developing even more sensitive, selective, and accurate graphene and GO-based biosensors for various applications can be seen globally. Figure 1 portrays the incorporation of biological molecules like antibodies, enzymes, DNA, cells, and proteins in the detection scheme of graphene (Bai et al., 2020).

Graphitic Carbon Nitride (g-C_3N_4)

In recent years, developing novel biosensing materials has substantially improved our ability to detect and monitor biological molecules and chemical substances. Graphitic carbon nitride (g-C_3 N_4) is one such substance that has gotten a lot of attention. This one-of-a-kind two-dimensional nanocomposite has various features and advantages, making it an outstanding option for biosensing applications. In this comprehensive review, we delve into the characteristics of g-C_3N_4, its applications in biosensing and present fresh instances of its use in various biosensor platforms.

Graphitic carbon nitride, often known as g-C_3N_4, is a two-dimensional nanocomposite of carbon and nitrogen atoms. Its structure is similar to that of graphite, with alternating carbon and nitrogen layers joined together via covalent bonds. This one-of-a-kind configuration provides various advantages (Elgrishi N et al., 2018). The bandgap of g-C3N4 can be modified by changing the synthesis conditions, allowing for the customization of its electrical characteristics to meet specific biosensing requirements. g-C_3N_4 is biocompatible and non-toxic, making it appropriate for interacting with biological systems without causing harm. It has been used in electrochemical biosensors, where its high conductivity and surface area aid in immobilizing biomolecules such as enzymes, antibodies, and DNA probes. As a result, highly sensitive and selective biosensors for diverse analytes are created. A g-C_3N_4-modified electrode for glucose detection that combines biocompatibility with the catalytic activity of glucose oxidase for precise detection (Elshafie et al., 2020).

Applications:

Researchers have used the photoluminescent characteristics of g-C_3N_4 to develop luminescent biosensors. The functionalization of g-C_3N_4 with specific recognition elements enables biomolecule identification via luminescence intensity or wavelength variations. A g-C_3N_4-based luminous biosensor for detecting DNA hybridization events, where the binding of DNA strands causes a change in luminescence intensity. In Field effect transistor(FET) based biosensors, g-C_3N_4 can be the active material. The interaction of target analytes with the g-C_3N_4 surface causes observable changes in electrical conductivity, allowing for the sensitive detection of various compounds.

A g-C_3N_4 FET biosensor, for instance, can monitor protein-protein interactions, which is important for understanding disease processes and therapeutic development (Zou et al., 2020). g-C_3N_4 can be used as a SERS substrate, increasing the Raman signal of molecules adsorbed on its surface. This feature has been used to develop highly sensitive biosensors for various applications—a graphitic nitride SERS biosensor with improved sensitivity and specificity for detecting low-concentration biomarkers in complicated biological samples. Researchers created g-C_3N_4 quantum dots that can permeate cell membranes and detect intracellular pH changes. These nanosensors provide insights into biological processes and have potential applications in drug delivery and disease diagnosis (Vinoth et al., 2021). Portable and economical point-of-care diagnostic systems based on g-C3N4 biosensors are being investigated for resource-limited situations. These gadgets identify diseases such as malaria, HIV, and COVID-19 promptly and accurately. g-C_3N_4 is a promising new biosensing material with the potential to revolutionize the way we diagnose diseases and monitor our health (Pourmadadi et al., 2022).

M-Xene

MXenes are a type of two-dimensional (2D) nanomaterial that was discovered in 2011. They comprise a transition metal carbide or nitride and at least one surface termination group. MXenes contain several features that make them excellent for use in biosensors. MXenes have extremely large surface areas, allowing them to immobilize various biomolecules such as enzymes, antibodies, and DNA probes (Xie et al., 2023). MXenes have strong electrical conductivity, making them ideal for electrochemical biosensors. MXenes feature unique optical properties, such as fluorescence quenching, which can be utilized to construct fluorescent biosensors and are generally biocompatible, which means they can be employed safely in biological systems. This chapter M-Xene nanocomposites as biosensing materials, demonstrating their versatility and promise. M-Xene incorporation into polymer matrices can increase biosensor mechanical stability and biocompatibility. These nanocomposites have applications in implantable biosensors and wearable health monitoring devices.

M-xenes have been coupled with metal nanoparticles such as gold or silver to improve biosensors' plasmonic properties. These nanocomposites are very effective in surface-enhanced Raman spectroscopy (SERS) biosensing, where they may detect trace analytes. Due to their superior electrical characteristics, M-Xenes and carbon nanotubes work together to improve biosensor performance. These composites have been utilized with high sensitivity to identify neurotransmitters and cancer biomarkers (Yoon et al., 2020).

Applications:

Figure 2. Diagrammatic presentation of Mxene nanocomposite as biosensors (Riazi et al., 2021)

A M-Xene/graphene oxide nanocomposite was created for DNA detection. The vast surface area of nanocomposite enabled efficient DNA immobilization, resulting in a highly sensitive and selective biosensor for genetic analysis.

A glucose biosensor based on an M-Xene/polymer nanocomposite was developed for continuous glucose monitoring in diabetic patients. The biocompatible polymer matrix ensured long-term stability, while M-Xene gave great sensitivity (Bilal et al., 2023).

Using SERS, researchers developed an M-Xene/gold nanoparticle composite for detecting tiny levels of pesticides in environmental samples. This biosensor demonstrated outstanding sensitivity and selectivity (Guo et al., 2020). A biosensor integrating M-Xenes and carbon nanotubes was developed to monitor neurotransmitters in the brain. This nanocomposite enabled neuroscience research by allowing for the quick and accurate detection of neurochemicals (Rajarathinam T et al., 2023)

An immunosensor based on an M-Xene and MoS_2 composite was developed to detect cancer biomarkers. The synergistic qualities of the composite increased the detection limit, making it useful for early cancer diagnosis. Figure 2 presents emerging applications of MXene-based sensors that are critically explored, including moisture, motion, gas, humidity detection, and pressure distribution mapping. Potential uses of MXene-based sensors in the food industry for monitoring food materials and manufacturing plants are highlighted.

Bimetallic Nanomaterials

Bimetallic nanomaterials comprising two separate metallic elements have gotten much attention due to their unique electrical, catalytic, and optical capabilities. Because of these characteristics, bimetallic nanoparticles are good candidates for biosensing applications. This chapter explores the synthesis methods, characteristics, and applications of bimetallic nanoparticles in biosensors. Bimetallic nanostructures have such huge surface areas that they may immobilize a wide range of biomolecules, such as enzymes, antibodies, and DNA probes (Kannan et al., 2021. Bimetallic nanoparticles conduct electricity well; they are ideal for electrochemical biosensors. Bimetallic nanoparticles offer distinct optical properties, such as fluorescence quenching, that can be exploited to create fluorescent biosensors. Bimetallic nanoparticles are generally biocompatible, which means they can be used safely in biological systems. Bimetallic nanomaterials can exhibit synergistic effects when the two metals combine to produce qualities superior to the individual metals. A schematic diagram of the use of bimetallic nanomaterials in biosensing applications has been shown in Figure 3.

Applications:

Gold-palladium nanoparticles are used to develop biosensors for detecting glucose, cholesterol, and other biomarkers. A gold-palladium nanomaterial-based biosensor, for instance, has been created for non-invasive glucose monitoring in saliva. Gold-silver nanoparticles are used to develop biosensors for detecting DNA, cancer cells, and other biomolecules. For instance, a gold-silver nanomaterial-based

Figure 3. Diagrammatic presentation of bimetallic nanomaterials in biosensing *(Stephanie et al., 2021)*

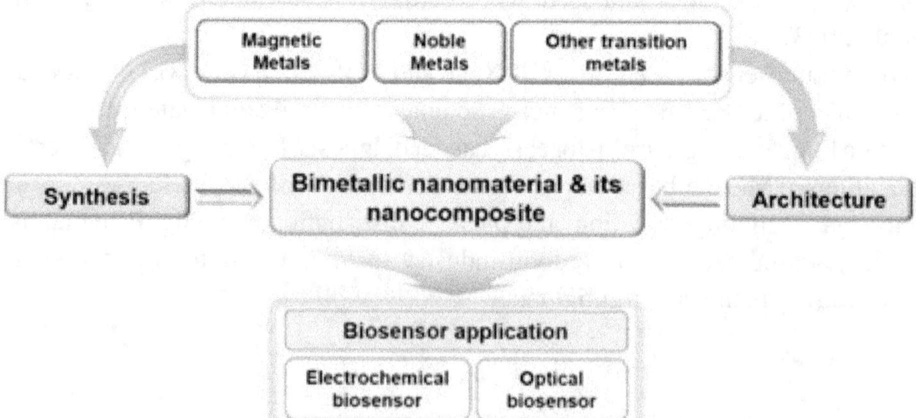

biosensor for detecting DNA from cancer cells has been developed (Zhaoxue et al., 2022). Gold-platinum nanoparticles are used to produce biosensors for detecting hydrogen peroxide, nitric oxide, and other chemicals. A gold-platinum nanomaterial-based biosensor, for instance, has been created to detect hydrogen peroxide in sweat (Patel V et al.,2020). Pt-Ir bimetallic nanoparticles have demonstrated outstanding electrocatalytic activity. They have been used in the construction of amperometric biosensors for the detection of neurotransmitters and glucose.

Pd-Pt bimetallic nanoparticles are recognized for their catalytic solid activity towards hydrogen peroxide, making them excellent for enzyme-based biosensors used in glucose monitoring and cancer biomarker detection (Rick et al., 2015). Due to its distinctive color change capabilities in the presence of specific analytes, Ag-Cu bimetallic nanoparticles have been used in colorimetric biosensors. While bimetallic nanoparticles are promising for biosensing applications, long-term stability, repeatability, and scalability must be addressed. Future research could concentrate on creating novel synthesis procedures, investigating new bimetallic combinations, and improving the incorporation of bimetallic nanomaterials into portable and point-of-care biosensing devices.

Chitosan

Chitosan is a chitin-derived natural biopolymer with various advantages as a biosensing material, including biocompatibility, ease of modification, and cost-effectiveness. This review delves into chitosan-based nanocomposites and shows how they can be used in biosensing. Chitosan is made by deacetylating chitin, a naturally abundant biopolymer found in the exoskeletons of crustaceans and insects. The chemical structure of chitosan is characterized by linear polymer chains comprised of linked D-glucosamine and N-acetyl-D-glucosamine units. This distinct structure confers biocompatibility, biodegradability, and film-forming ability to chitosan (Ye et al., 2016). The advantages of chitosan in biosensing applications have been schematically shown in Figure 4.

Applications:

Gold nanoparticles are widely used in biosensing due to their unique optical features. When mixed with chitosan, they combine to generate a stable nanocomposite that can be functionalized with biorecognition elements such as antibodies or DNA probes. These chitosan-AuNP nanocomposites have been used to detect various biomolecules, including proteins and DNA—chitosan-AuNP-based electrochemical immunosensors for detecting cancer biomarkers (Lee et al., 2010). This biosensor is based on immobilizing antibodies on a chitosan film. Antibodies are proteins that bind to specific antigens, such as proteins on the surface of foodborne pathogens.

Figure 4. Diagrammatic presentation of advantages of chitosan-based materials in biosensing applications
(Amirthalingam et al., 2021)

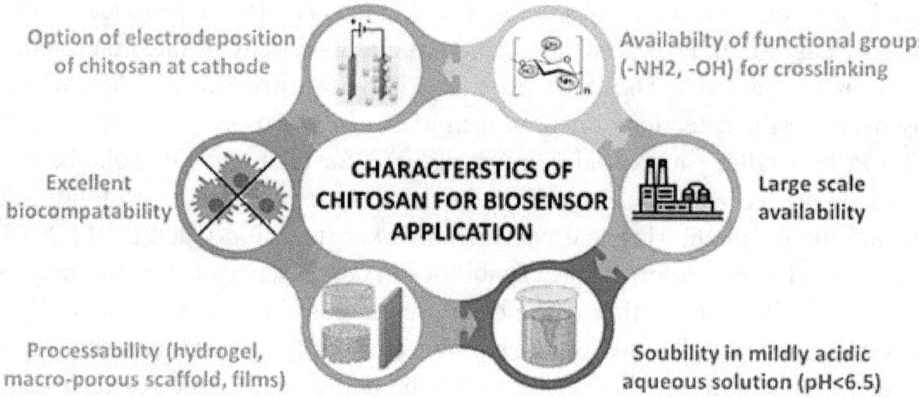

When a sample containing the foodborne pathogen is introduced to the biosensor, the antibodies bind to the foodborne pathogen. This binding event can then be detected using various techniques, such as electrochemical impedance or fluorescence spectroscopy. Chitosan-based foodborne pathogen biosensors are more sensitive and selective than typical biosensors. They are also more biocompatible and can be used to construct in situ foodborne pathogen biosensors. Carbon nanotubes have great surface area and strong electrical conductivity, making them ideal for improving biosensor performance. Chitosan can be used as a matrix to disseminate and immobilize carbon nanotubes (CNTs), increasing sensitivity and selectivity in biosensing applications (Tsujimura et al., 2014).

To generate multifunctional biosensors, chitosan can be mixed with various nanoparticles such as metal oxides (e.g., TiO_2, ZnO), magnetic nanoparticles, or silica nanoparticles (Roy et al., 2015). These hybrid nanocomposites improve performance, stability, and adaptability in biosensing applications. Chitosan can be used as a scaffold to immobilize enzymes, increasing their stability and activity. Biosensors that use enzyme-chitosan nanocomposites to detect specific substrates and metabolic products have been developed—for instance, chitosan-enzyme-based biosensors for glucose monitoring in diabetes management (Jiang et al., 2019).

Limitations:

The difficulty is that chitosan is not very soluble in water. This can make processing chitosan and creating chitosan-based biosensors problematic. Despite these obstacles, chitosan remains a potential biosensor material. With more research and development, chitosan-based biosensors can potentially revolutionize how we diagnose and monitor diseases.

Polymers and Biopolymers (Polypyrrole, Polyaniline)

Polypyrrole (PPy) is an electrically conducting polymer with high stability and biocompatibility. Given these characteristics, it is an excellent option for biosensing applications. Polymers and biopolymers are generally biocompatible, meaning they can be safely used in biological systems (Gourlay et al.,1978).

Polymers and biopolymers are chemically stable and can withstand various pH and temperature conditions. Electrochemical polymerization, chemical synthesis, and self-assembly can all be used to create PPy-based biosensors. Due to its simplicity and fine control over film thickness, electrochemical polymerization is one of the most extensively utilized processes (Moulahoum et al.,2022). Several practical applications of PPy-based biosensors show their adaptability and usefulness. For instance, PPy-based biosensors have been used for glucose monitoring in diabetes treatment, neurotransmitter detection in neurochemical investigations, and heavy metal environmental monitoring.

Polyaniline (PANI) is another conducting polymer with outstanding features such as electrical conductivity, simplicity of production, and environmental durability, making it ideal for biosensing applications. Chemical oxidation, electropolymerization, and mixing with other materials can all be used to create PANI-based biosensors (Lanzalaco et al., 2020).

Applications:

A chitosan-based glucose biosensor for implantable blood glucose detection has been created. The biosensor is built around glucose oxidase enzymes immobilized on a chitosan scaffold. Gluconic acid and hydrogen peroxide are produced when glucose oxidase enzymes catalyze glucose oxidation. An electrochemical sensor is then used to detect the hydrogen peroxide.

A silk fibroin-based DNA biosensor for cancer cell detection has been developed. The biosensor works by immobilizing DNA probes on a silk fibroin scaffold. The DNA probes complement the particular DNA sequence seen in cancer cells. Cancer cells bind to the DNA probes on the biosensor surface when present (Zhang et al., 2023). A PPy-based glucose biosensor for non-invasive glucose measurement in saliva has been created. The biosensor works by immobilizing glucose oxidase enzymes on a PPy scaffold. Gluconic acid and hydrogen peroxide are produced when glucose oxidase enzymes catalyze glucose oxidation. An electrochemical sensor is then used to detect the hydrogen peroxide.

A PANI-based DNA biosensor for detecting specific DNA sequences has been created. The biosensor works by immobilizing DNA probes on a PANI scaffold. The

DNA probes are complementary to the particular DNA sequence being identified. When the target DNA sequence is present, it binds to the DNA probes of the biosensor.

Carbon Nanofibers

Carbon nanofibers (CNFs) have emerged as a notable class of nanomaterials with diverse features that make them highly adaptable to various applications, including biosensing. CNFs are generally biocompatible, making them ideal for interacting with biological systems and minimizing adverse reactions in biosensing applications. CNFs are mechanically robust and can survive various environmental conditions, ensuring biosensor stability and endurance (Ayesha et al., 2022). Electrospinning enables the creation of tunable CNF-based nanofiber mats. These mats can be used in biosensors to immobilize biomolecules and detect analytes. Nanoporous membranes, for instance, can guide the growth of CNFs into well-defined shapes, allowing the production of ordered arrays for biosensing applications (Datsyuk et al., 2019).

Applications:

CNF-based biosensors have been used to identify diseases and biomarkers early, such as glucose for diabetes monitoring, cardiac biomarkers for heart disease diagnosis, and cancer markers for early cancer detection. For instance, a gold-based electrochemical biosensor was created for the ultrasensitive detection of Prostate-specific antigen (PSA), a biomarker for prostate cancer, providing a low-cost and quick diagnostic tool (Akbari Jonous et al., 2019). CNF-based biosensors detect environmental contaminants and poisons in water and air, such as heavy metals, pesticides, and pathogens. A CNF-based biosensor, for instance, was developed for the fast and selective detection of mercury ions (Hg^{2+}) in water, highlighting its potential for environmental monitoring and remediation (Ahmadian et al., 2021).

In neuroscience research, CNF-based biosensors monitor neurotransmitters such as dopamine and serotonin, providing insights into brain function and diseases. For instance, CNF-based microelectrode arrays have been used to track dopamine release in real-time, providing vital insight into neurodegenerative illnesses. CNF-based biosensors are integrated into wearable devices to measure physiological factors such as glucose levels, lactate levels, and pH. A CNF-based flexible biosensor, for instance, was incorporated into a wearable patch for non-invasive glucose monitoring, improving the quality of life for diabetes patients (Liu et al., 2014).

Molybdenum Sulphide

Molybdenum sulfide (MoS_2), a layered transition metal dichalcogenide, has lately emerged as a promising material for biosensing applications. This research examines the many elements of MoS2-based nanocomposites in biosensing, beginning with their distinguishing characteristics. MoS_2 is a two-dimensional structure with a high surface area composed of stacked layers of molybdenum atoms sandwiched between two sulfur atoms (Navid et al., 2022).

Due to their high surface area and strong electron transport capabilities, MoS2 is suitable for electrochemical biosensors such as glucose sensors, DNA sensors, and enzyme-based biosensors. MoS_2 nanocomposites can be used in surface plasmon resonance (SPR) and fluorescence-based biosensors to detect biomolecules in a label-free and extremely sensitive manner. MoS_2 FETs have shown potential in the ultrasensitive detection of proteins and nucleic acids, allowing for quick and real-time monitoring (Yu et al., 2022).

Graphene, a single sheet of carbon atoms, has been widely employed in biosensors due to its superior electrical conductivity and biocompatibility. Graphene-based biosensors have been used to detect a variety of analytes, including glucose, DNA, and heavy metals. QDs are semiconductor nanoparticles with distinct optical characteristics. Due to their tunable fluorescence and size-dependent electronic characteristics, they have been used in biosensors. QD-based biosensors can be used for imaging, immunoassays, and DNA detection (Mohammad et al., 2022). Gold and silver nanoparticles, for example, have been added into biosensor systems to improve sensitivity via localized surface plasmon resonance (LSPR) and catalytic effects. These nanoparticle-enhanced biosensors have been used to detect biomolecules, environmental contaminants, and infections. Figure 5 describes the prevailing and potential applications of MoS_2 nanostructures in optical, electrochemical, and electronic biosensors.

Metals

Metals have been frequently included in nanocomposite biosensors to improve sensitivity, selectivity, and stability due to their unique physical and chemical properties. Metals have a vast surface area, allowing them to immobilize various biomolecules, including enzymes, antibodies, and DNA probes.Metals are excellent electrical conductors, making them ideal for electrochemical biosensors. Gold has unique optical characteristics that can be exploited to construct optical biosensors. Metals are generally biocompatible, which means they can be employed safely in biological systems (Zhou et al., 2009).

Nanocomposites of zinc oxide (ZnO) have a wide bandgap and outstanding piezoelectric characteristics. They are employed in acoustic wave biosensors to detect sensitive biomolecules, such as DNA and proteins. Tin Oxide Nanocomposites have gas-sensing properties and are used in environmental monitoring to detect gases such as carbon dioxide, methane, and volatile organic compounds. Superparamagnetic iron oxide nanoparticles have been integrated into magnetic biosensors to quickly

Figure 5. Pictorial presentation of nanostructured MoS2-based advanced biosensors (Barua et al.,2017)

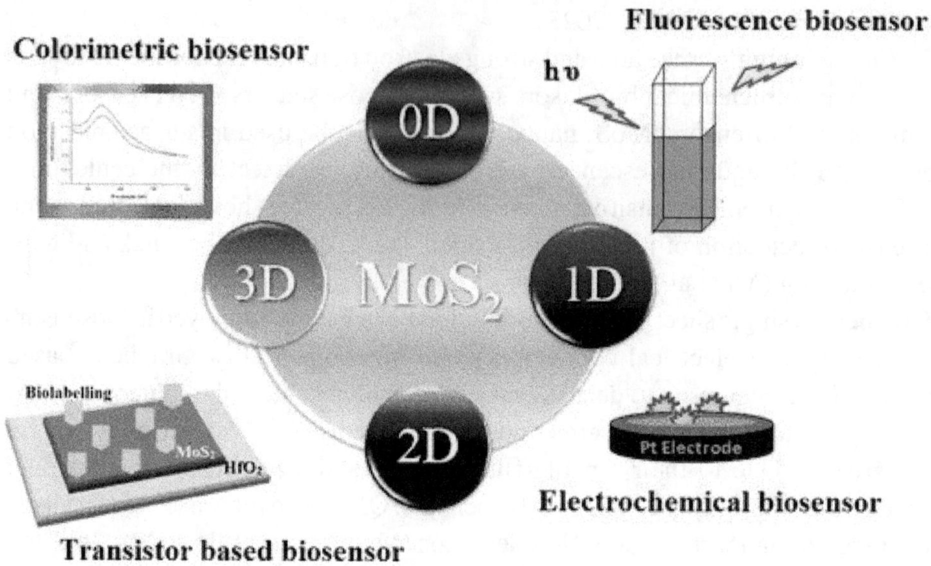

and effectively separate biomolecules and cells in applications such as immunoassays and DNA purification (Qurashi et al., 2015).

Applications

This biosensor immobilizes glucose oxidase enzymes with gold nanoparticles. Glucose oxidase enzymes convert glucose into hydrogen peroxide, which is then detected by an electrochemical sensor. This biosensor is very sensitive to glucose and selective for it (N Arunadevi et al., 2022)

This biosensor immobilizes DNA probes using silver nanoparticles. The hybridization of the DNA probes with the target DNA sequence is measured using an electrochemical sensor. This biosensor is highly sensitive and DNA selective. This biosensor is based on immobilizing antibodies with platinum nanoparticles. An electrochemical sensor detects the target protein after the antibodies bind to it. This biosensor detects highly sensitive and selective proteins (Salehabadi et al., 2018).

This biosensor uses a gold-coated electrode to immobilize antibodies. The antibodies bind to the target cells, which are subsequently detected with the help of an electrochemical sensor. This biosensor is highly sensitive and cell-selective.

Figure 6. Diagrammatic display highlighting the application areas of biosensors in different facets of food safety

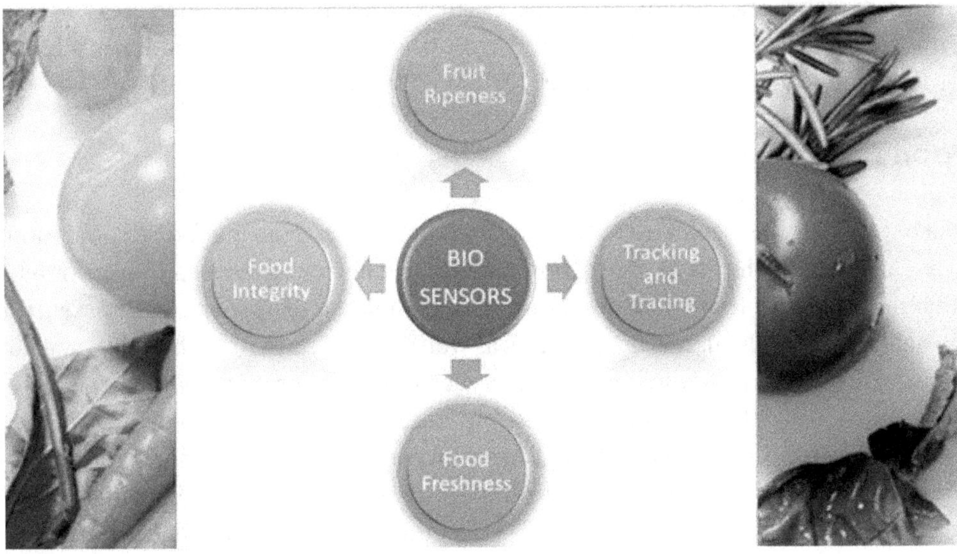

MAJOR GENERAL USES OF DIFFERENT BIOSENSORS

Food Packaging

In recent decades, the top consumer concern has been food poisoning. Although there are many ways (physical, chemical, and biological) for food to get contaminated throughout the year, infectious bacteria are the most dangerous. An organic polymer known as a biopolymer is created by covalently bonding monomers of the organic molecule. It is biodegradable, broken into smaller pieces in the soil by naturally occurring organisms, and leaves behind organic by-products like carbon dioxide and water that are safe for the ecosystem. To create bio-nanocomposites, biopolymers are combined with nanosized particles, such as carbon nanotubes, nitrocellulose, silver, zinc, copper, magnesium, and gold nanoparticles (Othman, 2014). Figure. 6 shows a clear overview of the different food safety sectors where these biosensors are commercially used.

A biosensor is an analytical tool that may transform input signals into a continuous output signal. The transducer, along with the receptor, makes up the biosensor. Receptors turn physical or chemical information into energy, which the transducer then converts into an electrical signal that can be used as an analytical signal. Clark and Lyons invented the biosensor in the 1960s. Due to their adequate microstructure, sensitivity, precision, stability, and fabrication costs, these biosensors are unsuitable for food packaging applications even if some have been utilized in environmental and biomedical applications. Some biosensors, including gas detection sensors, fluorescent and microfluidic sensors, electrochemical or imprinted biosensors, immunosensors, and thermal biosensors, have been described and tested specifically for food applications (Park, Kim, Lee, & Jang, 2015).

Food Freshness

The biosensor is chosen to track target metabolites created during food degradation to evaluate the freshness of foods. For instance, a calorimetric-based biosensor has been invented to follow meat and fish freshness via the identification of essential nitrogen compounds such as ammonium, as well as nitrogen oxide, which are created from the amination of ketones and aldehydes during microbial food metabolism (Pospiskova et al. 2013); a type of glucose biosensor that is improved with L-cysteine, as well as a nano-gold solution, had been invented to determine food freshness. These biosensors, which include calorimetric-based and glucose biosensors, may be considered and combined with food packaging components to create smart food packaging (Smiddy et al., 2002). By monitoring fundamental nitrogen constituents or levels of glucose as a freshness indicator, they can measure the freshness of meat, fish, and seafood. One of the primary markers created during adenine nucleotide breakdown in mammalian tissues is xanthanin. To assess the amounts of xanthanin in meat, a xanthine-immobilized xanthanin biosensor was also created. In this situation, the xanthan biosensor can be taken into consideration for the packaging of fish and meat to carry out the identification of xanthan during protein breakdown in meat or fish, so confirming the freshness of foods in packaged goods like ham, luncheon meat, red sausage, etc. (Devi, Yadav, Nehra, Yadav, & Pundir, 2013; Dolmacı, Çete, Arslan, & Yas¸ar, 2012).

Prevention of Food Contamination

The contamination of food is a global health hazard. Foods can become polluted in various ways, including physically by heavy metals like arsenic, mercury, cadmium, and lead, chemically by pesticides, toxins, or medication compounds (Hua, Yu, Liu, & Xianyu, 2021), and microbiologically by contaminants found in food (Kotsiri, Vidic, & Vantarakis, 2022). A colorimetric sensor was previously utilized to detect aflatoxin B1 chemicals, and lately, a fluorescence-based biosensor that employs DNA molecules to capture Hg^{+2} ions has been designed (Jokar, Safaralizadeh, Hadizadeh, Rahmani, & Kalani, 2016). To identify pesticides, a brand-new dual-responsive electrochemical/fluorescent biosensor was recently created (Chen et al., 2020). The potential for using these biosensors with intelligent packaging for food to track physical and chemical impurities in food exists. Food safety monitoring is essential for microbial contamination in foods. To detect microbial contamination in food, various biosensors have been developed.

For instance, one type of color-based sensor that is based on acoustic transduction that can detect color changes when in touch with bacteria is the optical biosensor. Salmonella groups B, D, E, E. coli, and E. coli 0157:H7 are among the microbiological pollutants that it has been reported to identify (Kuswandi et al., 2011). To identify the food mentioned above pathogens, this optical biosensor may find value in packaging for meals. Since it can identify contaminated bacteria like E. coli 0157:H7 by changing color when meat and fish are ruined, nano-sphere silica conjugated with a fluorescent dye has already been reported in previous studies. It may be appropriate for packing meat and dairy foods (Farooq et al., 2013). These electrochemical-based biosensors can detect the state of food potential of packaging for spoiling. It has been claimed that conducting polymer-based sensors can identify food spoilage gas generated during bacterial metabolism. Such conductive polymers are created by adding conducting nanoparticles to a matrix of polymers, where the resistance changes depending on how much gas escapes (Ahuja, Ahmad, & Kumar, 2007). These conducting-based biosensors may be employed with polymers for packaging to create intelligent sensors for food packaging.

Pesticide Detection and Environmental Remediation

Disposable amperometric enzymatic (acetylcholinesterase) biosensors have been suggested employing a cysteamine self-assembled monolayer on gold screen-printed electrodes for the identification of organophosphorous insecticides employing paraoxon as the prototype. The disposable biosensors displayed a linear spectrum of up to 40 ppb, a detection threshold of 2 ppb, and a sensitivity of 113 uA.mM. cm^2. Due to the exceptionally directed enzyme immobilization utilizing the self-assembled monolayer, the result of the analysis was reasonable. Recovery rates of $97 \pm 5\%$ ($n = 3$) were recorded after tests in river water samples spiked with 10 ppb of paraoxon, proving the efficiency of such enzymatic biosensors (Arduini, 2013). Additionally, the utilization of disposable screen-printed electrodes eliminates the need for laborious processes like reactivating the immobilized enzymes using pralidoxime iodide (PAM) and obidoxime solution or the use of a regenerative enzymatic membrane that would be required for subsequent use of the biosensors.

Employing hydrolase as well as an even nanocomposite made of magnetic Fe_3O_4 (average diameter of 120 nm) and gold nanoparticles (diameter 13 nm), a sensitive and selective enzymatic biosensor was used to measure the concentration of methyl parathion, another organophosphorus insecticide. The biggest benefit of utilizing hydrolase is that there is no poisoning from the organophosphorus insecticide, and the biosensor can potentially be reused for continuous testing (Zhao, 2013). Another enzyme-based biosensor (acetylcholinesterase) developed using amperometry has been suggested to detect methyl parathion in specimens of natural water taken from a polluted lake, with boundaries of detection between 1.5 and 1.8 ng mL^{-1}. This biosensor utilized a graphite working electrode and macroalgae (Cladophoropsis membranacea) to support functional immobilization (Nunes, 2014). The cost of the biosensor was decreased by using hydrolyzed macroalgae to form the immobilization matrix because it required less enzyme and substrate. On the other hand, it was using macroalgae as the immobilization support had a synergistic impact that increased the enzyme's stability and the biosensor's sensitivity (Nunes, 2014).

An electrolytic biosensor that uses a carbon paste electrode alongside reticulated hollow spherical forms of nickel cobaltite ($NiCo_2S_4$) material was developed, and it now has an increased limit of detection for methyl parathion (0.42 pg mL^{-1}) with a broad linear range(1.0 pgmL^{-1} to 10 ngmL^{-1}). The $NiCo_2S_4$ material's enhanced electron migration rate, which can produce abundant reaction active sites and strong electronic conductivity, may cause a lower detection limit (Peng, 2017).

Acetamiprid was found in actual environmental specimens, like water samples and fresh surface soil samples, using colorimetric and impedimetric apta-sensors, respectively. A greater linear range (50 fM to 10 μM) and lower limit of detection (17 fM) were reported using the impedimetric apta-sensor compared to the colorimetric apta-sensor, which had a linear range between 75 nM and 7.5 μM and the threshold for detection of 5 nM. The acetamiprid aptamer was supported at the electrode surface of that biosensor by a composite made of gold nanoparticles, multi-walled carbon nanotubes (MWCNT), and reduced graphene oxide nanoribbons. This composite may have increased the transfer of electrons and improved the analytical efficacy of the biosensor (Fei, 2015).

An apta-sensor constructed using tiny silver particles attached to nit-ogen-doped graphite oxide nanocomposite developed for identifying acetamiprid in sewage samples achieved a comparable identification limit (33 fM). The good electrical characteristics and substantial surface areas of nitrogen-doped graphene and silver nanoparticles, which exhibit more efficient electron transfer and high loading capacity than the nanomaterials alone, may cause the low detection limit. Additionally, the nanocomposite supports the immobilization of aptamers, which encourages amplifying their response signal and further enhances the susceptible detection of acetamiprid. By comparing the various outcomes the selectivity of the apta-sensor was also evaluated. The response signals to the other pesticides were insignificant, suggesting the strong selectivity of the APTA sensor to acetamiprid. Ecological samples, including crop specimens, brine or ocean water samples, and pesticide samples, were analyzed. A throwaway and label-free electrochemical immunosensor made from a field-effect transistor with single-walled carbon nanotubes

Figure 7. Pictorial presentation of biological and biomedical utilization of electrochemical biosensors
(Maduraiveeran, 2017)

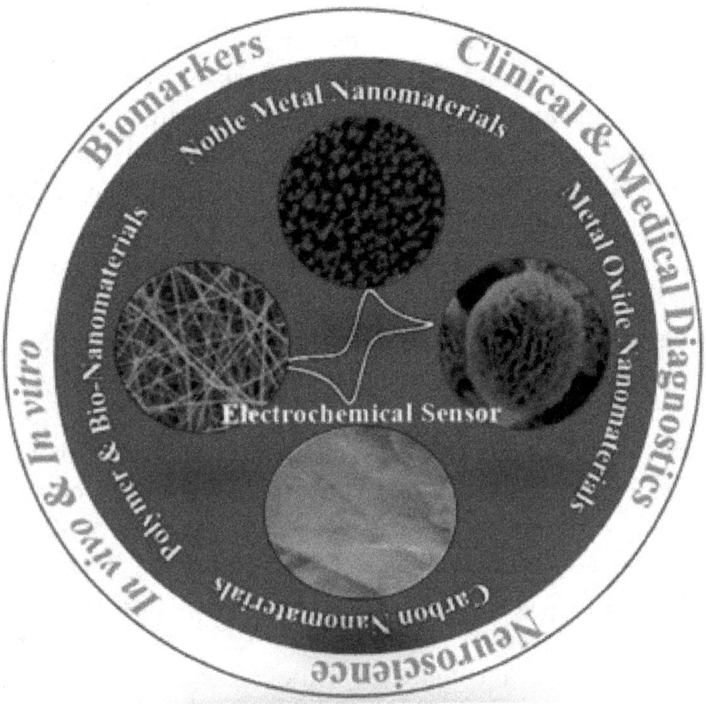

(SWCNT) and a label-free electrochemical (voltammetric) immunosensor based on gold nanoparticles were employed. The minimum level of detection for the FET immunosensor is 0.001 ng mL^{-1} as opposed to the voltammetric immunosensor's 0.016 ng mL^{-1}, and both values are lower than the permissible limits (0.1 ng mL^{-1}) (Madianos, 2018).

Clinical Analysis

The most significant nanomaterials used in biological and biomedical fields are depicted schematically in Figure. 7. Due to their remarkable affinity, effectiveness, and specificity, comparable to or even better than antibodies, aptamers-based biosensors have attracted a lot of attention.

Li et al. have come up with a protein biosensor employing a nanocomposite of CNTs, aptamer, and horseradish peroxidase (HRP) for thrombin that has a monitoring limit of 0.05 ppm (Li et al., 2015). For the detection of thrombin, the amperometric apta-sensor was recently developed using a variety of sensing substrates of nanocomposites, including aptamer-Au NPs-SWCNTs, ferrocene-aptamer-CNTs, aptamer-MWCNTs-ionic liquid-chitosan, alkaline phosphatase-aptamer-SWCNTs, etc. Shin et al. (2016) created the microfluidic aptamer-based electrochemical biosensor for detecting cardiac organoids. Aptamers are highly sensitive to the creatine kinase (CK)-MB (CK-MB) biomarker produced by defective cardiac tissue and were used to functionalize a gold-based microelectrode. This biosensor was used to detect minute amounts of CK-MB released by cardiac organoids in response to medication, which was in good agreement with the thrashing features and cell viability examinations.

The aptamer structure, which has recently been used to develop biosensors because of its excellent selectivity, affordable price, facile synthesizing, reuse ability, excellent affinity, and flexibility, is remarkably comparable to the configuration of the DNA triplex. One recent example is the melamine biosensor developed by Fu et al. on an indium tin oxide (ITO) electrode surface (Fu et al., 2016). With a broad functional span of 1.0 nM to 0.5 M, this sensor showed a low detection threshold of 0.43 nM.

By combining a ferrocene-labeled signal probe (Fc-sP) as well as a methylene blue-altered internal orientation probe (MB-rP) in a single hairpin-like DNA, Deng et al. showed a dual signal-tagged and hairpin framework DNA(dhDNA)-based ratiometric probe for the identification of Mucin 1 (MUC1) (Deng et al., 2017). Mucin 1 (MUC1) is regarded as a model tumor marker. This is important for preliminary diagnosis, differentiating evaluation, therapeutic consequence tracking, and ongoing inspections for patients with tumors or carcinomas. Electrochemical biosensors that utilize hemagglutinin (HA), which include nanosensors and immunosensors, were created for clinical purposes. A biosensor that utilizes neuraminidase (NA) enzyme activity evaluation for identifying the presence of the influenza A virus was developed (Anik et al., 2016).

To achieve excellent sensitivity and specificity, it has proven possible to use DNA-facilitating amplified sensors, enzyme-based amplifiers of signals, and nanomaterial-based signal enhancement. Devices are more suited for point-of-care (POC) diagnostics, DNA/enzyme amplification approaches, and new electroanalytical procedures using electrochemical immunosensors that utilize useful nanotechnology.

CONCLUSIVE REMARKS

Integrating nanocomposites into biosensors can enhance sensitivity, selectivity, stability, and response time, leading to more efficient and sustainable sensing platforms. In several industries, including sensor technology, there is an increased demand for sustainable and environmentally friendly materials to battle the environment's escalating problems. This chapter highlighted that effective biosensors and green nanocomposites, which blend biodegradable polymers with nanoparticles derived from renewable resources, may provide a solution. Chitosan, cellulose, starch, and proteins are biopolymers with unique qualities like biocompatibility, decomposition, and non-toxicity. It is possible to use green nanocomposites and biopolymers in various biosensing technologies, including enzyme, immunological, and DNA sensors. The creation of indirect electro-sensing techniques that ought to be competent to enhance the selectivity afforded by biological receptors or biomimetic receptors paired with nanocomposite materials still faces several challenges, including the reproducibility and mechanization of the remodeling procedures. However, few electrochemical sensor applications exist in extensive environmental and food contamination investigations. Creating novel materials and technologies, such as inexpensive paper-based sensors or perceptive nanoparticles, may make it easier to manufacture these products automatically and, as a result, use them regularly. Although the indirect detection of the aforementioned methodologies demonstrated potential, more work needs to be put into their technology transfer and deployment as analytical instruments to address health, ecological, and food-associated analytical issues. Future developments in biosensor technology include wearable technologies, embedded systems, a wide variety of analytes, microfluidics, and a greater focus on entire cell and tissue biosensors. Real-time on-the-spot testing for safety and efficacy evaluations in fields, homes, businesses, and marketplaces will soon be available to farmers and end users—even on smartphones and other portable electronic devices. Although certain specialized applications for which sensors are available on the market, they are insufficient to address the vast majority of current and growing issues in agriculture. Improvements in detection limit, particularity, sensitivity electrode design, matrix production, prices, and field-use viability are still highly desirable. By integrating sustainability into the design, synthesis, and application of nanocomposite-based biosensors, we can contribute to more environmentally friendly and efficient sensing technologies that align with a sustainable future.

ACKNOWLEDGMENT

The author (S. Roy) would like to acknowledge 'Scheme for Transformational and Advanced Research in Sciences (STARS)' (MoE-STARS/STARS-2/2023-0175) by the Ministry of Education for promoting translational India-centric research in sciences implemented and managed by Indian Institute of Science (IISc), Bangalore, for their support.

REFERENCES

Ahmadian, A., Shafiee, A., Aliahmad, N., & Agarwal, M. (2021). Overview of nano-fiber mats fabrication via electrospinning and morphology analysis. *Textiles*, *1*(2), 206–226. doi:10.3390/textiles1020010

Ahuja, T., Ahmad, I., & Kumar, D. (2007). Biomolecular immobilization on conducting polymers for biosensing applications. *Biomaterials*, *28*(5), 791–805. doi:10.1016/j.biomaterials.2006.09.046 PMID:17055573

Akbari Jonous, Z., Shayeh, J. S., Yazdian, F., Yadegari, A., Hashemi, M., & Omidi, M. (2019, February 7). An electrochemical biosensor for prostate cancer biomarker detection using graphene oxide-gold nanostructures. *Engineering in Life Sciences*, *19*(3), 206–216. doi:10.1002/elsc.201800093 PMID:32625003

Al-Arjan, W. S., Aslam Khan, M. U., Nazir, S., Abd Razak, S. I., & Abdul Kadir, M. R. (2020). Development of arabinoxylan-reinforced apple pectin/graphene oxide/nano-hydroxyapatite based nanocomposite scaffolds with controlled release of drug for bone tissue engineering: In-vitro evaluation of biocompatibility and cytotoxicity against MC3T3-E1. *Coatings*, *10*(11), 1120. doi:10.3390/coatings10111120

Alinejadian, N., Kollo, L., & Odnevall, I. (2022). Progress in additive manufacturing of MoS2-based structures for energy storage applications – A review. *Materials Science in Semiconductor Processing*, *139*, 106331. doi:10.1016/j.mssp.2021.106331

Amirthalingam, S., & Rangasamy, J. (2021). Chitosan-Based Biosensor Fabrication and Biosensing Applications. In R. Jayakumar & M. Prabaharan (Eds.), *Chitosan for Biomaterials III. Advances in Polymer Science* (Vol. 287). Springer., doi:10.1007/12_2021_85

Anik, U., Tepeli, Y., & Diouani, M. F. (2016). Fabrication of electrochemical model influenza A virus biosensor based on the measurements of neuraminidase enzyme activity. *Analytical Chemistry*, *88*(12), 6151–6153. doi:10.1021/acs.analchem.6b01720 PMID:27281347

Araújo, R. G., González-González, R. B., Martinez-Ruiz, M., Coronado-Apodaca, K. G., Reyes-Pardo, H., Morreeuw, Z. P., Oyervides-Muñoz, M. A., Sosa-Hernández, J. E., Barceló, D., Parra-Saldívar, R., & Iqbal, H. M. N. (2022). Expanding the Scope of Nanobiocatalysis and Nanosensing: Applications of Nanomaterial Constructs. *ACS Omega*, *7*(37), 32863–32876. doi:10.1021/acsomega.2c03155 PMID:36157779

Arduini, F., Guidone, S., Amine, A., Palleschi, G., & Moscone, D. (2013). Acetylcholinesterase biosensor based on self-assembled monolayer-modified gold-screen printed electrodes for organophosphorus insecticide detection. *Sensors and Actuators. B, Chemical*, *179*, 201–208. doi:10.1016/j.snb.2012.10.016

Arunadevi, N. (2022). Metal nanocomposites for advanced futuristic biosensing applications. *Materials Letters*, *309*, 131320. doi:10.1016/j.matlet.2021.131320

Bai, Y., Xu, T., & Zhang, X. (2020, January 3). Graphene-Based Biosensors for Detection of Biomarkers. *Micromachines*, *11*(1), 60. doi:10.3390/mi11010060 PMID:31947894

Barua, S., Dutta, H. S., Gogoi, S., Devi, R., & Khan, R. (2017). Nanostructured MoS2-Based Advanced Biosensors: A Review. In ACS Applied Nano Materials (Vol. 1, Issue 1, pp. 2–25). American Chemical Society (ACS). https://doi.org/doi:10.1021/acsanm.7b00157

Beatriz, F. M. R., Marcelo, M. S., Daniela, S. F., Devaney, R. D. C., & Glaucia, M. M. S. (2020). Graphene oxide-based nanomaterial interaction with human breast cancer cells. *Journal of Biomedical Materials Research. Part A*, *108*(4), 863–870. doi:10.1002/jbm.a.36864 PMID:31846174

Bilal, M., Singh, A. K., Iqbal, H. M., & Boczkaj, G. (2023). Enzyme-conjugated MXene nanocomposites for biocatalysis and biosensing. *Chemical Engineering Journal*, *474*, 145020. doi:10.1016/j.cej.2023.145020

Chen, H., Shao, S., Yu, Y., Huang, Y., Zhu, X., Zhang, S., Fan, J., Yin, G. Y., Chi, B., Wan, M., & Mao, C. (2020). A dual responsive biosensor for blood lead detection. *Analytica Chimica Acta*, *1093*, 131–141. doi:10.1016/j.aca.2019.09.062 PMID:31735206

D'Souza, A. A., Kumari, D., & Banerjee, R. (2017). *Nanocomposite biosensors for point-of-care—evaluation of food quality and safety*. Nanobiosensors., doi:10.1016/B978-0-12-804301-1.00015-1

Datsyuk, V., Trotsenko, S., Peikert, K., Höflich, K., Wedel, N., Allar, C., Sick, T., Deinhart, V., Reich, S., & Krcmar, W. (2019). Polystyrene nanofibers for nonwoven porous building insulation materials. *Engineering Reports*, *1*(2), e12037. doi:10.1002/eng2.12037

Deng, C., Pi, X., Qian, P., Chen, X., Wu, W., & Xiang, J. (2017). High-performance ratiometric electrochemical method based on the combination of signal probe and in Devi, R., Yadav, S., Nehra, R., Yadav, S., & Pundir, C. S. (2013). Electrochemical biosensor based on gold-coated iron nanoparticles/chitosan composite bound xanthine oxidase for detection of xanthine in fish meat. *Journal of Food Engineering*, *115*(2), 207–214. doi:10.1016/j.jfoodeng.2012.10.014

Deng, Z., Zhao, L., Zhou, H., Xu, X., & Zheng, W. (2022). Recent advances in electrochemical analysis of hydrogen peroxide towards in vivo detection. *Process Biochemistry (Barking, London, England)*, *115*, 57–69. doi:10.1016/j.procbio.2022.01.025

Dey, A., Abdul Moyez, Sk., Kumar Mandal, M., & Roy, S. (2016). Fabrication of solar cells using extracted biomolecules from tea leaves and hybrid perovskites. Materials Today: Proceedings. 3(10),part A, 3498-3504,. http://dx.doi.org/ doi:10.1016/j.matpr.2016.10.033

Dey, A., Das, B. C., Biswas, A. B., Biswas, P., Dhar, A., Roy, S., & Moyez Sk, A. (2017). Graphene Co-Doped TiO2 Nanocomposites for Photocatalysis and Photovoltaics Applications. *Indian Journal of Science and Technology*, *10*(31), 1–6. doi:10.17485/ijst/2017/v10i31/113878

Elgrishi, N., Rountree, K. J., McCarthy, B. D., Rountree, E. S., Eisenhart, T. T., & Dempsey, J. L. (2018). A Practical Beginner's Guide to Cyclic Voltammetry. *Journal of Chemical Education*, *95*(2), 197–206. doi:10.1021/acs.jchemed.7b00361

Elshafie, M., Younis, S. A., Serp, P., & Gad, E. A. M. (2020). Preparation characterization and non-isothermal decomposition kinetics of different carbon nitride sheets. *Egypt. J. Pet.*, *29*(1), 21–29. doi:10.1016/j.ejpe.2019.09.003

Ezhil Vilian, A. T. (2017). Krishnan Giribabu, Sang Rak Choe, Rethinasabapathy Muruganantham, Hoomin Lee, Changhyun Roh, Yun Suk Huh, Young-Kyu Han. A spick-and-span approach to the immobilization of horseradish peroxidase on Au nanospheres incorporated with a methionine/graphene biomatrix for the determination of endocrine disruptor bisphenol A. *Sensors and Actuators. B, Chemical*, *251*, 804–812. doi:10.1016/j.snb.2017.05.122

Farooq, A., Al-jowder, R., Narayanaswamy, R., Azzawi, M., Roche, P. J. R., & Whitehead, D. E. (2013). Gas detection using quenching fluorescence of dye-immobilized silica nanoparticles. *Sensors and Actuators. B, Chemical*, *183*, 230–238. doi:10.1016/j.snb.2013.03.058

Fei, A., Liu, Q., Huan, J., Qian, J., Dong, X., Qiu, B., Mao, H., & Wang, K. (2015). Label-free impedimetric apta-sensor for detection of femtomole level acetamiprid using gold nanoparticles decorated multiwalled carbon nanotube-reduced graphene oxide nanoribbon composites. *Biosensors & Bioelectronics*, *70*, 122–129. doi:10.1016/j.bios.2015.03.028 PMID:25797851

Fu, C., Liu, C., Li, Y., Guo, Y., Luo, F., Wang, P., Guo, L., Qiu, B., & Lin, Z. (2016). Homogeneous electrochemical biosensor for melamine based on DNA triplex structure and exonuclease III-assisted recycling amplification. *Analytical Chemistry*, *88*(20), 10176–10182. doi:10.1021/acs.analchem.6b02753 PMID:27665764

Ge, Y., Liu, P., Xu, L., Qu, M., Hao, W., Liang, H., Sheng, Y., Zhu, Y., & Wen, Y. (2022). A portable wireless intelligent electrochemical sensor based on layer-by-layer sandwiched nanohybrid for terbutaline in meat products. *Food Chemistry*, *371*, 131140. doi:10.1016/j.foodchem.2021.131140 PMID:34583185

Gourlay, S. J., Rice, R. M., Hegyeli, A. F., Wade, C. W. R., Dillon, J. G., Jaffe, H., & Kulkarni, R. K. (1978). Biocompatibility Testing of Polymers: In Vivo Implantation Studies. *Journal of Biomedical Materials Research*, *12*(2), 219–232. doi:10.1002/jbm.820120207 PMID:649628

Guo, Q., Zhang, X., Zhao, F., Song, Q., Su, G., Tan, Y., Tao, Q., Zhou, T., Yu, Y., Zhou, Z., & Lu, C. (2020). Protein-Inspired Self-Healable Ti3C2MXenes/Rubber-Based Supramolecular Elastomer for Intelligent Sensing. *ACS Nano*, *14*(3), 2788–2797. doi:10.1021/acsnano.9b09802 PMID:32045216

Gupta, V., Biswas, D., & Roy, S. (2022). A Comprehensive Review of Biodegradable Polymer-Based Films and Coatings and Their Food Packaging Applications. *Materials (Basel)*, *15*(17), 5899. doi:10.3390/ma15175899 PMID:36079280

Haleem, A., Javaid, M., Singh, R. P., Suman, R., & Rab, S. (2020). Biosensors applications in the medical field: A brief review. *Sensors International*, *2*, 100100. doi:10.1016/j.sintl.2021.100100

Hua, Z., Yu, T., Liu, D., & Xianyu, Y. (2021). Recent advances in gold nanoparticle-based biosensors for food safety detection. *Biosensors & Bioelectronics*, *179*, 113076. Advance online publication. doi:10.1016/j.bios.2021.113076 PMID:33601132

Jiang, Y., & Wu, J. (2019). Recent development in chitosan nanocomposites for surface-based biosensor applications. In ELECTROPHORESIS (Vol. 40, Issues 16–17, pp. 2084–2097). Wiley. https://doi.org/ doi:10.1002/elps.201900066

Jokar, M., Safaralizadeh, M. H., Hadizadeh, F., Rahmani, F., & Kalani, M. R. (2016). Design and evaluation of an APTA-nanosensor to detect Acetamiprid in vitro and silico. *Journal of Biomolecular Structure & Dynamics*, *34*(11), 2505–2517. doi:10.1080/07391102.2015.1123188 PMID:26609886

Kannan, P., & Maduraiveeran, G. (2021, December 31). Bimetallic Nanomaterials-Based Electrochemical Biosensor Platforms for Clinical Applications. *Micromachines*, *13*(1), 76. doi:10.3390/mi13010076 PMID:35056240

Kausar, A. (2022). Potential of Polymer/Fullerene Nanocomposites for Anticorrosion Applications in the Biomedical Field. *Journal of Composites Science*, *6*(12), 394. doi:10.3390/jcs6120394

Kotsiri, Z., Vidic, J., & Vantarakis, A. (2022). Applications of biosensors for bacteria and virus detection in food and water–A systematic review. *Journal of Environmental Sciences (China)*, *111*, 367–379. doi:10.1016/j.jes.2021.04.009 PMID:34949365

Kuswandi, B., Wicaksono, Y., Jayus, Abdullah, A., Heng, L. Y., & Ahmad, M. (2011). Smart packaging: Sensors for monitoring of food quality and safety. *Sensing and Instrumentation for Food Quality and Safety*, *5*(3–4), 137–146. doi:10.1007/s11694-011-9120-x

Lanzalaco, S., & Molina, B. G. (2020, May 24). Polymers and Plastics Modified Electrodes for Biosensors: A Review. *Molecules (Basel, Switzerland)*, *25*(10), 2446. doi:10.3390/molecules25102446 PMID:32456314

Lee, Y., Garcia, M. A., Frey Huls, N. A. F., & Sun, S. (2010). Synthetic Tuning of the Catalytic Properties of Au-Fe_3O_4 Nanoparticles. *Angewandte Chemie International Edition*, *49*(7), 1271–1274. doi:10.1002/anie.200906130 PMID:20077449

Li, J., Wang, J., Guo, X., Zheng, Q., Peng, J., Tang, H., & Yao, S. (2015). Carbon nanotubes labeled with aptamer and horseradish peroxidase as a probe for highly sensitive protein biosensing by post=electropolymerization of insoluble precipitates on electrodes. *Analytical Chemistry*, *87*(15), 7610–7617. doi:10.1021/acs.analchem.5b00640 PMID:26176595

Liu, D., Zhang, X., & You, T. (2014). Electrochemical Performance of Electrospun Free-Standing Nitrogen-Doped Carbon Nanofibers and Their Application for Glucose Biosensing ACS Appl. *ACS Applied Materials & Interfaces*, *6*(9), 6275–6280. doi:10.1021/am501713g PMID:24707890

Madianos, L., Tsekenis, G., Skotadis, E., Patsiouras, L., & Tsoukalas, D. (2018). A highly sensitive impedimetric apta-sensor for the selective detection of acetamiprid and atrazine based on microwires formed by platinum nanoparticles. *Biosensors & Bioelectronics*, *101*, 268–274. doi:10.1016/j.bios.2017.10.034 PMID:29096365

Maduraiveerana, G., Sasidharana, M., & Ganesan, V. (2017). *Electrochemical sensor and biosensor platforms based on advanced nanomaterials for biological and biomedical applications, Biosensors and Bioelectronics*. Elsevier., doi:10.1016/j.bios.2017.12.031

Malhotra, B. D., & Ali, M. A. (2018). *Nanomaterials in Biosensors: Fundamentals and Applications*. Nanomaterials for Biosensors., doi:10.1016/B978-0-323-44923-6.00001-7

Mandal, A., Maitra, S., Roy, S., Hazra, B., Ray, K., & Kargupta, K. (2023). Selective photo-reduction of CO2 to methanol using Cu-doped 1D-Bi2S3/rGO nanocomposites under visible light irradiation. New Journal of Chemistry. [RSC]. *New Journal of Chemistry*, *47*(3), 1422–1434. doi:10.1039/D2NJ03892G

Mehmandoust, M., Erk, N., Karaman, C., & Karaman, O. (2022). An electrochemical molecularly imprinted sensor based on CuBi2O4/rGO@MoS2 nanocomposite and its utilization for highly selective and sensitive for linagliptin assay. *Chemosphere*, *291*, 132807. doi:10.1016/j.chemosphere.2021.132807 PMID:34762887

Moulahoum, H., Ghorbanizamani, F., Guler Celik, E., & Timur, S. (2022). Nano-Scaled Materials and Polymer Integration in Biosensing Tools. *Biosensors (Basel)*, *12*(5), 301. doi:10.3390/bios12050301 PMID:35624602

Naresh, V., & Lee, N. (2021). A Review on Biosensors and Recent Development of Nanostructured Materials-Enabled Biosensors. *Sensors (Basel)*, *21*(4), 1109. doi:10.3390/s21041109 PMID:33562639

Nunes, G. S., Lins, J. A. P., Silva, F. G. S., Araujo, L. C., Silva, F. E. P. S., Mendonça, C. D., Badea, M., Hayat, A., & Marty, J. L. (2014). Design of a macroalgae amperometric biosensor; application to the rapid monitoring of organophosphate insecticides in an agroecosystem. *Chemosphere*, *111*, 623–630. doi:10.1016/j.chemosphere.2014.05.028 PMID:24997974

Othman, S. H. (2014). Bio-nanocomposite materials for food packaging applications: Types of biopolymer and nano-sized filler. *Agriculture and Agricultural Science Procedia*, *2*, 296–303. doi:10.1016/j.aaspro.2014.11.042

Park, Y. W., Kim, S. M., Lee, J. Y., & Jang, W. (2015). Application of biosensors in smart packaging. *Molecular & Cellular Toxicology*, *11*(3), 277–285. doi:10.1007/s13273-015-0027-1

Patel, V., Kruse, P., & Selvaganapathy, P. R. (2020). Solid State Sensors for Hydrogen Peroxide Detection. *Biosensors (Basel)*, *11*(1), 9. doi:10.3390/bios11010009 PMID:33375685

Peña-Bahamonde, J., Nguyen, H.N., Fanourakis, S.K. *et al.* Recent advances in graphene-based biosensor technology with applications in life sciences.

Peng, L., Dong, S., Wei, W., Yuan, X., & Huang, T. (2017). Synthesis of reticulated hollow spheres structure NiCo2S4 and its application in organophosphate pesticides biosensor. *Biosensors & Bioelectronics*, *92*, 563–569. doi:10.1016/j.bios.2016.10.059 PMID:27836591

Pospiskova, K., Safarik, I., Sebela, M., & Kuncova, G. (2013). Magnetic particle–based biosensor for biogenic amines using an optical oxygen sensor as a transducer. *Mikrochimica Acta*, *180*(3–4), 311–318. doi:10.1007/s00604-012-0932-0

Pourmadadi, M., Rajabzadeh-Khosroshahi, M., Saeidi Tabar, F., Ajalli, N., Samadi, A., Yazdani, M., Yazdian, F., Rahdar, A., & Díez-Pascual, A. M. (2022, October 26). Two-Dimensional Graphitic Carbon Nitride (g-C3N4) Nanosheets and Their Derivatives for Diagnosis and Detection Applications. *Journal of Functional Biomaterials*, *13*(4), 204. doi:10.3390/jfb13040204 PMID:36412845

Qin, M., Yuan, W., Zhang, X., Cheng, Y., Xu, M., Wei, Y., Chen, W., & Huang, D. (2022). Di Huang. Preparation of PAA/PAM/MXene/TA hydrogel with antioxidant, healable ability as a strain sensor. *Colloids and Surfaces. B, Biointerfaces*, *214*, 112482. doi:10.1016/j.colsurfb.2022.112482 PMID:35366577

Qurashi, A., Rather, J. A., Yamazaki, T., Sohail, M., De Wael, K., Merzougui, B., & Hakeem, A. S. (2015). Swift electrochemical detection of paraben an endocrine disruptor by In2O3 nanobricks. *Sensors and Actuators. B, Chemical*, *221*, 167–171. doi:10.1016/j.snb.2015.06.026

Rabiee, N., Khatami, M., Soufi, G. J., Fatahi, Y., Iravani, S., & Varma, R. S. (2021). Diatoms with Invaluable Applications in Nanotechnology, Biotechnology, and Biomedicine: Recent Advances. *ACS Biomaterials Science & Engineering*, *7*(7), 3053–3068. doi:10.1021/acsbiomaterials.1c00475 PMID:34152742

Rajangam, B., Daniel, D. K., & Krastanov, A. I. (2018). Progress in enzyme inhibition based detection of pesticides. *Engineering in Life Sciences*, *18*(1), 4–19. doi:10.1002/elsc.201700028 PMID:32624856

Rajarathinam, T., Kang, M., Hong, S., & Chang, S.-C. (2023). Nanocomposite-Based Electrochemical Sensors for Neurotransmitters Detection in Neurodegenerative Diseases. *Chemosensors (Basel, Switzerland)*, *11*(2), 103. doi:10.3390/chemosensors11020103

Riazi, H., Taghizadeh, G., & Soroush, M. (2021). MXene-Based Nanocomposite Sensors. In ACS Omega (Vol. 6, Issue 17, pp. 11103–11112). American Chemical Society (ACS). https://doi.org/ doi:10.1021/acsomega.0c05828

Rick, J., Tsai, M.-C., & Hwang, B. (2015). Biosensors Incorporating Bimetallic Nanoparticles. *Nanomaterials (Basel, Switzerland)*, *6*(1), 5. doi:10.3390/nano6010005 PMID:28344262

Roy, S., Han, G. S., Shin, H., Lee, J. W., Mun, J., Shin, H., & Jung, H. S. (2015). Low Temperature Synthesis of Rutile TiO2 Nanocrystals and Their Photovoltaic and Photocatalytic Properties. Journal of Nanoscience and Nanotechnology. *American Scientific Publishers.*, *6*, 4516–4521. doi:10.1166/jnn.2015.9778 PMID:26369074

Salehabadi, A., & Enhessari, M. (2018). Application of (mixed) metal oxides-based nanocomposites for biosensors. *Materials for Biomedical Engineering*, 357-396. https://doi.org/ doi:10.1016/B978-0-08-102814-8.00013-5

Smiddy, M., Papkovskaia, N., Papkovsky, D. B., & Kerry, J. P. (2002). Use of oxygen sensors for the non-destructive measurement of the oxygen content in a modified atmosphere and vacuum packs of cooked chicken patties; the impact of oxygen content on lipid oxidation. *Food Research International*, *35*(6), 577–584. doi:10.1016/S0963-9969(01)00160-0

Song, M., Lin, X., Peng, Z., Xu, S., Jin, L., Zheng, X., & Luo, H. (2021). Materials and Methods of Biosensor Interfaces With Stability. In Frontiers in Materials (Vol. 7). Frontiers Media SA. doi:10.3389/fmats.2020.583739

Stephanie, R., Kim, M. W., Kim, S. H., Kim, J. K., Park, C. Y., & Park, T. J. (2021, February). Recent advances in bimetallic nanomaterials and their nanocomposites for biosensing applications. *Trends in Analytical Chemistry*, *135*, 116159. doi:10.1016/j.trac.2020.116159

Thakur, M., Wang, B., & Verma, M. L. (2022). Development and applications of nano biosensors for sustainable agricultural and food industries: Recent developments, challenges, and perspectives. *Environmental Technology & Innovation*, *26*, 102371. doi:10.1016/j.eti.2022.102371

Tsujimura, S., Murata, K., & Akatsuka, W. (2014). Exceptionally High Glucose Current on a Hierarchically Structured Porous Carbon Electrode with "Wired" Flavin Adenine Dinucleotide-Dependent Glucose Dehydrogenase. *Journal of the American Chemical Society, 136*(41), 14432–14437. doi:10.1021/ja5053736 PMID:25244161

Vinoth, S., Shalini Devi, K., & Pandikumar, A. (2021). A comprehensive review on graphitic carbon nitride-based electrochemical and biosensors for environmental and healthcare applications. *Trends in Analytical Chemistry, 140*, 116274. doi:10.1016/j.trac.2021.116274

Wang, J., Shen, B., Lan, M., Kang, D., & Wu, C. (2020). Carbon nanotubes (CNTs) production from catalytic pyrolysis of waste plastics: The influence of catalyst and reaction pressure. *Catalysis Today, 351*, 50–57. doi:10.1016/j.cattod.2019.01.058

Wang, L., Zhang, Y., Wu, A., & Wei, G. (2017). Designed graphene-peptide nanocomposites for biosensor applications: A review. *Analytica Chimica Acta, 985*, 24–40. doi:10.1016/j.aca.2017.06.054 PMID:28864192

Xie, K., Wang, J., Xu, S., Hao, W., Zhao, L., Huang, L., & Wei, Z. (2023). Application of Two-Dimensional MXene materials in sensors. *Materials & Design, 228*, 111867. doi:10.1016/j.matdes.2023.111867

Yao, B., Yao, J., Fan, Z., Zhao, J., Zhang, K., & Huang, W. (2022). Rapid Advances of Versatile MXenes for Electrochemical Enzyme-Based Biosensors, Immunosensors, and Nucleic Acid-Based Biosensors. *ChemElectroChem, 9*(11), e202200103. Advance online publication. doi:10.1002/celc.202200103

Ye, J.-S., Hong, B.-D., Wu, Y.-S., Chen, H.-R., & Lee, C.-L. (2016). Heterostructured Palladium-Platinum Core-Shell Nanocubes for Use in a Nonenzymatic Amperometric Glucose Sensor Microchim. *Mikrochimica Acta, 183*(12), 3311–3320. doi:10.1007/s00604-016-1976-3

Yoon, J., Shin, M., Lim, J., Lee, J.-Y., & Choi, J.-W. (2020). Recent Advances in MXene Nanocomposite-Based Biosensors. *Biosensors (Basel), 10*(11), 185. doi:10.3390/bios10110185 PMID:33233574

Zhang, X., Tan, X., Wang, P., & Qin, J. (2023). Application of Polypyrrole-Based Electrochemical Biosensor for the Early Diagnosis of Colorectal Cancer. *Nanomaterials (Basel, Switzerland), 13*(4), 674. doi:10.3390/nano13040674 PMID:36839042

Zhao, Y., Zhang, W., Lin, Y., & Du, D. (2013). The vital function of Fe3O4@Au nanocomposites for hydrolase biosensor design and its application in detection of methyl parathion. *Nanoscale*, *5*(3), 1121–1126. doi:10.1039/c2nr33107a PMID:23280070

Zhou, X., Huang, X., Qi, X., Wu, S., Xue, C., Boey, F. Y. C., Yan, Q., Chen, P., & Zhang, H. (2009). In Situ Synthesis of Metal Nanoparticles on Single-Layer Graphene Oxide and Reduced Graphene Oxide Surfaces. *The Journal of Physical Chemistry. C, Nanomaterials and Interfaces*, *113*(25), 10842–10846. doi:10.1021/jp903821n

Zou, R., Teng, X., Lin, Y., & Lu, C. (2020). Graphitic carbon nitride-based nanocomposites electrochemiluminescence systems and their applications in biosensors. *Trends in Analytical Chemistry*, *132*, 116054. doi:10.1016/j.trac.2020.116054

Chapter 7
Nanocellulose Composites as Chemo-/Bio-Sensing Agents and Polymeric Matrixes:
A Nanotechnological Application

Paula Tatiana Peña González
Universidad Santo Tomás, Bucaramanga, Colombia

Inés Hernández Celi
Universidad Santo Tomás, Bucaramanga, Colombia

Franci Nathalie Gómez Jaimes
Universidad de Santander, Colombia

Carlos Andrés Martínez Bonilla
Universidad Santo Tomás, Bucaramanga, Colombia

ABSTRACT

This chapter highlights the nanocellose-based composite sensors as powerful tools in multiple sensing/detecting applications and polymeric matrixes. The worldwide need for rapid, precise/accurate, selective/specific detection positions the nanosensors as crucial devices for sensing purposes. The authors discuss the basic sensor conception, a basic architecture applied in different spatial scales and research areas, such as environmental sciences, medicine, and optoelectronics. Chemo- and biosensors are also described in the chapter. The use of nanocellulose as supporting or embedding material in sensors and polymeric matrixes gives the attributes of this noble material, which, combined with active and sensitive materials at the nanoscale, assemble new opportunities in the sensors field.

DOI: 10.4018/979-8-3693-1094-6.ch007

Copyright © 2024, IGI Global. Copying or distributing in print or electronic forms without written permission of IGI Global is prohibited.

INTRODUCTION

For decades, the use of petroleum-based products, wasteful consumption, and the evident environmental challenge has promoted the development of *sustainable alternatives*. All those alternatives keep in common characteristics such as renewability, biocompatibility, biodegradability, and environmental friendliness. In this sense, nature brings the possibility of a broad range of *sustainable alternatives* like chemical compounds that have been employed as support materials for diverse anthropic activities to improve living conditions or life quality. Among these compounds, we can find cellulose, a natural polymer with thousands of applications and outstanding properties that has maintained its position as an essential building block material in fields such as engineering, material science, electronics, energy, and, nowadays, nanotechnology.(Kim et al., 2015) Many articles, reviews, and book chapters have described in detail the properties, structure, and composition of materials obtained from higher plants, mainly composed of hemicellulose, lignin, and cellulose. (Tayeb et al., 2018) All those reports agree that cellulose is the most abundant biopolymer (polysaccharide, ~ 3000 glucose units) in nature, represents a large fragment of available carbon on Earth, is an essential component of the cell walls composed of β-1,4-glucan chains, confers mechanical properties to derived-plant fibers and plants itself, its structure is hierarchical with a helical arrangement of microfiber and nanofibers with two clearly defined non/crystalline domains as a result of the biosynthesis process. (Heinze, 2015) Additionally, it can be obtained from nature by physical, chemical, or biological methods, and it works as the source of materials for the conception of high-performance composite materials. (Nechyporchuk & Belgacem, 2016) In this field, these new composites appear due to nanocellulose *(NC)* properties and their possibilities, opening the door to developing engineering materials that improve current composites with the sustainable concept always in mind. (Kumar et al., 2022) The term NC is mainly referred to as nanocellulose nanocrystals (CNCs), cellulose nanofibrils (CNFs), and bacterial cellulose (BC), three kinds of materials with length or widths in the nanoscale (1-100 nm) which confer nanoscale advantages such as high-specific surface area, high aspect ratio, unique mechanical behavior, high strength, flexibility, and low density. (Singh & Kumar, 2017) This chapter intends to describe the conjugation of nanocellulose properties in chemical/biological sensing platforms and polymeric matrixes as an example of the nanotechnological scope of these composites. In order to understand the opportunities of nanosensors, we will focus on the basic conception of sensors, sensing agents, recent progress in nanocellulose-based sensors, chemosensing, and biosensing.

Figure 1. Schematic representation of sensor basic architecture

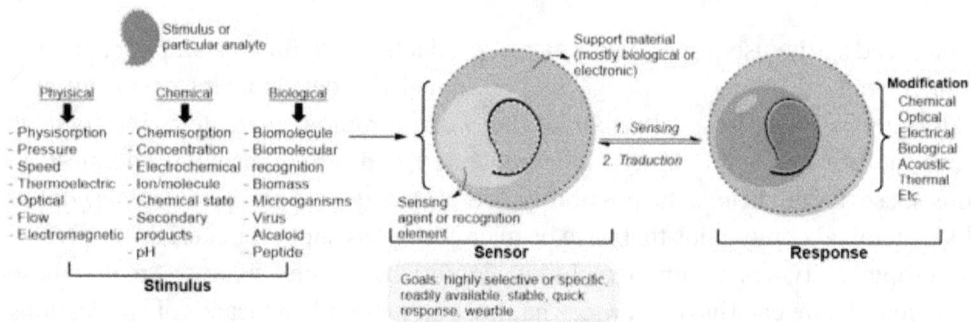

SENSORS: BASICS AND CONCEPTION

The term sensor (from Latin *sensorius*) is referred to perceive or feel and is part of the glossary commonly used in biotechnology and microelectronics. (Ohashi & Dai, 2006) In general, a sensor is a self-contained device that acts in response to a physical or chemical stimulus (including biological) such as concentration, light, pressure, magnetism, target species, or a particular analyte. It transmits a signal resulting from the interaction (direct or indirect). *Sensing* is the act of obtaining information by the sensor, a process that involves two steps: recognition and, in some cases, amplification, which brings insight into the parameter or analyte, whether semi-qualitative, qualitative, or quantitative. (C. W. Lee et al., 2019) A sensory system can be simplified into three essential components, a *stimulus* (exogenous or endogenous), either physical or chemical, a *sensor* accompanied by a detection element (sensing agent) on its sensory surface; and a *response* or traducción with its mechanism as a signal that brings information to the system to analyze, and finally decide (Figure 1). (Basabe-Desmonts et al., 2007)

Sensor's conception emerges from the ability to visualize analytes in real-time without complex equipment requirements. Nature is an excellent example of sensor diversity and development with a delicate and precise sensory system that responds to internal or external environmental stimuli. In nature, it is mandatory to be able to *sense* and adjust in real time to external conditions to ensure survival. For example, neuronal cells (mammalian mostly) are recognized as one of the most sophisticated and complex signaling and sensory processing systems in any living matter. Therefore, a nature-based approach allows deep learning from the sensing domains at the macro, micro, and nanoscale (Figure 2). Nowadays, many researchers transfer nature sensor technology to daily life devices in order to improve environmental, biological, or chemical sensors. (Liu et al., 2017)

Figure 2. Representative examples of biosensors in nature

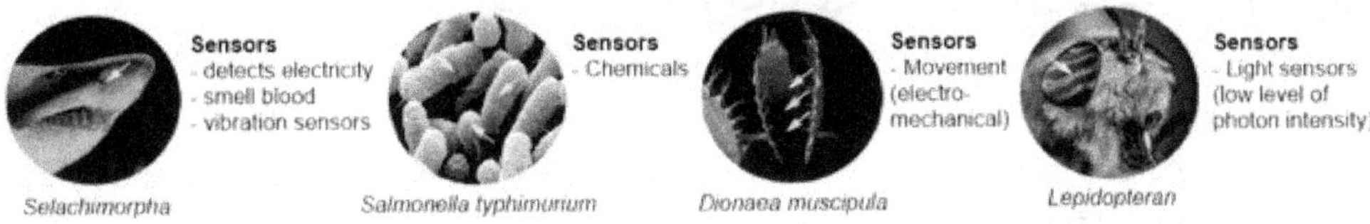

In a simplified scheme, sensors can be classified according to the nature of their sensing agent in chemosensors or biosensors. A small group corresponds to the physical sensors, devices able to measure physical properties (temperature, speed, amount of light, voltage, radiation, etc.). (dos Santos de Abreu et al., 2020) A chemosensor is mainly associated with agents or receptors from nonbiological chemical entities (in supramolecular chemistry). (Ikeda et al., 2010) According to this definition, molecular sensors and nanomaterial-based sensors may be classified as chemosensors if they include a device that traduces information from one kind of system to another. (Gooding & Gaus, 2016) In some cases, no physical distinction between the sensing agent and transducer can be made. If the transducer device is missing, such species should be considered advanced analytical reagents or synthetic receptors corresponding to synthetic or biomimetic materials.

On the other hand, a biosensor involves using sensing agents based on biological fragments or structures such as proteins, lipids (membrane components), peptides, oligonucleotides, nucleic acids, or RNA/DNA fragments. Biochemical o biological reactions as a signal of specific recognitions always accompany biosensors. (Anker et al., 2009) Typically, biosensors coupled with chemo/biosensors have emerged in the last 20 years as *nano*sensors by conjugating nanomaterials properties in sensing devices. A nanosensor is a submicroscopic assembly composed of nanomaterials, sensing agents, and support materials. In most cases, nanomaterials are both support material for bonded biological fragments in biosensors and sensing agents involved in the signaling process through superficial electronic interactions. (Bogue & Bogue, 2008) Specific organic molecules are also used as sensing agents in sensors. (Larson et al., 2012) However, nanomaterial-based sensors exhibit improved performances in terms of stability, photobleaching, aggregation tendency, surface functionalization, and mainly ultrasensitive detection. The following section will describe nanomaterials-based sensing agents, types, characteristics, and mechanisms.

SENSING AGENTS: OVERVIEW AND MECHANISMS IN NANOSENSORS

In sensors and their subclassifications, a sensing agent always works as a recognition element in the detection process. This element is vital in the sensing/measurement process because it responds to the analyte's presence directly or indirectly.(Rodrigues et al., 2017) We can generally find nanofibers, nanotubes, nanobelts, nanoprobes, nanowires, nanograins, thick films, and quantum dots as common recognition elements in nanosensors. (Javaid et al., 2021) According to their configuration and disposition, these recognition elements can be found in nanosensors at nanoscopic dimensions; however, sensors that perform measurements at the nanoscale could also be nanosensors, even if they do not necessarily have nanoscale dimensions. Additionally, nanosensors can be classified as active or passive sensors, requiring an energy source.

In the nanosensors field, the detection mechanism is as diverse as the type of nanoparticle involved in the recognition system. (Abdel-Karim et al., 2020) For accurate detection of ultra-low levels of analytes, nanoparticles need an ultra-high response, even for trace or single molecule detection. Most sensing mechanisms are often based on luminescence (fluorescence or phosphorescence) intensity changes. For quantum dots, fluorescence-quenching and enhancement effects account for the main sensing mechanisms. (Frigerio et al., 2012) Metallic nanoparticles made of gold and silver (rods, wires, stars, triangles, etc.) interact strongly with light waves, even if the light wavelength is larger than the particle. This type of interaction, "plasma oscillations" driven by external electromagnetic fields, leads to strong resonances at specific wavelengths finding application in sensing devices due to particle size, shape, and the dependence of the local dielectric environment. These plasmonic bands are tunable in metallic nanoparticles throughout the visible and near-infrared region. In that sense, the affectation of these specific bands in metallic nanoparticles corresponds to a detection mechanism in the sensor. Surface-enhanced Raman scattering (SERS) is an optical property in nanomaterials. (Vo-Dinh et al., 2015) This effect has been used to detect molecules such as environmental pollutants, pharmaceuticals, and biomolecules. SERS has become a vibrational technique to enhance (by orders of magnitude) the intensity of the Raman signatures of molecules (analyte) near metal nanoscale surfaces. In this case, SERS enhancement could be described by electromagnetic (EM) or chemical (CHEM) mechanisms. In EM, the incident local electric field increases once the analyte is absorbed due to visible light absorption by the metallic nanoparticle. In contrast, CHEM enhancement results from electronic resonance/charge transfer between the analyte molecule and metal surface, which leads to an increase in the polarizability of the molecule that can be detected as a response. (Kneipp et al., 2008)

Another mechanism in metallic nanoparticles in surface-enhanced fluorescence (SEF) relies on the nanoparticle's degradation -photoexcitation- to induce an altered response to analyte presence. (Guo et al., 2015) In SEF, the optimum fluorescence enhancement signal occurs when the analyte is 10–100 nm away from the metal surface, unlike the SERS mechanism, where proximity is needed. Such mechanisms produce an increase in quantum yield, lifetime decrease, increase in photo-stability, and energy transfer enriched. Aggregation-based colorimetric sensing is a strategy for molecular detection using metallic nanoparticles. In this case, the surface-charged nanoparticle negative surface is active for metal ions detection, resulting in aggregation changing their visible color. (Kozawa et al., 2020) In fact, these colorimetric appear as a combination of the nanoparticle solutions' absorption and scattering of light. For example, metal cations could be bound to the nanoparticle through the combination of nonspecific electrostatic interactions and complexation to the negatively charged groups on the nanoparticle surface, which accompanied aggregation. (Wang et al., 2018)

The hybrid nanoparticle is a new approach to take advantage of the physicochemical properties of two or more nanomaterials, tunning the response and behavior of their single components for parallel applications. This hybrid sensor opens the possibility of sensing and synchronously sequestering contaminants. (Tümay et al., 2021)

For non-metallics recognition agents, carbon nanoparticles arise as "less toxic" frameworks that act as platforms for tailored nanosensors. (Luo et al., 2020) Carbon nanotubes (CNT), carbon quantum dots (CQDs), fullerene, and nanographite are nanomaterials used in sensing applications. Optical, electrochemical, and SET mechanisms are involved in biosensors using CNTs. In this case, a biological molecule is attached covalently to the CNTs surface, and biorecognition is the mechanism accepted with their molecular target. (Qian et al., 2014)

NANOCELLULOSE-BASED CHEMOSENSING AND BIOSENSING: APPLICATIONS

Currently, sensors require a non-toxic, stable, and biocompatible supporting material. In this scenario, nanocellulose fibers have become a raw material applied effectively in many applications. (Nguyen et al., 2019) Nanocellulose properties will be exposed in Section 5.2.

Wearable devices, a new generation of sensors, use nanocellulose for its mechanical properties like high mechanical strength, toughness, and excellent flexibility. This new sensor can continuously monitor physiological signals associated with health conditions. (Wan et al., 2021) Additionally, they can sense external factors that could influence people's well-being in everyday life in real-time. In a profound development, researchers are using the nanocelluse-based sensor as "internal" devices, which means implanting the vascular as drug delivery platforms for drug release into the bloodstream. Despite remarkable results, the human body does not produce any cellulolytic enzyme, leaving the excretory system as the biological control to avoid stable nanoparticles in the internal body, which could cause inflammatory responses or tumors. (Li et al., 2022)

In nanocellulose-based chemosensors, multi stimuli materials have been developed, including modified nanocellulose modified with dual fluorescent for pH-sensing. In another example, pyrene-modified nanocellulose was demonstrated to be a selective chemosensor for Fe^{3+} and could differentiate between Fe^{3+} and Fe^{2+}. (Zhang et al., 2012) Spherical nanocellulose-based multifunctional material, a type of chemosensor, shows excellent results acting as a naked-eye sensor, rapid adsorbent, and antimicrobial agent. (Ram et al., 2018) Fluorescent quenching was the proposed mechanism in the detection of Cr^{6+} ions. In this example, the detection limits of Cr^{6+} were less than 30 ppb from 100 ppm, below maximal drinking water concentration. A regeneration process can be done using 0.1 N NaOH as a desorbing agent for chemosensor reuse. An optical plasmonic chemosensor made of copper nanoparticles and flexible nanocellulose film via *in situ* embedding was applied for the optical sensing of cyanide (CN^-), a high-impact mining pollutant. (Pouzesh et al., 2019) Lab-on-nanopaper, a differential approach, allows the visual sensing of human serum albumin (HAS) in human blood serums. This nanopaper was prepared by embedding curcumin within transparent bacterial cellulose. In this chemosensor, curcumin color changes due to the inhibitory effect of HSA during the molecule degradation in alkaline solutions. This change can be easily monitored visually (naked eye/smartphone camera) or spectroscopically. (Naghdi et al., 2019)

In nanocellulose-based biosensors, bonded biomolecules like peptides function as biorecognition agents to detect enzymes like destructive protease associated with wound dressing. (Bacakova et al., 2019) Nanocomposite made of Au nanoparticles and nanocellulose has been synthesized by a single-step technique in aqueous media. This biosensor displays exceptional biocompatibility and can trap an enzyme named horseradish peroxidase (HRP) and conserves its biological function. HRP enzyme biosensors can also detect hydrogen peroxide at low concentrations ($< 1 \mu M$). (Alle et al., 2022) A layer arrangement of nanocellulose fibers was employed to immobilize antibodies. At first, the layers have a carboxylated function and are further activated by covalent coupling with two specific proteins. Using a physical absorption mechanism, this activated layer architecture can trap an antibody recognized as antihuman immunoglobulin G. (38, 39)

Nanocellulose is also utilized in electrochemical biosensors to perceive DNA hybridization. (Aljabali et al., 2020; Shalauddin et al., 2019; Zaid et al., 2020) In genomics, hybridization is when two complementary single-stranded DNA and/or

RNA molecules bond to form a double-stranded molecule. It also plays an essential role in a procedure called a polymerase chain reaction, known as PCR, where we amplify specific gene regions, which is used in clinical testing. In this biosensor, DNA was embedded in nanocellulose, promoting the agglomeration in much larger sizes related to the original nanocellulose. (43,44) Tumor cells are always associated with folate receptors opening the opportunity for appropriate monitoring to avoid unwanted growth. In this field, a folic acid-embedded nanocellulose biosensor was designed to travel up to the folate receptor of the tumor cells of rat and human brain cells. Experiments show that the cellular binding of the fluorescein isothiocyanate on the folic acid-embedded nanocellulose by the folate receptor was much greater than the folic acid alone. (Do et al., 2021; Karimian et al., 2022)

CELLULOSE, NANOCELLULOSE, AND NANOCELLULOSE-BASED POLYMERIC MATRIXES

Since prehistory, man has found various applications for cellulosic materials, from the generation of fire through its ignition to the use of this material in the manufacture of clothing, where they were also used primarily in the various ancient scripts allowing the subsequent manufacture of paper, opening multiple research fields. In recent years, cellulose has gained interest in the industry due to its role as an innovative green chemistry material for a series of end-use applications. It has been used as a support for catalysts, matrices to capture molecular hosts, separation filters, thermal insulation materials, coating or reinforcing agents in flexible screens, filler in the polymeric matrix, and due to its various properties, in sensing applications.

Many publications related to cellulose as a natural polymeric matrix (~490,000), evidencing an exponential growth in the number of publications over the years (Figure 3). Of the pioneering publications in this field, those of the French chemist Anselme Payen stand out, through a series of investigations, determined the molecular formula of this polymer $[(C_6H_{10}O_5)n]$. (Fisher, 1989) Currently, most of the research on this topic is based on cellulose's production, isolation, and applications as support in polymeric matrices.

The development of numerous supporting materials from cellulose has generated interest in the industry due to the various advantages such as stability, biocompatibility, and high-water absorption. (A. Sharma et al., 2019) That is why with the growing development of nanotechnology and the manufacture of new materials, nanocellulose has acquired a significant interest due to its improved properties and the possibility of being used as a replacement for cellulose in the manufacture of reinforced composites for high-performance applications. (Ruiz et al., 2017; A. Sharma et al., 2019) Among the enhanced properties of this nanomaterial are the following:

Figure 3. Number of publications in cellulose, nanocellulose, and sensors (Web of science)

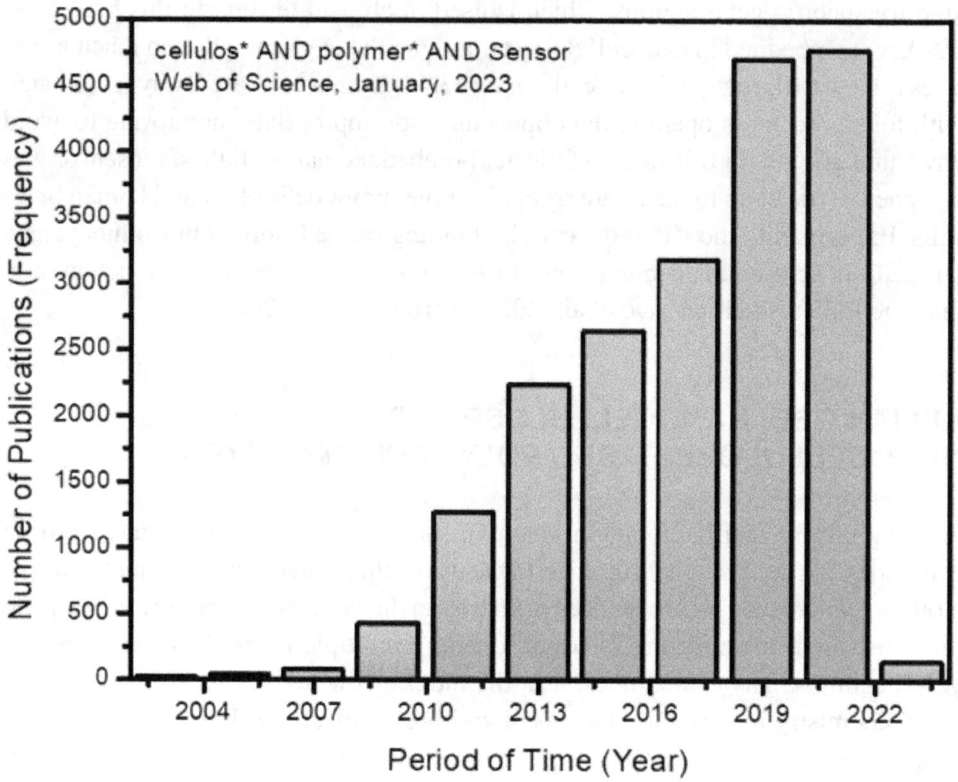

- Large surface area
- High rigidity, lightness, thermal and mechanical stability
- Optical transparency and reactive surface
- High porosity
- Biocompatibility and biodegradability

Two approaches can obtain nanocellulose: bottom-up (from the bottom up) through biological processes and top-down (from top to bottom) through chemical, physical or enzymatic processes (Figure 4). (Daud & Lee, 2017) According to the method and the raw material used in producing this nanomaterial, it is given a characteristic morphology and properties, which will determine its applications in industry. (Ruiz et al., 2017)

Due to the gained importance of the use of nanocellulose as a reinforcing agent, three polymeric matrices reinforced with this nanomaterial have been widely studied: natural matrices (for example, starch, cellulose, etc.), biopolymers (for example,

Figure 4. Methods for obtaining nanocellulose

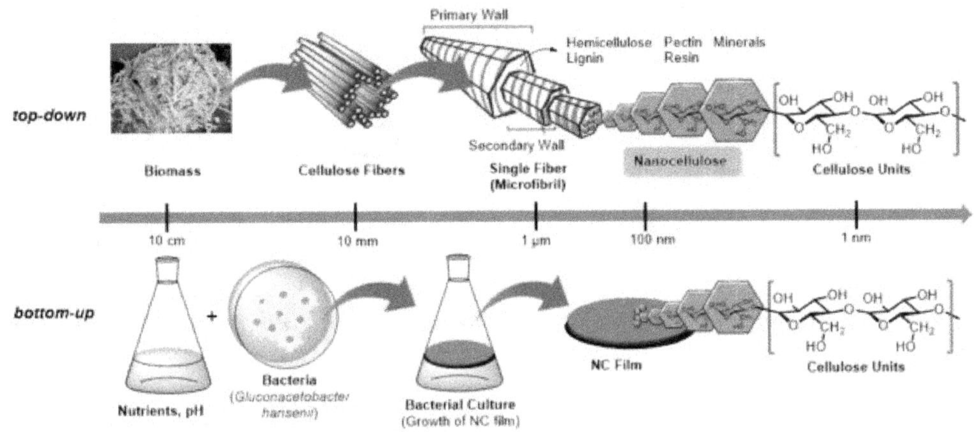

polylactic acid -PLA) and polycaprolactone (PCL). (Yan et al., 2016) Also, synthetic matrices (polypropylene, polystyrene, or polyethylene) have been explored, where this last reinforced matrix uses a hybrid approach arising from the adaptation of the properties of the renewable polymer, which is based on cellulose, and synthetic polymers; and due to the synergy of these two materials, unique intrinsic properties such as biodegradability, flexibility, and improved physical-mechanical properties are obtained. (K. Y. Lee & Bismarck, 2016; Thakur & Thakur, 2014; Zinge & Kandasubramanian, 2020)

Nanocellulose Structure

Cellulose has been classified as a homopolymer with a linear branching of glucose monomers linked by β–(1→4) glycosidic bonds (Figure 5), with a degree of polymerization variable between 10,000 to 15,000, depending on their source of origin. (Atalla et al., 2017; Mishra et al., 2018) Its structure comprises three hydroxyl groups, the primary in the C-6 position and the second in C-2 and C-3, which are hydrophilic. It has a syndiotactic configuration, where its linear structure causes intracellular unions by hydrogen bonds between the hydroxyl groups and the oxygen atoms of the nearby rings to balance the cellulose chains. (Hon, 1994; Mishra et al., 2018; Tshikovhi et al., 2020)

Cellulose fibers or crystals with at least one dimension at the nanoscale are known as nanocellulose and are obtained from various natural renewable sources such as plants, bacteria, and algae. (Mishra et al., 2018; Ramos et al., 2016) It can be classified into three main types: nanocrystalline cellulose, nanofibrillated cellulose, and bacterial nanocellulose (Figure 6); These three types have a similar chemical

Figure 5. Cellulose unit (a); steric configuration of cellulose (b)

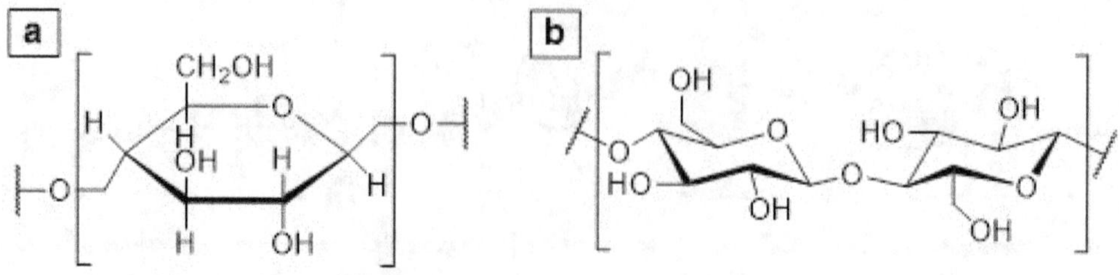

Figure 6. SEM images of (a) Cellulose nanofibers; (b) Cellulose nanocrystals; (c) Bacterial cellulose
Source: Tan et al. (2019)

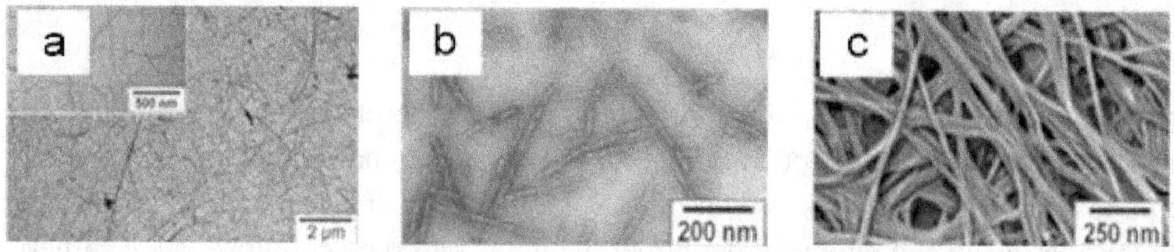

composition. However, they differ in their morphology, crystallinity, particle size, and some properties, which will confer their subsequent applications. (Phanthong et al., 2018)

Crystalline nanocellulose, known for its high strength, is extracted from cellulose fibers from sources such as plants, marine tunicates, and algae through a chemical process that involves acid hydrolysis in order to extract the crystalline region of cellulose, where 54 to 88% of its structure is made up of this region. (Phanthong et al., 2018; Zinge & Kandasubramanian, 2020) This nanocellulose has a short rod shape with a diameter between 2 to 20 nm and a length of 100 to 500 nm. (Phanthong et al., 2018) Due to its crystallinity, it is less flexible. Its surface contains a large volume of hydroxyl groups, displaying properties such as good transparency, high purity, high resistance, and ultrafine structure. (Zinge & Kandasubramanian, 2020)

Nanofibrillated cellulose is extracted from cellulose fibers by mechanical methods, and its source is plants, where the wood pulp is mainly found. (Phanthong et al., 2018; Zinge & Kandasubramanian, 2020) They are in the form of long fibers with a diameter between 1 and 100 nm and a length between 500 and 2,000 nm, where their fibers´ structure is composed of crystalline and amorphous regions. (Phanthong et al., 2018) This type of nanocellulose is characterized by being long and flexible due to the amorphous regions that contain it, where its complex structure of band-shaped fibers recognizes it and extended masses of nanofibers that have areas of anisotropic and reciprocal isotropic. (Phanthong et al., 2018; Zinge & Kandasubramanian, 2020)

Bacterial nanocellulose is not extracted from cellulose fibers; it is obtained from accumulating low molecular weight sugars produced by acetic acid type (AAB) bacteria such as *acetobacter*, *agrobacterium*, *alcaligenes*, *pseudomonas*, *rhizobium*, or *sarcina*. (Phanthong et al., 2018) The primary source being *acetobacter xylinum* (also known as *glucanoacetobacter xylinus*). (Zinge & Kandasubramanian, 2020) They are torsioned ribbons with a diameter between 20 to 100 nm and lengths given in micrometers. (Phanthong et al., 2018) This type of nanocellulose is characterized by its high purity because it does not have other components, such as lignin, hemicellulose, or pectin from cellulose fibers. It also has a high molecular weight, high water retention capacity (greater than 90%), high crystallinity, biocompatibility, and superior mechanical properties. (Zinge & Kandasubramanian, 2020)

The final properties and the morphology of the nanocellulose depend mainly on the crystalline and amorphous regions' content, which is additionally influenced by the extraction and treatment method used to break the amorphous regions of the cellulose (mechanical, chemical, physical, or enzymatic). Additionally, the length of the nanocellulose is also affected by the source of extraction; that extracted from vegetable cellulose has a length between 100 and 500 nm, while when it is obtained from marine tunicates and algae, its length is between 100 nm to several micrometers. (Ramos et al., 2016)

Nanocellulose Properties

In the last decade, nanocellulose has been a center of interest in research due to its promising properties and great potential in various applications. (Mondal, 2017) This nanomaterial combines the properties of cellulose with those acquired at the nanoscale, mainly enhanced by its large surface area. (Gopakumar et al., 2016) The main properties of nanocellulose are described in Figure 7.

Generally, the physical and mechanical properties of the various types of nanocellulose (Table 1) are affected by the type of raw material and the extraction process used; for example, the cellulose that is extracted from plant sources has fine fibers and a higher crystalline composition. (Thomas et al., 2020) Additionally, the pretreatment performed on the cellulosic material can also affect these properties, thereby promoting the accessibility of various hydroxyl groups, crystallinity, and internal surface area increase. (Farooq et al., 2020)

Due to the increase in research on nanocellulose, it has been shown that the mechanical properties and average modulus (~ 100 GPa) that make it up are better than those of the original lignocellulosic biomass because it has a more complex morphological structure with a greater crystalline component and absence of the

Figure 7. Properties of nanocellulose
Sources: Gopakumar et al. (2016); Loelovich (2016); Mondal (2017); Tshikovhi et al. (2020)

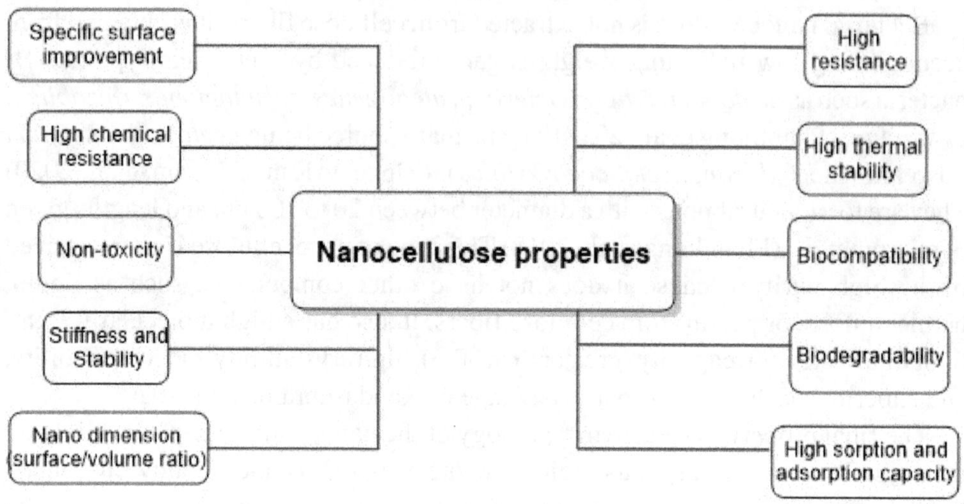

Table 1. Mechanical and physicochemical properties according to the type of nanocellulose

Mechanical Properties			
Property	*Crystalline Nanocellulose*	*Nanofibrated cellulose*	*Bacterial Nanocellulose*
Modulus axial (GPa)	140 - 160	30 - 40	80 - 110
Modulus transversal (GPa)	15 - 30	10 - 15	15 – 20
TS axial (GPa)	8 - 10	0.8 - 1	1.5 – 1.7
TS transversal (GPa)	c. 1	c. 0.1	c. 0.1
Physicochemical Properties			
Property	***Crystalline Nanocellulose***	***Nanofibrated cellulose***	***Bacterial Nanocellulose***
Sorption (%) [a]	10 - 12	17 - 20	10 – 12
Viscosity (mPa s) [b]	c. 10^3	c. 10^4	Gel
Conc. of gel-forming (%) [c]	5 - 7	2 - 3	< 1
Percolation threshold (vol. %)	1.5 – 1.7	0.7 - 1	0.3 – 0.7

[a]Maximum sorption of water from a saturated atmosphere
[b]viscosity of 1% aqueous dispersion at a shear rate 1 s^{-1}
[c]Minimum concentration of aqueous dispersion to form a stable gel
Source: (Ioelovich, 2016)

Table 2. Mean modulus for the different hierarchical cellulosic structures obtained from lignocellulosic biomass. (Puglia et al., 2016)

Structure	Component	Young's Modulus (GPa)
	Biomass	~20
	Single fiber network	40
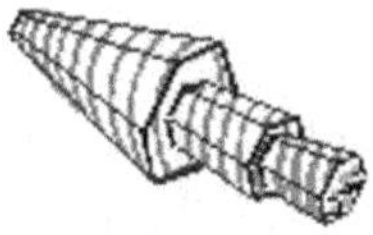	Microfibril	70
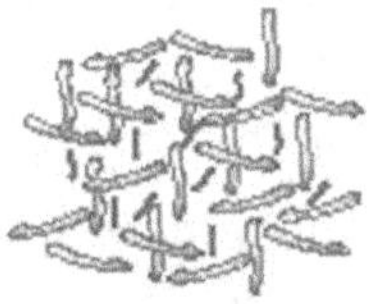	Cellulose nanocrystal	50 - 143
	Nanofibrillated cellulose	145 - 150

amorphous phase of cellulose. (Mondal, 2017; Puglia et al., 2016) Table 2 shows the average Young's Modulus of the various cellulosic structures obtained from lignocellulosic biomass.

Among other properties of nanocellulose, there are thermal stability and viscosity. The first depends on the flexible structure and the elimination of lignocellulosic materials with low thermal stability, such as hemicellulose and lignin. In the case of viscosity, this is determined by the rheological behavior acquired by the carboxyl functional group of the nanocellulose gel, where increasing shearing reduces the viscosity of the gels, and the increase in this is due to mainly to the increase in the specific area and the number of links by hydrogen bonds. (Mondal, 2017; Puglia et al., 2016)

The properties described above allow the use of nanocellulose in applications in the biomedical field, the energy sector, the environment, etc. (Mondal, 2017) Additionally, it has been used as a filler in the polymeric matrix because it improves the mechanical, physical, and thermophysical, and barrier properties of biopolymeric materials and films of, for example, carboxymethylcellulose (CMC), chitosan, agar, among others. (Farooq et al., 2020) Additionally, it has been shown that nanocomposites from nanocellulose are lighter than those from the pure polymer, that is, from cellulose, because they show an improvement in their rheological, thermal, mechanical, barrier, and biodegradation properties, allowing the use of this material in reinforcement applications (such as coating, nanocomposites, additives, etc.) with other synthetic or natural polymers. (Farooq et al., 2020; Ferreira et al., 2018)

Nanocellulose and Polymeric Matrix

Paper has been one of the most common applications of cellulosic materials dating back 5,000 years ago when the Egyptians invented papyrus by adhering to materials that had cellulose fibers. (Robles et al., 2017) Although the study of nanocellulose dates back to the 1960s, when its isolation was carried out using a top-down approach, it stagnated and only accelerated after 1999. (Tan et al., 2019) with investigations focused on the applications proposed mainly for their environmentally friendly nature, remarkable mechanical reinforcement properties, nanoporous structure, and specific surface area. (Gopakumar et al., 2016)

Currently, nanocellulose is used as a reinforcing agent in the polymeric matrix, a continuous phase of compounds used to hold a reinforcing agent in place, where its properties determine most of the degradation processes. (Zinge & Kandasubramanian, 2020) This matrix can be classified as thermosetting or thermoplastic, where thermosetting polymers are characterized by their dimensional/thermal stability (for example, epoxides, phenolics, and polyamides), and thermoplastic polymers are charactemarkedability to be molded, cast, and remolded without altering their physical properties (for example, polyesters, polyphenyl sulfides, etc.). (Saba et al., 2017)

Research and incorporation of nanocellulose into the polymer matrix are booming because it entails a series of positive aspects, such as an improvement in the material's biocompatibility, mechanical and thermal properties, lower cost, simple processability, recyclability, and easy accessibility. (Ferreira et al., 2018; Zinge & Kandasubramanian, 2020) Figure 8 summarizes the different polymer matrices used in coupling with nanocellulose to prepare various nanocomposites.

Non-biodegradable polymers have an essential role in improving the quality of human life, being used for more than 90 years in various applications and daily activities. These polymers are mainly characterized by their resistance to biological degradation after their elimination, which is why one objective of using

Figure 8. Biodegradable and non-biodegradable polymer matrices are used to form nanocellulose-based nanocomposites
Source: Ramos et al. (2016)

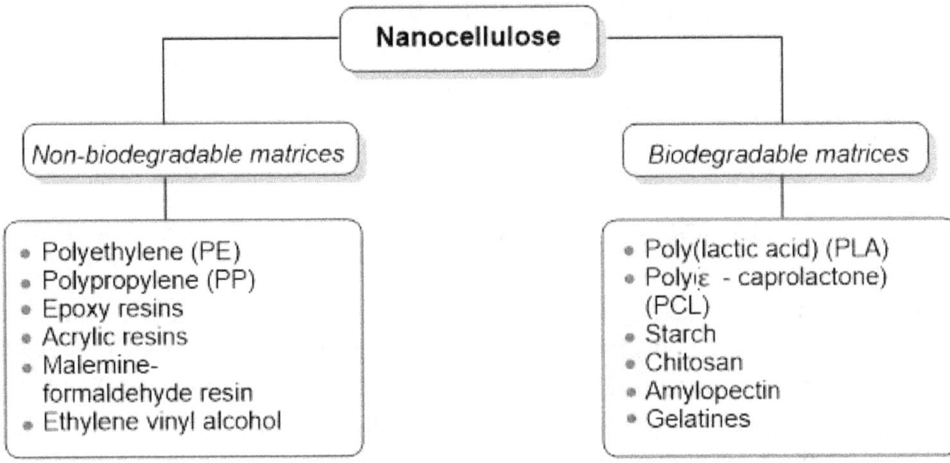

nanocellulose as a reinforcing material is the development of more environmentally friendly nanocomposite materials. (Ramos et al., 2016) It has been shown that nanocomposites based on nanocellulose as reinforcement have a higher Young's modulus and resistance to breakage, reporting increases of 60% and 32%, respectively, when compared with the pure synthetic polymer. Additionally, improvements have been reported in the thermal stability of these nanocomposites by adding up to 5% by weight of nanocellulose, similarly improving the barrier properties of the nanomaterial. (Ramos et al., 2016) Finally, nanocellulose has been found to improve the hardness and strength of these nanocomposites.

Despite an extensive list of advantages, the substitution of synthetic polymers for reinforced polymers with nanocellulose has a series of drawbacks, such as the low dispersion of the nanomaterial in the polymer matrix due to the presence of hydroxyl groups on the surface of the nanocellulose causing it to behave as an anisotropic material, being incompatible with nonpolar or hydrophobic polymeric matrices such as polylactic acid (PLA), polyethylene (PE) or polypropylene (PP). (Mondal, 2017; Ramos et al., 2016; Robles et al., 2017) However, methodologies have been studied to modify the structural characteristics of nanocellulose to promote its dispersion in the polymer matrix, improving the compatibility between both materials. These methodologies are mainly based on physical modifications (such as coating with surfactants or macromolecules) or chemical modifications (such as the derivation or grafting of polymers), which promote greater use of the properties obtained from the synergy between both materials. (H. Kargarzadeh et al., 2018)

Nanocellulose properties are comparable with other reinforcing materials when used as reinforcing material for the polymer matrix. For example, the tensile strength of nanocellulose crystals (~7.5 GPa) can be compared with the data reported for glass fibers and steel wires (being 4.8 and 4.1 GPa, respectively), evidencing in this way that nanocellulose is a promising material to be used as reinforcement in polymer matrices, providing improvements in their properties and reducing the environmental impact generated by synthetic polymers. (Ramos et al., 2016)

BACTERIAL NANOCELLULOSE

Today there is a vast demand for nanocellulose in the world market in industries such as textiles, food, and pharmaceuticals, among others, generating profits of USD 54.19 million in 2014 and USD 87.5 million in 2016, estimated for 2023 production of approximately USD 699.6 million with an annual growth rate of 33.8%; where additionally an annual production between 1010 and 1012 tons of cellulose is highlighted, with around 6x109 tons going to the textile and paper industries. (Farooq et al., 2020; Tan et al., 2019; Tshikovhi et al., 2020)

Specifically, bacterial nanocellulose was found value in the global market in 2016 at USD 207.36 million, forecasting for the year 2020 a reach of USD 497.76 million by its leading manufacturers such as Celluforce, University of Maine, US Forest Service, American Process, Seven Biotech Products Industry, Cellulose Lab, Bowil Biotech Ltd., etc. (Charreau et al., 2020; C. Sharma & Bhardwaj, 2019)

Several patents have arisen on incorporating nanocellulose in various compounds in the globalized world. Figure 9 shows the impact of nanocellulose on developing new materials and the research areas in which it has the most significant influence.

Figure 9. Percentage of patents according to the area in which nanocellulose is used (a), development of nanocellulose distributed in study areas (b)
Sources: Blanco et al. (2018) and Tan et al. (2019)

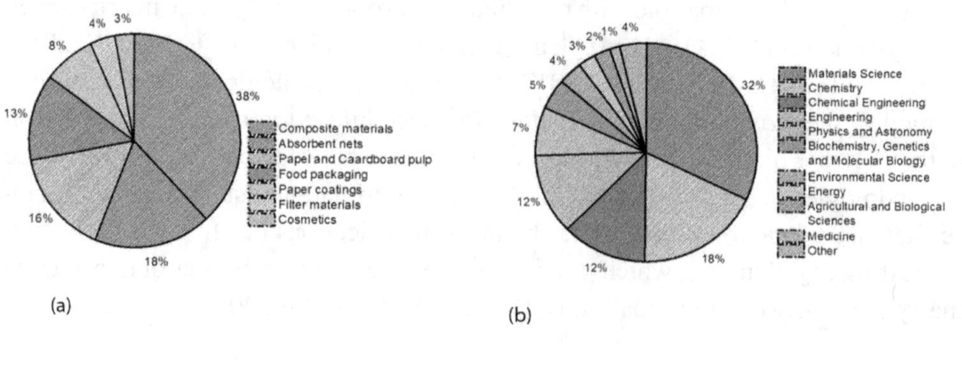

Nanocellulose is mainly considered a promising reinforcing agent in polymeric matrices, as mentioned above, due to its adaptability, flexibility, and processability. The use of bacterial nanocellulose as a support material demonstrated increases in the traction of the nanocomposites of 44% in the parallel measurement and 68% in the off-axis resistance; additionally, it increases the interfacial adhesion between the cellulose and the polymeric matrix, leading to an improvement in the mechanical properties and a reduction in the adsorption of water, it improves the thermodynamic stability, the crystallinity, and Young's modulus, increasing the latter to about 68%. (Blanco et al., 2018)

However, there are several disadvantages, such as moisture absorption, processing temperature limitation, the hydrophilic nature of the nanomaterial, and poor wettability, problems solved by various pretreatment techniques that allow obtaining materials reinforced with high-performance nanocellulose. (Blanco et al., 2018) These polymers reinforced by nanocellulose can be helpful as ecological materials in packaging, agriculture, membranes, etc., where nanocellulose provides biocompatibility, biodegradability, and increased tensile strength. (Mondal, 2017; Ramos et al., 2016)

Added to these last drawbacks, it is currently not recognized whether bacterial nanocellulose can be used to replace or compete with other traditional sources of cellulose since research on its production in commercial uses and fermentation bioprocesses has yet to be established studied enough (Sharma, C. & Bhardwaj, 2019). The main problem with the production of this polymer on an industrial scale is that it is neither efficient nor profitable because the production cost is high and it takes long periods (Jedrzejczak-Krzepkowska et al., 2016; C. Sharma & Bhardwaj, 2019)

Nanocellulose Production, Modification, And Applications

The technology and methods used to produce bacterial nanocellulose differ significantly from those used to produce crystalline nanocellulose and cellulose nanofibers. (Blanco et al., 2018) Commonly this production is given by Gram-negative bacteria of the *glucanacetobacter xylinus* type, previously known as *acetobacter xylinum*, in a process composed mainly of 4 stages:

1) Activation of monosaccharides through the development of sugar nucleotides.
2) Polymerization that leads to the assembly of sugar units repeatedly.
3) Addition of acyl groups.
4) Excretion of bacterial nanocellulose from the cell's cytoplasm through the cell membrane.

Table 3. Bacterial nanocellulose production methods (Blanco et al., 2018; C. Sharma & Bhardwaj, 2019)

Production Method	Advantages	Disadvantages
Static culture	Simple process No need for complex instruments Low energy costs. Easy implementation in production on site	Laborious Extensive culture times: 5 to 20 days Fermentation cannot be monitored or controlled It is not possible to apply it in large-scale production
Batch static, intermittent feeding technology	Simple process Improved production compared to static crops It can be used for large-scale production	Fermentation cannot be monitored Cellulose forms as a film and sometimes as a suspension of crosslinked cellulose
Cell-free extract technology	Simple process. It is applicable for large-scale production in a short time Increased production output	It is not possible to control the fermentation parameters
Agitated culture: agitator/ impeller	Profitable. Applies to large-scale production Overcomes some limitations of static cultures, such as diffusion, control capacity, and expansion	The cellulose takes the shape of an irregular sphere Shearing results in crop mutation, which reduces productivity There are instabilities in the crop evidenced in the loss of productivity
Bioreactor-based production: rotating disc reactor, airlift reactor, biofilm reactor with plastic composite supports, drip bed reactor	High productivity Less labor required Short production times depending on the reactor Easy scaling	Generally, there are no disadvantages in the production of bioreactors. If the right conditions are maintained, high production of bacterial nanocellulose can be reached

In the static culture of *glucanoacteboacter xylinus*, the cellulose fibers that form from the glucose units' polymerization accumulate to create a surface mat at the air-liquid interface, called a pellicle, which the bacteria grow, to finally excrete nanocellulose using a wide variety of substrates. (C. Sharma & Bhardwaj, 2019)

The production method influences the structure and mechanical properties of bacterial nanocellulose, which varies according to this nanomaterial's final use. Various methods for manufacturing bacterial nanocellulose range from static cultures, such as batch and fed batches, to continuous culture methods using bioreactors. (C. Sharma & Bhardwaj, 2019) Table 3 shows the currently used bacterial nanocellulose production methods with their advantages and disadvantages.

All types of nanocellulose, including bacterial nanocellulose, have hydroxyl groups on their surface, due to which the nanocellulose can be physically or chemically modified. It is indicated that this modification or functionalization improves the thermal and mechanical properties and increases the compatibility with nonpolar

Table 4. Bacterial nanocellulose applications according to its properties

Properties	Applications	Observation
Absorption and resistance.	Paper-based dressing	-
Moisture resistance: ability to withstand extreme temperatures	Adhesives on medical samples.	Medical labels that require refrigeration
Inertness: ability to avoid reaction between the chemicals in the material and the packaged product	Packaging of medicines and surgical products	-
Support at extreme temperatures	Packaging of sterile equipment and instruments	-
Biocompatibility	Manufacture of implants, masks, wound dressings, etc.	-
Biodegradability	Manufacture of scaffolds	The scaffolds must have the characteristic of degrading after being introduced into the body.
Mechanical strength	Support materials	As implants in bone tissue

polymeric matrices. (Islam & Rahman, 2018; H. Kargarzadeh et al., 2018; Hanieh Kargarzadeh et al., 2017) The changes to the bacterial nanocellulose mainly occur when it is coated on the fibers, allowing a final uniform distribution of cellulose in the polymer matrix, and providing improved interfacial adhesion through mechanical interlocking. Some authors, such as Pommet et al., reported a method in which bacterial nanocellulose was placed on natural hemp and sisal fibers to improve their adhesion to bio-based polymeric matrices, forming robust nanocomposites. (A. Sharma et al., 2019)

Recent studies have shown various applications of bacterial nanocellulose in different industrial sectors, where reinforcement applications stand out due to their potential and mechanical properties. (Robles et al., 2017) According to the property that nanocellulose has, it is given a series of applications. Table 4 highlights some applications of bacterial nanocellulose according to its properties.

In general, there is a wide variety of potential markets for materials based on nanocellulose in high-volume pulp and paper, plastics, automotive, food, coating, and support industries, and low-volume applications in sectors such as medicine, cosmetics, and pharmaceuticals. Due to its improved properties, novel applications have been studied in organic LEDs, electronics, photovoltaics, enzyme immobilization, and sensor technology. (Farooq et al., 2020) In the area of medicine, bacterial nanocellulose has been used in wound healing and tissue curtains, where, for example, polyvinyl alcohol was reinforced with this nanomaterial in order to form a material with mechanical properties comparable to those of cardiovascular tissue. (Tan et al., 2019)

Several review articles have recently focused on using compounds reinforced with bacterial nanocellulose. Due to its properties, applications have arisen as filtration membranes, reinforcing material in polymeric matrices, paper manufacturing, adsorption products, speaker diaphragms, hydrocarbon recovery, and a variety of medical applications as pads used in the topical application of drugs, artificial skin, and scaffolds to promote tissue formation *in vivo* or *ex vivo*. (Charreau et al., 2020) Additionally, nanocomposites from the reinforcement of polymeric matrices with bacterial nanocellulose offer promising opportunities in automotive, biomedical, aeronautical, and food applications because they are lighter than synthetic polymers. (Ferreira et al., 2018)

CONCLUSION

The research that is carried out around cellulose and nanocellulose as promising materials in various applications focuses mainly on pretreatment, manufacturing, and characterization methodologies, making use of processes that involve lower production cost and high effectiveness, obtaining; as a result, materials that have remarkable properties such as high biodegradability and biocompatibility.

Bacterial nanocellulose has become a promising support material for different polymer matrices. However, it has a series of disadvantages regarding its large-scale production, where the improvement of production techniques and modification of the surface of This nanomaterial will make it stand out in the future as a promising raw material in several applications in various areas of knowledge, primarily replacing current synthetic polymers. In Colombia, the large-scale production of bacterial nanocellulose is not yet carried out; however, it is considered that due to the large amount of agro-industrial waste materials produced by the country, an effective culture medium can be counted on for the production of bacterial nanocellulose, so it is necessary to carry out more selective research in this field.

Finally, the challenge surrounding the manufacture of this nanomaterial is to focus its production methods on less expensive techniques that involve shorter times and allow obtaining higher production yields; to be able to use this nanomaterial in the production of new, more environmentally friendly compounds.

REFERENCES

Abdel-KarimR.RedaY.Abdel-FattahA. (2020). Review-Nanostructured Materials-Based Nanosensors. https://doi.org/ doi:10.1149/1945-7111/ab67aa

Aljabali, A. A. A., Obeid, M. A., Al Zoubi, M. S., Charbe, N. B., Chellappan, D. K., Mishra, V., Dureja, H., Gupta, G., Prasher, P., Dua, K., Elnashar, R. M., Tambuwala, M. M., & Barhoum, A. (2020). In A. Barhoum (Ed.), *Nanocelluloses in Sensing Technology BT - Handbook of Nanocelluloses: Classification, Properties, Fabrication, and Emerging Applications* (pp. 1–30). Springer International Publishing., doi:10.1007/978-3-030-62976-2_44-1

Alle, M., Bandi, R., Sharma, G., Dadigala, R., Lee, S.-H., & Kim, J.-C. (2022). Gold nanoparticles spontaneously grown on cellulose nanofibrils as a reusable nanozyme for colorimetric detection of cholesterol in human serum. *International Journal of Biological Macromolecules, 201*, 686–697.. doi:10.1016/j.ijbiomac.2022.01.158 PMID:35104471

Anker, J. N., Hall, W. P., Lyandres, O., Shah, N. C., Zhao, J., & Van Duyne, R. P. (2009). Biosensing with plasmonic nanosensors. *Nanoscience and Technology: A Collection of Reviews from Nature Journals, 7*(June), 308–319. doi:10.1142/9789814287005_0032

Atalla, R. H., Atalla, R. S., & Agarwal, U. P. (2017). The Nanostructures of Native Celluloses, Their Transformations upon Isolation, and Their Implications for Production of Nanocelluloses. In *ACS Symposium Series* (Vol. 1251, pp. 1–18). 10.1021/bk-2017-1251.ch001

Bacakova, L., Pajorova, J., Bacakova, M., Skogberg, A., Kallio, P., Kolarova, K., & Svorcik, V. (2019). Versatile application of nanocellulose: From industry to skin tissue engineering and wound healing. *Nanomaterials (Basel, Switzerland), 9*(2), 164. doi:10.3390/nano9020164 PMID:30699947

Basabe-Desmonts, L., Müller, T. J. J., & Crego-Calama, M. (2007). Design of fluorescent materials for chemical sensing. *Chemical Society Reviews, 36*(6), 993–1017. doi:10.1039/b609548h PMID:17534482

Blanco, A., Monte, M. C., Campano, C., Balea, A., Merayo, N., & Negro, C. (2018). Nanocellulose for industrial use: Cellulose nanofibers (CNF), cellulose nanocrystals (CNC), and bacterial cellulose (BC). In M. Hussain (Ed.), *Handbook of Nanomaterials for Industrial Applications* (pp. 74–126). Elsevier Inc., doi:10.1016/B978-0-12-813351-4.00005-5

Bogue, R., & Bogue, R. (2008). Nanosensors: A review of recent progress. *Sensor Review, 28*(1), 12–17. doi:10.1108/02602280810849965

Charreau, H., Cavallo, E., & Foresti, M. L. (2020). Patents involving nanocellulose: Analysis of their evolution since 2010. *Carbohydrate Polymers, 237*(December 2019), 116039. doi:10.1016/j.carbpol.2020.116039

Dai, L., Wang, Y., Zou, X., Chen, Z., Liu, H., & Ni, Y. (2020). Ultrasensitive Physical, Bio, and Chemical Sensors Derived from 1-, 2-, and 3-D Nanocellulosic Materials. *Small, 16*(13), 1906567. doi:10.1002/smll.201906567 PMID:32049432

Daud, J., & Lee, K.-Y. (2017). Surface Modification of Nanocellulose. In Hanieh Kargarzadeh, I. Ahmad, S. Thomas, & A. Dufresne (Eds.), Handbook of Nanocellulose and Cellulose (pp. 101–122). Wiley-VCH. doi:10.1002/9783527689972.ch3

Do, T. T. A., Grijalvo, S., Imae, T., Garcia-Celma, M. J., & Rodríguez-Abreu, C. (2021). A nanocellulose-based platform towards targeted chemo-photodynamic/photothermal cancer therapy. *Carbohydrate Polymers, 270*, 118366. doi:10.1016/j.carbpol.2021.118366 PMID:34364611

dos Santos de Abreu, D., Strauss, M., & Santhiago, M. (2020). Flexible cellulose-based devices for monitoring physical parameters. In *Comprehensive Analytical Chemistry* (1st ed., Vol. 89). Elsevier B.V., doi:10.1016/bs.coac.2020.01.002

Farooq, A., Patoary, M. K., Zhang, M., Mussana, H., Li, M., Naeem, M. A., Mushtaq, M., Farooq, A., & Liu, L. (2020). Cellulose from sources to nanocellulose and an overview of synthesis and properties of nanocellulose/zinc oxide nanocomposite materials. *International Journal of Biological Macromolecules, 154*, 1050–1073. doi:10.1016/j.ijbiomac.2020.03.163 PMID:32201207

Ferreira, F. V., Dufresne, A., Pinheiro, I. F., Souza, D. H. S., Gouveia, R. F., Mei, L. H. I., & Lona, L. M. F. (2018). How do cellulose nanocrystals affect the overall properties of biodegradable polymer nanocomposites: A comprehensive review. *European Polymer Journal, 108*(July), 274–285. doi:10.1016/j.eurpolymj.2018.08.045

Fisher, C. H. (1989). In R. B. Seymour, H. F. Mark, L. Pauling, C. H. Fisher, G. A. Stahl, L. H. Sperling, C. S. Marvel, C. E. Carraher, & R. B. Seymour (Eds.), *Anselm Payen Pioneer in Natural Polymers and Industrial Chemistry BT - Pioneers in Polymer Science* (pp. 47–61). Springer Netherlands., doi:10.1007/978-94-009-2407-9_5

Frigerio, C., Ribeiro, D. S. M., Rodrigues, S. S. M., Abreu, V. L. R. G., Barbosa, J. A. C., Prior, J. A. V., Marques, K. L., & Santos, J. L. M. (2012). Application of quantum dots as analytical tools in automated chemical analysis: A review. *Analytica Chimica Acta, 735*, 9–22. doi:10.1016/j.aca.2012.04.042 PMID:22713912

Gooding, J. J., & Gaus, K. (2016). Single-Molecule Sensors: Challenges and Opportunities for Quantitative Analysis. *Angewandte Chemie International Edition*, *55*(38), 11354–11366. doi:10.1002/anie.201600495 PMID:27444661

Gopakumar, D. A., Thomas, S., & Grohens, Y. (2016). Nanocelluloses as Innovative Polymers for Membrane Applications. In D Puglia, E. Fortunati, & J. M. Kenny (Eds.), Multifunctional Polymeric Nanocomposites Based on Cellulosic Reinforcements (Issue 1, pp. 253–275). William Andrew Publishing. doi:10.1016/B978-0-323-44248-0.00008-0

Guo, L., Jackman, J. A., Yang, H. H., Chen, P., Cho, N. J., & Kim, D. H. (2015). Strategies for enhancing the sensitivity of plasmonic nanosensors. *Nano Today*, *10*(2), 213–239. doi:10.1016/j.nantod.2015.02.007

Heinze, T. (2015). Cellulose: Structure and properties. *Advances in Polymer Science*, *271*, 1–52. doi:10.1007/12_2015_319

Hon, D. (1994). Cellulose: A random walk along its historical path. *Cellulose (London, England)*, *1*(1), 1–25. doi:10.1007/BF00818796

Huang, Z., Raghuwanshi, V. S., & Garnier, G. (2017). Functionality of immunoglobulin G and immunoglobulin M antibody physisorbed on cellulosic films. *Frontiers in Bioengineering and Biotechnology*, *5*(AUG), 41. doi:10.3389/fbioe.2017.00041 PMID:28770196

Ikeda, M., Ochi, R., & Hamachi, I. (2010). Supramolecular hydrogel-based protein and chemosensor array. *Lab on a Chip*, *10*(24), 3325–3334. doi:10.1039/c004908e PMID:20862441

Ioelovich, M. (2016). Nanocellulose—fabrication, structure, properties, and application in the area of care and cure. In A. Grumezescu (Ed.), *Fabrication and Self-Assembly of Nanobiomaterials* (pp. 243–288). William Andrew Publishing. doi:10.1016/B978-0-323-41533-0.00009-X

Islam, M. N., & Rahman, F. (2018). Production and modification of nanofibrillated cellulose composites and potential applications. In G. Koronis & A. Silva (Eds.), *Green Composites for Automotive Applications* (pp. 115–141). Elsevier Ltd. doi:10.1016/B978-0-08-102177-4.00006-9

Javaid, M., Haleem, A., Singh, R. P., Rab, S., & Suman, R. (2021). Exploring the potential of nanosensors: A brief overview. *Sensors International*, *2*(October), 100130. doi:10.1016/j.sintl.2021.100130

Jedrzejczak-Krzepkowska, M., Kubiak, K., Ludwicka, K., & Bielecki, S. (2016). Bacterial NanoCellulose Synthesis, Recent Findings. In M. Gama, F. Dourado, & S. Bielecki (Eds.), *Bacterial Nanocellulose: From Biotechnology to Bio-Economy* (pp. 19–46). Elsevier B.V. doi:10.1016/B978-0-444-63458-0.00002-0

Kargarzadeh, H., Huang, J., Lin, N., Ahmad, I., Mariano, M., Dufresne, A., Thomas, S., & Gałęski, A. (2018). Recent developments in nanocellulose-based biodegradable polymers, thermoplastic polymers, and porous nanocomposites. *Progress in Polymer Science*, *87*, 197–227. doi:10.1016/j.progpolymsci.2018.07.008

Kargarzadeh, H., Mariano, M., Huang, J., Lin, N., Ahmad, I., Dufresne, A., & Thomas, S. (2017). Recent developments on nanocellulose reinforced polymer nanocomposites: A review. *Polymer*, *132*, 368–393. doi:10.1016/j.polymer.2017.09.043

Karimian, A., Yousefi, B., Sadeghi, F., Feizi, F., Najafzadehvarzi, H., & Parsian, H. (2022). Synthesis of biocompatible nanocrystalline cellulose against folate receptors as a novel carrier for targeted delivery of doxorubicin. *Chemico-Biological Interactions*, *351*, 109731. doi:10.1016/j.cbi.2021.109731 PMID:34728188

Kim, J.-H., Shim, B. S., Kim, H. S., Lee, Y.-J., Min, S.-K., Jang, D., Abas, Z., & Kim, J. (2015). Review of nanocellulose for sustainable future materials. *International Journal of Precision Engineering and Manufacturing-Green Technology*, *2*(2), 197–213. doi:10.1007/s40684-015-0024-9

Kneipp, J., Kneipp, H., & Kneipp, K. (2008). SERS—A single-molecule and nanoscale tool for bioanalytics. *Chemical Society Reviews*, *37*(5), 1052–1060. doi:10.1039/b708459p PMID:18443689

Kozawa, D., Cho, S. Y., Gong, X., Nguyen, F. T., Jin, X., Lee, M. A., Lee, H., Zeng, A., Xue, G., Schacherl, J., Gibson, S., Vega, L., & Strano, M. S. (2020). A Fiber Optic Interface Coupled to Nanosensors: Applications to Protein Aggregation and Organic Molecule Quantification. *ACS Nano*, *14*(8), 10141–10152. doi:10.1021/acsnano.0c03417 PMID:32667777

Kumar, S., Ngasainao, M. R., Sharma, D., Sengar, M., Gahlot, A. P. S., Shukla, S., & Kumari, P. (2022). Contemporary nanocellulose-composites: A new paradigm for sensing applications. *Carbohydrate Polymers*, *298*, 120052. doi:10.1016/j.carbpol.2022.120052 PMID:36241259

Larson, K. K., He, M., Teichert, J. F., Naganawa, A., & Bode, J. W. (2012). Chemical sensing with shapeshifting organic molecules. *Chemical Science (Cambridge)*, *3*(6), 1825–1828. doi:10.1039/c2sc20238g

Lee, C. W., Suh, J. M., & Jang, H. W. (2019). Chemical Sensors Based on Two-Dimensional (2D) Materials for Selective Detection of Ions and Molecules in Liquid. *Frontiers in Chemistry*, *7*, 708. doi:10.3389/fchem.2019.00708 PMID:31803712

Lee, K. Y., & Bismarck, A. (2016). Bacterial NanoCellulose as Reinforcement for Polymer Matrices. In Miguel Gama, F. Dourado, & S. Bielecki (Eds.), Bacterial Nanocellulose: From Biotechnology to Bio-Economy (pp. 109–122). Elsevier B.V. doi:10.1016/B978-0-444-63458-0.00006-8

Li, S., Chen, H., Liu, X., Li, P., & Wu, W. (2022). Nanocellulose as a promising substrate for advanced sensors and their applications. *International Journal of Biological Macromolecules*, *218*, 473–487. doi:10.1016/j.ijbiomac.2022.07.124 PMID:35870627

Liu, Y., He, K., Chen, G., Leow, W. R., & Chen, X. (2017). Nature-Inspired Structural Materials for Flexible Electronic Devices. *Chemical Reviews*, *117*(20), 12893–12941. doi:10.1021/acs.chemrev.7b00291 PMID:28991450

Luo, X., Han, Y., Chen, X., Tang, W., Yue, T., & Li, Z. (2020). Carbon dots derived fluorescent nanosensors as versatile tools for food quality and safety assessment: A review. *Trends in Food Science & Technology*, *95*, 149–161. doi:10.1016/j.tifs.2019.11.017

Mishra, R. K., Sabu, A., & Tiwari, S. K. (2018). Materials chemistry and the futurist eco-friendly applications of nanocellulose: Status and prospect. *Journal of Saudi Chemical Society*, *22*(8), 949–978. doi:10.1016/j.jscs.2018.02.005

Mondal, S. (2017). Preparation, properties and applications of nanocellulosic materials. *Carbohydrate Polymers*, *163*, 301–316. doi:10.1016/j.carbpol.2016.12.050 PMID:28267510

Naghdi, T., Golmohammadi, H., Vosough, M., Atashi, M., Saeedi, I., & Maghsoudi, M. T. (2019). Lab-on-nanopaper: An optical sensing bioplatform based on curcumin embedded in bacterial nanocellulose as an albumin assay kit. *Analytica Chimica Acta*, *1070*, 104–111. doi:10.1016/j.aca.2019.04.037 PMID:31103163

Nechyporchuk, O., Belgacem, M. N., & Bras, J. (2016). Production of cellulose nanofibrils: A review of recent advances. *Industrial Crops and Products*, *93*, 2–25. doi:10.1016/j.indcrop.2016.02.016

Nguyen, L. H., Naficy, S., Chandrawati, R., & Dehghani, F. (2019). Nanocellulose for Sensing Applications. *Advanced Materials Interfaces*, *6*(18), 30–33. doi:10.1002/admi.201900424

Ohashi, T., & Dai, L. (2006). C60 and carbon nanotube sensors. Carbon Nanotechnology: *Recent Developments in Chemistry, Physics, Materials Science and Device Applications*, 525–575. doi:10.1016/B978-044451855-2/50018-8

Phanthong, P., Reubroycharoen, P., Hao, X., Xu, G., Abudula, A., & Guan, G. (2018). Nanocellulose: Extraction and application. *Carbon Resources Conversion, 1*(1), 32–43. doi:10.1016/j.crcon.2018.05.004

Pouzesh, M., Nekouei, S., Ferdosi Zadeh, M. A., Keshtpour, F., Wang, S., & Nekouei, F. (2019). Fabrication of stable copper nanoparticles embedded in nanocellulose film as a bionanocomposite plasmonic sensor and thereof for optical sensing of cyanide ion in water samples. *Cellulose (London, England), 26*(8), 4945–4956. doi:10.1007/s10570-019-02405-0

Puglia, D. Fortunati, E., & Kenny, J. (2016). Multifunctional Polymeric Nanocomposites Based on Cellulosic Reinforcements (First Edit). William Andrew Publishing. doi:10.1016/C2014-0-04565-0

Qian, Z. S., Shan, X. Y., Chai, L. J., Ma, J. J., Chen, J. R., & Feng, H. (2014). DNA nanosensor based on biocompatible graphene quantum dots and carbon nanotubes. *Biosensors & Bioelectronics, 60,* 64–70. doi:10.1016/j.bios.2014.04.006 PMID:24768864

Ram, B., Chauhan, G. S., Mehta, A., Gupta, R., & Chauhan, K. (2018). Spherical nanocellulose-based highly efficient and rapid multifunctional naked-eye Cr(VI) ion chemosensor and adsorbent with mild antimicrobial properties. *Chemical Engineering Journal, 349*(Vi), 146–155. doi:10.1016/j.cej.2018.05.085

Ramos, M., Valdés, A., & Garrigós, M. C. (2016). Multifunctional Applications of Nanocellulose-Based Nanocomposites. In D. Puglia, E. Fortunati, & J. M. Kenny (Eds.), *Multifunctional Polymeric Nanocomposites Based on Cellulosic Reinforcements* (pp. 177–204). William Andrew Publishing. doi:10.1016/B978-0-323-44248-0.00006-7

Robles, E., Labidi, J., Halász, K., & Csóka, L. (2017). Key issues in reinforcement involving nanocellulose. In M. Jawadi, S. Boufi, & K. H.P.S (Eds.), Cellulose-Reinforced Nanofibre Composites: Production, Properties and Applications (pp. 401–425). Woodhead Publishing. doi:10.1016/B978-0-08-100957-4.00018-8

Rodrigues, S. S. M., Ribeiro, D. S. M., Soares, J. X., Passos, M. L. C., Saraiva, M. L. M. F. S., & Santos, J. L. M. (2017). Application of nanocrystalline CdTe quantum dots in chemical analysis: Implementation of chemo-sensing schemes based on analyte-triggered photoluminescence modulation. *Coordination Chemistry Reviews, 330*, 127–143. doi:10.1016/j.ccr.2016.10.001

Ruiz, C., Soriano, L., & Valcârcel, M. (2017). Nanocellulose as analyte and analytical tool: Opportunities and challenges. *Trends in Analytical Chemistry*, *87*, 1–18. doi:10.1016/j.trac.2016.11.007

Saba, N., Jawaid, M., & Sultan, M. T. H. (2017). Thermal properties of oil palm biomass based composites. In M. Jawadi, P. Tahir, & N. Saba (Eds.), *Lignocellulosic Fibre and Biomass-Based Composite Materials: Processing, Properties and Applications* (Vol. 1, pp. 95–122). Woodhead Publishing., doi:10.1016/B978-0-08-100959-8.00006-8

Shalauddin, M., Akhter, S., Basirun, W. J., Bagheri, S., Anuar, N. S., & Johan, M. R. (2019). Hybrid nanocellulose/f-MWCNTs nanocomposite for the electrochemical sensing of diclofenac sodium in pharmaceutical drugs and biological fluids. *Electrochimica Acta*, *304*, 323–333. doi:10.1016/j.electacta.2019.03.003

Sharma, A., Thakur, M., Bhattacharya, M., Mandal, T., & Goswami, S. (2019). Commercial application of cellulose nano-composites – A review. *Biotechnology Reports (Amsterdam, Netherlands)*, *21*, e00316. doi:10.1016/j.btre.2019.e00316 PMID:30847286

Sharma, C., & Bhardwaj, N. K. (2019). Bacterial nanocellulose: Present status, biomedical applications and future perspectives. *Materials Science and Engineering C*, *104*, 109963. doi:10.1016/j.msec.2019.109963 PMID:31499992

Singh, R., & Kumar, S. (2017). Green technologies and environmental sustainability. *Green Technologies and Environmental Sustainability,* 1–492. doi:10.1007/978-3-319-50654-8

Tan, K., Heo, S., Foo, M., Chew, I. M., & Yoo, C. (2019). An insight into nanocellulose as soft condensed matter: Challenge and future prospective toward environmental sustainability. *The Science of the Total Environment*, *650*(1), 1309–1326. doi:10.1016/j.scitotenv.2018.08.402 PMID:30308818

Tayeb, A. H., Amini, E., Ghasemi, S., & Tajvidi, M. (2018). Cellulose Nanomaterials-Binding Properties and Applications: A Review. *Molecules (Basel, Switzerland)*, *23*(10), 2684. doi:10.3390/molecules23102684 PMID:30340374

Thakur, V. K., & Thakur, M. K. (2014). Processing and characterization of natural cellulose fibers/thermoset polymer composites. *Carbohydrate Polymers*, *109*, 102–117. doi:10.1016/j.carbpol.2014.03.039 PMID:24815407

Thomas, P., Duolikun, T., Rumjit, N. P., Moosavi, S., Lai, C. W., Bin Johan, M. R., & Fen, L. B. (2020). Comprehensive review on nanocellulose: Recent developments, challenges and future prospects. *Journal of the Mechanical Behavior of Biomedical Materials*, *110*(May), 103884. doi:10.1016/j.jmbbm.2020.103884 PMID:32957191

Tshikovhi, A., Mishra, S. B., & Mishra, A. K. (2020). Nanocellulose-based composites for the removal of contaminants from wastewater. *International Journal of Biological Macromolecules*, *152*, 616–632. doi:10.1016/j.ijbiomac.2020.02.221 PMID:32097743

Tümay, S. O., Şanko, V., Şenocak, A., & Demirbas, E. (2021). A hybrid nanosensor based on novel fluorescent iron oxide nanoparticles for highly selective determination of Hg2+ ions in environmental samples. *New Journal of Chemistry*, *45*(32), 14495–14507. doi:10.1039/D1NJ02908H

Vo-Dinh, T., Liu, Y., Fales, A. M., Ngo, H., Wang, H. N., Register, J. K., Yuan, H., Norton, S. J., & Griffin, G. D. (2015). SERS Nanosensors and Nanoreporters: Golden Opportunities in Biomedical Applications. *Wiley Interdisciplinary Reviews. Nanomedicine and Nanobiotechnology*, *7*(1), 17–33. doi:10.1002/wnan.1283 PMID:25316579

Vuoriluoto, M., Orelma, H., Zhu, B., Johansson, L.-S., & Rojas, O. J. (2016). Control of Protein Affinity of Bioactive Nanocellulose and Passivation Using Engineered Block and Random Copolymers. *ACS Applied Materials & Interfaces*, *8*(8), 5668–5678. doi:10.1021/acsami.5b11737 PMID:26844956

Wan, C., Zhang, L., Yong, K.-T., Li, J., & Wu, Y. (2021). Recent progress in flexible nanocellulosic structures for wearable piezoresistive strain sensors. *Journal of Materials Chemistry. C, Materials for Optical and Electronic Devices*, *9*(34), 11001–11029. doi:10.1039/D1TC02360H

Wang, F., Lu, Y., Chen, Y., Sun, J., & Liu, Y. (2018). Colorimetric Nanosensor Based on the Aggregation of AuNP Triggered by Carbon Quantum Dots for Detection of Ag+ Ions. *ACS Sustainable Chemistry & Engineering*, *6*(3), 3706–3713. doi:10.1021/acssuschemeng.7b04067

Yan, L., Kasal, B., & Huang, L. (2016). A review of recent research on the use of cellulosic fibres, their fibre fabric reinforced cementitious, geo-polymer and polymer composites in civil engineering. *Composites. Part B, Engineering*, *92*, 94–132. doi:10.1016/j.compositesb.2016.02.002

Zaid, M. H. M., Abdullah, J., Yusof, N. A., Wasoh, H., Sulaiman, Y., Noh, M. F. M., & Issa, R. (2020). Reduced Graphene Oxide/TEMPO-Nanocellulose Nanohybrid-Based Electrochemical Biosensor for the Determination of Mycobacterium tuberculosis. *Journal of Sensors, 4051474*, 1–11. Advance online publication. doi:10.1155/2020/4051474

Zha, R., Shi, T., He, L., & Zhang, M. (2022). Nanoengineering and green chemistry-oriented strategies toward nanocelluloses for protein sensing. *Advances in Colloid and Interface Science*, *308*, 102758. doi:10.1016/j.cis.2022.102758 PMID:36037672

Zhang, L., Li, Q., Zhou, J., & Zhang, L. (2012). Synthesis and photophysical behavior of pyrene-bearing cellulose nanocrystals for Fe 3+ sensing. *Macromolecular Chemistry and Physics*, *213*(15), 1612–1617. doi:10.1002/macp.201200233

Zinge, C., & Kandasubramanian, B. (2020). Nanocellulose based biodegradable polymers. *European Polymer Journal*, *133*(May), 109758. doi:10.1016/j.eurpolymj.2020.109758

ADDITIONAL READING

Cao, P. F., Yan, Y. H., Mangadlao, J. D., Rong, L. H., & Advincula, R. (2016). Star-like Copolymer Stabilized Noble-metal Nanoparticle Powders. *Nanoscale*, *8*(14), 7435–7442. doi:10.1039/C5NR07000G PMID:26659728

Chen, F., Li, X., Hihath, J., Huang, Z., & Tao, N. (2006). Effect of Anchoring Groups on Single-Molecule Conductance: Comparative Study of Thiol-, Amine-, and Carboxylic-Acid-Terminated Molecules. *Journal of the American Chemical Society*, *128*(49), 15874–15881. doi:10.1021/ja065864k PMID:17147400

Huang, H., Liu, X., Hu, T., & Chu, P. K. (2010). Ultra-sensitive Detection of Cysteine by Gold Nanorod Assembly. *Biosensors & Bioelectronics*, *25*(9), 2078–2083. doi:10.1016/j.bios.2010.02.003 PMID:20197234

Lee, J. S., Ulmann, P. A., Han, M. S., & Mirkin, C. A. (2008). A DNA–Gold Nanoparticle-Based Colorimetric Competition Assay for the Detection of Cysteine. *Nano Letters*, *8*(2), 529–533. doi:10.1021/nl0727563 PMID:18205426

Sudeep, P. K., Joseph, S. T. S., & Thomas, K. G. (2005). Selective Detection of Cysteine and Glutathione Using Gold Nanorods. *Journal of the American Chemical Society*, *127*(18), 6516–6517. doi:10.1021/ja051145e PMID:15869256

Chapter 8
Nanocomposites for Advanced Biosensing Applications

Nawal Alghamdi
https://orcid.org/0000-0001-5994-9737
University of Tabuk, Saudi Arabia

ABSTRACT

Nanocomposites have an essential role in a number of different applications these days. Therefore, they have been utilised in the biosensing field. The integration of nanocomposites in different types of biosensors for detection or mentoring of different analytes results in improved performance biosensors. In this chapter, the role of nanocomposites in biosensors will be studied, with a number of examples, of such improvements in these sensors' performance. An overview of the biosensors will be given first, along with key characteristics to quantity the enhancing in their performance. After that, a detailed discussion of nanocomposites as materials and their roles in biosensor development will be included.

INTRODUCTION

There is an increased demand for effective biosensors nowadays for a number of reasons, such as outbreaks and the spread of infections and diseases worldwide. Additionally, biosensors are applied for the monitoring of human health and medical diagnosis, they are also significantly needed in the detection of different contaminations that may be present in water, food, and the environment. All of these needs require a development in biosensor applications to achieve effective sensing such as highly sensitive and selective biosensors with short response time and at reasonable cost. The performance of biosensors has been improved recently after the

DOI: 10.4018/979-8-3693-1094-6.ch008

Copyright © 2024, IGI Global. Copying or distributing in print or electronic forms without written permission of IGI Global is prohibited.

advent of nanotechnology and, more precisely, the incorporation of nanocomposites in biosensors for various applications (X. Wang et al., 2020). Nanocomposites are mixtures of different materials and phases where at least one nanostructure material, such as nanoparticles, is included. The unique properties of nanocomposites, such as their high surface area, high mechanical stability, and electrical conductivity, have improved the performance of such sensors, which results in a favourable and stable family of biosensors (Huang et al., 2021). Various nanocomposites have been utilised in this application for different purposes, such as enhancement in the sensitivity and selectivity of the overall sensors or the production of fast-responding sensors. Nanocomposites with different nanostructure fillings are utilised, such as graphene, metal oxides, and carbon nanotubes. The employment of these materials has brought a number of preferred characteristics and advantages to the biosensing field. This chapter describes the significant role of the incorporation of nanocomposites in the bio-sensing field. The application of nanotechnology in biosensors has improved their overall performance in different ways, such as their sensitivity, selectivity, and response time. Various types of nanomaterials have been used, including nanocomposites (Kazemi et al., 2021). This chapter will emphasise the role of nanocomposites and their related nanostructure materials in the development of biosensors, with an emphasis on the advantages of using different nanocomposites in biosensors. The first part of this chapter will include explanations and discussions of the biosensors, including their structure, working principle, types, and characterization factors. The performance characteristics of biosensors will provide a clear observation of the function of the nanocomposites in developing the biosensors. The second part of this chapter will include a detailed discussion about nanocomposites, their types and properties, and their important role in the improvement of biosensors, with several examples of some of their applications from recent literature and research.

BIOSENSORS

Structure and Components

First, let us start by defining sensors and accordingly biosensors. A sensor is a device that helps detect variations in the quantity of a certain property and translate these changes into readable signals. For example, if there is variation in physical property (e.g., heat) or electrical property (e.g., electric voltage), such changes can be detected by a sensor. In the case of biosensors, they incorporate biological elements into their design and use them to convert the biological response into an electrical signal (Ramesh et al., 2023a; Naresh & Lee, 2021). Biosensors cover different applications, for example, in healthcare and clinical diagnosis (Ding et al., 2023; Mills & Nestorova, 2023). Also, biosensors are used in environmental monitoring, like detecting environmental pollutants (Alghamdi et al., 2019, 2020). Another important biosensor application is monitoring the quality of food and water (Alqahtani et al., 2020). The main elements in the structure of any biosensor can be identified as follows: the biological recognition element, the transducer element, and the signal processor (Suresh et al., 2014). The biological recognition consists of two components: analytes and bio-receptors. The analyte is the target substance or molecule to be detected, and the bio-receptor is a biological element that could recognise the target molecule and bind to it. The process between the target molecule and the bio-receptor is called biorecognition, where the interaction occurs between the analyte and bio-receptor (Kalakonda et al., 2023; Naresh & Lee, 2021) (Bansi Dhar Malhotra & Chandra Mouli Pandey, 2017). The second part of a biosensor is the transducer, which is a key element as it translates the biorecognition event between the analyte and bioreceptor into measurable electrical signals. A transducer basically converts energy from one type to another, so in this case, it transforms the energy produced from biorecognition into another type of energy that can be a readable signal. Transducers have several types depending on the process to obtain the signals, for example, electrochemical, optical, and electronic transducers (Huang et al., 2021; Naresh & Lee, 2021). The last part of the biosensor is the signal processor, which works by processing the electric signal produced by the transducer into an expressive signal or parameter that can be shown on a display (Naresh & Lee, 2021; Suresh et al., 2014.). Figure 1 below describes the biosensor component and the working principles.

Figure 1. The main components in the sensor. The binding between analytes and bioreceptors (biorecognition) results in signals that are translated by the transducer

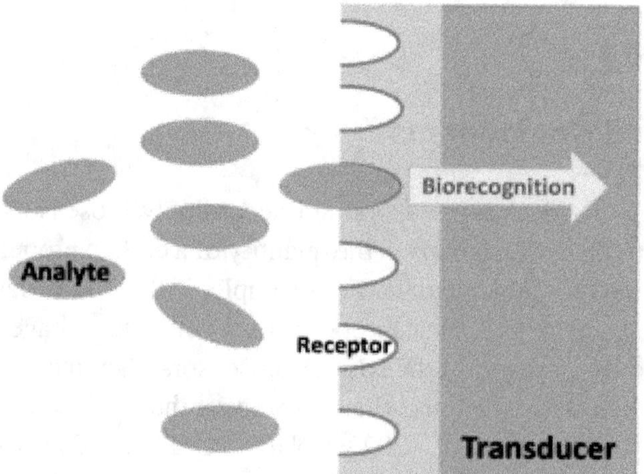

Types of Bio-Sensors

Biosensors could be categorised according to different criteria, but commonly they are classified based on the types of transducers and also the types of bioreceptors.

Classification of Biosensors Based on Bioreceptors

The interaction between analytes and bioreceptors could lead to a biochemical reaction product; in this case, the biosensors are called catalytic biosensors, for example, enzyme receptors (Naresh & Lee, 2021; Rocchitta et al., 2016). In the other case, the analytes are permanently attached to the bioreceptors without any biochemical reactions, such as antibodies; this type is called non-catalytic biosensors (Ahmad et al., 2021; Naresh & Lee, 2021). A brief description of some of the types of biosensors based on their bioreceptors is included below:

Enzyme Based Biosensors

As the known role of enzymes is to catalyse biochemical reactions, so in the enzyme based biosensors, the biorecognition process here results from the induced changes in the presence of a specific analyte. A common example of such sensors is the glucose biosensor (Rocchitta et al., 2016).

Antibody Based Biosensors

The strong interaction between antigens and antibodies enables the use of antibodies as bioreceptors. These types of sensors are often called immunosensors such as those used in the detection of coronavirus disease (Ahmad et al., 2021).

Nanoparticles Based Biosensors

Recently, different nanomaterials have been used in biosensors, mainly as bioreceptors. The involvement of nanoparticles has helped with very good achievements in the biosensing field. As the topic of this book chapter applies, we will discuss this further in the second part of this chapter. Interestingly, nanomaterials can also be used as transducers for biosensors (Ramesh et al., 2023a).

Classification of Biosensors Based on Transducers

Different types of transducers are integrated into biosensors, and according to the working principle of these transducers, biosensors can be grouped accordingly. The most common types of transducers are electrochemical, optical, thermal, and piezoelectric. A brief description of some biosensor types based on their transducers is included below (Curulli, 2023; Huang et al., 2021; Naresh & Lee, 2021).

Electrochemical Biosensors

Electrochemical biosensors are the most common type of biosensors; they can be divided into sub-categories (potentiometric, amperometric, voltametric, conductometric, and impedimetric biosensors). depending on the nature of the transduction process (e.g., a potential or impedance change). In these sensors, the interaction between the analyte and receptor, which are somehow attached to the transducer, will result in an electrochemical reaction that will translate the transducer into a readable signal (Huang et al., 2021; Kalakonda et al., 2023; Naresh & Lee, 2021).

Optical Biosensors

In this type of biosensor, a change in the optical properties is a result of the interaction between the analyte and the bioreceptor. Examples of such optical changes can be changes in optical absorption, emission, reflection, or refraction (Kalakonda et al., 2023; Naresh & Lee, 2021).

Thermal Biosensors

These sensors are used when the biorecognition reaction involves temperature change. So the thermal transducers (e.g., thermistors) in this case will measure the energy released or absorbed. The thermal signals recorded from the target analyte-receptor interactions are proportional to the amount of analyte (Sivarajasekar et al., 2019; Bansi Dhar Malhotra & Chandra Mouli Pandey, 2017).

Characterization of Biosensors

In order to determine the effectiveness of any sensor, a number of characteristics need to be considered, as they give an indicator of how good the sensor's performance is. As we discuss the performance of biosensors and the role of the implantation of nanocomposites in improving these sensors functions, we need to know how we measure such enhancement. Significant sensor characteristics are listed below with their definitions (Bansi Dhar Malhotra & Chandra Mouli Pandey, 2017; Naresh & Lee, 2021):

Selectivity

Selectivity is mainly a property of the bio-receptor in the biosensor, which indicates how well this biosensor is at detecting the target analyte when mixed with other analytes. If the biosensor (specifically the bioreceptor) is capable of detecting a specific substance in the presence of other substances that may be similar to the target substance in their nature, in this case, we could describe this biosensor as selective.

Sensitivity

Biosensor sensitivity could be described as the ability of a biosensor to detect a very low concentration of the analyte; the lower the detection limit, the more sensitive the biosensor is to such an analyte. This limit is commonly called the limit of detection (LOD) of a sensor.

Response Time

The time taken to show almost all the results of this sensor, or 95% of the results, means shorter response times give faster biosensors.

Reproducibility

Reproducible sensors reproduce almost the same result or reading each time when repeating the measurements.

Stability

The stability of a sensor means that it still possesses the same characteristics over time (e.g., the same sensitivity and response time). In other words, stability can describe the ageing of a sensor, which may be a result of some component degradation, such as bioreceptor degradation.

NANOCOMPOSITES IN BIOSENSORS

The advent of nanotechnology improves the performance of many devices and technologies; one of these applications is sensor technology, generally, and more specifically, biosensors. The use of nanotechnology through the implantation of nanomaterials in the biosensor architecture has brought numerous advantages and overcome a number of challenges in the biosensors. Various types of nanomaterials can be used, such as nano-particles, nano-tubes and nano-wires, nano-pores, self-adhesive monolayers, and nanocomposites (Huang et al., 2021; Ramesh et al., 2023a). Some of challenges that have been overcome by nanostructure incorporation include increased efficiency in biorecognition binding, improved sensitivity, and lowering the limit of detection (LOD) (Huang et al., 2021). Also, it improved the selectivity of biosensors and shortened the response time (Akgöl et al., 2021). In electrochemical biosensors, it aids the sensing process by speeding the electron transfer after reactions to the sensing electrodes (Huang et al., 2021). In this part of the chapter, nanocomposite-based biosensors will be particularly discussed by mentioning the properties of nanocomposites and how these properties improve the performance of the biosensors, with a number of examples of such applications.

Properties of Nanocomposites in Biosensing

Nanocomposites are composite materials that have one or more phases on the nanometer scale (1–100 nm), such as nanoparticles or nanotubes (R. P. Singh et al., 2020a). In other words, they are a mixture of nanomaterials with matrix materials (Figure 2). Nanocomposites are mainly categorised based on the matrix material type and the existence of polymer material in the composite mixture into polymeric based nanocomposites and nonpolymeric based materials. Non-polymeric based nanocomposites (also called inorganic nanocomposites) include metal nanocomposites, ceramic nanocomposites, and ceramic-ceramic nanocomposites. Examples of metal matrices are Al, Fe and Mg, while Al_2O_3/SiO_2, Al_2O_3/TiO_2 are examples of ceramic matrices (Din, 2019; R. P. Singh et al., 2020a). The non-polymeric nanocomposites have some advantages, such as boosted strength and hardness and enhanced magnetic and eclectic properties. On the other hand, polymeric matrices nanocomposites could include for example vinyl polymers, condensation polymers, and polyolefins. A number of advantages have been reported for the polymeric type of nanocomposites, including improving the thermal properties and the recyclability. Despite these merits, there is still a challenge in producing unform nanocomposites from the dispersion of nanostructure materials in the polymeric matrix, which results in poor mechanical properties (Sen, 2020; Vroman & Tighzert, 2009). The unique properties of nanocomposites allow them to be used in different applications. For example, they are involved in green energy production, water treatment, supercapacitors, and biosensors (Singh et al., 2020). The chemical, physical, and electrical properties of nanocomposites materials are more special and useful in different fields, especially when compared with their related bulk materials. The small particles have a large surface area, which results in a high surface-to-volume ratio and high surface energies. The interactions at the interfaces at such low nanoscale dimensions are improved, resulting in more active reactions, which make the nanocomposites emerge as candidates to overcome limitations in different applications (Din, 2019; Huang et al., 2021; Sen, 2020).

In biosensors, nanocomposites and nanostructures are significantly involved in the development of these sensors performance. The large surface area of nanostructures leads to a larger surface area of the biosensors that are covered by bioreceptors, which results in an increase in the absorption of biomolecules. Such practice aids the biorecognition process and results in improving the sensitivity of the biosensor (Huang et al., 2021; Zhang et al., 2021). Additionally, the small dimension of nanoscale materials is considered to be at the same scale as the biological process, which simplifies the sensing process and also the functionalization of the sensing surfaces for different molecules (Bisht et al., 2021). The integration of nanocomposites in particle-conductive nanocomposites also uncomplicates the detection by reducing the need for any chemical mediators in the electron transport process between the activity part and the electrode part of the biosensor, which leads to improved speedy response and accuracy. Such merit occurs in some biosensors, especially biochemical biosensors (Huang et al., 2021; T. Wang et al., 2013). The high surface energy of nanoparticles compared with bulk material increases their surface activity, and therefore nanoparticles are known for their catalytic activity and drive reactions in the biosensing process, especially metal nanoparticles, which are used as catalysts in this case. Also, nanoparticles could be employed as reactants in different procedures in the operative biosensors. Other benefits of nanocomposites include providing straight attachment between enzymes and the surface of sensing electrodes and their common biocompatibility properties (Huang et al., 2021). The capability of nanomaterials to be functionalized has extended their unique properties in the overall performance of biosensors, such as durable, sensitive, selective, and fast-responsive biosensors (Kausar et al., 2023) For example, the application of conductive polymer-based nanocomposites in biosensor construction leads to a number of excellent properties such as improved conductivity, good electro-catalytic activity, and high absorption capacity (Naseri et al., 2018).

Figure 2. The formation of a nanocomposite by mixing matrix material with nanomaterial (here nanoparticles)

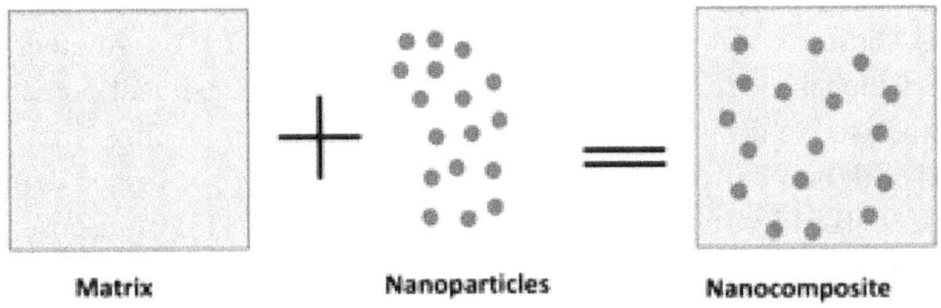

Nanocomposites Advantages in Biosensing

As shown in the previous section, the unique properties that nanocomposites acquire due to their nanoscale dimensionality make them very attractive materials in biosensing applications. As examples of these benefits, nanocomposites enhance the sensitivity of the biosensors and also give a relatively rapid response. There are many nanocomposites that have been investigated in biosensors, and they have exhibited desirable performance in the field. In this section, we will discuss a number of these nanocomposites materials and types and show the advantages that these materials bring to the biosensors world. Nanocomposites could be classified based on the filler type (the content of nanomaterials) or based on the type of matrix, as mentioned before. Here, we will discuss more about the classification of nanocomposites based on the nanostructure used in the composites formation. Mainly, nanocomposites could include metal nanoparticles M-NPs, metal oxide nanoparticles MO-NPs, magnetic nanoparticles, quantum dots QDs, carbon nanotube CNT, graphene and graphene oxide, or hybrid nanocomposites where more than one nanostructure is incorporated. We will discuss below the employment and advantages of each type in biosensing.

Graphene Based Nanocomposites

Graphene as a nanomaterial has a number of outstanding characteristics that make it get great attention in many applications. Properties such as high thermal and electric conductivity, a high surface to area ratio, and excellent mechanical properties make it appealing for sensing and biosensing applications (Kazemi et al., 2021). Multifunctional nanocomposites can be created by using graphene as filler in polymeric matrices. Such composites can be utilised in biosensors due to the great properties they gain. For example, due to their high electric conductivity, the electron transfer will be higher, resulting in better sensing performance (Akgöl et al., 2021; Kausar et al., 2023).

Here will show some cases from the literature when graphene-based nanocomposites are used in biosensors and correlated to performance improvement. A graphene-based optical biosensor has been used to detect cancer cells in living tissue and shows a very high sensitivity (Wan et al., 2021). Another type of biosensor, which is the field effect transistor FETs (electrochemical-potentiometric sensor), was used to detect different viruses. The detection of Zika virus, which is associated with non-structural protein 1 (ZKV NS1), was performed using FETs. Also, the COVID-19 virus was detected by graphene-based FET biosensors, which showed a low limit of detection and rapid response (Ramesh et al., 2023; Seo et al., 2020). A polypropylene/graphene/polyaniline nanocomposite biosensor was used to detect the volatile gases, which are sulphur compounds (e.g. H_2S) and ammonia. Such sensors show a high sensing response due to the highly conductive composite. Also, the sensing response increases by about 250% by using graphene filler nanocomposites compared to only polyaniline sensors (Kausar et al., 2023) (Wu et al., 2022). A paper-based biosensor fabricated with graphene nanocomposite (graphene, polyvinylpyrrolidone, polyaniline, G/PVP/PANI) is used to detect cholesterol in real samples from individuals. Such electrode sensors show brilliant electro-catalytic activity for hydrogen peroxide oxidation, which is created from an enzymatic reaction between cholesterol and its oxidase (Li et al., 2019; Naseri et al., 2018). In addition, graphene based electrochemical biosensors have been utilised for different drug detection. The development of such composites in drug detection has been reviewed between 2010 and 2020, and this review shows a great improvement in the electrochemical biosensing performance (Jiang et al., 2021). However, the great performance of graphene-based nanocomposite biosensors could be disturbed by the fact that pure and high-quality graphene is still a challenge (Kausar et al., 2023).

Graphene Oxide Based Nanocomposites

Graphene oxide (GO) is one of the best graphene derivatives and shows similar outstanding properties, making it an important candidate for the improvement of biosensors. In addition to the unique chemical, thermal, electrical, and high surface area, the surfaces of graphene oxide (GO) surfaces have oxygen-containing functional groups that assist different biomolecule attachment, which aids the biorecognition process in biosensors (Gosai et al., 2021; Kausar et al., 2023; Sharma et al., 2016). Promising nanocomposites in the biosensing field can be made by mixing graphene oxide (GO) with other polymer matrices to obtain mechanically strong functionalized sensors and transducers for ultra-sensitive detection of various biomolecules (Gosai et al., 2021). For example, the composites of polyaniline and graphene oxide have improved their sensing performance towards a number of gases and toxic ions, including hydrogen and methane gases and lead Pb and Cd cadmium ions (Biswas & Oh, 2019; Dongyang Wang et al., 2018; Kausar et al., 2023). Another nanocomposite containing graphene oxides was prepared to sense the presence of hydrogen H_2 in ambient air at room temperature. Such a sensor detects H2 at a low concentration below 2% with a response time of 72 sec (Mohammadi et al., 2020). A nanocomposite fabricated from polyaniline, natural rubber, and reduced graphene oxide was prepared to be implanted in various biosensors for health monitoring. This composite has an eco-friendly and simpler technique and shows improved sensing performance (Zhou et al., 2016). In glucose sensors, nanocomposites made from reduced graphene oxide (rGO) and polypyrrole PPY matrix help to hold the enzymic bioreceptor, which is glucose oxidase, and also assist in the transfer of electrons to the collector. Also, it is reported that the incorporation of rGO in ZnO nanorod-based biosensors improved the sensitivity of this sensor by 16 folds and decreased the limit of detection LoD by 2.3 folds (Y. Wang et al., 2017; Gosai et al., 2021). A ternary nanocomposite of PEDOT as a polymer and graphene oxide with hemin (H-G/PEDOT) was included in a biosensor to detect H_2O_2. Such sensors result in a great improvement in the sensing performance, including a low limit of detection of 0.08 mM and high sensitivity, selectivity, stability, and reproducibility, and this enhancement is due to the properties of the GO and other components in the composites (Naseri et al., 2018; W. Wang et al., 2014).

Metal Nanoparticles Based Nanocomposites

Metal-based nanocomposites form by including metals or alloys of metal in matrices. A number of metal nanoparticles (NPs) are commonly used, such as gold, silver, platinum, palladium, and copper NPs (Yu & Li, 2019). The excellent properties of such nanoparticles in biosensing are inherited from their composites, including superior conductivity and stability, electrocatalytic properties, high sensitivity, and selectivity (Huang et al., 2021; Kausar et al., 2023; Suresh et al., 2014). Another advantage for the use of metallic nanoparticles is the ease of their process and fabrication compared with other nanostructure materials used in sensors, such as graphene (Huang et al., 2021). Below, we will discuss some advantages of each type of these metal NPs with examples of their applications in biosensors. Gold NPs are noble metal NPs that have great chemical stability and are broadly investigated in biosensing applications in different electrochemical and optical sensors. These nanoparticles are also known for their simple manufacturing methods, biocompatibility, good conductivity, and high catalytic activity (Ramesh et al., 2023; Vidotti et al., 2011). A selective biosensor for the tumour marker alpha fetoprotein (AFP) was formed by incorporation of the nanocomposites Au NPs/PEDOT/PEG matrix, such composites improve the binding between the AFP and its related antibody (Cui et al., 2016; Naseri et al., 2018). Another Au-NPs-based nanocomposite (PANI-AuNPs) was used in an immuno-biosensor to detect carcinoembryonic antigen (CEA). An improvement in the sensor performance was obtained, including enhanced stability, selectivity, and repeatability (Korzeniewska et al., 2020; Naseri et al., 2018). Silver nanoparticles Ag NPs are very similar to Au NPs in their advantages, including biocompatibility and electro-catalytic activity in catalysing electrochemical reactions. Ag NPs have been used in the detection of MERS coronavirus in a paper-based colorimetric biosensor; also, Ag NPs were used to fix hepatitis B antibodies on sensors of platinum electrodes, where a highly stable, sensitive, and selective sensor was obtained (Bisht et al., 2021; Huang et al., 2021). Copper Cu also has been extensively studied for biosensing applications due to its promising properties as a sensing material due to its desirable electric conductivity, good stability, and as well the electro-catalytical properties. Compared with noble metals such as Au and Cu, Cu considered cost-effective (Naresh & Lee, 2021). Cu NPs were used in composites with polyaniline as a matrix for the detection of chloroform, and a desired performing sensor was achieved (Liang et al., 2018a). A hydrogen peroxide sensor based on the nanocomposite of polyaniline (PANI) and Cu NPs on an electrode of glassy carbon showed improved electrocatalytic response compared with the same sensing electrode with this composite, which led to a lower limit of detection and good stability (Liang et al., 2018b; Naseri et al., 2018). Platinum nanoparticles (Pt NPs) and Palladium nanoparticles (Pd NPs) are other metal nanoparticles that have been investigated in biosensor applications and show excellent performance. They possess very similar characteristics as the previously mentioned NPs. The abundance of Pd metal compared with other metals makes it economical and engaging in sensing applications (Yu & Li, 2019; Naresh & Lee, 2021).

Metal Oxide Nanoparticles Based Nanocomposites

Metal oxide NPs have many advantages that make them dominant in different fields. Metal oxide-based nanocomposites have led to improvements in bio-sensing performances due to the attractive properties they hold. These NPs have been utilised in the detection of various biomarker analytes due to their excellent properties, including their strong catalytic activity due to the high number of catalytic sites as they form single atom catalysts (Kannan & Maduraiveeran, 2023). The oxygen atoms linked to the transition metal atom are correlated to the presence of different forms and variable stoichiometry materials (Ramesh et al., 2023). Also, they have high organic capturing capability, low cost, high hardness, biocompatibility, and a high level of crystallinity, with the merit of NPs having a large surface-to-volume ratio (Kannan & Maduraiveeran, 2023). The most commonly used metal oxides in nanocomposites for biosensing applications are tin oxide SnO2, nickel oxide NiO, zinc oxide ZnO, and titanium oxide TiO_2 (Kausar et al., 2023). Some examples of different metal oxides NPs based sensors in biosensors will be included below. Different composites for tin oxide SnO_2 have been studied for the detection of hydrocarbons and NOx gases. The inclusion of these sensing materials in the gas sensors shows high sensitivity, selectivity, and reproducibility characteristics (Kausar et al., 2023; Kulkarni et al., 2018; Su & Lu, 2020). Also, a biosensor based on polypyrrole/tin oxide and polypyrrole/zinc oxide composites was used to detect the presence of ammonia, and an improvement in thermal stability and electrical conductivity was observed (Singh et al., 2021). In glucose sensing, nanocomposites of NiO/CuO/PANI and NiCo2O4/polyaniline were integrated into different electrochemical sensors. Both composites result in highly selective and sensitive biosensors (Ghanbari &

Babaei, 2016; Naseri et al., 2018). Mixed metal oxide nanocomposites of NiO, ZnO, and Fe3O4 were included in electrochemical biosensors to detect dopamine at a pH of 7. Such sensors showed a rapid electron transfer process, which enhanced the electrochemical response of dopamine oxidation and indicated desirable selectivity (Fayemi et al., 2018; Kanan & Maduraiveeran, 2023).

Carbon nanotube based nanocomposites

Carbon nanotubes CNTs are the most widely examined nanotube materials in various applications, such as biosensors, drug delivery, and diagnostics. CNTs are one dimension hollow cylindrical tubes formed as one or more rolled up graphene layers that define whether these CNTs are single-walled (SWCNTs) or double-walled (DWCNTs), or multi-walled (MWCNTs). SWCNTs have a low diameter of 1-2 nm, whereas MWCNTs diameters can reach 100 nm (Curulli, 2023; Ramesh et al., 2023). In the biosensors field, CNTs have various unique properties that make them attractive in this field. These properties include their high thermal stability, mechanical strength, electrical conductivity, surface to volume ratio and significant electrocatalytic activity. Also, they contribute to the sensing process by their excellent capability to capture different analytes, such as ions, and other molecules, such as gaseous molecules. In addition, due to their high electrical conductivity, they have fast electron and charge transmission properties and also improve the electrochemical reactivity of essential biomolecules (Ferrier & Honeychurch, 2021; Kausar et al., 2023; Naresh & Lee, 2021; Suresh et al., 2014.). The CNTs based biosensors are known for their promising sensing features such as sensitivity, selectivity, and fast response. The CNTs polymeric based nanocomposites play a significant role in improving the mechanical properties and porosity as well as the electrical conductivity. These composites can be integrated into different families of biosensors, such as electrochemical and optical sensors (Ferrier & Honeychurch, 2021). As an example of CNTs based nanocomposites in biosensors, an immune electrochemical biosensor based on CNTs and conductive polymers was used for the detection of hepatitis B (Hu et al., 2011). Also, another electrochemical sensor based on another composite of poly(allylamine)/CNT was studied for detecting dengue virus (Silva et al., 2015). A sensitive detection of HIV and other contagious diseases was achieved by applying the nanocomposites of MIPs and MWCNT (MIP=Molecularly imprinted polymers) in different biosensors (Ferrier & Honeychurch, 2021; Meskher et al., 2022). A nanocomposites of single-walled carbon nanotubes (SWCNT) with polytyrosine (Polytyr) (SWCNT/Polytyr) was investigated as an electrode material in a biosensor for the detection of nicotinamide adenine dinucleotide (NADH). Such biosensors provide excellent electrocatalytic activity and also good sensing performance, which makes such a composite mixture promising in other sensing applications as it is later used to detect ethanol (Ates et al., 2017; Eguílaz et al., 2016).

Other nanostructure-based nanocomposites

As a number of different nanocomposites explained above, there are other types of nanocomposites that have been extensively investigated in the literature in the biosensing field and their appealing contribution to the development of the performance of biosensors. These nanocomposites include quantum dot based nanocomposites, magnetic nanoparticles based nanocomposites, nanorods based nanocomposites, and nanowires based nanocomposites. Another commonly used type is hybrid nanocomposite, which is formed by the inclusion of more than one nanostructure (Sen, 2020).

Applications of Nanocomposites based biosensors

Nowadays, there is a great necessity for biosensors with significantly improved performance, as the number of diseases and outbreaks has increased lately and the world population has also increased. Therefore, there will be an expanded demand for more effective biosensors for monitoring essentials in human life and the environment, such as health, water, and food. Different nanomaterials have recently been applied in biosensor applications, and they show an enhancement in biosensor performance, as discussed in the previous section. Some of these materials are nanocomposites with nanostructured materials as fillers, such as gold and silver nanoparticles, quantum dots, carbon nanomaterials (such as CNTs and graphene), and also hybrid nanocomposites where more than one type of nanofiller is included in the composite (D'Souza et al., 2016; Ferrier & Honeychurch, 2021; Naseri et al., 2018; Wang et al., 2020). Nanocomposites have been broadly utilised in many sensing applications, including gas sensors, chemical sensors, and biosensors. In the biosensing application, nanocomposites play a significant role in the bio-recognition process, the transduction process, and generating signals by different means. For example, there are a number of factors that can vary the sensor features, such as the nanofiller material type and amount, the type of polymer used, and the processing technique for the composite (Kausar et al., 2023; Wang et al., 2020). The incorporation of nanocomposites in biosensors shows improvements in the performance and efficiency of such sensors. Also, they help with reducing the sensor sizes, shrinking the consumed time for response to a few minutes and even seconds compared with other analytical techniques (D'Souza et al., 2016). Generally, nanocomposites-based biosensors have been used in a wide range of applications, including health care, water and food quality monitoring, disease diagnosis, drug delivery, and environmental monitoring. For example, as reported in the previous section, glucose biosensors have been exploited for the observation of diabetes disease and to help patients keep their blood sugar at the required levels (Naresh & Lee, 2021; Ramesh et al., 2023).

Some of the applications of nanocomposites-based sensors in the biosensing field are listed below.

- Biomedical mentoring biosensors: When the biosensors are used to monitor the health of the patients and the progress of their condition to help them to In some cases, wearable devices are used (Ramesh et al., 2023).
- Cancer detection biosensors: In this application, the target analytes for the biosensors are called cancer biomarkers, such as proteins, DNA, RNA, and enzymes that are caused by genetic change. Detecting such biomarkers helps in identifying the appropriate therapy (Hasan et al., 2021).
- Diagnostic biosensors: The integration of nanomaterials in the biosensors could lead to precise detection of the target biomarkers in a short time. For example, glucose sensors are vital to checking the glucose level to prevent any associated health problems with its fluctuation (Ghanbari & Babaei, 2016; Ramesh et al., 2023).
- Water and food biosensors: This family of biosensors is used to detect contamination in food and water to ensure the healthy life for humans and other creatures. These contaminations could include, for example, heavy metals (such as mercury Hg and lead Pd) or pesticides. Also, these sensors could help examine food quality through the detection of some markers linked to nutritional damage and the detection of food allergens. (D'Souza et al., 2016)

FUTURE RESEARCH DIRECTIONS

Further investigation of nanocomposites in the biosensing field is very promising. As nanocomposites are attractive in various research areas, they can therefore be studied further in the biosensing field. There are great opportunities for researchers to try new and different nanocomposites to be integrated into the biosensor structures to achieve superior-performance biosensors.

CONCLUSION

The biosensors are such important tools to monitor for example human heath, water quality and the spread of toxic gases. The investigation of the role of nanocomposites in biosensors, or as they are called in this chapter, nanocomposites-based biosensors, demonstrates several advantages. These biosensors have been obtained faster responses, good sensitivity and selectivity. Such merits are significant and as much as needed nowadays due to the changes in the world. The advancement in the nanocomposites materials field is promising, especially here in biosensing applications, and may allow for further studies and investigations for more and new nanocomposites materials in the biosensing field as well as in sensing different target analytes.

REFERENCES

Akgöl, S., & Ulucan-Karnak, F. kuru, C. İ., & Kuşat, K. (2021). The usage of composite nanomaterials in biomedical engineering applications. In Biotechnology and Bioengineering (Vol. 118, Issue 8, pp. 2906–2922). John Wiley and Sons Inc. doi:10.1002/bit.27843

Al Ahmad, M., Mustafa, F., Panicker, N., & Rizvi, T. A. (2021). Optical detection of sars-cov-2 utilizing antigen-antibody binding interactions. *Sensors (Basel)*, *21*(19), 6596. doi:10.3390/s21196596 PMID:34640920

Alghamdi, N., Alqahtani, Z., & Grell, M. (2019). Sub-nanomolar detection of cesium with water-gated transistor. *Journal of Applied Physics*, *126*(6), 064502. doi:10.1063/1.5108730

Alghamdi, N., Alqahtani, Z., Zhou, C., Sano, N., Conte, M., & Grell, M. (2020). Sensing aromatic pollutants in water with catalyst-sensitized water-gated transistor. *Chemical Papers*, *74*(12), 4169–4180. doi:10.1007/s11696-020-01212-3

Alqahtani, Z., Alghamdi, N., & Grell, M. (2020). Monitoring the lead-and-copper rule with a water-gated field effect transistor. *Journal of Water and Health*, *18*(2), 159–171. doi:10.2166/wh.2020.186 PMID:32300089

Ates, M., Eker, A. A., & Eker, B. (2017). Carbon nanotube-based nanocomposites and their applications. In Journal of Adhesion Science and Technology (Vol. 31, Issue 18, pp. 1977–1997). Taylor and Francis Ltd. doi:10.1080/01694243.2017.1295625

Bisht, A., Mishra, A., Bisht, H., & Tripathi, R. M. (2021). Nanomaterial Based Biosensors for Detection of Viruses Including SARS-CoV-2: A Review. In Journal of Analysis and Testing (Vol. 5, Issue 4, pp. 327–340). Nonferrous Metals Society of China. doi:10.1007/s41664-021-00200-0

Biswas, M. R. U. D., & Oh, W. C. (2019). Comparative study on gas sensing by a Schottky diode electrode prepared with graphene-semiconductor-polymer nanocomposites. *RSC Advances*, *9*(20), 11484–11492. doi:10.1039/C9RA00007K PMID:35520227

Cui, M., Song, Z., Wu, Y., Guo, B., Fan, X., & Luo, X. (2016). A highly sensitive biosensor for tumor maker alpha fetoprotein based on poly(ethylene glycol) doped conducting polymer PEDOT. *Biosensors & Bioelectronics*, *79*, 736–741. doi:10.1016/j.bios.2016.01.012 PMID:26774088

Curulli, A. (2023). Functional Nanomaterials Enhancing Electrochemical Biosensors as Smart Tools for Detecting Infectious Viral Diseases. In Molecules (Vol. 28, Issue 9). MDPI. doi:10.3390/molecules28093777

D'Souza, A. A., Kumari, D., & Banerjee, R. (2016). Nanocomposite biosensors for point-of-care—evaluation of food quality and safety. In *Nanobiosensors* (pp. 629–676). Elsevier. doi:10.1016/B978-0-12-804301-1.00015-1

Din, S. H. (2019). Nano-Composites and their Applications: A review. *Characterization and Application of Nanomaterials*, *2*(1). doi:10.24294/can.v2i1.875

Ding, Y., Sun, Y., Liu, C., Jiang, Q. Y., Chen, F., & Cao, Y. (2023). SERS-Based Biosensors Combined with Machine Learning for Medical Application. In ChemistryOpen (Vol. 12, Issue 1). John Wiley and Sons Inc. doi:10.1002/open.202200192

Eguílaz, M., Gutierrez, F., González-Domínguez, J. M., Martínez, M. T., & Rivas, G. (2016). Single-walled carbon nanotubes covalently functionalized with polytyrosine: A new material for the development of NADH-based biosensors. *Biosensors & Bioelectronics*, *86*, 308–314. doi:10.1016/j.bios.2016.06.003 PMID:27387261

Fayemi, O. E., Adekunle, A. S., Kumara Swamy, B. E., & Ebenso, E. E. (2018). Electrochemical sensor for the detection of dopamine in real samples using polyaniline/NiO, ZnO, and Fe3O4 nanocomposites on glassy carbon electrode. *Journal of Electroanalytical Chemistry (Lausanne, Switzerland)*, *818*, 236–249. doi:10.1016/j.jelechem.2018.02.027

Ferrier, D. C., & Honeychurch, K. C. (2021). Carbon nanotube (CNT)-based biosensors. In Biosensors (Vol. 11, Issue 12). MDPI. doi:10.3390/bios11120486

Ghanbari, Kh., & Babaei, Z. (2016). Fabrication and characterization of non-enzymatic glucose sensor based on ternary NiO/CuO/polyaniline nanocomposite. *Analytical Biochemistry*, *498*, 37–46. doi:10.1016/j.ab.2016.01.006 PMID:26778527

Gosai, A., Khondakar, K. R., Ma, X., & Ali, M. A. (2021). Application of functionalized graphene oxide based biosensors for health monitoring: Simple graphene derivatives to 3D printed platforms. In Biosensors (Vol. 11, Issue 10). MDPI. doi:10.3390/bios11100384

Hasan, M. R., Ahommed, M. S., Daizy, M., Bacchu, M. S., Ali, M. R., Al-Mamun, M. R., Saad Aly, M. A., Khan, M. Z. H., & Hossain, S. I. (2021). Recent development in electrochemical biosensors for cancer biomarkers detection. *Biosensors & Bioelectronics: X*, *8*, 100075. doi:10.1016/j.biosx.2021.100075

Hu, Y., Zhao, Z., & Wan, Q. (2011). Facile preparation of carbon nanotube-conducting polymer network for sensitive electrochemical immunoassay of Hepatitis B surface antigen in serum. *Bioelectrochemistry (Amsterdam, Netherlands)*, *81*(2), 59–64. doi:10.1016/j.bioelechem.2011.01.005 PMID:21458390

Huang, X., Zhu, Y., & Kianfar, E. (2021). Nano Biosensors: Properties, applications and electrochemical techniques. In Journal of Materials Research and Technology (Vol. 12, pp. 1649–1672). Elsevier Editora Ltda. doi:10.1016/j.jmrt.2021.03.048

Jiang, W., Huang, L., Zhang, D., Wang, Y., & Pan, G. (2021). Graphene-Based Composites for Electrochemical Sensor Fabrication and Their Application for Drug Detection. *International Journal of Electrochemical Science*, *16*(3), 1–13. doi:10.20964/2021.03.69

Kalakonda, S. N., Bammidi, R., Edubilli, H., Medapati, S., Boyina, S. L. D., & Prasad, V. V. S. R. (2023). Biosensors – An Insight into the Electrochemical and Optical Biosensors. *International Journal of Pharmaceutical Investigation*, *13*(3), 402–412. doi:10.5530/ijpi.13.3.051

Kannan, P., & Maduraiveeran, G. (2023). Metal Oxides Nanomaterials and Nanocomposite-Based Electrochemical Sensors for Healthcare Applications. In Biosensors (Vol. 13, Issue 5). MDPI. doi:10.3390/bios13050542

Kausar, A., Ahmad, I., Zhao, T., Aldaghri, O., Ibnaouf, K. H., & Eisa, M. H. (2023). Multifunctional Polymeric Nanocomposites for Sensing Applications—Design, Features, and Technical Advancements. In Crystals (Vol. 13, Issue 7). Multidisciplinary Digital Publishing Institute (MDPI). doi:10.3390/cryst13071144

Kazemi, F., Naghib, S. M., Zare, Y., & Rhee, K. Y. (2021). Biosensing Applications of Polyaniline (PANI)-Based Nanocomposites: A Review. In Polymer Reviews (Vol. 61, Issue 3, pp. 553–597). Taylor and Francis Ltd. doi:10.1080/15583724.2020.1858871

Korzeniewska, E., Szczęsny, A., & Kasprowicz, D. (2020). IDEs structures created in the physical vacuum deposition process on textile substrates. *Journal of Physics: Conference Series*, *1534*(1), 012004. doi:10.1088/1742-6596/1534/1/012004

Kulkarni, S., Patil, P., Mujumdar, A., & Naik, J. (2018). Synthesis and evaluation of gas sensing properties of PANI, PANI/SnO2 and PANI/SnO2/rGO nanocomposites at room temperature. *Inorganic Chemistry Communications*, *96*, 90–96. doi:10.1016/j.inoche.2018.08.008

Li, Y., He, R., Niu, Y., & Li, F. (2019). Paper-Based Electrochemical Biosensors for Point-of-Care Testing of Neurotransmitters. In Journal of Analysis and Testing (Vol. 3, Issue 1, pp. 19–36). Nonferrous Metals Society of China. doi:10.1007/s41664-019-00085-0

Liang, J., Wei, M., Wang, Q., Zhao, Z., Liu, A., Yu, Z., & Tian, Y. (2018a). Sensitive Electrochemical Determination of Hydrogen Peroxide Using Copper Nanoparticles in a Polyaniline Film on a Glassy Carbon Electrode. *Analytical Letters*, *51*(4), 512–522. doi:10.1080/00032719.2017.1343832

Liang, J., Wei, M., Wang, Q., Zhao, Z., Liu, A., Yu, Z., & Tian, Y. (2018b). Sensitive Electrochemical Determination of Hydrogen Peroxide Using Copper Nanoparticles in a Polyaniline Film on a Glassy Carbon Electrode. *Analytical Letters*, *51*(4), 512–522. doi:10.1080/00032719.2017.1343832

Malhotra, B. D., & Pandey, C. M. (2017). *Biosensors: Fundamentals and Applications*. https://www.ebsco.com/terms-of-use

Meskher, H., Mustansar, H. C., Thakur, A. K., Sathyamurthy, R., Lynch, I., Singh, P., Han, T. K., & Saidur, R. (2022). Recent trends in carbon nanotube (CNT)-based biosensors for the fast and sensitive detection of human viruses: a critical review. In Nanoscale Advances (Vol. 5, Issue 4, pp. 992–1010). Royal Society of Chemistry. doi:10.1039/D2NA00236A

Mills, D. K., & Nestorova, G. G. (2023). Biosensor Development and Innovation in Healthcare and Medical Applications. In Sensors (Vol. 23, Issue 5). MDPI. doi:10.3390/s23052717

Mohammadi, M. M., Kumar, A., Liu, J., Liu, Y., Thundat, T., & Swihart, M. T. (2020). Hydrogen Sensing at Room Temperature Using Flame-Synthesized Palladium-Decorated Crumpled Reduced Graphene Oxide Nanocomposites. *ACS Sensors*, *5*(8), 2344–2350. doi:10.1021/acssensors.0c01040 PMID:32786377

Naresh, V., & Lee, N. (2021). A review on biosensors and recent development of nanostructured materials-enabled biosensors. In Sensors (Switzerland) (Vol. 21, Issue 4, pp. 1–35). MDPI AG. doi:10.3390/s21041109

Naseri, M., Fotouhi, L., & Ehsani, A. (2018). Recent Progress in the Development of Conducting Polymer-Based Nanocomposites for Electrochemical Biosensors Applications: A Mini-Review. *Chemical Record (New York, N.Y.)*, *18*(6), 599–618. doi:10.1002/tcr.201700101 PMID:29460399

Ramesh, M., Janani, R., Deepa, C., & Rajeshkumar, L. (2023a). Nanotechnology-Enabled Biosensors: A Review of Fundamentals, Design Principles, Materials, and Applications. In Biosensors, 13(1). MDPI. doi:10.3390/bios13010040

Ramesh, M., Janani, R., Deepa, C., & Rajeshkumar, L. (2023b). Nanotechnology-Enabled Biosensors: A Review of Fundamentals, Design Principles, Materials, and Applications. In Biosensors, 13(1). MDPI. doi:10.3390/bios13010040

Rocchitta, G., Spanu, A., Babudieri, S., Latte, G., Madeddu, G., Galleri, G., Nuvoli, S., Bagella, P., Demartis, M. I., Fiore, V., Manetti, R., & Serra, P. A. (2016). Enzyme biosensors for biomedical applications: Strategies for safeguarding analytical performances in biological fluids. In Sensors (Switzerland), 16(6). MDPI AG. doi:10.3390/s16060780

Sen, M. (2020). *Nanocomposite Materials*. www.intechopen.com

Seo, G., Lee, G., Kim, M. J., Baek, S. H., Choi, M., Ku, K. B., Lee, C. S., Jun, S., Park, D., Kim, H. G., Kim, S. J., Lee, J. O., Kim, B. T., Park, E. C., & Kim, S. (2020). Rapid Detection of COVID-19 Causative Virus (SARS-CoV-2) in Human Nasopharyngeal Swab Specimens Using Field-Effect Transistor-Based Biosensor. *ACS Nano*, *14*(4), 5135–5142. doi:10.1021/acsnano.0c02823 PMID:32293168

Sharma, D., Kanchi, S., Sabela, M. I., & Bisetty, K. (2016). Insight into the biosensing of graphene oxide: Present and future prospects. *Arabian Journal of Chemistry*, *9*(2), 238–261. doi:10.1016/j.arabjc.2015.07.015

Silva, M. M. S., Dias, A. C. M. S., Silva, B. V. M., Gomes-Filho, S. L. R., Kubota, L. T., Goulart, M. O. F., & Dutra, R. F. (2015). Electrochemical detection of dengue virus NS1 protein with a poly(allylamine)/carbon nanotube layered immunoelectrode. *Journal of Chemical Technology and Biotechnology*, *90*(1), 194–200. doi:10.1002/jctb.4305

Singh, P., Kushwaha, C. S., Singh, V. K., Dubey, G. C., & Shukla, S. (2021). Chemiresistive sensing of volatile ammonia over zinc oxide encapsulated polypyrrole based nanocomposite. *Sensors and Actuators. B, Chemical*, *342*, 130042. doi:10.1016/j.snb.2021.130042

Singh, R. P., Singh, P., & Singh, K. R. B. (2020a). Introduction to Composite Materials. In *Composite Materials* (pp. 1–28). CRC Press. doi:10.1201/9781003080633-1

Singh, R. P., Singh, P., & Singh, K. R. B. (2020b). Introduction to Composite Materials. In *Composite Materials* (pp. 1–28). CRC Press. doi:10.1201/9781003080633-1

Sivarajasekar, N., Dharshni, Ra., & Dharshini, Rd. (2019). Liquid-liquid extraction for wastewater treatment View project THERMAL BIOSENSORS AND THEIR APPLICATIONS. *American International Journal of Research in Science*, *50*. http://www.iasir.net

Su, P.-G., & Lu, P.-H. (2020). Electrical and Humidity-Sensing Properties of Impedance-Type Humidity Sensors that Were Made of Ag Microwires/PPy/SnO2 Ternary Composites. *Chemosensors (Basel, Switzerland)*, *8*(4), 92. doi:10.3390/chemosensors8040092

Suresh, S., Mathan, P., Sagadevan, S., & Periasamy, M. (2014). RECENT TRENDS IN NANOBIOSENSORS AND THEIR APPLICATIONS-. *RE:view*.

Vidotti, M., Carvalhal, R. F., Mendes, R. K., Ferreira, D. C. M., & Kubota, L. T. (2011). Biosensors based on gold nanostructures. *Journal of the Brazilian Chemical Society*, *22*(1), 3–20. doi:10.1590/S0103-50532011000100002

Vroman, I., & Tighzert, L. (2009). Biodegradable polymers. In Materials, 2(2), 307–344. doi:10.3390/ma2020307

Wan, Q., Han, L., Guo, Y., Yu, H., Tan, L., Zheng, A., & Chen, Y. (2021). Graphene-based biosensors with high sensitivity for detection of Ovarian cancer cells. *Molecules (Basel, Switzerland)*, *26*(23), 7265. doi:10.3390/molecules26237265 PMID:34885851

Wang, D., Huang, B., Liu, J., Guo, X., Abudukeyoumu, G., Zhang, Y., Ye, B.-C., & Li, Y. (2018). A novel electrochemical sensor based on Cu@Ni/MWCNTs nanocomposite for simultaneous determination of guanine and adenine. *Biosensors & Bioelectronics*, *102*, 389–395. doi:10.1016/j.bios.2017.11.051 PMID:29174972

Wang, T., Zhu, H., Zhuo, J., Zhu, Z., Papakonstantinou, P., Lubarsky, G., Lin, J., & Li, M. (2013). Biosensor based on ultrasmall MoS2 nanoparticles for electrochemical detection of H2O2 released by cells at the nanomolar level. *Analytical Chemistry*, *85*(21), 10289–10295. doi:10.1021/ac402114c PMID:24067077

Wang, W., Xu, G., Cui, X. T., Sheng, G., & Luo, X. (2014). Enhanced catalytic and dopamine sensing properties of electrochemically reduced conducting polymer nanocomposite doped with pure graphene oxide. *Biosensors & Bioelectronics*, *58*, 153–156. doi:10.1016/j.bios.2014.02.055 PMID:24632460

Wang, X., Li, F., & Guo, Y. (2020). Recent Trends in Nanomaterial-Based Biosensors for Point-of-Care Testing. *Frontiers in Chemistry*, *8*, 586702. doi:10.3389/fchem.2020.586702 PMID:33195085

Wang, Y., Qing, X., Zhou, Q., Zhang, Y., Liu, Q., Liu, K., Wang, W., Li, M., Lu, Z., Chen, Y., & Wang, D. (2017). The woven fiber organic electrochemical transistors based on polypyrrole nanowires/reduced graphene oxide composites for glucose sensing. *Biosensors & Bioelectronics*, *95*, 138–145. doi:10.1016/j.bios.2017.04.018 PMID:28437640

Wu, G., Lee, D., Cha, Y. L., Kim, W., Kim, D.-J., Du, H., & Zhang, X. (2022). *Polyaniline/Graphene-Functionalized Flexible Waste Mask Sensors for Ammonia and Volatile Sulfur Compound Monitoring*. American Chemical Society. doi:10.1021/acsami.2c15443

Yu, L., & Li, N. (2019). Noble metal nanoparticles-based colorimetric biosensor for visual quantification: A mini review. In Chemosensors, 7(4). MDPI Multidisciplinary Digital Publishing Institute. doi:10.3390/chemosensors7040053

Zhang, C., Gholipour, G., & Mousavi, A. A. (2021). State-of-the-Art Review on Responses of RC Structures Subjected to Lateral Impact Loads. *Archives of Computational Methods in Engineering*, *28*(4), 2477–2507. doi:10.1007/s11831-020-09467-5

Zhou, Z., Zhang, X., Wu, X., & Lu, C. (2016). Self-stabilized polyaniline@graphene aqueous colloids for the construction of assembled conductive network in rubber matrix and its chemical sensing application. *Composites Science and Technology*, *125*, 1–8. doi:10.1016/j.compscitech.2016.01.016

ADDITIONAL READING

Kaur, G., Kaur, A., & Kaur, H. (2021). Review on nanomaterials/conducting polymer based nanocomposites for the development of biosensors and electrochemical sensors. *Polymer-Plastics Technology and Materials*, *60*(5), 504–521. doi:10.1080/25740881.2020.1844233

Mohammadpour-Haratbar, A., Zare, Y., & Rhee, K. Y. (2022). Electrochemical biosensors based on polymer nanocomposites for detecting breast cancer: Recent progress and future prospects. *Advances in Colloid and Interface Science*, *309*, 102795. doi:10.1016/j.cis.2022.102795 PMID:36242876

Tao, C., & Rouhi, J. (2023). A biosensor based on graphene oxide nanocomposite for determination of carcinoembryonic antigen in colorectal cancer biomarker. *Environmental Research*, *238*, 117113. doi:10.1016/j.envres.2023.117113 PMID:37696325

KEY TERMS AND DEFINITIONS

Biocompatibility: Ability to be within different living systems without causing any damages.

Biorecognition: The process where the receptors in the biosensor bind to the target analytes and give signals that can be measured.

Biosensor: A sensor that is used to detect biological signals.

Catalytic activity: increase in the chemical reaction rate at the presence of the catalyst.

Electrochemical reaction: The chemical reaction that results in a change in the electrical properties.

Limit of Detection: The lower concentration of the analyte that a sensor is able to detect.

Nanocomposites: Mixtures of nanostructure materials and other matrix material.

Chapter 9
Applications of Bionanocomposites in Food Packaging

Mukhtar Iderawumi Abdulraheem
https://orcid.org/0000-0002-6745-378X
Henan Agricultural University, China

Gholamreza Abdi
Persian Gulf University, Iran

Nafeesa Farooq Khan
University of Kashmir, India

Abdul Wahab
Chinese Academy of Sciences, Shanghai, China

Kasahun Gudeta
Adama Science and Technology University, Ethiopia

Chandan Singh
Dr. Harisingh Gour Vishwavidyalaya, India

Oluwadamilola Oluwatoyin Hazzan
Chinese Academy of Sciences, Xiamen, China

ABSTRACT

Although bio-based nanocomposite materials have various packaging options, the future is still evolving. At this point, multifunctional intelligent packaging will replace simple traditional packaging. As the use of synthetic plastics decreases, so does the demand for biodegradable packaging materials. Because of its potential to alleviate environmental issues associated with conventional packaging materials, biodegradable nanocoating for food packaging has seen a significant increase in interest in recent years. This innovative technology offers several perspectives and benefits, including improved food safety, extended shelf life, reduced waste, and enhanced sustainability. The objectives of this chapter are to highlight the latest advancements and research in the field of biodegradable nanocoating and its potential to revolutionize the food packaging industry as well as the advanced development teams about the benefits and potential applications of biodegradable nanocoating in food packaging.

DOI: 10.4018/979-8-3693-1094-6.ch009

Copyright © 2024, IGI Global. Copying or distributing in print or electronic forms without written permission of IGI Global is prohibited.

INTRODUCTION

The current global crisis in agriculture is worrisome, with the depletion of natural resources, food loss and waste, population growth, and climate change being all contributing factors. It is concerning to note that UN projections estimate the global population could reach 9.15 billion by 2050, an increase of 2.25 billion (Iderawumi and Yusuff 2021). Thus, we must start implementing sustainable solutions to ensure the future of our food production and supply. Furthermore, global agricultural output and consumption are expected to rise by 60% by 2050 (Kim and Lee 2022; Pawlak and Kołodziejczak 2020). Agricultural production may be expanded to satisfy this need by adopting advanced technology, large farm sizes, and improved cultivar technologies (Ahmad et al. 2023; Panel 2019). However, a complete and sustainable technique for raising agricultural yields while reducing related environmental pollutants has yet to be adequately described (Sharma et al. 2022). Logistics systems, such as food packaging and market services, must be digitized to achieve this. However, using synthetic plastic in food packaging can negatively impact the environment and climate (Chisenga et al. 2020). Therefore, the demand for eco-friendly packaging materials is rising as a sustainable alternative.

Bio-nanocomposites for food packaging and contact surfaces are advancing rapidly (Alsufyani and M'Sakni N 2023). Inorganic nanoparticles in polymers and composites will drive progress for the next ten years, offering significant benefits for food safety and sustainability (Sorrentino et al. 2007; Wahab et al. 2023). Plastic is the most sought-after component for food packaging due to its fantastic features, which can fit every packaging element. However, the complexities of dealing with plastic waste forced the company to focus on developing creative alternative bio-sourced polymers (Moshood et al. 2022). Hence, using bio-sourced plastic packaging material in food packaging has enormous potential (Sid et al. 2021). Packaging technologies have recently significantly improved product shelf life, food quality, safety, and freshness (Barage et al. 2022; Vasile and Baican 2021). Active and smart materials used in food packaging include biodegradable polymers, nanocomposite materials, nanostructured materials, and tailored nanoparticles (Ahari and Soufiani 2021). Furthermore, the suggested procedures and equipment allow for rapid in-situ analysis and the production of environmentally safe and biodegradable food packaging (Jung et al. 2022).

Biodegradable nanocoatings help promote perishable foods' shelf-life by providing a protective barrier against oxygen, moisture, and light. These coatings can slow deterioration by minimizing oxidative reactions and inhibiting microbial growth. As a result, packaged food's freshness and quality can be maintained for a longer period, reducing food waste and improving overall sustainability (Dehghani et al. 2018). Traditional packaging materials often contribute to environmental pollution due to their non-biodegradable nature. Biodegradable nanocoatings offers an eco-friendly alternative by utilizing renewable resources and enabling the decomposition of packaging materials after use (Kandeepan 2021). This reduces the accumulation of plastic waste in landfills and minimizes the environmental impact.

To keep food fresh and safe, biodegradable nanocoatings are essential for packaging with coatings, made from biopolymers such as starch, cellulose, chitosan, and proteins, which form a strong barrier against moisture loss, gas exchange, and microbial contamination (Ezati and Rhim 2020). This results in high-quality packaged food and staying fresh for longer (Tan et al. 2021). One of the best things about these nanocoatings is that they significantly improve food safety by preventing harmful substances from entering the food through packaging materials (Wahab et al. 2023). Unlike traditional packaging materials, such as plastics, which can contain additives or chemicals that pose health risks, biodegradable nanocoatings act as a shield, greatly reducing the risk of contamination (Al-Tayyar et al. 2020). Enhanced sustainability is an essential aspect of biodegradable nanocoatings in food packaging. Traditional packaging materials, such as plastic, require fossil fuels and emit greenhouse gases. On the other hand, biodegradable nanocoatings can be made from renewable resources and thus have a lower carbon footprint. By using these coatings, the food industry can contribute to a more sustainable future by reducing its reliance on nonrenewable resources and reducing climate change (Tan et al. 2021).

Biodegradable nanocoating has emerged in recent years as a significant solution to the environmental issues associated with traditional food packaging materials (Ahmad et al. 2023). These nanocoatings offer several advantages, including enhanced barrier properties, extended shelf life, and reduced waste generation. As the demand for sustainable packaging solutions continues to grow, the prospects of biodegradable nanocoating in food packaging appear promising (Zhang and Sablani 2021). The main benefits of using biodegradable nanocoatings in food packaging are that they are environmentally friendly, economical, efficient, and effective (Wahab et al. 2023). Biodegradable nanocoatings break down naturally over time, unlike traditional plastic packaging, which can take hundreds of years to decompose, reducing waste in landfills and oceans (Din et al. 2020). Furthermore, using biodegradable nanocoatings helps extend food shelf life by acting as a barrier against moisture and oxygen, preventing spoilage, and keeping food fresher for extended periods. This is especially critical for perishable fruits, vegetables, and meats.

This chapter aims to give a comprehensive overview of the potential advantages, obstacles, and future possibilities of utilizing biodegradable nanocoatings in food packaging applications. It delves into the diverse perspectives surrounding this emerging technology and its impact on the food industry. Additionally, the chapter examines the benefits of incorporating biodegradable nanocoatings in food packaging to extend the shelf life of perishable food items. These coatings create safeguards and preserve the quality and freshness of food by preventing microbial growth and minimizing chemical reactions that lead to degradation.

Furthermore, the chapter sheds light on the difficulties of implementing biodegradable nanocoatings in food packaging. In this chapter, we delve into the latest innovations in nanomaterials, coating methods, and characterization techniques that pertain to biodegradable nanocoatings utilized in food packaging (Onyeaka et al. 2022). Our goal is to offer a thorough comprehension of the possible effects of this burgeoning technology on the food sector. We examine different viewpoints, benefits, and hurdles and scrutinize current research endeavors. The primary objective is to enrich the knowledge base and aid well-informed decision-making in sustainable food packaging by showcasing ongoing investigations and potential avenues for future exploration.

SYNTHETIC BIODEGRADABLE NANOCOATINGS MATERIALS FOR FOOD PACKAGING

Recently, there has been a surge of interest in utilizing synthetic biodegradable nanocoating materials for food packaging. This innovative material has the potential to enhance the safety and longevity of food products while simultaneously reducing their environmental footprint (Mohammad et al. 2022; Souza and Fernando 2016). Essentially, the primary function of these nanocoating is to create a barrier that shields the packaged food against external factors such as moisture and oxygen, which can significantly compromise its quality (Souza and Fernando 2016). This response will delve into the different kinds of synthetic biodegradable nanocoating materials used in food packaging, their unique attributes, and their potential benefits.

Methods and Types of Synthetic Biodegradable Nanocoatings

Synthetic biodegradable nanocoatings are materials that possess both biodegradability and nanoscale size. These coatings are often composed of synthetic polymers or other materials that can be degraded over time by natural processes such as microbes or enzymes (Table 1). Synthetic biodegradable nanocoatings come in a variety of forms, each with its own set of unique features and applications (Table 1).

Polylactic Acid (PLA): PLA is a widely used biodegradable polymer derived from renewable resources such as corn starch or sugarcane (Maraveas 2020; Perera et al. 2023; Taib et al. 2023). It boasts remarkable film-forming capabilities and can be quickly transformed into nanocoatings, which offer superb defense against oxygen and water vapor (Maraveas 2020; Sun et al. 2018). As a result, these nanocoatings are ideal for safeguarding food during packaging. Additionally, PLA is biocompatible and non-toxic, ensuring the food's safety (Maraveas 2020; Svagan et al. 2012). Additionally, various other polymers and fillers have been added to enhance the performance and decrease the cost of the final product, and integrating nanofillers such as talc, silica, and nanoadditives has enhanced food products (Taib et al. 2023; Wang et al. 2022). However, incorporating nanomaterials may result in substantial costs, which can be mitigated by including raw materials in the mixture.

Polyhydroxyalkanoates (PHA): PHA refers to a series of biodegradable polymers produced by various microorganisms during fermentation (Fitriani et al. 2023). These versatile polymers can be tailored to meet specific application requirements and offer unique features (Nilsen-Nygaard et al. 2021; Vasile 2018). PHA nanocoatings are particularly effective at preventing the permeation of gases and moisture, making them an excellent choice for food packaging (Fitriani et al. 2023; Westlake et al. 2023). PHA synthesis requires excess carbon and limited nitrogen in the substrate, producing hydroxyalkyl bacterial metabolites due to low oxygen and nutrient conditions (Costa et al. 2019). Moreover, PHA is biocompatible and can be decomposed by microorganisms in the environment (Agarwal et al. 2023).

Polybutylene Succinate (PBS): PBS is a remarkable type of polyester that is both eco-friendly and biodegradable (Westlake et al. 2023). Made from renewable resources such as succinic acid and 1,4-butanediol, it boasts impressive mechanical properties and can effectively create a barrier against moisture and gases (Hong et al. 2021; Su and Kang 2020). When used as nanocoatings, PBS has been scientifically shown to prevent spoilage caused by microbial growth or oxidation, extending the shelf life of a wide range of food products (Westlake et al. 2023).

Polyvinyl Alcohol (PVA): PVA is a synthetic polymer that dissolves in water and can be altered to create biodegradable nanocoatings (Perera et al. 2023). These coatings have excellent barrier properties against oxygen and can effectively regulate the release of active ingredients like antimicrobials and antioxidants (Fitriani et al. 2023; Vasile and Baican 2021). Nevertheless, PVA's sensitivity to moisture restricts its use in certain food packaging situations.

Chitosan: a biopolymer derived from chitin, a natural polysaccharide found in crustacean exoskeletons, boasts impressive film-forming properties and has garnered significant attention for its potential applications in food packaging (Al-Naamani et al. 2016; De Silva et al. 2017; Hong et al. 2021). Notably, chitosan nanocoatings possess antimicrobial qualities that effectively inhibit the growth of spoilage-causing microorganisms on the surface of packaged food (Ezati and Rhim 2020; Perera et al. 2023).

Cellulose-based Polymers: Derived from wood, cotton, or agricultural residues, cellulose is an exceptional material for creating polymers used in food packaging (Gadhave et al. 2018). These polymers, which are both transparent and mechanically robust, improve the shelf life of food products when used as films or coatings

Table 1. Comparison and summary of various nanocoatings technologies

Biodegradable polymers	Sources	Properties	Uses	Advantages	Limitations	References
PLA	renewable resources	Biodegradable and bio-based polymer suitable for food packaging	For products like fresh produce, deli items, and bakery goods	Good mechanical properties, transparency, and is heat sealable	Unstable and can degrade when exposed to water, relatively short shelf-life, high cost and melting point of around 190-210°C	(Perera et al. 2023; Shaikh et al. 2021)
PHA	Synthesized by bacteria	Produced by microorganisms under certain conditions and have barrier properties against oxygen and moisture	For items that require good barrier properties, such as snack foods and perishable goods	Biodegradability, biocompatibility, and potential applications	Limited availability and affordability, expensive, and not compatible with certain food products	(Nilsen-Nygaard et al. 2021)
PBS	Petroleum	It is compostable under industrial conditions	For items like disposable cutlery, cups, and trays.	good mechanical properties, flexibility, and biodegradability	Expensive materials to produce, short shelf life due to susceptibility to degradation from heat and melting point of around 150-170°C	(Fitriani et al. 2023)
PVA	Synthetic polymer	Modified to improve its barrier properties and mechanical strength	For applications such as water-soluble pouches for single-dose products and edible films for items like cheese slices	attractive option for single-use plastic products	Unstable, expensive materials to produce, relatively short shelf life and melting point of around 210-230°C	(Shaikh et al. 2021; Westlake et al. 2023)
Chitosan	Crustacean shells	It has antimicrobial properties and can form films with good barrier properties	Extend the shelf life of perishable products due to their antimicrobial properties	Biocompatible, biodegradable, and non-toxic	Expensive materials to produce, short shelf life, unstable and melting point of around 150-180°C	(Hong et al. 2021)
Cellulose-based polymer	Cellulose	Biodegradability and renewable sourcing.	Used in food packaging for items like pouches, wraps, and coatings	Renewable, biodegradable, and have excellent mechanical properties	Expensive to produce, short shelf life and melting point of around 250-280°C	(Fitriani et al. 2023)
Starch-based polymer	Starch from plant	good film-forming properties and can be modified to improve their barrier properties	For applications like trays, containers, and films.	Biodegradable, renewable, and exhibit good film-forming properties	Expensive, short shelf life and melting point of around 150-180°C	(Gadhave et al. 2018; Shaikh et al. 2021)

(Agarwal et al. 2023; Westlake et al. 2023). Nonetheless, they may have specific water resistance and processability limitations.

Starch-based Polymers: Polymers derived from starch and plant-based sources, such as corn, potatoes, or cassava, are frequently utilized in food packaging due to their exceptional mechanical strength and water vapor barrier properties (Gadhave et al. 2018; Perera et al. 2023). However, they may not excel in the oxygen barrier because their properties can be enhanced by mixing with other biopolymers or additives. Starch-based polymers are biodegradable and compostable when subjected to the appropriate conditions.

In conclusion, each of these polymers has distinct features that make them suited for a variety of food packaging applications. While PLA provides transparency and heat scalability, PHA provides outstanding oxygen and moisture barrier characteristics. PBS is biodegradable and has high mechanical qualities, whilst PVA's water solubility makes it suitable for some packaging applications. Because of its antibacterial qualities, chitosan is perfect for increasing the shelf life of perishable items, whilst cellulose- and starch-based polymers provide renewable sources and biodegradability.

Biodegradation of Biodegradable Nanocoatings

The natural breakdown of organic materials is known as biodegradation, a sustainable alternative to conventional coating. Biodegradable nanocoatings are being researched for various uses, such as packaging, electronics, textiles, and biomedical devices (Bajpai et al. 2018). The degradation process of nanocoatings is influenced by various factors such as the material type, the specific enzymes or microorganisms involved in the breakdown, and the surrounding environmental conditions (Kuswandi 2017; Remya et al. 2022). The choice of natural (cellulose, chitosan, starch, and proteins) or synthetic polymers derived from renewable resources like polylactic acid (PLA), polyhydroxyalkanoates (PHA), and polycaprolactone (PCL) are typical considerations based on their degradability and compatibility with the underlying substrate (Sun et al. 2018; Svagan et al. 2012). The mechanisms of biodegradation nanocoating in food packaging is a complex process that involves for nanocoatings involves preparing a biodegradable, adsorption it onto food packaging materials, allowing it to interact with food, undergoing compostable (Figure 1)

The biodegradation rate is influenced by various factors, such as chemical structure, environmental conditions, and additives (Din et al. 2020). Crystalline or crosslinked polymers take longer to degrade, whereas additives can either hinder or accelerate the process. Creating biodegradable nanocoatings necessitates meticulous attention to environmental factors, including temperature, pH levels, and moisture (Panou and Karabagias 2023). While outdoor coatings must endure diverse conditions, they

Figure 1. Mechanisms of action of biodegradable nanocoatings in food packaging

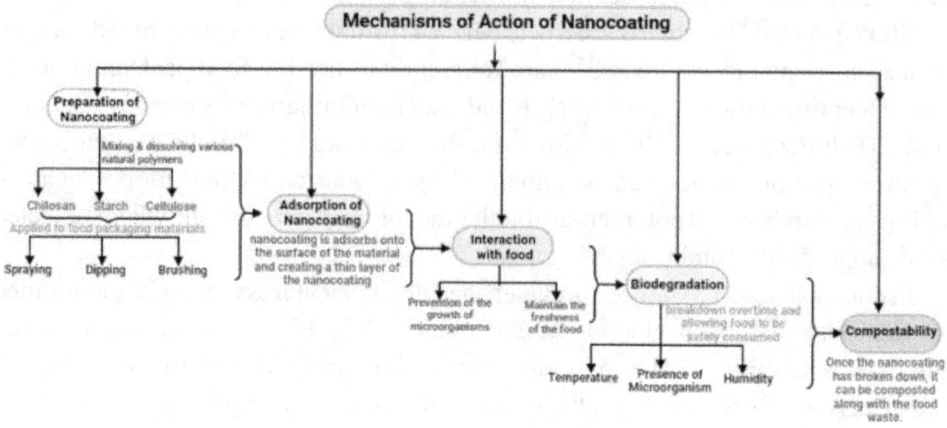

must also facilitate biodegradation. It is vital to consider the existence of certain enzymes and microorganisms that can decompose the coating material (Cagri et al. 2004). The use of biodegradable nanocoatings offers several advantages over non-biodegradable coatings. One of the most significant benefits is the reduction of waste accumulation in landfills, contributing to a more sustainable waste management system (Shaikh et al. 2021). Additionally, biodegradable nanocoatings can help minimize pollution by minimizing the ingestion of toxic materials into the environment during degradation. Moreover, these coatings offer potential health and safety benefits compared to traditional coatings that often contain toxic substances such as heavy metals or volatile organic compounds (VOCs).

Biodegradable nanocoatings made from natural or renewable materials are safer and more environmentally friendly. When it comes to biodegradable nanocoatings, finding the right balance between durability and biodegradability is crucial for their specific use. The biodegradation process is influenced by multiple factors, making it essential to carefully approach these coatings' creation for optimal performance and sustainability. The benefits of such coatings are manifold and include better waste management, reduced pollution, and improved human health and safety.

Reinforced Biodegradable Materials: Biocomposites and Bio-Nanocomposites

There has been a notable increase in interest in using biodegradable materials like biocomposites and bio-nanocomposites. These materials can achieve an optimal balance of biodegradability and mechanical strength through reinforcing agents (Kovačević et al. 2021). Specifically, biocomposites comprise a biopolymer matrix reinforced with natural fibers or particles such as flax, hemp, jute, sisal, bamboo, or kenaf. In order to produce biocomposites, natural fibers are treated to ensure compatibility with a biopolymer matrix before being mixed and shaped through compression molding or extrusion. These materials boast excellent mechanical properties, are environmentally friendly, and can be disposed of with minimal environmental impact (Fortunati et al. 2012).

Bio-nanocomposites are advanced materials that contain minuscule reinforcing agents at the nanoscale within the biopolymer matrix (Alsufyani and M'Sakni N 2023). These agents come in nanoparticles, nanofibers, or nanoclays and can significantly boost the material's mechanical properties by improving its strength, stiffness, and toughness. Multiple techniques can be utilized to integrate these reinforcements, ultimately leading to superior mechanical characteristics. Biodegradable materials, such as biocomposites and bio-nanocomposites, possess enhanced barrier properties, flame retardancy, and thermal stability, making them superior to non-biodegradable materials. They are highly suitable for food packaging and agricultural films, providing advantages for mechanical performance and environmental sustainability (Mir et al. 2018).

APPLICATIONS OF BIODEGRADABLE NANOCOATINGS IN FOOD PACKAGING

Biodegradable nanocoatings offer numerous applications that can address various challenges related to food preservation, safety, and sustainability. Here are some critical applications of biodegradable nanocoatings in food packaging:

- Barrier protection: Biodegradable nanocoatings can be applied to packaging material to create a barrier against oxygen, moisture, and contaminants, which preserves the freshness, flavor, and quality of food for a longer period of time (Amini et al. 2014).
- Extended shelf life: Biodegradable nanocoatings enhance the durability and handiness of food products by preventing the penetration of oxygen and moisture. This can decrease food waste and provide healthier food options for consumers.
- Antimicrobial properties: Engineered biodegradable nanocoatings can inhibit the growth of bacteria, molds, and fungi on food surfaces, enhancing food safety and preventing foodborne illnesses (Al-Tayyar et al. 2020; Cagri et al. 2004).

- Active packaging: Active compounds like antioxidants, antimicrobials, and oxygen scavengers can be gradually released from biodegradable nanocoatings to enhance food quality, safety, and shelf life (Naqvi et al. 2022).
- Smart packaging: Nanocoatings can detect changes in temperature, oxygen levels, or moisture content, allowing for intelligent packaging solutions that provide instant information on food quality, freshness, and potential spoilage.
- Sustainable packaging: Biodegradable nanocoatings provide an eco-friendly option for packaging, reducing the environmental impact of food packaging compared to non-biodegradable materials.
- Brand differentiation: Food companies can use biodegradable nanocoatings to differentiate their products and enhance brand reputation by offering sustainable food packaging due to the prevailing information on the environmental impact of packaging.

It is important to note that biodegradable nanocoatings require further research and development to optimize their performance, ensure safety, and address scalability and cost-effectiveness challenges (Amini et al. 2014). Biodegradable nanocoatings have numerous advantages for food preservation, safety, and health, and they enhance the freshness of food items, minimize food waste, preserve nutritional value and freshness, reduce the reliance on chemical additives, elevate food hygiene, offer antimicrobial benefits, detect spoilage, limit exposure to harmful substances, and improve the overall quality of food.

Sustainable Packaging Through Using Biodegradable Nanocoatings

With traditional packaging materials' harmful and adverse environmental effects, there is a growing need for innovative solutions that prioritize sustainability and waste reduction. Biodegradable nanocoatings are an exciting option for sustainable packaging in the future (Ashfaq et al. 2022). This eco-friendly and functional protective layer comprises natural polymers or bio-based materials that can decompose into non-toxic substances when exposed to environmental conditions. Integrating biodegradable nanocoatings into packaging makes it possible to prolong the lifespan of products while improving their barrier properties, thus boosting overall product performance and minimizing environmental harm (Shaikh et al. 2021; Zhang and Sablani 2021). Integrating biodegradable nanocoatings into packaging materials can significantly reduce waste by enabling them to decompose naturally over time. Furthermore, these coatings aid in segregation and recycling various components of packaging materials, thereby promoting a circular economy (Honarvar et al. 2016). In addition to their positive impact on waste reduction and recyclability, biodegradable nanocoatings offer a wide range of benefits for packaging materials (Danushika et al. 2021).

These coatings provide superior barrier properties, including moisture and gas permeability resistance, crucial in packaged food products. By extending the shelf life of goods, biodegradable nanocoatings can reduce food waste and promote sustainability (Zhang and Sablani 2021). Additionally, these coatings can be tailored to meet specific packaging requirements, with manufacturers able to adjust the composition and thickness of the coating to achieve a variety of properties, such as enhanced strength, flexibility, or UV resistance (Guillard et al. 2018). This flexibility allows for the creation packaging solutions that are both environmentally friendly and functional for various industries. While biodegradable nanocoatings offer a promising solution for sustainable packaging, obstacles still need to be addressed. One of the main concerns is cost-effectiveness, as the current production of biodegradable nanocoatings may be more expensive than traditional coatings. Despite this, the cost, economic scale, and technological advancements are expected to decrease, making it a more viable option for widespread use.

Another challenge is ensuring compatibility with existing packaging manufacturing processes, which may require modifications or investments in new equipment (Honarvar et al. 2016). Packaging manufacturers and coating suppliers must collaborate closely to overcome these compatibility issues and promote the adoption of biodegradable nanocoatings on a larger scale. In conclusion, the path to achieving sustainable packaging in the future lies in using biodegradable nanocoating technology (Guillard et al. 2018). This innovative solution holds great promise for minimizing waste, streamlining recycling efforts, and tailoring packaging to satisfy unique requirements. Despite potential obstacles, persistent exploration and advancement will undoubtedly pave the way for a more environmentally conscious packaging sector.

Biodegradable Nanocoatings as Key to Eco-Friendly Food Packaging

As the public becomes more conscious of the detrimental effects of non-biodegradable packaging on the environment, the demand for sustainable alternatives is growing. Biodegradable nanocoatings are a promising solution for eco-friendly food packaging (Honarvar et al. 2016). This type of coating consists of a thin layer of biodegradable material that is applied to the surface of food packaging. Biodegradable nanocoatings are often made from natural polymers derived from renewable resources, such as cellulose, chitosan, or starch (De Silva et al. 2017). Compared to traditional packaging materials, it provides several benefits, making it an excellent option for reducing waste and promoting a more sustainable future. Using biodegradable nanocoatings in food packaging is an eco-friendly measure that curbs the production

Table 2. Biodegradable nanocoatings in food packaging

Polymer	Types	Active agent	Mechanism/Effects	References
Synthetic polymers	Bacterial polyester poly(3-hydroxybutyrate-co-3-hydroxyvalerate) (PHBV)	ZnO NPs	Enhanced the microbial activities, mechanical and thermal properties	(Díez-Pascual and Díez-Vicente 2014)
	Polylactide (PLA)	Nanoclay	Improved the oxygen barriers	(Svagan et al. 2012)
	PLA	CNC	Promote oxygen barriers and reduce water permeability	(Fortunati et al. 2012)
Chitosan	Chitosan/PVA	TiO2	Enhance antimicrobial activities, mechanical properties and prevent particles migration	(Lian et al. 2016)
	Chitosan coated on PE lm	Nano ZnO	Promote solubility and antimicrobial activities	(Al-Naamani et al. 2016)
	Chitosan	CNC	Effective in barrier properties and tensile strength	(Khan et al. 2012)
	Chitosan	ZnO NPs	Enhance antimicrobial activities and mechanical properties	(Sanuja et al. 2015)
	Chitosan	MgO	Enhanced the thermal, stability, moisture barrier and mechanical properties	(De Silva et al. 2017)
Cellulose	Cellulose nanocrystals	CNC	Enhanced the thermal, stability, moisture barrier and mechanical properties	(Kirk et al. 2018)
	Cellulose	AgNPs and CNC	Promote antibacterial, antioxidant and mechanical properties	(Wu et al. 2019)

of non-biodegradable waste and conserves our natural resources. This innovative approach also plays a crucial role in prolonging the shelf life of food by inhibiting spoilage and preserving freshness, thereby fostering sustainable practices in the food industry. It is worth noting that the assurance of safe food is an integral part of our daily existence, and significant research has been devoted to this area, as indicated in Table 2. It is noteworthy to mention that, apart from the progress in packaging technology, like the use of active materials, visual cues such as alterations in color remain a primary means of gauging the freshness of food (Motelica et al. 2020). As previously discussed, some active food packaging materials can also be sensors to evaluate the food's freshness (Yang et al. 2020).

Biodegradable nanocoatings are a sustainable packaging solution with a notably lower carbon footprint than traditional packaging materials. Manufacturing requires less energy and resources, resulting in fewer carbon emissions. Moreover, using natural and non-toxic substances ensures the safety of the packaged food, making it a safer option for consumers. Traditional packaging materials, on the other hand, may contain harmful chemicals that pose health risks. Sustainable packaging solutions can be achieved using biodegradable nanocoatings on paper, cardboard, and plastic (Moshood et al. 2022). However, the challenge lies in the cost of large-scale production. With increased demand and technological advancements, costs may likely be reduced. Biodegradable nanocoatings can serve as an eco-friendly solution for food packaging. However, these coatings must meet stringent safety and quality standards to comply with regulatory requirements. Rigorous testing and certification processes are necessary to ensure food quality is not compromised. Although challenging, this technology can minimize waste, extend shelf life, reduce carbon footprints, increase safety, and provide flexibility. Continued exploration and advancement in this field will pave the way for a more sustainable future in the food industry.

SMART AND INTELLIGENT BIODEGRADABLE NANOCOATINGS IN FOOD PACKAGING

There is a growing trend in the food industry to adopt sustainable and eco-friendly packaging solutions using innovative technology, which is the key driver of the shift in the use of smart and intelligent biodegradable nanocoatings in food packaging (Ashfaq et al. 2022; Wahab et al. 2023). Certain foods are highly perishable and can decline rapidly under varying storage conditions. So, various indicators, such as time, temperature, pH, and leak indicators, have been created to monitor food quality (Amin et al. 2022) from the changes and alterations in the composition of the packaged environments, such as gaseous composition, microbial spoilage, and chemical (Fig. 2). Biodegradable packaging technology utilizes sophisticated and dynamic ingredients to ensure optimal product preservation (Amin et al. 2022; Ezati and Rhim 2020). The electronic smart packaging sector, valued at $1.45 billion, is projected to expand significantly because experts are exploring using environmentally friendly materials to replace conventional electronic components (Schaefer and Cheung 2018). The freshness indicator is a visual cue that signals whether the product is safe to eat by changing color because time and temperature sensors indicate how much time has passed since opening and how long the food can be safely consumed by modifying color (Fuertes et al. 2016).

Bio-based, active, and intelligent packaging solutions have been created for direct and indirect food uses. A study conducted by (Jancikova et al. 2019) concluded that dry-leaf rosemary extract (at a concentration of 20%) proved highly effective

Figure 2. Various techniques and classification of bio-based polymer

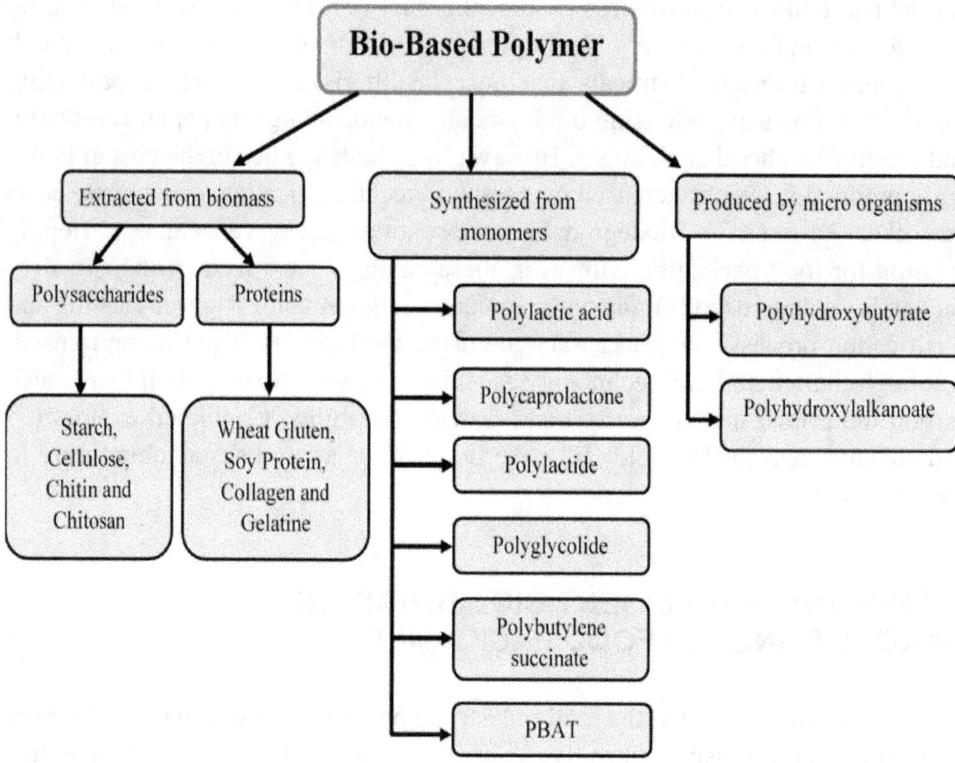

in inhibiting 88% of DPPH radicals, making it a promising candidate for smart packaging applications.

Minimizing food spoilage and reducing food waste are significant challenges that require attention, especially incorporating active materials in food packaging to extend the product's shelf life (Primožič et al. 2021). Traditional packaging materials, such as polyethylene, polypropylene, and polyethylene terephthalate, are non-degradable and limit the penetration of oxygen and water molecules (Chinaglia et al. 2018). However, advancements in functional materials allow for the development of active food packaging that includes scavengers, absorbers, emitters, and coatings. These agents can be added to conventional non-degradable packaging and work with biodegradable components to reduce waste further (Girigoswami et al. 2022). For more information on these agents, please refer to Table 3.

Detecting particular gases or pathogens contributing to food spoilage is vital, and nanosensors are critical in this process (Patel et al. 2020). These sensors are responsible for monitoring the changes that occur within the food, guaranteeing that its quality and safety (Girigoswami et al. 2022). Nanomaterials are employed in various

Table 3. Active food packaging to extend shelf life and packaging

Component	Functions	Agents	References
Moisture absorbers	Reducing microbial growth	Organic, polyvinyl alcohol, clay	(Gaikwad et al. 2019)
Oxygen scavengers	Protection again oxidation of fat	Enzyme based, metallic, organic and inorganic, Ag, SiO_2,	(Gaikwad et al. 2018)
CO_2 emitters	Accumulation of damage by microbial actions	Ascorbic, sodium bicarbonate and citric acid	(Han et al. 2018)
Ethylene scavengers	Reducing the ripening of vegetables and fruits	ZnO nanoparticles, polyvinyl chloride	(Sadeghi et al. 2021)

ways to improve packaging functionality, including integrity indicators, freshness indicators, time and temperature indicators, and radio frequency identification (Mohammadi and Jafari 2020). The innovative technology of smart packaging has many potential uses that ensure and prolong the quality, durability, and medication and facilitate the secure transportation of goods through integrated security measures (Schaefer and Cheung 2018). From a consumer standpoint, these capabilities are highly valued, but there are distinctive obstacles to buying a sustainable product, such as non-availability, poor quality, and complex processing methods (Ketelsen et al. 2020). In today's hyper-connected world, tracking and monitoring purchased items through linked applications has become crucial for businesses to improve customer satisfaction and loyalty (Kovačević et al. 2021). As (Pacquit et al. 2007) noted, smart packaging can identify supply chain inefficiencies, minimize errors and costs, enhance product functionality, and drive higher profit margins. However, it should be noted that replacing oxygen, CO_2, and water vapor within food packaging can vary depending on the food and packaging utilized. Fortunately, nanocomposite materials, particularly polymers, have effectively addressed and overcome these complexities (Abbaspour et al. 2015). In fact, (Mills 2005) utilized nano-sized titanium dioxide (TiO_2) or tin dioxide particles to design a photo-indicator intelligent ink that changes color in response to oxygen-level shifts.

Additionally, the packaging industry has been exploring numerous nanostructures and nanomaterials, such as nanoclay particulates, due to their accessibility, low cost, ease of processing, and exceptional performance for detecting oxygen levels within the packaging (Sarfraz et al. 2020). Graphene nanosheets and carbon nanotubes are developing as promising carbon-based nanomaterials (Arora and Padua 2010). Polymeric substrate-based flexible or static films and bottles made from various materials, such as glass and plastic, are combined with vacuum-coated thin inorganic layers to fulfill the demand for material packaging that can block oxygen and aromas.

Although aluminum foil and metallization were once the preferred materials and methods, new materials are now used more frequently.

CHALLENGES AND LIMITATIONS

The challenges associated with using biodegradable nanocoatings in food packaging are its cost. Currently, it is more expensive than traditional plastic packaging. However, as technology advances and more companies adopt this alternative, costs are expected to decrease over time. Its limited availability of biodegradable materials is also another challenge (Wang et al. 2022). Though several options are available, they may only be suitable for some types of food packaging. Additionally, biodegradable nanocoatings' durability and shelf life are still being studied to ensure that they can effectively protect food products. Despite these challenges, researchers and manufacturers actively work to improve the technology and find new solutions to overcome these limitations. The technical, regulatory and economic challenges must be overcome to implement these coatings in the food packaging industry successfully.

Technical Challenges

- Durability and Shelf Life: Food packaging has to endure various handling processes, transportation, and storage conditions without compromising its protective properties. The nanocoatings should be able to maintain their integrity and barrier performance throughout the product's shelf life.
- Compatibility with Different Packaging Materials: Ensuring compatibility with different packaging materials such as plastics, paper, and metal. Each of these materials has unique surface properties that require specific coating formulations for optimal adhesion and performance. Therefore, developing nanocoatings that can adhere effectively to a wide range of packaging materials is crucial for successful implementation.
- Barrier Performance: Food packaging protects food from external factors like moisture, oxygen, and light. Biodegradable nanocoatings must provide an effective barrier against these factors to maintain the quality and safety of packaged food. Achieving comparable or superior barrier performance to conventional non-biodegradable coatings is a significant technical challenge that requires extensive research and development.

Regulatory Challenges

- Safety and Toxicity: To ensure the safety of consumers, it is crucial to thoroughly assess the use of biodegradable nanocoatings in food packaging. Regulatory bodies require comprehensive toxicological evaluations to determine any possible risks associated with these coatings. The evaluation must consider the potential migration of nanomaterials from the coating to the food and any adverse effects on human health (Onyeaka et al. 2022). Manufacturers face a significant challenge in meeting the strict safety standards of regulatory authorities.
- Labeling and Consumer Perception: It is essential to have clear labeling on biodegradable nanocoatings in food packaging. Unfortunately, there is currently no standard labeling system for these coatings. To increase transparency and consumer acceptance, it is necessary to develop clear and informative labeling guidelines that accurately communicate the benefits and potential risks associated with biodegradable nanocoatings.
- Regulatory Framework: The rules and regulations of nanotechnology and biodegradable materials are continuously changing (Wang et al. 2022). To use biodegradable nanocoatings in food packaging, following the existing regulations, which may not have specific guidelines for these new materials, is necessary. Therefore, creating a clear regulatory framework that considers biodegradable nanocoatings' unique properties and potential risks is crucial for their successful implementation.

Economic Challenges

- Cost-Effectiveness: Implementing biodegradable nanocoatings in food packaging poses a significant economic challenge because the procedures involved in creating these coatings can incur substantial costs, especially compared to their non-biodegradable counterparts (Wang et al. 2022). To increase the accessibility of biodegradable nanocoatings, manufacturers must focus on optimizing production efficiency and reducing expenses.
- Market Demand and Consumer Acceptance: As environmental concerns become more prevalent, the need for sustainable packaging options grows (Iderawumi and Yusuff 2021). Whether consumers will accept biodegradable nanocoatings in food packaging depends on their safety, efficacy, and affordability. Establishing trust in these emerging technologies and addressing any skepticism is paramount to ensuring their successful integration into the marketplace (Panel 2019).
- Infrastructure and Supply Chain: Manufacturers must make infrastructure and supply chain changes to use biodegradable nanocoatings in food packaging. They must have the right equipment, facilities, and expertise to apply these

coatings properly (Wyser et al. 2016). They also need to ensure a reliable and sustainable supply chain by improving the availability of biodegradable raw materials.

Implementing biodegradable nanocoatings in food packaging presents challenges that must be addressed. Technical obstacles include guaranteeing compatibility with various packaging materials, achieving comparable barrier performance, and maintaining durability throughout the product's shelf life. Regulatory challenges include conducting safety evaluations, adhering to labeling requirements, and establishing a well-defined regulatory framework. Economic hurdles encompass cost-effectiveness, market demand, consumer acceptance, and adapting infrastructure and supply chains.

FUTURE PROSPECTS AND MARKET TRENDS OF BIODEGRADABLE NANOCOATINGS IN FOOD PACKAGING

The prospects of biodegradable nanocoatings in food packaging are promising due to increasing consumer demand for sustainable products, stringent regulations on conventional packaging materials, and market trends indicating significant growth in the biodegradable packaging market. At the same time, there are challenges to overcome; ongoing research and development efforts address scalability, cost-effectiveness, and safety concerns associated with these coatings (Lindh et al. 2016). Biodegradable nanocoatings have significant solutions to the environmental issues associated with traditional food packaging materials and have several advantages, including enhanced barrier properties, extended shelf life, and reduced waste generation. As the demand for sustainable packaging solutions continues to grow, the future prospects of biodegradable nanocoatings in food packaging appear promising (Nilsen-Nygaard et al. 2021). In terms of application, biodegradable nanocoatings can be used in various food packaging materials, including films, trays, bottles, and containers (Alsufyani and M'Sakni N 2023). These coatings provide an effective barrier against moisture, oxygen, and other external factors that can degrade food products' quality and shelf life. Additionally, they can be tailored to specific requirements, such as antimicrobial properties, to prevent bacterial growth and maintain food safety (Al-Tayyar et al. 2020).

The adoption of biodegradable nanocoatings in food packaging also offers economic benefits. Though the initial cost of implementing these coatings may be higher than traditional materials, the long-term savings can be significant. Extended shelf life and reduced food waste translate into cost savings for manufacturers, retailers, and consumers (Sarfraz et al. 2020). Furthermore, the growing demand for sustainable packaging solutions presents an opportunity for companies to differentiate themselves in the market and gain a competitive edge (Wyser et al. 2016). However, certain challenges must to be addressed for the widespread adoption of biodegradable nanocoatings in food packaging. One of the key challenges is scalability and cost-effectiveness. Currently, producing biodegradable nanocoatings on a large scale is still relatively expensive compared to conventional materials (Mangaraj et al. 2019).

Another factor contributing to the future prospects of biodegradable nanocoatings is the stringent regulations imposed by governments and regulatory bodies on conventional packaging materials. Many countries have implemented strict regulations to reduce plastic waste and promote sustainable packaging practices (Sarfraz et al. 2020). Biodegradable nanocoatings align with these regulations offering a more environmentally friendly alternative. However, it should be noted that the key drivers for adopting of biodegradable nanocoatings in food packaging are increasing consumer awareness and demand for eco-friendly products (Wyser et al. 2016). Consumers have recently become more conscious of their purchasing decisions' environmental impact and actively seek sustainable alternatives. Biodegradable nanocoatings provide a viable solution by reducing the reliance on non-biodegradable materials such as plastics.

CONCLUSION

In conclusion, using biodegradable nanocoatings in food packaging is an important step towards reducing the negative impact of traditional packaging on the environment. Not only does it offer numerous benefits, such as increased shelf life and reduced waste, but it also addresses the growing concern of plastic pollution. However, it is important to acknowledge that there are still challenges and limitations associated with using biodegradable nanocoatings in food packaging. As we continue to develop and improve this technology, we must prioritize sustainability and responsible manufacturing practices (Mukhtar Iderawumi et al. 2023). There is a need to know what role individuals, researchers, organizations, or industries can plays in promoting the use of biodegradable nanocoatings in food packaging and giving encouragement to companies to adopt more sustainable practices in food packaging. By working together, we can create a better future for our planet and ensure that future generations can enjoy a healthy and thriving environment through the use of biodegradable nanocoatings in food packaging

REFERENCES

Abbaspour, A., Norouz-Sarvestani, F., Noori, A., & Soltani, N. (2015). Aptamer-conjugated silver nanoparticles for electrochemical dual-aptamer-based sandwich detection of staphylococcus aureus. *Biosensors & Bioelectronics*, *68*, 149–155. doi:10.1016/j.bios.2014.12.040 PMID:25562742

Agarwal, A., Shaida, B., Rastogi, M., & Singh, N. B. (2023). Food Packaging Materials with Special Reference to Biopolymers-Properties and Applications. *Chemistry Africa*, *16*(1), 117–144. doi:10.1007/s42250-022-00446-w

Ahari, H., & Soufiani, S. P. (2021). Smart and active food packaging: Insights in novel food packaging. *Frontiers in Microbiology*, *12*, 657233. doi:10.3389/fmicb.2021.657233 PMID:34305829

Ahmad, A., Qurashi, A., & Sheehan, D. (2023). Nano packaging – Progress and future perspectives for food safety, and sustainability. *Food Packaging and Shelf Life*, *35*, 100997. doi:10.1016/j.fpsl.2022.100997

Al-Naamani, L., Dobretsov, S., & Dutta, J. (2016). Chitosan-zinc oxide nanoparticle composite coating for active food packaging applications. *Innovative Food Science & Emerging Technologies*, *38*, 231–237. doi:10.1016/j.ifset.2016.10.010

Al-Tayyar, N. A., Youssef, A. M., & Al-Hindi, R. (2020). Antimicrobial food packaging based on sustainable Bio-based materials for reducing foodborne Pathogens: A review. *Food Chemistry*, *310*, 125915. doi:10.1016/j.foodchem.2019.125915 PMID:31841936

Alsufyani, T., & M'Sakni, N. H. (2023). Biodegradable Bio-Composite Film Reinforced with Cellulose Nanocrystals from Chaetomorpha linum into Thermoplastic Starch Matrices. *Polymers*, *15*(6), 1542. doi:10.3390/polym15061542 PMID:36987321

Amin, U., Khan, M. K. I., Maan, A. A., Nazir, A., Riaz, S., Khan, M. U., Sultan, M., Munekata, P. E. S., & Lorenzo, J. M. (2022). Biodegradable active, intelligent, and smart packaging materials for food applications. *Food Packaging and Shelf Life*, *33*, 100903. doi:10.1016/j.fpsl.2022.100903

Amini, S. M., Gilaki, M., & Karchani, M. (2014). Safety of nanotechnology in food industries. *Electronic Physician*, *6*, 962–968. doi:10.14661/2014.962-968 PMID:25763176

Arora, A., & Padua, G. W. (2010). Review: Nanocomposites in Food Packaging. *Journal of Food Science*, *75*(1), R43–R49. doi:10.1111/j.1750-3841.2009.01456.x PMID:20492194

Ashfaq, A., Khursheed, N., Fatima, S., Anjum, Z., & Younis, K. (2022). Application of nanotechnology in food packaging: Pros and Cons. *Journal of Agriculture and Food Research*, *7*, 100270. doi:10.1016/j.jafr.2022.100270

Bajpai, V. K., Kamle, M., Shukla, S., Mahato, D. K., Chandra, P., Hwang, S. K., Kumar, P., Huh, Y. S., & Han, Y.-K. (2018). Prospects of using nanotechnology for food preservation, safety, and security. *Yao Wu Shi Pin Fen Xi*, *26*(4), 1201–1214. doi:10.1016/j.jfda.2018.06.011 PMID:30249319

Barage, S., Lakkakula, J., Sharma, A., Roy, A., Alghamdi, S., Almehmadi, M., Hossain, M. J., Allahyani, M., & Abdulaziz, O. (2022). Nanomaterial in Food Packaging: A Comprehensive Review. *Journal of Nanomaterials*, *6053922*, 1–12. doi:10.1155/2022/6053922

Cagri, A., Ustunol, Z., & Ryser, E. T. (2004). Antimicrobial edible films and coatings. *Journal of Food Protection*, *67*(4), 833–848. doi:10.4315/0362-028X-67.4.833 PMID:15083740

Chinaglia, S., Tosin, M., & Degli-Innocenti, F. (2018). Biodegradation rate of biodegradable plastics at molecular level. *Polymer Degradation & Stability*, *147*, 237–244. doi:10.1016/j.polymdegradstab.2017.12.011

Chisenga, S. M., Tolesa, G. N., & Workneh, T. S. (2020). Biodegradable Food Packaging Materials and Prospects of the Fourth Industrial Revolution for Tomato Fruit and Product Handling. *International Journal of Food Sciences*, *8879101*, 1–17. doi:10.1155/2020/8879101 PMID:33299850

Costa, S. S., Miranda, A. L., de Morais, M. G., Costa, J. A. V., & Druzian, J. I. (2019). Microalgae as source of polyhydroxyalkanoates (PHAs) — A review. *International Journal of Biological Macromolecules*, *131*, 536–547. doi:10.1016/j.ijbiomac.2019.03.099 PMID:30885732

Danushika, C. M., Rohan, S. D., & Renuka, N. L. (2021). Novel Acumens into Biodegradation: Impact of Nanomaterials and Their Contribution. In M. Kamila Cabral (Ed.), *M Kassio Ferreira, S Rodrigo Nogueira de*. Biodegradation Technology of Organic and Inorganic Pollutants. IntechOpen.

De Silva, R. T., Mantilaka, M. M. M. G. P. G., Ratnayake, S. P., Amaratunga, G. A. J., & de Silva, K. M. N. (2017). Nano-MgO reinforced chitosan nanocomposites for high performance packaging applications with improved mechanical, thermal and barrier properties. *Carbohydrate Polymers*, *157*, 739–747. doi:10.1016/j.carbpol.2016.10.038 PMID:27987986

Dehghani, S., Hosseini, S. V., & Regenstein, J. M. (2018). Edible films and coatings in seafood preservation: A review. *Food Chemistry*, *240*, 505–513. doi:10.1016/j.foodchem.2017.07.034 PMID:28946304

Díez-Pascual, A. M., & Díez-Vicente, A. L. (2014). ZnO-reinforced poly(3-hydroxybutyrate-co-3-hydroxyvalerate) bionanocomposites with antimicrobial function for food packaging. *ACS Applied Materials & Interfaces*, *6*(12), 9822–9834. doi:10.1021/am502261e PMID:24846876

Din, M. I., Ghaffar, T., Najeeb, J., Hussain, Z., Khalid, R., & Zahid, H. (2020). Potential perspectives of biodegradable plastics for food packaging application-review of properties and recent developments. *Food Additives & Contaminants. Part A, Chemistry, Analysis, Control, Exposure & Risk Assessment*, *37*(4), 665–680. doi:10.1080/19440049.2020.1718219 PMID:32049609

Ezati, P., & Rhim, J.-W. (2020). pH-responsive chitosan-based film incorporated with alizarin for intelligent packaging applications. *Food Hydrocolloids*, *102*, 105629. doi:10.1016/j.foodhyd.2019.105629

Fitriani, F., Bilad, M. R., Aprilia, S., & Arahman, N. (2023). Biodegradable Hybrid Polymer Film for Packaging: A Review. *Journal of Natural Fibers*, *20*(1), 2159606. doi:10.1080/15440478.2022.2159606

Fortunati, E., Peltzer, M., Armentano, I., Torre, L., Jiménez, A., & Kenny, J. M. (2012). Effects of modified cellulose nanocrystals on the barrier and migration properties of PLA nano-biocomposites. *Carbohydrate Polymers*, *90*(2), 948–956. doi:10.1016/j.carbpol.2012.06.025 PMID:22840025

Fuertes, G., Soto, I., Vargas, M., Valencia, A., Sabattin, J., & Carrasco, R. (2016). Nanosensors for a Monitoring System in Intelligent and Active Packaging. *Journal of Sensors*, *7980476*, 1–8. doi:10.1155/2016/7980476

Gadhave, R. V., Das, A., Mahanwar, P. A., & Gadekar, P. T. (2018). Starch based bio-plastics: The future of sustainable packaging. *Open Journal of Polymer Chemistry*, *8*(2), 21–33. doi:10.4236/ojpchem.2018.82003

Gaikwad, K. K., Singh, S., & Ajji, A. (2019). Moisture absorbers for food packaging applications. *Environmental Chemistry Letters, 17*(2), 609–628. doi:10.1007/s10311-018-0810-z

Gaikwad, K. K., Singh, S., & Lee, Y. S. (2018). Oxygen scavenging films in food packaging. *Environmental Chemistry Letters, 16*(2), 523–538. doi:10.1007/s10311-018-0705-z

Girigoswami, A., Thirumalai, A., Karthick, H., Pallavi, P., Gowtham, P., & Girigoswami, A. (2022). Nanotechnology driven improvement of smart food packaging. *Materials Research Innovations.* doi:10.1080/14328917.2022.2114667

Guillard, V., Gaucel, S., Fornaciari, C., Angellier-Coussy, H., Buche, P., & Gontard, N. (2018). The next generation of sustainable food packaging to preserve our environment in a circular economy context. *Frontiers in Nutrition, 5*, 121. doi:10.3389/fnut.2018.00121 PMID:30564581

Han, J. W., Ruiz-Garcia, L., Qian, J. P., & Yang, X. T. (2018). Food Packaging: A Comprehensive Review and Future Trends. *Comprehensive Reviews in Food Science and Food Safety, 17*(4), 860–877. doi:10.1111/1541-4337.12343 PMID:33350114

Honarvar, Z., Hadian, Z., & Mashayekh, M. (2016). Nanocomposites in food packaging applications and their risk assessment for health. *Electronic Physician, 8*(6), 2531–2538. doi:10.19082/2531 PMID:27504168

Hong, L. G., Yuhana, N. Y., & Zawawi, E. Z. E. (2021). Review of bioplastics as food packaging materials. *AIMS Materials Science, 8*(2), 166–184. doi:10.3934/matersci.2021012

Iderawumi, A. M., & Yusuff, M. A. (2021). Effects of nanoparticles on improvement in quality and shelf life of fruits and vegetables. *J Plant Biol Crop Res, 4*, 2–7.

Jancikova, S., Jamróz, E., Kulawik, P., Tkaczewska, J., & Dordevic, D. (2019). Furcellaran/gelatin hydrolysate/rosemary extract composite films as active and intelligent packaging materials. *International Journal of Biological Macromolecules, 131*, 19–28. doi:10.1016/j.ijbiomac.2019.03.050 PMID:30853584

Jung, Y., Min, J., Choi, J., Bang, J., Jeong, S., Pyun, K. R., Ahn, J., Cho, Y., Hong, S., Hong, S., Lee, J., & Ko, S. H. (2022). Smart paper electronics by laser-induced graphene for biodegradable real-time food spoilage monitoring. *Applied Materials Today, 29*, 101589. doi:10.1016/j.apmt.2022.101589

Kandeepan, G. (2021). Biodegradable nanocomposite packaging films for meat and meat products: A review. *Journal of Packaging Technology and Research, 5*(3), 143–166. doi:10.1007/s41783-021-00123-2

Ketelsen, M., Janssen, M., & Hamm, U. (2020). Consumers' response to environmentally-friendly food packaging - A systematic review. *Journal of Cleaner Production, 254*, 120123. doi:10.1016/j.jclepro.2020.120123

Khan, A., Khan, R. A., Salmieri, S., Le Tien, C., Riedl, B., Bouchard, J., Chauve, G., Tan, V., Kamal, M. R., & Lacroix, M. (2012). Mechanical and barrier properties of nanocrystalline cellulose reinforced chitosan based nanocomposite films. *Carbohydrate Polymers, 90*(4), 1601–1608. doi:10.1016/j.carbpol.2012.07.037 PMID:22944422

Kim, M.-Y., & Lee, K. H. (2022). Electrochemical sensors for sustainable precision agriculture—A review. *Frontiers in Chemistry, 10*, 848320. doi:10.3389/fchem.2022.848320 PMID:35615311

Kirk, K. A., Othman, A., & Andreescu, S. (2018). Nanomaterial-functionalized Cellulose: Design, Characterization and Analytical Applications. *Analytical Sciences, 34*(1), 19–31. doi:10.2116/analsci.34.19 PMID:29321453

Kovačević, Z., Flinčec Grgac, S., & Bischof, S. (2021). Progress in Biodegradable Flame Retardant Nano-Biocomposites. *Polymers*. PMID:33673607

Kuswandi, B. (2017). Environmental friendly food nano-packaging. *Environmental Chemistry Letters, 15*(2), 205–221. doi:10.1007/s10311-017-0613-7

Lian, Z., Zhang, Y., & Zhao, Y. (2016). Nano-TiO2 particles and high hydrostatic pressure treatment for improving functionality of polyvinyl alcohol and chitosan composite films and nano-TiO2 migration from film matrix in food simulants. *Innovative Food Science & Emerging Technologies, 33*, 145–153. doi:10.1016/j.ifset.2015.10.008

Lindh, H., Williams, H., Olsson, A., & Wikström, F. (2016). Elucidating the indirect contributions of packaging to sustainable development: A terminology of packaging functions and features. *Packaging Technology & Science, 29*(4-5), 225–246. doi:10.1002/pts.2197

Mangaraj, S., Yadav, A., Bal, L. M., Dash, S. K., & Mahanti, N. K. (2019). Application of biodegradable polymers in food packaging industry: A comprehensive review. *Journal of Packaging Technology and Research, 3*(1), 77–96. doi:10.1007/s41783-018-0049-y

Maraveas, C. (2020). Production of Sustainable and Biodegradable Polymers from Agricultural Waste. *Polymers*, *12*(5), 1127. doi:10.3390/polym12051127 PMID:32423073

Mills, A. (2005). Oxygen indicators and intelligent inks for packaging food. *Chemical Society Reviews*, *34*(12), 1003–1011. doi:10.1039/b503997p PMID:16284666

Mir, S. A., Dar, B. N., Wani, A. A., & Shah, M. A. (2018). Effect of plant extracts on the techno-functional properties of biodegradable packaging films. *Trends in Food Science & Technology*, *80*, 141–154. doi:10.1016/j.tifs.2018.08.004

Mohammad, Z. H., Ahmad, F., Ibrahim, S. A., & Zaidi, S. (2022). Application of nanotechnology in different aspects of the food industry. *Discover Food*, *2*(1), 12. doi:10.1007/s44187-022-00013-9

Mohammadi, Z., & Jafari, S. M. (2020). Detection of food spoilage and adulteration by novel nanomaterial-based sensors. *Advances in Colloid and Interface Science*, *286*, 102297. doi:10.1016/j.cis.2020.102297 PMID:33142210

Moshood, T. D., Nawanir, G., Mahmud, F., Mohamad, F., Ahmad, M. H., & AbdulGhani, A. (2022). Sustainability of biodegradable plastics: New problem or solution to solve the global plastic pollution? *Current Research in Green and Sustainable Chemistry*, *5*, 100273. doi:10.1016/j.crgsc.2022.100273

Motelica, L., Ficai, D., Ficai, A., Oprea, O. C., Kaya, D. A., & Andronescu, E. (2020). Biodegradable Antimicrobial Food Packaging: Trends and Perspectives. *Foods*, *9*(10), 1438. doi:10.3390/foods9101438 PMID:33050581

Mukhtar Iderawumi, A., Jiandong, H., Shakeel, A., Linze, L., & Syed Muhammad Zaigham Abbas, N. (2023). Advances in the Use of Organic and Organomineral Fertilizers in Sustainable Agricultural Production. In H. Khalid Rehman (Ed.), *Organic Fertilizers*. IntechOpen.

Naqvi, S. M. Z. A., Haq, S. I. U., Iderawumi, A. R. M., Anwar, A., Mujtaba, G., Khan, N. M., Islam, A., Awais, M., Mustafa, T., & Iqbal, J. (2022). GREEN SYNTHESIS OF SILVER NANOPARTICLES AND ANTI-OXIDANT ACTIVITY IN PLANTS UNDER SEMIARID CONDITION—A REVIEW. *Pakistan Journal of Weed Science Research*, *28*(3), 319–330. doi:10.28941/pjwsr.v28i3.1064

Nilsen-Nygaard, J., Fernández, E. N., Radusin, T., Rotabakk, B. T., Sarfraz, J., Sharmin, N., Sivertsvik, M., Sone, I., & Pettersen, M. K. (2021). Current status of biobased and biodegradable food packaging materials: Impact on food quality and effect of innovative processing technologies. *Comprehensive Reviews in Food Science and Food Safety*, *20*(2), 1333–1380. doi:10.1111/1541-4337.12715 PMID:33547765

Onyeaka, H., Passaretti, P., Miri, T., & Al-Sharify, Z. T. (2022). The safety of nanomaterials in food production and packaging. *Current Research in Food Science, 5*, 763–774. doi:10.1016/j.crfs.2022.04.005 PMID:35520272

Pacquit, A., Frisby, J., Diamond, D., Lau, K. T., Farrell, A., Quilty, B., & Diamond, D. (2007). Development of a smart packaging for the monitoring of fish spoilage. *Food Chemistry, 102*(2), 466–470. doi:10.1016/j.foodchem.2006.05.052

Panel, M. (2019). Byte by byte: Policy innovation for transforming Africa's food system with digital technologies, Dakar. Dakar, Senegal. *International Food Policy Research Institute of Bonn, 9780896296848.*

Panou, A., & Karabagias, I. K. (2023). Biodegradable Packaging Materials for Foods Preservation: Sources, Advantages, Limitations, and Future Perspectives. *Coatings, 13*(7), 1176. doi:10.3390/coatings13071176

Patel, G., Pillai, V., Bhatt, P., & Mohammad, S. (2020). Application of nanosensors in the food industry. In B. Han, V. K. Tomer, T. A. Nguyen, A. Farmani, & P. Kumar Singh (Eds.), *Nanosensors for Smart Cities*. Elsevier. doi:10.1016/B978-0-12-819870-4.00020-7

Pawlak, K., & Kołodziejczak, M. (2020). The role of agriculture in ensuring food security in developing countries: Considerations in the context of the problem of sustainable food production. *Sustainability (Basel), 12*(13), 5488. doi:10.3390/su12135488

Perera, K. Y., Jaiswal, A. K., & Jaiswal, S. (2023). *Biopolymer-Based Sustainable Food Packaging Materials: Challenges, Solutions, and Applications*. Foods.

Primožič, M., Knez, Ž., & Leitgeb, M. (2021). (Bio)nanotechnology in Food Science-Food Packaging. *Nanomaterials (Basel, Switzerland), 11*(2), 292. doi:10.3390/nano11020292 PMID:33499415

Remya, R. R., Julius, A., Suman, T. Y., Mohanavel, V., Karthick, A., Pazhanimuthu, C., Samrot, A. V., & Muhibbullah, M. (2022). Role of Nanoparticles in Biodegradation and Their Importance in Environmental and Biomedical Applications. *Journal of Nanomaterials, 6090846*, 1–15. doi:10.1155/2022/6090846

Sadeghi, K., Lee, Y., & Seo, J. (2021). Ethylene Scavenging Systems in Packaging of Fresh Produce: A Review. *Food Reviews International, 37*(2), 155–176. doi:10.1080/87559129.2019.1695836

Sanuja, S., Agalya, A., & Umapathy, M. J. (2015). Synthesis and characterization of zinc oxide–neem oil–chitosan bionanocomposite for food packaging application. *International Journal of Biological Macromolecules*, *74*, 76–84. doi:10.1016/j.ijbiomac.2014.11.036 PMID:25499891

Sarfraz, J., Gulin-Sarfraz, T., Nilsen-Nygaard, J., & Pettersen, M. K. (2020). Nanocomposites for Food Packaging Applications: An Overview. *Nanomaterials (Basel, Switzerland)*, *11*(1), 10. doi:10.3390/nano11010010 PMID:33374563

Schaefer, D., & Cheung, W. M. (2018). Smart Packaging: Opportunities and Challenges. *Procedia CIRP*, *72*, 1022–1027. doi:10.1016/j.procir.2018.03.240

Shaikh, S., Yaqoob, M., & Aggarwal, P. (2021). An overview of biodegradable packaging in food industry. *Current Research in Food Science*, *4*, 503–520. doi:10.1016/j.crfs.2021.07.005 PMID:34401747

Sharma, V., Tripathi, A. K., & Mittal, H. (2022). Technological revolutions in smart farming: Current trends, challenges & future directions. *Computers and Electronics in Agriculture*, *201*, 107217. doi:10.1016/j.compag.2022.107217

Sid, S., Mor, R. S., Kishore, A., & Sharanagat, V. S. (2021). Bio-sourced polymers as alternatives to conventional food packaging materials: A review. *Trends in Food Science & Technology*, *115*, 87–104. doi:10.1016/j.tifs.2021.06.026

Sorrentino, A., Gorrasi, G., & Vittoria, V. (2007). Potential perspectives of bio-nanocomposites for food packaging applications. *Trends in Food Science & Technology*, *18*(2), 84–95. doi:10.1016/j.tifs.2006.09.004

Souza, V. G. L., & Fernando, A. L. (2016). Nanoparticles in food packaging: Biodegradability and potential migration to food—A review. *Food Packaging and Shelf Life*, *8*, 63–70. doi:10.1016/j.fpsl.2016.04.001

Su, S., & Kang, P. M. (2020). Systemic Review of Biodegradable Nanomaterials in Nanomedicine. *Nanomaterials (Basel, Switzerland)*, *10*(4), 656. doi:10.3390/nano10040656 PMID:32244653

Sun, J., Shen, J., Chen, S., Cooper, M. A., Fu, H., Wu, D., & Yang, Z. (2018). Nanofiller Reinforced Biodegradable PLA/PHA Composites: Current Status and Future Trends. *Polymers*, *10*(5), 505. doi:10.3390/polym10050505 PMID:30966540

Svagan, A. J., Åkesson, A., Cárdenas, M., Bulut, S., Knudsen, J. C., Risbo, J., & Plackett, D. (2012). Transparent films based on PLA and montmorillonite with tunable oxygen barrier properties. *Biomacromolecules*, *13*(2), 397–405. doi:10.1021/bm201438m PMID:22229499

Taib, N.-A. A. B., Rahman, M. R., Huda, D., Kuok, K. K., Hamdan, S., Bakri, M. K. B., Julaihi, M. R. M. B., & Khan, A. (2023). A review on poly lactic acid (PLA) as a biodegradable polymer. *Polymer Bulletin, 80*(2), 1179–1213. doi:10.1007/s00289-022-04160-y

Tan, C., Han, F., Zhang, S., Li, P., & Shang, N. (2021). Novel bio-based materials and applications in antimicrobial food packaging: Recent advances and future trends. *International Journal of Molecular Sciences, 22*(18), 9663. doi:10.3390/ijms22189663 PMID:34575828

Vasile, C. (2018). Polymeric Nanocomposites and Nanocoatings for Food Packaging: A Review. *Materials (Basel), 11*(10), 1834. doi:10.3390/ma11101834 PMID:30261658

Vasile, C., & Baican, M. (2021). Progresses in food packaging, food quality, and safety—Controlled-release antioxidant and/or antimicrobial packaging. *Molecules (Basel, Switzerland), 26*(5), 1263. doi:10.3390/molecules26051263 PMID:33652755

Wahab, A., Munir, A., Saleem, M. H., AbdulRaheem, M. I., Aziz, H., Mfarrej, M. F. B., & Abdi, G. (2023). Interactions of Metal-Based Engineered Nanoparticles with Plants: An Overview of the State of Current Knowledge, Research Progress, and Prospects. *Journal of Plant Growth Regulation, 42*(9), 5396–5416. doi:10.1007/s00344-023-10972-7

Wang, J., Euring, M., Ostendorf, K., & Zhang, K. (2022). Biobased materials for food packaging. *Journal of Bioresources and Bioproducts, 7*(1), 1–13. doi:10.1016/j.jobab.2021.11.004

Westlake, J. R., Tran, M. W., Jiang, Y., Zhang, X., Burrows, A. D., & Xie, M. (2023). Biodegradable biopolymers for active packaging: Demand, development and directions. *Sustainable Food Technology, 1*(1), 50–72. doi:10.1039/D2FB00004K

Wu, Z., Deng, W., Luo, J., & Deng, D. (2019). Multifunctional nano-cellulose composite films with grape seed extracts and immobilized silver nanoparticles. *Carbohydrate Polymers, 205*, 447–455. doi:10.1016/j.carbpol.2018.10.060 PMID:30446127

Wyser, Y., Adams, M., Avella, M., Carlander, D., Garcia, L., Pieper, G., Rennen, M., Schuermans, J., & Weiss, J. (2016). Outlook and Challenges of Nanotechnologies for Food Packaging. *Packaging Technology & Science, 29*(12), 615–648. doi:10.1002/pts.2221

Yang, Y.-N., Lu, K.-Y., Wang, P., Ho, Y.-C., Tsai, M.-L., & Mi, F.-L. (2020). Development of bacterial cellulose/chitin multi-nanofibers based smart films containing natural active microspheres and nanoparticles formed in situ. *Carbohydrate Polymers*, *228*, 115370. doi:10.1016/j.carbpol.2019.115370 PMID:31635728

Zhang, H., & Sablani, S. (2021). Biodegradable packaging reinforced with plant-based food waste and by-products. *Current Opinion in Food Science*, *42*, 61–68. doi:10.1016/j.cofs.2021.05.003

Chapter 10
Ethical and Environmental Implications Associated With the Application of Nanotechnology

Nnabuk Okon Eddy
https://orcid.org/0000-0001-7704-3082
Independent Researcher, Nigeria

Rajni Garg
https://orcid.org/0000-0002-9629-8141
Galgotias College of Engineering and Technology, India

Richard Alexis Ukpe
Federal University, Otuoke, Nigeria

Nyeneime William Akpanudo
Akwa Ibom State University, Nigeria

Hillary Abugu
https://orcid.org/0000-0002-6045-8449
University of Nigeria, Nigeria

Rishav Garg
https://orcid.org/0000-0001-6565-9234
Galgotias College of Engineering and Technology, India

Ansar Anjum
Galgotias College of Engineering and Technology, India

Balu Anand
https://orcid.org/0000-0003-3984-0237
Sir Theagaraya College, India

ABSTRACT

Nanotechnology is gaining wider acceptability in the global system because of the wider scope of its applications and the vast expectations for future demands. However, some research documents have reported on the potential toxicity of some nanoparticles to life, but the current status in our society seems to be much more concerned with the advantages than the considerations of both factors. It seems indicative that a critical review of the toxicological consequences of nanoparticles

DOI: 10.4018/979-8-3693-1094-6.ch010

Copyright © 2024, IGI Global. Copying or distributing in print or electronic forms without written permission of IGI Global is prohibited.

should move forward in Pari Pasu with researchers working on their existing and newer applications. However, this cannot be achieved without sound ethical considerations. The ethical concern should embrace opposition to the use of nanoparticles in the fabrication of lethal autonomous weapons, unbiased evaluation of hazards and risks, nonmaleficence, etc. This literature-based review examines nanoparticle ethics and ecotoxicity.

INTRODUCTION

Nanoparticles are currently dominating the materials world concerning applications, market price, demand and supply. Compared to different classes of hybrid, composites and reinforced materials, nanoparticles are gradually dominating the global market. A comparative survey of the global market values for nanoparticles and some polymer materials are shown in Table 1.

From the presented information, the market price for polymers and metals is currently higher but the expected growth rate for nanoparticles may gradually be multiplied more than other materials. The implication is that nanoparticles are gradually reducing the market value for other materials and may dominate more shortly if their applications are not properly regulated.

Technical considerations of ethical issues must be able to define the scope, positive and negative impacts and allowance for the well-being of the future global society. Concerning nanotechnology, ethical issues have been reported by Schulte and Salamanca-Buentello (2007) to include issues related to environmental impact, equity, metaphysical applications concerning the human machine, privacy and

Table 1. Global market values for nanoparticles and polymers

Materials	Market Value ($)	Growth (%)for 2009	Growth (%)for 2023	Projected growth (%) for 2030	Reference
Nanoparticles	10.88 billion	14.60	14.80	22.00	Grand View Research (2023), BCC Research ()2023
Polymers	132.10 billion	5.60	4.69	4.90	Grand View Research (2023), SkyQuest (2023)
Metals	551.03 billion	3.80	7.40	2.60	The Business Research Company (2023).
Polymeric nanoparticles blend	610.40 million	3.20	10.00	10.61	MNR (2023)

security. Specific ethical considerations have also been documentedfor nanoparticles, concerning opposition to their involvement in military applications especially, the fabrication of lethal autonomous weapons in addition to the fearful opinion of self-replication ad infinitum in grey goo scenario, which defines possible catastrophic events that can be caused by molecular nanoparticles arising from uncontrolled self-replicating machines (nanobot) tending to consume all matters in the earth while developing more to themselves (Garrick, 2008). Issues related to transparency in the provision of information, viable research and obligations aimed at avoidingharm from mistakesthrough the handling of nanoparticles are also relevant to the discussion of ethical issues. The overall benefit of ethical issues is the protection of the entire components of the environment including the biotic and abiotic components.

Nanoparticles are defined by particles within the dimension of 0 and 100 nm with possible sub-classification covering porosity (including microporous mesoporous and microporous with particle size (x) in the ranges of 0-2 nm, $0<x<2$ nm and $2 < x<50$ nm respectively) (Eddy *et al.*, 2022a), dimensional appearance (such as 0-dimension, 1-dimension and 2-dimension) (Chakraborty *et al.*, 2019), organic and inorganic (Khan *et al.*, 2022), carbon-based nanoparticles (Bhatia, 2016), etc.

Considering the wider scope of current and expected future applications of nanoparticles, envisaging risk factors (toxicity) and future environmental health, several studies need to be conducted on the ethical issues that may result from nanoparticle production, applications and consequences.Therefore, the present review is designed to appraise the various levels of issues in nanotechnology that need to be regulated and also access the existing regulations concerning the production, handling, applications and disposal of nanoparticles, where necessary.

REVIEW OF SOME APPLICATIONS OF NANOPARTICLES

In Table 2, some applications of metal, metal oxide and carbon-based nanoparticles are presented. The Tables reveal that nanoparticles have a broader scope of applications ranging from medical, agricultural, pharmaceutical, industrial and research. Most of them seem to be useful in the treatment of water, as antibacterial agents and other listed areas. It is worth pointing out that in most cases, where nanoparticles can do better, enhanced efficiency would be expected from the corresponding nanocomposite. Therefore, although we have not presented information on the application of nanocomposites or hybrid/advanced nanoparticles, their outstanding performance is often better than those of the constituents' nanoparticles.

One of the current concerns in the application of nanoparticles is their roles in the military, which may include both positive and negative applications. The positive applications is in the development of sensors for detecting radioactivity, chemicals

Table 2. Applications of some nanoparticles

Nanoparticles	Applications	References
AgNPS	Antimicrobial, antineoplastic, antibacterial, anticancer, antioxidant, wastewater treatment, agriculture, drug delivery, photocatalyst, electronics, etc	Algotiml*et al.* (2022), Dhaka *et* al. (2023), Hasan *et al.* (2022), Vidyasagar *et al.* (2023)
AuNPS	Component of optical microscope, immunoassays, biosensors, phototherapy, drug delivery, DNA and antigens, genomics, organic catalysis, antimicrobial, photothermal therapy, wastewater treatment, plasmon fluorescence, etc	Bustanji*et*al. (2023), Daraee*et al.* (2016),Silvestre *et al.* (2023), Yao *et al.* (2023),
CuNPS	Catalyst for organic reactions and photocatalyst, dental applications, antibacterial, antimicrobial, antimicrobial, antibacterial, antiviral, antibiotic, anticancer, etc	Al-Hakkani*et al.* (2020), Crisan*et*al. (2022), Gwande*et al.* (2016), Palani *et al.* (2023)
CaONPS	Wastewater treatment, drug delivery, firefighting, catalysis, agriculture,	Eddy *et al.* (2023d), Emeka *et al.* (2023), Kelle *et al.* (2023), Kumari *et al.* (2023)
CaO_2NP	Anticancer, antitumor, antimicrobial, antibacterial, photocatalysis, etc.	Wu *et al.* (2023)
Carbon-based nanoparticles	Water treatment, gene therapy, biosensors, tissue regeneration, pharmaceuticals, detectors, electronics, sports equipment, armour vehicles, drug delivery, anticancer, imagining, biosensors, agriculture, fluorescence quenchers, etc	Holmannova*et*al. (2022), Sridharan *et al.* (2022),
ZnONPS	Agriculture, water treatment, labelling, sensors, drug delivery, delivery of genes, food additives, agriculture, etc	Lakshmipriya and Gopinath (2021), Odoemelam*et al.* (2023), Zhou *et al.* (2023)
MgONPS	Photocatalysts, refractory fibre, fuel additives, biosensors, water treatment, catalysis of organic reactions, semiconductors, dehydrating agents, refractory materials, materials improvement such as soundproofing, heat resistance, corrosion inhibitor, magnetic rod antenna, antistatic agent, etc	Ramezani Farani *et al.* (2023),
SiO_2NPS	Biomedicine, paper making, drainage aid, binding agent for polymers and concrete, water treatment, agriculture, water treatment, components of semiconductors, plastic, rubbers, in drug delivery, advanced catalysis, etc	Guo *et* al. (2018), Nay *et* al. (2022), Rastogi *et al.* (2019),
TiO_2NPS	Pharmaceuticals, food products, textiles, cosmetics, toothpaste, coating, ink, paper, plastics, etc	Jabeen (2018), Waghmode*et al.* (2019)
Al_2O_3NPS	Biomedical, wear protection, ballistics, electronics, regulator and seal discs, textile, antimicrobial, sensors, immunotherapy, etc	Gudkov*et*al. (2022), Hassanpour *et al.* (2018),

and biological weapons. They can also provide useful functions in aerospace such as solid fuel cells, surveillance cameras, monitors, etc.

Toxicity of Nanoparticleswhat

One of the basic challenges facing nanoparticles is their toxicity (Kumah *et al.* 2023). Nanoparticles id said to be toxic when it is harmful or poisonous to any component of the environment. Different forms of products have been listed as toxic but toxicity can generally be grouped into the following,

(i) Chemical toxicity: This form of toxicity arises from chemicals such as those used in the preparation of some products (Eddy *et* al., 2022a)
(ii) Biological toxicity: Mostly associated with the pathogen that can penetrate organisms and interact to cause diseases and even death at times (Kakoty*et al.*, 2022)
(iii) Physical toxicity: Physical toxicity can be accepted based onsubstances exerting toxic remark due to their physical nature (Sukhanova*et al.*, 2018)
(iv) Radioactive toxicity embraces all forms of radiation that do sustain or activate toxicity after interaction with the environment through absorption, reflection, scattering, or transmittance of ionizing radiation with a wavelength that matches the theta, delta, beta, alpha and gamma (Ray, 2017)

The toxicity of nanoparticles defines their tendency to act as a poison (Eddy *et al.*, 2023a). The toxicity of nanoparticles can be affected by several factors which may include dosages, exposure/entrance route, surface properties, particle size, level of aggregation, transmembrane diffusivity,immunogenicity, excretion pathway and other factors summarized below (Kumah *et al.*, 2023).

(i) The nature or characteristics of the toxicant: Some toxicants are highly penetrative to a wide range of target systems (Malhotra*et al.*, 2020), some are selective, and some can be transported at ease while others may require direct contact. Some are gaseous and others are solid or liquid.
(ii) Solubility: The score point for the activity of a toxicant can also be achieved through its solubility in a medium, especially an aqueous medium (Souza*et al.*, 2021). Solubility can enhance the tendency toward transportation, absorption and penetration (OdauduandAkinsiku, 2022). For example, the toxicity of lead and other heavy metals to aquatic organisms is an active process when the heavy metal is present in the aquatic medium (Essien and Eddy, 2015)
(iii) Adsorption/absorption tendency/capacity: Most toxicants can only be effective in such capacity if the part of the ecosystem can absorb or adsorb them (Ogoko*et al.*, 2023). For example, some toxicants can penetrate the stem, the leaf and the fruit of a given plant through foliar absorption (Eddy and Ekop, 2007).

Some toxicants containing compounds can be secreted by humans because the system cannot absorb or adsorb them (Behzadi*et al.*, 2017)

(iv) The residence time: The magnitude of the impact generated by a toxicant can vary with the listed and other factors but the severity has a strong correlation with the period of contact (Eddy *et al.*, 2022a). Consequently, highly toxic contaminants can lead to death at a time measured as short and yet less toxic ones can also lead to death if a prolonged period of contact is ensured.

(v) Dosage: All forms of contamination and toxicity depend on dosage, which is a direct consequence, expected for the dependency on the progress of a reaction with concentration. The interplay of toxicity levels can be worse encouraging when concentration-time dependency are mutually operational towards the achievement of negative impact (Zheng and Zhou, 2018)

(vi) Exposure route: The severity of a toxicant can be considered harsh depending on the exposure route. For example, the absorption of a toxic compound by the leaf or root of a plant operates under different mechanisms and will subsequently release a diverse measure of toxicity (ref). Such could also account for the differential severity of toxicity for a given substance absorbed/absorbed by the skin compared to the oral or nasal intake routes (Clement *et al.*, 2014).

Eddy *et* al. (2023c) reported that nanoparticles can also exhibit neurotoxicity, hepatoxicity (referring to nanoparticles acting as xenobiotics to impaired liver functioning), chronic toxicity, nephrotoxicity (toxicity associated with effective deterioration of the kidney function), acute and chronic toxicity. Some documented evidence also supports these classes of toxicity about nanoparticles including nephrotoxicity (Iavicoli*et al.*, 2016), neurotoxicity associated with PbONPS (Mei *et al.*, 2023), hepatoxicity by metallic nanoparticles (Vilas-Boas and Vinken, 2021), acute toxicity associated with silica nanoparticles (Almanaa*et al.*, 2022) and chronic toxicity (Almanaa, *et* al., 2022), as well as chronic and acute toxicity of AgNPS (Panacek*et al.*, 2011).The different forms of toxicity listed above are demonstrated in Figure 1.

Based on observed nanotoxicity data, several consequences ranging from mild to death have been ascribed to the toxicity of nanoparticles (Abbasi *et al.*,.2023). For example, Kumah *et* al. (2023) observed that the toxicity of nanoparticles to biological organisms includes the generation of oxidative stress, damage to DNA, apoptosis, cell death and induction of inflammatory responses. Impacts on the endocrine, respiratory, nervous, immune, carcinogenicity and immune systems have also been affirmed (Xuan *et al.*, 2023).

The particle size of nanoparticles can influence their mobility and reactivity including the extent of interaction with other systems. Studies conducted by Park *et al.* (2011) on the toxicity of various sizes of silver nanoparticles (AgNPS)(namely

Figure 1. Various levels of nanotoxicity

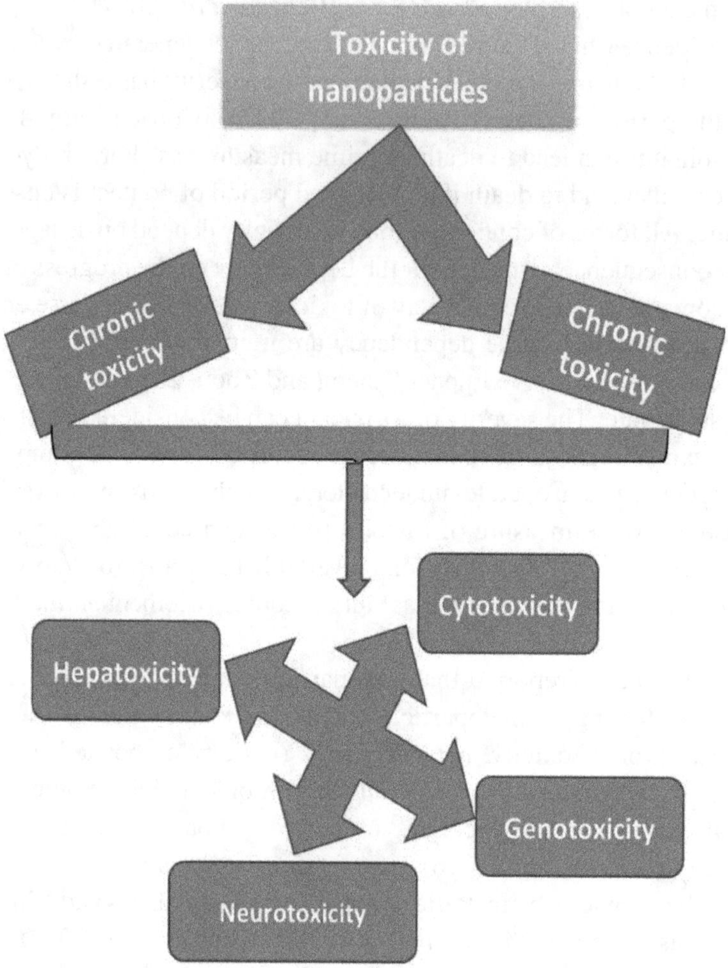

20, 80 and 110 nm) indicated that the toxicity is size-dependent. Also, their study compared in vitro bioassays concerning inflammation, genotoxicity, cytotoxicity and developmental toxicity. The study also led to the observation that the AgNPS have a higher potential to indued the listed toxic effects compared to silver ions. Also, the effect was more significant as the particle size decreased and they concluded that the toxicity also varied with the cell and with the period of exposure. A combination of particle size and charge has also been investigated concerningthe toxicity of nanoparticles, especially cytotoxicity. For example, Huang *et al.* (2017) reported that the cytotoxicity of iron nanoparticles, having similar size but different charges showed observable incrementswith an increase in thesurface charge. The influence of particle size on the toxicity of nanoparticles can also be correlated with the following reasons,

(i) The specific surface area of nanoparticles increases as the particle size decreases indicating that nanoparticles with smaller particle sizes tend to have higher levels of toxicity
(ii) Smaller particle size nanoparticles can penetrate cells and block their biological or biochemical functions more rapidly than larger particles
(iii) The reactivity of nanoparticles increases as the surface area increases and with a decrease in the particle size. Therefore, toxicity driven by reactivity would be more severed with nanoparticles having larger surface area and smaller particle size.
(iv) The particle size of nanoparticles can also affect the production rate of reactive oxygen species and hence toxicity. Therefore, smaller particle sizes are known for their capacity to produce more ROS than larger particle-size nanoparticles.

Some studies have also shown that the level of toxicity of nanoparticles can be diverse considering the site of attack. In this regard, Rosário*et al.* (2022), observed differential distribution of AgNPS in the liver, lungs, kidney and blood. The retention time was also observed to varywhile the major excretion route was through the urine. On this note, it can be stated that the organs that show a higher retention time will develop higher toxicity. Some metal nanoparticles have been confirmed to be more toxic than others. In Table 3, the toxicity of some common nanoparticles (NPS) is recorded. It is seen that different nanoparticles can exert different toxic impacts ranging from mild to chronic impact. For example, PbONPS and CdSNP can exhibit chronic toxic impacts that can result in death. The reported impact are highlighted to cover animals and man as well as some lower organism and plants

Some studies have shown that the initial mechanism concerning the toxicity of nanoparticles (NPS) is the generation of free radicals and reactive oxygen species via oxidative stress and the consequence manifestation of genotoxicity, cellular dysfunction, inflammation, etc.

Carcinogenicity occurs when nanoparticles are inhaled, which defines the tendency of nanoparticles to induce tumour (malignancy) when dermally applied, ingested or inhaled (Becker *et al.*, 2011). Most nanoparticles are carcinogenic (i.e. they tend to cause cancer) because of their large surface area and significant number of particles with nano dimension (Ogoko*et al.*, 202). Some reports have indicated that most nanoparticles are carcinogenic. For example, Magaye*et al.* (2012) reviewed the carcinogenic aspects of cobalt, nickel and copper nanoparticles. They concluded through a literature search that there is adequate evidence concerning the toxicity of the listed metal nanoparticles in experimental animals but evidence backing their

Table 3. Reported toxic impacts of some common nanoparticles

Nanoparticles (NPS)	Toxicity	Reference
AgNPS	Negative impact on photosynthetic pigment at higher concentrations, impaired fin development, higher cytotoxicity in monocytes and macrophages impaired proliferation and expression of human lymphocyte cells, and impairment of peripheral blood mononuclear cells (PBMCs). Toxic to mammalian skin and kidney, liver tissues	Muñoz-Carrillo *et al.* (2023), Noga *et al.* (2023) Ferdousn and Nemmar (2020)
Al_2O_3NPS	oxidative stress, apoptosis, increase in white blood cell count. Decrease in immunoglobulin G, antitrypsin, red blood cell count, maturation index ratio and total and differential bone marrow cell counts (MIR).	Alghriany, *et al.* (2022)
AuNPS	Decreases in body and absolute testicular weights, impaired level of testosterone hormone, low motility and sperm count increase in level of sperms of abnormal morphology	Abdulhaq *et al.* (2023)
CaNPS	Subarachnoid haemorrhage in the brain tissues. Hemorrhagic lung infarcts acute arterial plethora pathomorphological changes (lymphoid tissue hyperplasia), eosinophilia and hydropic and hyaline-drop dystrophy of parenchymal liver tissue	Zaitseva*et al.* (2019)
CuNP	Impaired kidney and liver, immunotoxicity, cytotoxicity, genotoxicity, ROS generation, oxidative stress, inflammation response associated with irritation, etc.	Ameh and Sayes (2019)
Carbon-based NPS	cell death, production of oxidative stress, DNA damage, apoptosis, and induction of inflammatory responses	Kumah *et al.* (2023)
MgONPS	Induction of DNA, oxidative and genotoxic damages, apoptotic impact	Mahmoud *et al.* (2016)\ Mangalampalli*et al.* (2017)
MnO_2NPS	Malfunctioning of the brain, spleen, liver and kidney	Fernández-Pampín*et al.* (2022)
SiO_2NPS	cell death, production of oxidative stress, DNA damage, apoptosis, and induction of inflammatory responses, malfunctioning of the lung, liver, spleen and other soft tissues.	Kumah *et al.* (2023)
TiO_2NPS	cell death, production of oxidative stress, DNA damage, apoptosis, and induction of inflammatory responses	Kumah *et al.* (2023)
ZnNPS	Decrease in cell viability, induce alteration in morphological factors, decreasing hatching rate by aquatic organisms,	Vaidiglesias*et al.* (2023)
Fe_2O_3NPS	Alteration of blood coagulation, cell lysis, inflammation, impaired cell viability, cardiac failure, generation of ROS	Wu *et al.* (2022)
PbONPS	Impairment of the brain, liver, blood poisoning, neurotoxicity effectssuch as memory impairment, hippocampus-dependent learning and antisocial behaviours	Sutunkova*et al.* (2020)
CdSNP	Damage to the epithelium of the gill of fish, loss of mucous cells, vacuolated cells. Significant degeneration of epithelial and mucous cells	Verma *et al.* (2020)

toxicity to humans is scanty. A similar report has also indicted ZnONPS, concerning its tendency to be carcinogenic (Barguilla*et al.*, 2020). Some evidence has also been reported concerning the carcinogenic potential of TiO_2NPS and some carbon-based nanoparticles on the experimental rate at a level that agrees with the European Union criteria for category 2 (Roller, 2009).

In genotoxicity, the property of interest is for nanoparticles to damagethe genetic information in cells with the resultant impact of mutation and the probable manifestation of cancer. Studies conducted on the toxicity of AgNPSbased on in-vitro and vivo studies showed that their tendency to initiate genotoxicity is most probable (Jaswal and Gupta, 2023).Ferdousn and Nemmar (2020) have also shown that AgNPS are toxic to the mammalian cell. Based on the observation reported by Sharma (2023), nanoparticles can be bound to DNA directly and generate indirect genotoxicity after their interaction with a nuclear protein or ROD. Such observations have been applied in the explanation of the mechanism of genotoxicity of some nanoparticles such as the genotoxicity of NiONPS and CoONPS which were confirmed by Fontaine *et al.* (2023) as a consequence of a significant increase in ROS and damage to DNA. Kohl *et al.* ((2020) also gave some explanations for the genotoxicity of nanoparticles based on in vitro and in vivo observations and highlighted that the outcome changes associated with in-vivo studies are related to induced DNA function through secondary genotoxicity, associated with inflammation. Ramos-Zúñiga *et al.* (2023) provided convincing reviewand experimental evidence that copper is needed by humans but at higher concentrations copper nanoparticles have a high genotoxicity tendency that is connected with their ability to generate hydroxyl radicals through the Fenton and Haber Weiss mechanism. They can persist and damage cells, and membrane, produces ROS, protein dysfunctioning, genotoxicity etc.

Copper nanoparticles (CuNPS) are also known fortheir cytotoxicity tendency, suggesting that as a chemotherapeutic agent, CuNPS can lead to toxicity in living cells (McGaw *et al.*, 2014). Oxidative stress responses have been studied regarding copper nanoparticle-induced effects as well as other pathways to cytotoxicity.One of the significant errors supporting theunderestimation of the toxicity of CuNPS is the fact that copper is regarded as a micronutrient. Given this conception in some quarters, we present some ecotoxicity data concerning CuNPS and some biological systems in Table 4 as representative data for some nanoparticles considered micronutrients. Such information justifies the fact that even the useful classes of nanoparticles may exert ecotoxicity depending on the affecting factors, such as dosage, entrance route, etc.

Apart from CuNPS, several organic, hybrid, inorganic and polymeric nanoparticles have significant chemotherapeutic properties and hence they also have inherent cytotoxicity (Rizi*et al.*, 2022) as shown in Table 4.However, some nanoparticles such as $Ca(OH)_2$NPS, ZnONPS, CaONPS, NaAlSiONPS and others have been found to exhibit a very low level of cytotoxicity even if they are also good chemotherapeutic agents (de Souza *et al.*, 2020; Eddy *et al.*, 2023b-c). Eddy *et al.* (2023d). The ecotoxicity of copper nanoparticles (CuNPs) has also been established for all the components of the ecosystem including soil microorganisms (Sharma *et al.*, 2021), and Enchytraeuscrypticus (Gomes *et al.*, 2015).

Lekamge*et al.* (2020) have also established that AgNPS are toxic to *Raphidocelissubcapitata* algae. The toxicity showed a positive incremental trend with the concentration of the algae. The activation of the glutathione transferase enzyme was also observed to be induced by the presence of AgNPS. Haase*et al.* (2011) have also indicated that AgNPs are toxic to human microphases and can be distributed to different organs to induce different levels of toxicity. Korani*et al.* (2015) have also confirmed that ANPS are toxic to humans and that the entrance routes are the skin, respiratory and gastric systems. Zhang *et al.* (2021) have verified that AgNPS can affect embryo and fetus development in man. They observed that AgNPS can create resorption growth retardation and abnormalities directed to the histological, central nervous system, respiratory, immune system and other disorders. Histopathological examination of AgNPSin mammals including humans has also shown that AgNPS can adversely affect humans (Iyiola*et al.*, 2018).

The ecotoxicity of AgNPS on plants is a significant sector to review since plants are the primary producers and through the food chain, other components of the ecosystem can be affected. Yan and Hen (2019) also observed that the phytotoxicity of AgNPS on plants has strong negative impacts on the morphological, physiological, cellular and molecular levels of plants. The absorption of AgNPS by *Arabidopsis thalian* was reported by Qian *et al.* (2013) to release adverse consequences such as low concentration of chlorophyll, delayed growth, alteration of the antioxidant-oxidant balance and homeostasis of water in the plant.

Literature supporting the ecotoxicity of NiNPSarenot scanty (Latvala *et al.*, 2016; Yokota *et al.* 2019). However, concerning TiO_2NPS literature is almost lacking on their ecotoxicity but platinum nanoparticles have been implicated in some research concerning their toxicity to the ecosystem.

Hlavkova*et al.* (2019) observed that the ecotoxicology of TiO_2NPs on some aquatic biocenosis showed increasing impact as the particle size decreased and also concluded that the toxicity significantly varies with an organism with bacteria showing higher sensitivity while *Daphnia*showed the lowest. Czubackaand Czerczak (2019) also confirmed that PtNPS can cause DNA damage and apoptosis with significant cytotoxicity. However, Labrador-Rached*et al.* (2018) reported that PtNPS may not record significant ecotoxicology at low concentrations.

Table 4. Ecotoxicology data for Cu-NPs

Nanoparticles	Toxicity	Target agent	Reference
Cu-NPs	Induced organ dames especially, the gills, kidney, brain, gonad, heart and liver	Some fishes	Malhotra*et al.* (2020)
Cu-NPs	Induced oxidative stress	crayfish	Yang *et al.* (2022)
Cu-NPs	In vivo toxicity was established especially concerning the pavement and cells of the fish gills	*Oncorhynchus Mykiss, Danio rerio and Pimephalespromelas,*	Song *et al.* (2015)
Cu-NPs	Reduction in the activities of glutathione, acetylcholinesterase, glutathione S- transferase and other enzymes	Tilapia fish	Al Ghais*et al.* (2019)
Cu-NPs	Disorderliness in energy metabolism and oxidative, reduced photosynthetic efficiency, oxidative stress, etc	*Paramecium bursaria–Chlorella*	Tan *et al.* (2022)
Cu-NPs	Induced inflammatory influx, the release of cytokine and damage to the lung	Mouse	Pietrofesa*et al.* (2021)
Cu-NPs	Denaturation of DNA, protein and protein, reduction in oxidative stress tolerance, enhanced production of reactive oxygen species, hemolysis, dead of the cell, etc	Some aquatic organism	Akalin*et al.* (2021)
Cu-NPs	Acute lethal toxicity was established	*Danio rerio*	Griffitt*et al.* (2007)
Cu-NPs	Increased metal load, inducement of oxidative stress and genotoxicity	*Labeorohita*	Aziz and Abdullah (2023)
Cu-NPs	Alteration in the activity of brain acetylcholinesterase, Na and K ions metal in gills	Crucian carp	Liu *et al.* (2017)
Cu-NPs	Damage to gills, epidermis and liver of Siberian sturgeon	Siberian sturgeon	Ostaszewska*et al.* (2016)
Cu-NPs	Exerted cytotoxicity on the kidney, liver and gill of the fishes	*Rutilus rutiluscaspicus and Caspian roach*	Aghamirkarimi*et al.* (2017)
Cu-NPs	Cytotoxicity was reported	Some aquatic organism	Kolukisaoğlu*et al.* (2022)
Cu-NPs	Genotoxicity, damage to the root, reduce photosynthetic capacity, etc.	*Cucumber plant*	Mosa*et al.* (2018)

Sources Of Toxicological And Ethical Issues

In Table 5, issues arising from various beneficial applications of nanoparticles are highlighted. It is evident from the presented results that the production route is the major source of issues that may lead to nanotoxicity.

Table 5. Some applications of nanoparticles and supporting properties

Application	Supporting properties	References	Potential source of issues
Adsorption	Particle size, surface area, porosity, surface area to volume ratio	Garg *et al.*, (2023)	Production, handling, application, disposal, inhalation, thermal contact,
Photocatalyst	Particle size, crystallite size, pore diameter, surface area, porosity, bandgap	Eddy et al. (2023e)	Production, handling
Pharmaceutical	Particle size, surface area, surface area to volume ratio	Zakaria *et al.* (2023)	Consumption, dermal contact, handling, reaction, etc
Chemotherapeutic agent	Anti-cancer property	Mousa *et al.* (2021)	Production, handling and migration to the cell, reaction, etc
Antimicrobial agent	Particle size, surface area,blockage of cells etc	Urnukhsaikhan*et al.* (2021)	Production, handling, applications, reaction, leakages, etc
Military applications	Particle size, toxicity, surface area, etc	Kumar and Dixit (2019a,b)	Production, degradation, leakage, explosion
Fire fighting	Particle size, surface area, surface area to volume ratio, porosity,	Sheng *et al.* (2023)	Production, handling, inhalation, dermal and tissue contact, etc
Agricultural	Particle size, reactivity, surface area, ionic character, etc	Yousef *et al.* (2023)	Production, ion exchange, reactions, absorption, leakages, etc
Industrial applications	Particle size, mechanical property, electrical property, optical property	Kumar *et al.* (2023)	Production, applications, disposal, degradation, etc

No matter the intended method of synthesis, the possibility of losing nanoparticles to the environment is very high if the conservation of the raw materials and the products are not properly guided. Also due to their small particle sizes, nanoparticles can be inhaled in the form of particulates. They can easily be leaked into the water system, aquatic animals can consume or absorb them and through the food chain, they will be bioaccumulated and biomagnified in plant and animals

Some nanoparticles are useful in agriculture, especially as soil stimulants. The risk factors include surface runoff and absorption by plants through the foliar absorption process.Since plants are the primary producers, once a risk factor has been established in plants, the entire ecosystem is also at risk. Intake from contaminated drinking water is also possible in addition to ingestion through food materials. The applications of nanoparticles in the military sector can lead to the contamination of the air, soil as well as the biotic and abiotic components of the environments. Such incidents may arise from explosion, discharge, disposal, etc. In agriculture, nanoparticles are used as pesticides, soil stimulants, fertilizers and other areas. Rainfall or irrigation can help to spread the nanoparticles away from the target to the non-target zone with associated consequences of contamination. Nanoparticles are strong antimicrobial agents and are often used in several remediation approaches such as antibacterial, antiviral, etc. In the course of application, nanoparticles canmiss their target and exertsome challenges to other components of the environment. The major applications of nanoparticles in environmental science are in adsorption and photodegradation. If the end product is not properly managed or there is a leakage of nanoparticles, the likelihood of nanotoxicity is most probable.

Issues and the Way Forward

From the above analysis,it can be stated that nanoparticles mayhave significant beneficial roles in different sectors of our society. However, the major challenges are listed to include the following,

(i) Method of synthesis and applications
(ii) Disposal of endproduct or unreacted materials
(iii) Leakage from the operational line
(iv) Migration by natural processes such as surface runoff, etc
(v) Improper handling
(vi) Military application
(vii) Lack of better substitution or replaceable materials
(viii) Degradation of nanoparticles or materials containing nanoparticles
(ix) Nature of application
(x) Improper personnel
(xi) Lack of adequate research

It is significant to state that one of the most serious issues concerning nanoparticles is the engagement of poor personnel within the production, application and disposal lines. This implies that ethical practices must first consider the employment and training of the right personnel. Issues arising from synthetic methods are currently receivingsolutions (especially at the pilot level) through the application of the green synthetic method. The green synthesis method is an environmentallyfriendly approach that can help to eliminate toxic compounds in the production lines, In several applications, the nanoparticles or materials/equipment containing the nanoparticles need to be disposed of if theyare expired or if their efficiency has reduced to a significant level. Consequently, efforts adopted in the management of waste from nanomaterials must be able toconserve environmental impact. Some studies have proposed re-cycling, resource recovery and reuse as the three principles that can be useful in the management of nanowastes (Eddy *et al.* (202b). It is also necessary that operational lines involving the applications of nanoparticles need to be periodically monitored to overcome possible leakage and where leakage occurs, technology for the recovery of the leak product should be made available. One of the basic issues in the nano industries is the absence of materials with comparable properties. This has led to the excessive engagement of nanoparticles in several sectors since the desired properties can hardly be met by other materials. It is the place of research to engage in future tasks that can lead to the production of new materials with comparative properties that are more environmentally friendly.

In most countries, there are no regulations concerning nanoparticles regarding their production, personnel requirements for handling, restriction, usage, transportation, military applications, etc. Perhaps the most popular and fundamental regulation concerning nanoparticles is the European Union regulation but the regulation has been widely criticized as overlooking the criteria of legislative authority, accountability, and due process. The most effective regulation concerning nanoparticles must ensure that all the factors needed to protect the present and future environment as well as the implementation plan and other issues.

CONCLUSION AND RECOMMENDATIONS

The study has employed available literature as a searchlight for issues concerning nanoparticles and the environment. We observed the growing economic and scientific markets as the major driving force for the proliferation of nanoparticles in the global market. In recent times, the market demand for nanoparticles has expanded excessively, indicating that if adequate research, ethical and alternative replacements are not considered, the world may gradually approach the grey goo scenario that has heightened fears among scientists, especially when most sectors are gradually replaced by nanomaterials.

Given the observed environmental issues attributed to nanotechnology and the risk of meeting worse issues, ethical consideration requires the re-strengthening of existing regulations, research for better replacement, engagement of right and trained personnel, employment of green synthetic routes and application of recovery, reuse and recycling principles of waste management.

ACKNOWLEDGEMENT

The authors sincerely acknowledge the Tertiary Education Trust fund of Nigeria for fully funding this work through the National Research Fund with Prof. NnabukOkon Eddy as the Principal Investigator(Grant number: TRTF/ES/DR&D-CE/NRF2020/ SETI/98/VOL.1).

REFERENCES

Abbasi, R., Shineh, G., Mobaraki, M., Doughty, S., & Tayebi, L. (2023). Structural parameters of nanoparticles affecting their toxicity for biomedical applications: A review. *Journal of Nanoparticle Research, 25*(3), 43. doi:10.1007/s11051-023-05690-w PMID:36875184

Abdulhaq, N. A., Elnady, D. A., Abo El-atta, H. M., El-Morsi, D. A., & Gad El-Hak, S. A. (2023). Assessment of reproductive toxicity of gold nanoparticles and its reversibility in male albino rats. *Toxicological Research*. doi:10.1007/s43188-023-00203-2

Al-Hakkani, M. F. (2020). Biogenic copper nanoparticles and their applications: A review. *SN Applied Sciences, 2*(3), 505. doi:10.1007/s42452-020-2279-1

Alghriany, A. A. I., Omar, H. E. M., Mahmound, A. M., & Atia, M. M. (2022). Assessment of the Toxicity of Aluminum Oxide and Its Nanoparticles in the Bone Marrow and Liver of Male Mice: Ameliorative Efficacy of Curcumin Nanoparticles. *ACS Omega, 7*(16), 13841–13852. doi:10.1021/acsomega.2c00195 PMID:35559158

Algotiml, R., Gab-Alla, A., Seoudi, R., Abulreesh, H. H., El-Readi, M. Z., & Elbanna, K. (2022). Anticancer and antimicrobial activity of biosynthesized Red Sea marine algal silver nanoparticles. *Scientific Reports, 12*(1), 2421. doi:10.1038/s41598-022-06412-3 PMID:35165346

Almanaa, T. N., Aref, M., Kakakhel, M. A., Elshopakey, G. E., Mahboub, H. H., Abdelazim, A. M., Kamel, S., Belali, T. M., Abomughaid, M. M., Alhujaily, M., Fahmy, E. M., Ezzat Assayed, M., Mostafa-Hedeab, G., & Daoush, W. M. (2022). Silica nanoparticle acute toxicity on male *rattus norvegicus* domestica: Ethological behavior, hematological disorders, biochemical analyses, hepato-renal function, and antioxidant-immune response. *Frontiers in Bioengineering and Biotechnology, 10*, 868111. doi:10.3389/fbioe.2022.868111 PMID:35464726

Ameh, T., & Sayes, C. M. (2019). The potential exposure and hazards of copper nanoparticles: A review. *Environmental Toxicology and Pharmacology*, *71*, 103220. doi:10.1016/j.etap.2019.103220 PMID:31306862

Baran, A. (2016). Nanotechnology: Legal and ethical issues. Economics and management. *Economics and Management*, *8*(1), 47–54.

Barguilla, I., Barszczewska, G., Annangi, B., Domenech, J., Velázquez, A., Marcos, R., & Hernández, A. (2020). MTH1 is involved in the toxic and carcinogenic long-term effects induced by zinc oxide and cobalt nanoparticles. *Archives of Toxicology*, *94*(6), 1973–1984. doi:10.1007/s00204-020-02737-y PMID:32377776

Becker, H., Herzberg, F., Schulte, A., & Kolossa-Gehring, M. (2011). The carcinogenic potential of nanomaterials, their release from products and options for regulating them. *International Journal of Hygiene and Environmental Health*, *214*(3), 231–238. doi:10.1016/j.ijheh.2010.11.004 PMID:21168363

Bhatia, S. (2016). *Nanoparticles Types, Classification, characterization, fabrication methods and drug delivery applications. in: natural polymer drug delivery systems*. Springer. doi:10.1007/978-3-319-41129-3_2

Bustanji, Y., Taneera, J., Semreen, M. H., Abu-Gharbieh, E., El-Huneidi, W., Faris, M. A. E., Alzoubi, K. H., Soares, N. C., Albustanji, B., Abuhelwa, A. Y., Abu-Zurayk, R., Alqudah, M. A., & AlKhatib, H. S. (2023). Gold nanoparticles and breast cancer: A bibliometric analysis of the current state of research and future directions. *OpenNano*, *12*, 100164. doi:10.1016/j.onano.2023.100164

Chakraborty, P., Das, T., & Saha-Dasgupta, T. (2019). 1.15 - MXene: A New Trend in 2D Materials Science. D. L. Andrews, R. H. Lipson, & T. Nann (eds.) Comprehensive Nanoscience and Nanotechnology (Second Edition). Academic Press. doi:10.1016/B978-0-12-803581-8.10414-X

Crisan, M. C., Teodora, M., & Lucian, M. (2022). Copper Nanoparticles: Synthesis and Characterization, Physiology, Toxicity and Antimicrobial Applications. *Applied Sciences (Basel, Switzerland)*, *12*(1), 141. doi:10.3390/app12010141

Daraee, H., Eatemadi, A., Abbasi, E., Aval, S. F., Kouhi, M., & Akbarzadeh, A. (2016). Application of gold nanoparticles in biomedical and drug delivery. *Artificial Cells, Nanomedicine, and Biotechnology*, *44*(1), 410–422. doi:10.3109/21691401.2014.955107 PMID:25229833 de Souza, G. L., Almeida, A. C., Danlas, N. O., Paula, A., Turrioni, S., & Moura, C. C. G. (2020). Cytotoxicity and Effects of a New Cacium Hydroxide Nanoparticle Material on Production of Reactive Oxygen Species by LPS-Stimulated Dental Pulp Cells. *Iranian Endodontic Journal, 15*(4), 227–235. doi:10.22037/iej.v15i4.28942 PMID:36704109

Dhaka, A., Mali, S. C., Sharma, S., & Trivedi, R. (2023). A review on biological synthesis of silver nanoparticles and their potential applications. *Results in Chemistry, 6*, 101108. doi:10.1016/j.rechem.2023.101108

Eddy, N. O., Edet, U. E., Oladele, J. O., Kelle, H. I., Ogoko. E. C., Odiongenyi, A. O., Ameh, P., Ukpe, R. A., Ogbodo, R., Garg, R. & Garg, R. (2023b). Synthesis and application of novel microporous framework of nanocomposite from trona for photocatalysed degradation of methyl orange dye. *Journal of Environmental Monitoring and Assessment*. doi:10.1007/s10661-023-12014-x

Eddy, N. O., Garg, R., Garg, R., Aikoye, A., & Ita, B. I. (2022a). Waste to resource recovery: Mesoporous adsorbent from orange peel for the removal of trypan blue dye from aqueous solution. *Biomass Conversion and Biorefinery*. doi:10.1007/s13399-022-02571-5

Eddy, N. O., Garg, R., Garg, R., Eze, S. I., Ogoko, E. C., Kelle, H. I., Ukpe, R. A., Ogbodo, R. and Chijoke, F. (2023a). Sol-gel synthesis, computational chemistry, and applications of CaO nanoparticles for the remediation of methyl orange contaminated water. *Advances in Nano Research*. https://doi.org/. doi:10.12989/anr.2023.15.1.000

Eddy, N. O., Odiongenyi, A. O., Garg, R., Ukpe, R. A., Garg, R., El Nemir, A., Ngwu, C. M., & Okop, I. J. (2023e). Quantum and experimental investigation of the application of *Crassostrea gasar* (mangrove oyster) shell–based CaO nanoparticles as adsorbent and photocatalyst for the removal of procaine penicillin from aqueous solution. *Environmental Science and Pollution Research International, 30*(23), 64036–64057. doi:10.1007/s11356-023-26868-8 PMID:37059957

Eddy, N. O., Odiongenyi, A. O., Garg, R., Ukpe, R. A., Garg, R., Ogoko, E. C., & Kelle, H. I. (2023d). *Ecotoxicological aspect of nanotechnology. In nanotechnology for sustainable Agriculture, Food and Environment*. Taylor and Francis. doi:10.1201/9781003397861-16

Eddy, N. O., Ukpe, R. A., Ameh, P., Ogbodo, R., Garg, R., & Garg, R. (2022). Theoretical and experimental studies on photocatalytic removal of methylene blue (MetB) from aqueous solution using oyster shell synthesized CaO nanoparticles (CaONP-O). *Environmental Science and Pollution Research International, 30*(34), 81417–81432. doi:10.1007/s11356-022-22747-w PMID:36057067

Eddy, N. O., Ukpe, R. A., Ameh, P., Ogbodo, R., Garg, R., & Garg, R. (2022b). Theoretical and experimental studies on photocatalytic removal of methylene blue (MetB) from aqueous solution using oyster shell synthesized CaO nanoparticles (CaONP-O). *Environmental Science and Pollution Research International, 30*(34), 81417–81432. doi:10.1007/s11356-022-22747-w PMID:36057067

Eddy, N. O., Ukpe, R. A., Garg, R., Garg, R., Odiongenyi, A. O., Ameh, P., Akpet, I. N., & Udo, E. S. (2023c). Review of in-depth knowledge on the application of oxides nanoparticles and nanocomposites of Al, Si and Ca as photocatalyst and antimicrobial agents in the treatment of contaminants in water. *Clean Technologies and Environmental Policy*. doi:10.1007/s10098-023-02603-2

Ferdous, Z., & Nemmar, A. (2020). Health Impact of Silver Nanoparticles: A Review of the Biodistribution and Toxicity Following Various Routes of Exposure. *International Journal of Molecular Sciences, 21*(7), 2375. doi:10.3390/ijms21072375 PMID:32235542

Fernández-Pampín, N., González Plaza, J. J., García-Gómez, A., Peña, E., Rumbo, C., Barros, R., Martel-Martín, S., Aparicio, S., & Tamayo-Ramos, J. A. (2022). Toxicology assessment of manganese oxide nanomaterials with enhanced electrochemical properties using human in vitro models representing different exposure routes. *Scientific Reports, 12*(1), 20991. doi:10.1038/s41598-022-25483-w PMID:36471154

Fontaine, M., Bartolami, E., Prono, M., Béal, D., Blosi, M., Costa, A. L., Ravagli, C., Baldi, G., Sprio, S., Tampieri, A., Fenoglio, I., Tran, L., Fadeel, B., & Carriere, M. (2023). Nanomaterial genotoxicity evaluation using the high-throughput p53-binding protein 1 (53BP1) assay. *PLoS One, 18*(9), e0288737. doi:10.1371/journal.pone.0288737 PMID:37713377

Garrick, B. J. (2008). Examples of Risks Having the Potential for Catastrophic Consequences. B. John Garrick, Quantifying and Controlling Catastrophic Risks. Academic Press. doi:10.1016/B978-0-12-374601-6.00007-8

Grand View Research. (2023). *Base Metal Mining Market Size, Share & Trends Analysis Report By Product (Aluminum, Copper, Zinc, Lead), By End-Use (Construction, Automotive), By Region, And Segment Forecasts*. GVR. htttps://www.grandviewresearch.com/industry-analysis/base-metal-mining-market

Grand View Research. (2023). *Nanomaterials Market Size, Share & Trends Analysis Report By Material (Gold, Silver, Iron, Copper), By Application (Aerospace, Automotive, Medical), By Region, And Segment Forecasts, 2023 – 2030*. https://www.google.com/search?q=Global+market+value+for+nanoparticles

Grand View Research. (2023). *Polymer Foam Market Size, Share & Trends Analysis Report By Type (Polystyrene, Polyurethane, Polyolefin, Melamine, Phenolic, PVC), By Application, By Region, And Segment Forecasts, 2022 – 2030*. https://www.grandviewresearch.com/industry-analysis/polymer-foam-market

Gudkov, S. V., Burmistrov, D. E., Smirnova, V. V., Semenova, A. A. & Lisitsyn, A. B. (2022). A mini review of antibacterial properties of Al_2O_3 nanoparticles. *Nanomaterials (Bassel), 12*(15), 2635, 30. . doi:10.3390/nano12152635

Guo, Z., Martucci, N. J., Liu, Y., Yoo, E., Tako, E. and Mahler, G. J. Silicon dioxide nanoparticles exposure affects small intestine function in an in vitro model. *Nanotoxicology, 12*(5), 485-508. . doi:10.1080/17435390.2018.1463407

Gwande, M. B., Goswami, A., Felpin, F., Asefa, T., Huang, X., Silva, R., Zou, X., Zboril, R., & Varma, R. S. (2016). Cu and Cu-based nanoparticles: Synthesis and applications in catalysis. *Chemical Reviews, 116*(6), 3722–3811. doi:10.1021/acs.chemrev.5b00482 PMID:26935812

Hasan, K. M. F., Xiaoyi, L., Shaoqin, Z., Horváth, P. G., Bak, M., Bejó, L., Sipos, G., & Alpár, T. (2022). Functional silver nanoparticles synthesis from sustainable point of view: 2000 to 2023 – A review on game changing materials. *Heliyon, 8*(12), e12322. doi:10.1016/j.heliyon.2022.e12322 PMID:36590481

Hassanpour, P., Panahi, Y., Ebrahimi-Kalan, A., Akbarzadeh, A., Davaran, S., Nasibova, A. N., Khallov, R., & Kavetskyy, T. (2018). Biomedical applications of aluminium oxide nanoparticles. *Micro & Nano Letters, 13*(9), 1227–1231. doi:10.1049/mnl.2018.5070

Holmannova, D., Borsky, P., Svadlakova, T., Borska, L., & Fiala, Z. (2022). Carbon Nanoparticles and Their Biomedical Applications. *Applied Sciences (Basel, Switzerland), 12*(15), 7865. doi:10.3390/app12157865

Huang, Y., Cambre, M., & Lee, H. (2017). The Toxicity of Nanoparticles Depends on Multiple Molecular and Physicochemical Mechanisms. *International Journal of Molecular Sciences, 18*(12), 2702. doi:10.3390/ijms18122702 PMID:29236059

Iavicoli, I., Fontana, L., & Nordberg, G. (2016). The effects of nanoparticles on the renal system. *Critical Reviews in Toxicology, 46*(6), 490–560. doi:10.1080/10408444.2016.1181047 PMID:27195425

Jabeen, F. (2018). Chapter 34 - Polymer–ceramic nanocomposites for controlled drug delivery, Editor(s): Inamuddin, Abdullah M. Asiri, Ali Mohammad, In Woodhead Publishing Series in Biomaterials, Applications of Nanocomposite Materials in Drug Delivery. Woodhead Publishing. doi:10.1016/B978-0-12-813741-3.00035-2

Jaswal, T. and Gupta, J. (2023). A review on the toxicity of silver nanoparticles on human health,Materials Today: *Proceedings, 81*(2), 859-863. . doi:10.1016/j.matpr.2021.04.266

Kelle, H. I., Ogoko, E. C., Akintola O & Eddy, N. O. (2023). Quantum and experimental studies on the adsorption efficiency of oyster shell–based CaO nanoparticles (CaONPO) towards the removal of methylene blue dye (MBD) from aqueous solution. *Journal: Biomass Conversion and Biorefinery*. . doi:10.1007/s13399-023-04947-7

Khan, Y., Sadia, H., Ali Shah, S. Z., Khan, M. N., Shah, A. A., Ullah, N., Ullah, M. F., Bibi, H., Bafakeeh, O. T., Khedher, N. B., Eldin, S. M., Fadhl, B. M., & Khan, M. I. (2022). Classification, Synthetic, and Characterization Approaches to Nanoparticles, and Their Applications in Various Fields of Nanotechnology: A Review. *Catalysts*, *12*(11), 1386. doi:10.3390/catal12111386

Kohl, Y., Rundén-Pran, E., Mariussen, E., Hesler, M., El Yamani, N., Longhin, E. M., & Dusinska, M. (2020). Genotoxicity of Nanomaterials: Advanced In Vitro Models and High Throughput Methods for Human Hazard Assessment—A Review. *Nanomaterials (Basel, Switzerland)*, *10*(10), 1911. doi:10.3390/nano10101911 PMID:32992722

Kumah, E. A., Fopa, R. D., Harati, S., Boadu, P., Zohoori, F. V., & Pak, T. (2023). Human and environmental impacts of nanoparticles: A scoping review of the current literature. *BMC Public Health*, *23*(1), 1059. doi:10.1186/s12889-023-15958-4 PMID:37268899

Kumar, N., & Dixit, A. (2019a). Nanotechnology-Driven Explosives and Propellants. In *Nanotechnology for Defence Applications*. Springer. doi:10.1007/978-3-030-29880-7_3

Kumar, N., & Dixit, A. (2019b). Nanotechnology-Enabled Management of Chemical, Biological, Radiological, and Nuclear Threats. In *Nanotechnology for Defence Applications*. Springer. doi:10.1007/978-3-030-29880-7_4

Kumar, R., Kumar, M., & Luthra, G. (2023). Fundamental approaches and applications of nanotechnology: A mini review. *Materials Today: Proceedings*. doi:10.1016/j.matpr.2022.12.172

Kumari, M., Sarkar, B., & Mukherjee, K. (2023). Nanoscale calcium oxide and its biomedical applications: A comprehensive review. *Biocatalysis and Agricultural Biotechnology*, *47*, 102506. doi:10.1016/j.bcab.2022.102506

Lakshmipriya, T., & Gopinath, S. C. B. (2021). 1 - Introduction to nanoparticles and analytical devices, Editor(s): Subash C.B. Gopinath, Fang Gang, Nanoparticles in Analytical and Medical Devices, Elsevier. doi:10.1016/B978-0-12-821163-2.00001-7

Magaye, R., Zhao, J., Bowman, L., & Ding, M. (2012). Genotoxicity of cobalt, nickel and copper based nanoparticles. *Experimental and Therapeutic Medicine*, *4*(4), 551–561. doi:10.3892/etm.2012.656 PMID:23170105

Mahmoud, A., Ezgi, Ö., Merve, A., & Özhan, G. (2016). In Vitro Toxicological Assessment of Magnesium Oxide Nanoparticle Exposure in Several Mammalian Cell Types. Int J Toxicol, 35(4):429-37. doi:10.1177/1091581816648624

Mangalampalli, B., Dumala, N., & Grover, P. (2017). Acute oral toxicity study of magnesium oxide nanoparticles and microparticles in female albino Wistar rats. *Regulatory Toxicology and Pharmacology*, *98*, 170–184. doi:10.1016/j.yrtph.2017.09.005 PMID:28899817

McGaw, L. J., Elgorashi, E. E., & Eloff, J. N. (2014). 8 - Cytotoxicity of African Medicinal Plants Against Normal Animal and Human Cells, Editor(s): Victor Kuete, Toxicological Survey of African Medicinal Plants, Elsevier. doi:10.1016/B978-0-12-800018-2.00008-X

Mei, L., Xie, R., Zhu, S., Deng, S., Xu, H., Fan, X., Yin, W., & Gu, Z. (2023). Neurotoxicity study of lead-based perovskite nanoparticles. *Nano Today*, *50*, 101830. doi:10.1016/j.nantod.2023.101830

MMR. (2023). Polym*eric Nanoparticles Market: Global Industry Analysis and Forecast (2022-2029)*. MMR. www.maximizemarketresearch.com/market-report/global-polymeric-nanoparticles-market/104820/#

Mousa, A. B., Moawad, R., & Abdallah, Y. (2021). Zinc oxide nanoparticles promise anticancer and antibacterial activity in ovarian cancer. *Pharmaceutical Research*. doi:10.1007/s11095-023-03505-0 PMID:37016170

Muñoz-Carrillo, M. G.,Garcidueñas-Piña, C., Pedroza-Jimenez,M. M., García, R. C., Carbajal-Hernández, A. L., Andrade,G. and Morales-Dominguez· J. F. (2023). *Toxicity of Silver Nanoparticles Synthesized from Opuntia ficus-indica on Danio rerio and Arabidopsis thaliana and Their Antimicrobial Activity*. https://doi.org/doi:10.1155/2023/6903787

Nay, A. A., Abd-Elhamid, A. I., Aly, A. A., & Brase, S. (2022). Recent progress in the applications of silica-based nanoparticles. *RSC Advances*, *12*(22), 13706–13726. doi:10.1039/D2RA01587K PMID:35530394

Noga, M., Milan, J., Frydrych, A., & Jurowski, K. (2023). Toxicological Aspects, Safety Assessment, and Green Toxicology of Silver Nanoparticles (AgNPs)—Critical Review: State of the Art. *International Journal of Molecular Sciences*, *24*(6), 5133. doi:10.3390/ijms24065133 PMID:36982206

Ogoko, E. C., Kelle, H. I., Akintola, O., & Eddy, N. O. (2023). Experimental and theoretical investigation of *Crassostrea* gigas (gigas) shells based CaO nanoparticles as a photocatalyst for the degradation of bromocresol green dye (BCGD) in an aqueous solution. *Biomass Conversion and Biorefinery*. doi:10.1007/s13399-023-03742-8

Palani, G., Trilaksana, H., Sujatha, R. M., Kannan, K., Rajendran, S., Kornirjrnko, K., Nykiel, M., & Uthayakuma, M. (2023). Silver nanoparticles for waste water management. *Molecules (Basel, Switzerland)*, *28*(8), 3520. doi:10.3390/molecules28083520 PMID:37110755

Panacek, A., Prucek, R., Safarowa, D., Dittrich, M., Richtrova, J., Benickova, K., Zhoril, R. & Kvitek, L. (2011). Acute and chronic toxicity effects of silver nanoparticles (nps) on drosophila melanogaster. *environmental Science & Technology*, *45* (11), 4974-4979, DOI: doi:10.1021/es104216b

Park, M. V. D. Z., Neigh, A. M., Vermeulen, J. P., de la Fonteyne, L. J. J., Verharen, H., Briedé, van Loveren, H., & de Jong, W. H. (2011). The effect of particle size on the cytotoxicity, inflammation, developmental toxicity and genotoxicity of silver nanoparticles. *Biomaterials*, *32*(36), 9810–9817. doi:10.1016/j.biomaterials.2011.08.085 PMID:21944826

Ramezani Farani, M., Farsadrooh, M., Zare, I., Gholami, A., & Akhavan, O. (2023). Green Synthesis of Magnesium Oxide Nanoparticles and Nanocomposites for Photocatalytic Antimicrobial, Antibiofilm and Antifungal Applications. *Catalysts*, *13*(4), 642. doi:10.3390/catal13040642

Ramos-Zúñiga, J., Bruna, N., & Pérez-Donoso, J. M. (2023). Toxicity Mechanisms of Copper Nanoparticles and Copper Surfaces on Bacterial Cells and Viruses. *International Journal of Molecular Sciences*, *10503*(13), 10503. doi:10.3390/ijms241310503 PMID:37445681

Rastogi, A., Tripathi, D.K., Yadav, S. *et al.* A(2019). pplication of silicon nanoparticles in agriculture. *3 Biotech* **9**, 90. . doi:10.1007/s13205-019-1626-7

Rizi, H. A., Shin, D. H., & Rizi, S. Y. (2022). Polymeric Nanoparticles in Cancer Chemotherapy: A Narrative Review. *Iranian Journal of Public Health*, *51*(2), 226–239. doi:10.18502/ijph.v51i2.8677 PMID:35866132

Rosário, F., Creylman, J., Verheyen, G., Van Miert, S., Santos, C., Hoet, P., & Oliveira, H. (2022). Impact of Particle Size on Toxicity, Tissue Distribution and Excretion Kinetics of Subchronic Intratracheal Instilled Silver Nanoparticles in Mice. *Toxics*, *10*(5), 260. doi:10.3390/toxics10050260 PMID:35622673

Schulte, P. A., & Salamanca-Buentello, F. (2007). Ethical and Scientific Issues of Nanotechnology in the Workplace. *Environmental Health Perspectives*, *115*(1), 5–1. doi:10.1289/ehp.9456 PMID:17366812

Sharma, P. (2023). Chapter 6 - Genotoxicity of the nanoparticles, Editor(s): Nar Singh Chauhan, Sarvajeet Singh Gill. In Nanomaterial-Plant Interactions, The Impact of Nanoparticles on Agriculture and Soil. Academic Press. doi:10.1016/B978-0-323-91703-2.00017-8

Sheng, Y., Li, Y., Peng, Y., Ma, L., Gao, W., Yan, C., & Liu, X. (2023). Effect of hydrophilic silica nanoparticles on foam properties of mixtures of silicone and hydrocarbon surfactants. *Chemical Engineering Science*, *278*, 118769. doi:10.1016/j.ces.2023.118769

Silvestre, O. F., Rao, A., & Liz-Marzán, L. M. (2023). Self-assembled colloidal gold nanoparticles as substrates for plasmon enhanced fluorescence. *European Journal of Materials*, *3*(1), 1. doi:10.1080/26889277.2023.2202676

Sodano, V. (2023). Nanotechnology and food system: Assessing the european union regulatory system. *European Journal of Development Studies*, *3*(5), 25–31. doi:10.24018/ejdevelop.2023.3.5.294

Sridharan, R., Monisha, B., Kumar, P. S., & Gayathri, K. V. (2022). Carbon nanomaterials and its applications in pharmaceuticals: A brief review. *Chemosphere*, *294*, 133731. doi:10.1016/j.chemosphere.2022.133731 PMID:35090848

Sutunkova, M. P., Solovyeva, S. N., Chemyshov, I. N., Kilnova, S. V., Gurvich, V. B., Shur, V. Y., Shishkine, E. V., Zubarev, I. V., Privalova, L. I., & Katsneison, B. A. (2020). Manifestation of Systemic Toxicity in Rats after a Short-Time Inhalation of Lead Oxide Nanoparticles. *International Journal of Molecular Sciences*, *21*(3), 690. doi:10.3390/ijms21030690 PMID:31973040

Urnukhsaikhan, E., Bold, B. E., Gunbileg, A., Sukhbaatar, N., & Mishig-Ochir, T. (2021). Antibacterial activity and characteristics of silver nanoparticles biosynthesized from *Carduus crispus*. *Scientific Reports*, *11*(1), 21047. doi:10.1038/s41598-021-00520-2 PMID:34702916

Vaidiglesias, V., Alba-Gonzalez, A., Fernandez-Bertolez, N., Touzani, A., Laffon, B., & Folgueira, M. (2023). Effects of zinc oxide nanoparticle exposure on human glial cells and zebra embryos. *International Journal of Molecular Sciences*, *24*(15), 12297. doi:10.3390/ijms241512297 PMID:37569675

Verma, Y., Rani, V., & Rana, S. V. S. (2020). Assessment of cadmium sulphide nanoparticles toxicity in the gills of a fresh water fish. *Environmental Nanotechnology, Monitoring & Management*, *13*, 100280. doi:10.1016/j.enmm.2019.100280

Vidyasagar, P., Patel, R. R., Singh, S. K., & Singh, M. (2023). Grenn synthesis of silver nanoparticles: Methods, biological applications, delivery and toxicity. *Materials Advances*, *4*(8), 1831–1849. doi:10.1039/D2MA01105K

Vilas-Boas, V., & Vinken, M. (2021, January). Hepatotoxicity induced by nanomaterials: Mechanisms and in vitro models. *Archives of Toxicology*, *95*(1), 27–52. doi:10.1007/s00204-020-02940-x PMID:33155068

Waghmode, M. S., Gunjal, A. B., Mulla, J. A., Patil, N. N., & Nawani, N. N. (2019). Studies on the titanium dioxide nanoparticles: Biosynthesis, applications and remediation. *SN Applied Sciences*, *1*(4), 310. doi:10.1007/s42452-019-0337-3

Wu, L., Wen, W., Wang, X., Huang, D., Cao, J., Qi, X., & Shen, S. (2022). Ultrasmall iron oxide nanoparticles cause significant toxicity by specifically inducing acute oxidative stress to multiple organs. *Particle and Fibre Toxicology*, *19*(1), 24. doi:10.1186/s12989-022-00465-y PMID:35351185

Wu, X., Han, X., Guo, Y., Liu, Q., Sun, R., Wen, Z., & Dai, C. (2023). Application prospect of calcium peroxide nanoparticles in biomedical field. *Reviews on Advanced Materials Science*, *62*(1), 20220308. doi:10.1515/rams-2022-0308

Xuan, L., Ju, Z., Skonieczna, M., Zhou, P., & Huang, R. (2020). Nanoparticles-induced potential toxicity on humans health: Application, toxicity mechanisms and evaluation models. *MedComm*, *4*(4), e327. doi:10.1002/mco2.327 PMID:37457660

Yao, L., Bojic, D., & Liu, M. (2023). Applications and safety of gold nanoparticles as therapeutic devices in clinical trials. *Journal of Pharmaceutical Analysis*, *13*(9), 960–967. doi:10.1016/j.jpha.2023.06.001 PMID:37842655

Yousef, H.A., Fahmy, H.M., Arafa, F. N. *et al.* (2023). Nanotechnology in pest management: advantages, applications, and challenges. *Int J Trop Insect Sci, 43,* 1387–1399. https://. doi:10.1007/s42690-023-01053-z

Zaitseva, N. V., Zemlyanova, M. A., Stepankov, M. S., & Ignatova, A. M. (2019). Studying and Assessing the Toxicity of Calcium Oxide Nanoparticles under One-Time Inhalation Exposure. *Nanotechnologies in Russia*, *14*(9-10), 497–503. doi:10.1134/S1995078019050148

Zakaria, M. Y., Abd El-Halim, S. M., Beshay, B. Y., Zaki, I., & Abourehab, M. A. S. (2023). 'Poly phenolic phytoceutical loaded nano-bilosomes for enhanced caco-2 cell permeability and SARS-CoV 2 antiviral activity': In-vitro and insilico studies. *Drug Delivery*, *30*(1), 1. doi:10.1080/10717544.2022.2162157 PMID:36587813

Zhou, X. Q., Hayat, Z., Zhang, D. D., Li, M. Y., Hu, S., Wu, Q., Cao, Y. F., & Yuan, Y. (2023). Zinc oxide nanoparticles: Synthesis, characterization, modification, and applications in food and agriculture. *Processes (Basel, Switzerland)*, *11*(4), 1193. doi:10.3390/pr11041193

Compilation of References

Abbasi, R., Shineh, G., Mobaraki, M., Doughty, S., & Tayebi, L. (2023). Structural parameters of nanoparticles affecting their toxicity for biomedical applications: A review. *Journal of Nanoparticle Research*, *25*(3), 43. doi:10.1007/s11051-023-05690-w PMID:36875184

Abbaspour, A., Norouz-Sarvestani, F., Noori, A., & Soltani, N. (2015). Aptamer-conjugated silver nanoparticles for electrochemical dual-aptamer-based sandwich detection of staphylococcus aureus. *Biosensors & Bioelectronics*, *68*, 149–155. doi:10.1016/j.bios.2014.12.040 PMID:25562742

Abdel-KarimR.RedaY.Abdel-FattahA. (2020). Review-Nanostructured Materials-Based Nanosensors. https://doi.org/ doi:10.1149/1945-7111/ab67aa

Abdulhaq, N. A., Elnady, D. A., Abo El-atta, H. M., El-Morsi, D. A., & Gad El-Hak, S. A. (2023). Assessment of reproductive toxicity of gold nanoparticles and its reversibility in male albino rats. *Toxicological Research*. doi:10.1007/s43188-023-00203-2

Aday, B., Yıldız, Y., Ulus, R., Eris, S., Sen, F., Kaya, M.(2016). One-pot, efficient and green synthesis of acridinedione derivatives using highly monodisperse platinum nanoparticles supported with reduced graphene oxide. *N J Chem., 40*(1), 748–54. doi: 02 098 K. doi:10.1039/C5NJ

Agarwal, A., Shaida, B., Rastogi, M., & Singh, N. B. (2023). Food Packaging Materials with Special Reference to Biopolymers-Properties and Applications. *Chemistry Africa*, *16*(1), 117–144. doi:10.1007/s42250-022-00446-w

Agnihotri, R., Gaur, S., & Albin, S. (2020). Nanometals in dentistry: Applications and toxicological implications-A systematic review. *Biological Trace Element Research*, *197*(1), 70–88. doi:10.1007/s12011-019-01986-y PMID:31782063

Ahari, H., & Soufiani, S. P. (2021). Smart and active food packaging: Insights in novel food packaging. *Frontiers in Microbiology*, *12*, 657233. doi:10.3389/fmicb.2021.657233 PMID:34305829

Ahmad, A., Qurashi, A., & Sheehan, D. (2023). Nano packaging – Progress and future perspectives for food safety, and sustainability. *Food Packaging and Shelf Life*, *35*, 100997. doi:10.1016/j.fpsl.2022.100997

Ahmadian, A., Shafiee, A., Aliahmad, N., & Agarwal, M. (2021). Overview of nano-fiber mats fabrication via electrospinning and morphology analysis. *Textiles*, *1*(2), 206–226. doi:10.3390/textiles1020010

Ahmadpour, A., Zare, M., Behjoomanesh, M., & Avazpour, M. (2015). Photocatalytic decolorization of methyl orange dye using nano-photocatalysts. *Advances in Environmental Technology*, *1*(3), 121–127. doi:10.22104/aet.2015.371

Ahmad, W., Jaiswal, K. K., & Soni, S. (2020). Green synthesis of titanium dioxide (TiO2) nanoparticles by using Mentha arvensis leaves extract and its antimicrobialproperties. *Inorganic and Nano-Metal Chemistry*, *50*(10), 1032–1038. doi:10.1080/24701556.2020.1732419

Ahuja, T., Ahmad, I., & Kumar, D. (2007). Biomolecular immobilization on conducting polymers for biosensing applications. *Biomaterials*, *28*(5), 791–805. doi:10.1016/j.biomaterials.2006.09.046 PMID:17055573

Akbari Jonous, Z., Shayeh, J. S., Yazdian, F., Yadegari, A., Hashemi, M., & Omidi, M. (2019, February 7). An electrochemical biosensor for prostate cancer biomarker detection using graphene oxide-gold nanostructures. *Engineering in Life Sciences*, *19*(3), 206–216. doi:10.1002/elsc.201800093 PMID:32625003

Akgöl, S., & Ulucan-Karnak, F. kuru, C. İ., & Kuşat, K. (2021). The usage of composite nanomaterials in biomedical engineering applications. In Biotechnology and Bioengineering (Vol. 118, Issue 8, pp. 2906–2922). John Wiley and Sons Inc. doi:10.1002/bit.27843

Akhtar, M. S., Panwar, J., & Yun, Y. S. (2013). Biogenic synthesis of metallic nanoparticles by plant extracts. *ACS Sustainable Chemistry & Engineering*, *1*(6), 591–602. doi:10.1021/sc300118u

Al Ahmad, M., Mustafa, F., Panicker, N., & Rizvi, T. A. (2021). Optical detection of sars-cov-2 utilizing antigen-antibody binding interactions. *Sensors (Basel)*, *21*(19), 6596. doi:10.3390/s21196596 PMID:34640920

Al Ghabban, A., Al Zubaidi, A. B., Jafar, M., & Fakhri, Z. (2018). Effect of Nano SiO2 and Nano CaCO3 on the Mechanical Properties, Durability and flowability of Concrete. *IOP Conference Series. Materials Science and Engineering*, *454*(1), 012016. doi:10.1088/1757-899X/454/1/012016

Alam, M. K., Karim, R., & Ahammad, A. S. (2023). Smart Nanomaterials and Their Applications. *Emerging Nanomaterials and Their Impact on Society in the 21st Century, 135*, 100-124. doi:10.21741/9781644902172-5

Alam, M. A., Samad, U. A., Khan, R., Alam, M., & Al-Zahrani, S. M. (2017). Anti-corrosive performance of epoxy coatings containing various nano-particles for splash zone applications. *Korean Journal of Chemical Engineering*, *34*(8), 2301–2310. doi:10.1007/s11814-017-0114-1

Alam, M. A., Samad, U. A., Sherif, E. S. M., Seikh, A., Al-Zahrani, S. M., Alharthi, N. H., & Alam, M. (2019). Synergistic effect of Ag and ZnO nanoparticles on polyaniline incorporated epoxy/2pack coatings for splash zone applications. *Journal of Coatings Technology and Research*, *16*(3), 835–845. doi:10.1007/s11998-018-00156-4

Al-Arjan, W. S., Aslam Khan, M. U., Nazir, S., Abd Razak, S. I., & Abdul Kadir, M. R. (2020). Development of arabinoxylan-reinforced apple pectin/graphene oxide/nano-hydroxyapatite based nanocomposite scaffolds with controlled release of drug for bone tissue engineering: In-vitro evaluation of biocompatibility and cytotoxicity against MC3T3-E1. *Coatings*, *10*(11), 1120. doi:10.3390/coatings10111120

Alghamdi, N., Alqahtani, Z., & Grell, M. (2019). Sub-nanomolar detection of cesium with water-gated transistor. *Journal of Applied Physics*, *126*(6), 064502. doi:10.1063/1.5108730

Alghamdi, N., Alqahtani, Z., Zhou, C., Sano, N., Conte, M., & Grell, M. (2020). Sensing aromatic pollutants in water with catalyst-sensitized water-gated transistor. *Chemical Papers*, *74*(12), 4169–4180. doi:10.1007/s11696-020-01212-3

Alghriany, A. A. I., Omar, H. E. M., Mahmound, A. M., & Atia, M. M. (2022). Assessment of the Toxicity of Aluminum Oxide and Its Nanoparticles in the Bone Marrow and Liver of Male Mice: Ameliorative Efficacy of Curcumin Nanoparticles. *ACS Omega*, *7*(16), 13841–13852. doi:10.1021/acsomega.2c00195 PMID:35559158

Algotiml, R., Gab-Alla, A., Seoudi, R., Abulreesh, H. H., El-Readi, M. Z., & Elbanna, K. (2022). Anticancer and antimicrobial activity of biosynthesized Red Sea marine algal silver nanoparticles. *Scientific Reports*, *12*(1), 2421. doi:10.1038/s41598-022-06412-3 PMID:35165346

Al-Hakkani, M. F. (2020). Biogenic copper nanoparticles and their applications: A review. *SN Applied Sciences*, *2*(3), 505. doi:10.1007/s42452-020-2279-1

Alinejadian, N., Kollo, L., & Odnevall, I. (2022). Progress in additive manufacturing of MoS2-based structures for energy storage applications – A review. *Materials Science in Semiconductor Processing*, *139*, 106331. doi:10.1016/j.mssp.2021.106331

Ali, S. M. (2018). Fabrication of a nanocomposite from an agricultural waste and its application as a biosorbent for organic pollutants. *International Journal of Environmental Science and Technology*, *15*(6), 1169–1178. doi:10.1007/s13762-017-1477-x

Aljabali, A. A. A., Obeid, M. A., Al Zoubi, M. S., Charbe, N. B., Chellappan, D. K., Mishra, V., Dureja, H., Gupta, G., Prasher, P., Dua, K., Elnashar, R. M., Tambuwala, M. M., & Barhoum, A. (2020). In A. Barhoum (Ed.), *Nanocelluloses in Sensing Technology BT - Handbook of Nanocelluloses: Classification, Properties, Fabrication, and Emerging Applications* (pp. 1–30). Springer International Publishing., doi:10.1007/978-3-030-62976-2_44-1

Alka, L., Anurag, V., Himanshi, J., Niti, Y., & Neha, K. (2014). NanotechnologyBased Cosmeceuticals. *ISRN Dermatology*, *2014*, 843687. doi:10.1155/2014/843687 PMID:24963412

Allaker, R. P. (2010). The use of nanoparticles to control oral biofilm formation. *Journal of Dental Research*, *89*(11), 1175–1186. doi:10.1177/0022034510377794 PMID:20739694

Alle, M., Bandi, R., Sharma, G., Dadigala, R., Lee, S.-H., & Kim, J.-C. (2022). Gold nanoparticles spontaneously grown on cellulose nanofibrils as a reusable nanozyme for colorimetric detection of cholesterol in human serum. *International Journal of Biological Macromolecules*, *201*, 686–697.. doi:10.1016/j.ijbiomac.2022.01.158 PMID:35104471

Almanaa, T. N., Aref, M., Kakakhel, M. A., Elshopakey, G. E., Mahboub, H. H., Abdelazim, A. M., Kamel, S., Belali, T. M., Abomughaid, M. M., Alhujaily, M., Fahmy, E. M., Ezzat Assayed, M., Mostafa-Hedeab, G., & Daoush, W. M. (2022). Silica nanoparticle acute toxicity on male *rattus norvegicus* domestica: Ethological behavior, hematological disorders, biochemical analyses, hepato-renal function, and antioxidant-immune response. *Frontiers in Bioengineering and Biotechnology*, *10*, 868111. doi:10.3389/fbioe.2022.868111 PMID:35464726

Al-Mutairi, N. H., Mehdi, A. H., & Kadhim, B. J. (2022). Nanocomposites Materials Definitions, Types and Some of Their Applications: A Review. *European Journal of Research Development and Sustainability*, *3*(2), 102–108.

Al-Naamani, L., Dobretsov, S., & Dutta, J. (2016). Chitosan-zinc oxide nanoparticle composite coating for active food packaging applications. *Innovative Food Science & Emerging Technologies*, *38*, 231–237. doi:10.1016/j.ifset.2016.10.010

Al-Nafori, M. K., Elshal, M. G., & Refai, W. M. (2017). The effect of incorporating gold and silver nanoparticles in orthodontic adhesive system on bond strength of orthodontic bracket. *EC Dental Science*, *11*(4), 119–131.

Alqahtani, Z., Alghamdi, N., & Grell, M. (2020). Monitoring the lead-and-copper rule with a water-gated field effect transistor. *Journal of Water and Health*, *18*(2), 159–171. doi:10.2166/wh.2020.186 PMID:32300089

Al-Shemy, M. T., El-Shafei, A. M., Al-Sayed, A., & Adel, A. M. (2023). Facile in-situ synthesis of nanocrystalline celluloses-silver bio-nanocomposite for chitosan based active packaging. *Egyptian Journal of Chemistry*, *66*(4), 335–352.

Alsufyani, T., & M'Sakni, N. H. (2023). Biodegradable Bio-Composite Film Reinforced with Cellulose Nanocrystals from Chaetomorpha linum into Thermoplastic Starch Matrices. *Polymers*, *15*(6), 1542. doi:10.3390/polym15061542 PMID:36987321

Al-Tayyar, N. A., Youssef, A. M., & Al-Hindi, R. (2020). Antimicrobial food packaging based on sustainable Bio-based materials for reducing foodborne Pathogens: A review. *Food Chemistry*, *310*, 125915. doi:10.1016/j.foodchem.2019.125915 PMID:31841936

Alvarez-Ayuso, E., Garcıa-Sánchez, A., & Querol, X. (2003). Purification of metal electroplating waste waters using zeolites. *Water Research*, *37*(20), 4855–4862. doi:10.1016/j.watres.2003.08.009 PMID:14604631

Amari, A., Alzahrani, F. M., Katubi, K. M., Alsaiari, N. S., Tahoon, M. A., & Ben Rebah, F. (2021). Clay-polymer nanocomposites: Preparations and utilization for pollutants removal. *Materials (Basel)*, *14*(6), 1365. doi:10.3390/ma14061365 PMID:33799810

Ameh, T., & Sayes, C. M. (2019). The potential exposure and hazards of copper nanoparticles: A review. *Environmental Toxicology and Pharmacology*, *71*, 103220. doi:10.1016/j.etap.2019.103220 PMID:31306862

Ameta, R., Solanki, M. S., Benjamin, S., & Ameta, S. C. (2018). Photocatalysis. In *Advanced Oxidation Processes for Waste Water Treatment* (pp. 135–175). Elsevier. doi:10.1016/B978-0-12-810499-6.00006-1

Amini, S. M., Gilaki, M., & Karchani, M. (2014). Safety of nanotechnology in food industries. *Electronic Physician*, *6*, 962–968. doi:10.14661/2014.962-968 PMID:25763176

Amin, U., Khan, M. K. I., Maan, A. A., Nazir, A., Riaz, S., Khan, M. U., Sultan, M., Munekata, P. E. S., & Lorenzo, J. M. (2022). Biodegradable active, intelligent, and smart packaging materials for food applications. *Food Packaging and Shelf Life*, *33*, 100903. doi:10.1016/j.fpsl.2022.100903

Amirthalingam, S., & Rangasamy, J. (2021). Chitosan-Based Biosensor Fabrication and Biosensing Applications. In R. Jayakumar & M. Prabaharan (Eds.), *Chitosan for Biomaterials III. Advances in Polymer Science* (Vol. 287). Springer., doi:10.1007/12_2021_85

An, C., Sun, C., Li, N., Huang, B., Jiang, J., Shen, Y., Wang, C., Zhao, X., Cui, B., Wang, C., Li, X., Zhan, S., Gao, F., Zeng, Z., Cui, H., & Wang, Y. (2022). Nanomaterials and nanotechnology for the delivery of agrochemicals: Strategies towards sustainable agriculture. *Journal of Nanobiotechnology*, *20*(1), 1–19. doi:10.1186/s12951-021-01214-7 PMID:34983545

Andelescu, A., Nistor, M. A., Muntean, S. G., & Rădulescu-Grad, M. E. (2018). Adsorption studies on copper, cadmium, and zinc ion removal from aqueous solution using magnetite/carbon nanocomposites. *Separation Science and Technology*, *53*(15), 2352–2364. doi:10.1080/01496395.2018.1457696

Anderson, D. S., Sydor, M. J., Fletcher, P., & Holian, A. (2016). Nanotechnology: The risks and benefits for medical diagnosis and treatment. *Journal of Nanomedicine & Nanotechnology*, *7*(4), e143.

Anik, U., Tepeli, Y., & Diouani, M. F. (2016). Fabrication of electrochemical model influenza A virus biosensor based on the measurements of neuraminidase enzyme activity. *Analytical Chemistry*, *88*(12), 6151–6153. doi:10.1021/acs.analchem.6b01720 PMID:27281347

Anker, J. N., Hall, W. P., Lyandres, O., Shah, N. C., Zhao, J., & Van Duyne, R. P. (2009). Biosensing with plasmonic nanosensors. *Nanoscience and Technology: A Collection of Reviews from Nature Journals*, *7*(June), 308–319. doi:10.1142/9789814287005_0032

Ansari, M. J., Unnisa, A., Singh, A., Verma, D., Kanaoujiya, R., & Gonzales, J. L. A. (2022). Application of porous nanomaterials for sustained and targeted drug release. *Advances in Science and Technology (Owerri, Nigeria)*, *117*, 105–110. https://www.scientific.net/AST.117.105. doi:10.4028/p-b2484c

Aragaw, T. A., Bogale, F. M., & Aragaw, B. A. (2021). Iron-based nanoparticles in wastewater treatment: A review on synthesis methods, applications, and removal mechanisms. *Journal of Saudi Chemical Society*, *25*(8), 101280. doi:10.1016/j.jscs.2021.101280

Araújo, R. G., González-González, R. B., Martinez-Ruiz, M., Coronado-Apodaca, K. G., Reyes-Pardo, H., Morreeuw, Z. P., Oyervides-Muñoz, M. A., Sosa-Hernández, J. E., Barceló, D., Parra-Saldívar, R., & Iqbal, H. M. N. (2022). Expanding the Scope of Nanobiocatalysis and Nanosensing: Applications of Nanomaterial Constructs. *ACS Omega*, *7*(37), 32863–32876. doi:10.1021/acsomega.2c03155 PMID:36157779

Arduini, F., Guidone, S., Amine, A., Palleschi, G., & Moscone, D. (2013). Acetylcholinesterase biosensor based on self-assembled monolayer-modified gold-screen printed electrodes for organophosphorus insecticide detection. *Sensors and Actuators. B, Chemical*, *179*, 201–208. doi:10.1016/j.snb.2012.10.016

Ariga, K. (2023). Molecular nanoarchitectonics: Unification of nanotechnology and molecular/materials science. *Beilstein Journal of Nanotechnology*, *14*(1), 434–453. Advance online publication. doi:10.3762/bjnano.14.35 PMID:37091285

Armarego, W. L. F. (2022). Nanomaterials. In editor(s): Wilfred L.F. Armarego, Purification of Laboratory Chemicals (Ninth Edition). Butterworth-Heinemann, 586-630. doi:10.1016/B978-0-323-90968-6.50005-9

Arora, A., & Padua, G. W. (2010). Review: Nanocomposites in Food Packaging. *Journal of Food Science*, *75*(1), R43–R49. doi:10.1111/j.1750-3841.2009.01456.x PMID:20492194

Arslan, S., Ekrikaya, S., Ildiz, N., Yusufbeyoglu, S., & Ocsoy, I. (2023). Evaluation of the antibacterial activity of dental adhesive containing biogenic silver nanoparticles decorated nanographene oxide nanocomposites (Ag@ nGO NCs) and effect on bond strength to dentine. *Odontology*, 1–14. doi:10.1007/s10266-023-00836-7 PMID:37436660

Arunadevi, N. (2022). Metal nanocomposites for advanced futuristic biosensing applications. *Materials Letters*, *309*, 131320. doi:10.1016/j.matlet.2021.131320

Ashfaq, A., Khursheed, N., Fatima, S., Anjum, Z., & Younis, K. (2022). Application of nanotechnology in food packaging: Pros and Cons. *Journal of Agriculture and Food Research*, *7*, 100270. doi:10.1016/j.jafr.2022.100270

Asif, A., & Hasan, M. Z. (2018). Application of nanotechnology in modern textiles: A review. *International Journal of Current Engineering and Technology*, *8*(2), 227–231.

Aslan, T., Dadi, Ş., Kafdag, O., Temur, N., Ildiz, N., Ocsoy, I., & Ustun, Y. (2023). Rational design of EDTA-incorporated nanoflowers as novel and effective endodontic disinfection against biofilms. *Odontology*, 1–9. doi:10.1007/s10266-023-00857-2 PMID:37787827

Asmathunisha, N., & Kathiresan, K. (2013). A review on biosynthesis of nanoparticles by marine organisms. *Colloids and Surfaces. B, Biointerfaces*, *103*, 283–287. doi:10.1016/j.colsurfb.2012.10.030 PMID:23202242

Aswini, R., Murugesan, S., & Kannan, K. (2021). Bio-engineered TiO_2 nanoparticles using Ledebouria revoluta extract: Larvicidal, histopathological, antibacterial and anticancer activity. *International Journal of Environmental Analytical Chemistry*, *101*(15), 2926–2936. doi:10.1080/03067319.2020.1718668

Atalla, R. H., Atalla, R. S., & Agarwal, U. P. (2017). The Nanostructures of Native Celluloses, Their Transformations upon Isolation, and Their Implications for Production of Nanocelluloses. In *ACS Symposium Series* (Vol. 1251, pp. 1–18). 10.1021/bk-2017-1251.ch001

Ates, M., Eker, A. A., & Eker, B. (2017). Carbon nanotube-based nanocomposites and their applications. In Journal of Adhesion Science and Technology (Vol. 31, Issue 18, pp. 1977–1997). Taylor and Francis Ltd. doi:10.1080/01694243.2017.1295625

Ates, M., & Topkaya, E. (2015). Nanocomposite film formations of polyaniline via TiO2, Ag, and Zn, and their corrosion protection properties. *Progress in Organic Coatings*, *82*, 33–40. doi:10.1016/j.porgcoat.2015.01.014

Ayan, S., Aranci-Ciftci, K., Ciftci, F., & Ustundag, C. B. (2023). Nanotechnology and COVID-19: Prevention, diagnosis, vaccine, and treatment strategies. In Frontiers in Materials (Vol. 9). Frontiers Media S.A. doi:10.3389/fmats.2022.1059184

Aziz, Z. A. A., Mohd-Nasir, H., Ahmad, A., Mohd. Setapar, S. H., Peng, W. L., Chuo, S. C., Khatoon, A., Umar, K., Yaqoob, A. A., & Mohamad Ibrahim, M. N. (2019). Role of Nanotechnology for Design and Development of Cosmeceutical: Application in Makeup and Skin Care. *Frontiers in Chemistry*, *7*, 739. doi:10.3389/fchem.2019.00739 PMID:31799232

Babapour, M., Hadi Dehghani, M., Alimohammadi, M., Moghadam Arjmand, M., Salari, M., Rasuli, L., Mubarak, N. M., & Ahmad Khan, N. (2022). Adsorption of Cr(VI) from aqueous solution using mesoporous metal-organic framework-5 functionalized with the amino acids: Characterization, optimization, linear and nonlinear kinetic models. *Journal of Molecular Liquids*, *345*, 117835. doi:10.1016/j.molliq.2021.117835

Bacakova, L., Pajorova, J., Bacakova, M., Skogberg, A., Kallio, P., Kolarova, K., & Svorcik, V. (2019). Versatile application of nanocellulose: From industry to skin tissue engineering and wound healing. *Nanomaterials (Basel, Switzerland)*, *9*(2), 164. doi:10.3390/nano9020164 PMID:30699947

Bacali, C., Baldea, I., Moldovan, M., Carpa, R., Olteanu, D. E., Filip, G. A., & Badea, F. (2020). Flexural strength, biocompatibility, and antimicrobial activity of a polymethyl methacrylate denture resin enhanced with graphene and silver nanoparticles. *Clinical Oral Investigations*, *24*(8), 2713–2725. doi:10.1007/s00784-019-03133-2 PMID:31734793

Badgar, K., Abdalla, N., El-Ramady, H., & Prokisch, J. (2022). Sustainable applications of nanofibers in agriculture and water treatment: A review. *Sustainability (Basel)*, *14*(1), 464. doi:10.3390/su14010464

Bahara, J., Ramezani, T., Mousavi, M., & Asadi-Samani, M. (2017). Antioxidant and anti-inflammatory activity of green synthesized silver nanoparticles using Salvia officinalis extract. *Annals of Tropical Medicine and Public Health, 10*(5), 1265–1270. doi:10.4103/ATMPH.ATMPH_174_17

Bai, H., & Li, Q. (2022). Electrodeposited Ni/TiN-SiC Nanocomposites on the Dumbbell: Reducing Sport Injuries. *Coatings, 12*(2), 177. doi:10.3390/coatings12020177

Bai, Y., Xu, T., & Zhang, X. (2020, January 3). Graphene-Based Biosensors for Detection of Biomarkers. *Micromachines, 11*(1), 60. doi:10.3390/mi11010060 PMID:31947894

Bajpai, V. K., Kamle, M., Shukla, S., Mahato, D. K., Chandra, P., Hwang, S. K., Kumar, P., Huh, Y. S., & Han, Y.-K. (2018). Prospects of using nanotechnology for food preservation, safety, and security. *Yao Wu Shi Pin Fen Xi, 26*(4), 1201–1214. doi:10.1016/j.jfda.2018.06.011 PMID:30249319

Bajwa, S. Z., Munawar, A., & Khan, W. S. (2017). Nanotechnology in medicine: Innovation to market. *Pharmaceutical Bioprocessing, 5*(2), 11–15.

Balhaddad, A. A., Kansara, A. A., Hidan, D., Weir, M. D., Xu, H. H., & Melo, M. A. S. (2019). Toward dental caries: Exploring nanoparticle-based platforms and calcium phosphate compounds for dental restorative materials. *Bioactive Materials, 4*, 43–55. doi:10.1016/j.bioactmat.2018.12.002 PMID:30582079

Banerjee, S., Dionysiou, D. D., & Pillai, S. C. (2015). Self-cleaning applications of TiO2 by photo-induced hydrophilicity and photocatalysis. *Applied Catalysis B: Environmental, 176–177*, 396–428. doi:10.1016/j.apcatb.2015.03.058

Banin, U., Waiskopf, N., Hammarström, L., Boschloo, G., Freitag, M., Johansson, E. M. J., Sá, J., Tian, H., Johnston, M. B., Herz, L. M., Milot, R. L., Kanatzidis, M. G., Ke, W., Spanopoulos, I., Kohlstedt, K. L., Schatz, G. C., Lewis, N., Meyer, T., Nozik, A. J., & Brudvig, G. W. (2020). Nanotechnology for catalysis and solar energy conversion. In Nanotechnology, 32(4). IOP Publishing Ltd. doi:10.1088/1361-6528/abbce8

Bankar, A., Joshi, B., Kumar, A. R., & Zinjarde, S. (2010). Banana peel extract mediated novel route for the synthesis of silver nanoparticles. *Colloids and Surfaces. A, Physicochemical and Engineering Aspects, 368*(1-3), 58–63. doi:10.1016/j.colsurfa.2010.07.024

Bapat, R. A., Chaubal, T. V., Dharmadhikari, S., Abdulla, A. M., Bapat, P., Alexander, A., & Kesharwani, P. (2020). Recent advances of gold nanoparticles as biomaterial indentistry. *International Journal of Pharmaceutics, 586*, 119596. doi:10.1016/j.ijpharm.2020.119596 PMID:32622805

Barage, S., Lakkakula, J., Sharma, A., Roy, A., Alghamdi, S., Almehmadi, M., Hossain, M. J., Allahyani, M., & Abdulaziz, O. (2022). Nanomaterial in Food Packaging: A Comprehensive Review. *Journal of Nanomaterials, 6053922*, 1–12. doi:10.1155/2022/6053922

Baran, A. (2016). Nanotechnology: Legal and ethical issues. Economics and management. *Economics and Management, 8*(1), 47–54.

Barber, D. J., & Freestone, I. C. (1990). *An Investigation Of The Origin Of The Colour Of The Lycurgus Cup By Analytical Transmission Electron Microscopy* (*Vol. 32*).

Barguilla, I., Barszczewska, G., Annangi, B., Domenech, J., Velázquez, A., Marcos, R., & Hernández, A. (2020). MTH1 is involved in the toxic and carcinogenic long-term effects induced by zinc oxide and cobalt nanoparticles. *Archives of Toxicology*, *94*(6), 1973–1984. doi:10.1007/s00204-020-02737-y PMID:32377776

Barreneche, C., Martín, M., Calvo-de la Rosa, J., Majó, M., & Fernández, A. I. (2019). Own-synthetize nanoparticles to develop nano-enhanced phase change materials (NEPCM) to improve the energy efficiency in buildings. *Molecules (Basel, Switzerland)*, *24*(7), 1232. doi:10.3390/molecules24071232 PMID:30934832

Barua, S., Dutta, H. S., Gogoi, S., Devi, R., & Khan, R. (2017). Nanostructured MoS2-Based Advanced Biosensors: A Review. In ACS Applied Nano Materials (Vol. 1, Issue 1, pp. 2–25). American Chemical Society (ACS). https://doi.org/ doi:10.1021/acsanm.7b00157

Basabe-Desmonts, L., Müller, T. J. J., & Crego-Calama, M. (2007). Design of fluorescent materials for chemical sensing. *Chemical Society Reviews*, *36*(6), 993–1017. doi:10.1039/b609548h PMID:17534482

Bashiri Rezaie, A., Montazer, M., & Mahmoudi Rad, M. (2018). Environmentally friendly low cost approach for nano copper oxide functionalization of cotton designed for antibacterial and photocatalytic applications. *Journal of Cleaner Production*, *204*, 425–436. doi:10.1016/j.jclepro.2018.08.337

Baskaran, X., Geo Vigila, A. V., Parimelazhagan, T., Muralidhara-Rao, D., & Zhang, S. (2016). Biosynthesis, characterization, and evaluation of bioactivities of leaf extract-mediated biocompatible silver nanoparticles from an early tracheophyte, Pteris tripartita Sw. *International Journal of Nanomedicine*, *11*, 5789–5806. doi:10.2147/IJN.S108208 PMID:27895478

Bathula, C., Subalakshmi, K., Kumar, A., Yadav, H., Ramesh, S., Shinde, S., & Kim, H. (2020). Ultrasonically driven green synthesis of palladium nanoparticles by Coleusamboinicus for catalytic reduction and Suzuki-Miyaura reaction. *Colloids and Surfaces. B, Biointerfaces*, *192*, 111026. doi:10.1016/j.colsurfb.2020.111026 PMID:32344163

Batır, G. G., Arık, M., Caldıran, Z., Turut, A., & Aydogan, S. (2018). Synthesis and Characterization of Reduced Graphene Oxide/Rhodamine 101 (rGO-Rh101) Nanocomposites and Their Heterojunction Performance in rGO-Rh101/p-Si Device Configuration. *Journal of Electronic Materials*, *47*(1), 329–336. doi:10.1007/s11664-017-5758-4

Bauerlein, E. (2003). Biomineralization of unicellular organisms: An unusual membrane biochemistry for the production of ınorganic nano-and microstructures. *Angewandte Chemie International Edition*, *42*(6), 614–641. doi:10.1002/anie.200390176 PMID:12574993

Beatriz, F. M. R., Marcelo, M. S., Daniela, S. F., Devaney, R. D. C., & Glaucia, M. M. S. (2020). Graphene oxide-based nanomaterial interaction with human breast cancer cells. *Journal of Biomedical Materials Research. Part A, 108*(4), 863–870. doi:10.1002/jbm.a.36864 PMID:31846174

Becker, H., Herzberg, F., Schulte, A., & Kolossa-Gehring, M. (2011). The carcinogenic potential of nanomaterials, their release from products and options for regulating them. *International Journal of Hygiene and Environmental Health, 214*(3), 231–238. doi:10.1016/j.ijheh.2010.11.004 PMID:21168363

Bezerra, P. C. S., Cavalcante, R. P., Garcia, A., Wender, H., Martines, M. A. U., Casagrande, G. A., Giménez, J., Marco, P., Oliveira, S. C., & Machulek, A. Jr. (2017). Synthesis, characterization, and photocatalytic activity of pure and N-, B-, or Ag-Doped TiO2. *Journal of the Brazilian Chemical Society, 28*(9), 1788–1802. doi:10.21577/0103-5053.20170040

Bhatia, S. (2016). *Nanoparticles Types, Classification, characterization, fabrication methods and drug delivery applications. in: natural polymer drug delivery systems*. Springer. doi:10.1007/978-3-319-41129-3_2

Bhaumik, M., Maity, A., Srinivasu, V. V., & Onyango, M. S. (2011). Enhanced removal of Cr(VI) from aqueous solution using polypyrrole/Fe3O4 magnetic nanocomposite. *Journal of Hazardous Materials, 190*(1–3), 381–390. doi:10.1016/j.jhazmat.2011.03.062 PMID:21497438

Bhunia, A. K. (2008). Biosensors and Bio-Based Methods for the Separation and Detection of Foodborne Pathogens. In Advances in Food and Nutrition Research (Vol. 54, pp. 1–44). Academic Press Inc. doi:10.1016/S1043-4526(07)00001-0

Biehler, E., Quach, Q., & Abdel-Fattah, T. M. (2023). Silver-Nanoparticle-Decorated Fused Carbon Sphere Composite as a Catalyst for Hydrogen Generation. *Energies, 16*(13), 5053. doi:10.3390/en16135053

Bilal, M., Singh, A. K., Iqbal, H. M., & Boczkaj, G. (2023). Enzyme-conjugated MXene nanocomposites for biocatalysis and biosensing. *Chemical Engineering Journal, 474*, 145020. doi:10.1016/j.cej.2023.145020

Binnig, G., Rohrer, H., Gerber, C., & Weibel, E. (1982). Tunneling through a controllable vacuum gap. *Applied Physics Letters, 40*(2), 178–180. doi:10.1063/1.92999

Bisht, A., Mishra, A., Bisht, H., & Tripathi, R. M. (2021). Nanomaterial Based Biosensors for Detection of Viruses Including SARS-CoV-2: A Review. In Journal of Analysis and Testing (Vol. 5, Issue 4, pp. 327–340). Nonferrous Metals Society of China. doi:10.1007/s41664-021-00200-0

Bissessur, R. (2020). Nanomaterials applications. In *Polymer Science and Nanotechnology* (pp. 435–453). Elsevier. doi:10.1016/B978-0-12-816806-6.00018-2

Biswas, M. R. U. D., & Oh, W. C. (2019). Comparative study on gas sensing by a Schottky diode electrode prepared with graphene-semiconductor-polymer nanocomposites. *RSC Advances, 9*(20), 11484–11492. doi:10.1039/C9RA00007K PMID:35520227

Blanco, A., Monte, M. C., Campano, C., Balea, A., Merayo, N., & Negro, C. (2018). Nanocellulose for industrial use: Cellulose nanofibers (CNF), cellulose nanocrystals (CNC), and bacterial cellulose (BC). In M. Hussain (Ed.), *Handbook of Nanomaterials for Industrial Applications* (pp. 74–126). Elsevier Inc., doi:10.1016/B978-0-12-813351-4.00005-5

Bogue, R., & Bogue, R. (2008). Nanosensors: A review of recent progress. *Sensor Review*, *28*(1), 12–17. doi:10.1108/02602280810849965

Boke, F., Gazioglu, C., Akkaya, S., & Akkaya, M. (2014). Relationship between orthodontic treatment and gingival health: A retrospective study. *European Journal of Dentistry*, *8*(03), 373–380. doi:10.4103/1305-7456.137651 PMID:25202219

Boström, M., & Micheletti, M. (2016). Introducing the Sustainability Challenge of Textiles and Clothing. *Journal of Consumer Policy*, *39*(4), 367–375. doi:10.1007/s10603-016-9336-6

Brittany, L., Carino, V., Kuo, J., Leong, L., & Ganesh, R. (2006). Adsorption of organic Compounds to metal oxide nanoparticles. In *Conference presentation is part of.* General Environmental.

Bustanji, Y., Taneera, J., Semreen, M. H., Abu-Gharbieh, E., El-Huneidi, W., Faris, M. A. E., Alzoubi, K. H., Soares, N. C., Albustanji, B., Abuhelwa, A. Y., Abu-Zurayk, R., Alqudah, M. A., & AlKhatib, H. S. (2023). Gold nanoparticles and breast cancer: A bibliometric analysis of the current state of research and future directions. *OpenNano*, *12*, 100164. doi:10.1016/j.onano.2023.100164

Cagri, A., Ustunol, Z., & Ryser, E. T. (2004). Antimicrobial edible films and coatings. *Journal of Food Protection*, *67*(4), 833–848. doi:10.4315/0362-028X-67.4.833 PMID:15083740

Celik, C., Ildiz, N., & Ocsoy, I. (2020). Building block and rapid synthesis of catecholamines-inorganic nanoflowers with their peroxidase-mimicking and antimicrobial activities. *Scientific Reports*, *10*(1), 2903. doi:10.1038/s41598-020-59699-5 PMID:32075999

Chakraborty, P., Das, T., & Saha-Dasgupta, T. (2019). 1.15 - MXene: A New Trend in 2D Materials Science. D. L. Andrews, R. H. Lipson, & T. Nann (eds.) Comprehensive Nanoscience and Nanotechnology (Second Edition). Academic Press. doi:10.1016/B978-0-12-803581-8.10414-X

Chang, H. S., Walsh, L. J., & Freer, T. J. (1999). The effect of orthodontic treatment onsalivary flow, pH, buffer capacity, and levels of mutans streptococci and lacto bacilli. *Australian Orthodontic Journal*, *15*(4), 229–234. https://search.informit.org/doi/10.3316/informit.986899452640633 PMID:11933357

Charreau, H., Cavallo, E., & Foresti, M. L. (2020). Patents involving nanocellulose: Analysis of their evolution since 2010. *Carbohydrate Polymers, 237*(December 2019), 116039. doi:10.1016/j.carbpol.2020.116039

Chaturvedi, S., Dave, P. N., & Shah, N. K. (2012). Applications of nano-catalyst in new era. *Journal of Saudi Chemical Society*, *16*(3), 307–325. doi:10.1016/j.jscs.2011.01.015

Chaudhary, L. S., Ghatmale, P. R., & Chavan, S. S. (2016). Review on: Application of nanotechnology in computer science. [IJSR]. *International Journal of Scientific Research, 5*, 1542–1545.

Chauhan, G., Madou, M. J., Kalra, S., Chopra, V., Ghosh, D., & Martinez-Chapa, S. O. (2020). Nanotechnology for COVID-19: Therapeutics and Vaccine Research. *ACS Nano, 14*(7), 7760–7782. doi:10.1021/acsnano.0c04006 PMID:32571007

Chavali, M. S., & Nikolova, M. P. (2019). Metal oxide nanoparticles and their applications in nanotechnology. *SN Applied Sciences, 1*(6), 607. doi:10.1007/s42452-019-0592-3

Chen, D., Cheng, Y., Zhou, N., Chen, P., Wang, Y., Li, K., Huo, S., Cheng, P., Peng, P., Zhang, R., Wang, L., Liu, H., Liu, Y., & Ruan, R. (2020a). Photocatalytic degradation of organic pollutants using TiO2-based photocatalysts: A review. *Journal of Cleaner Production, 268*, 121725. doi:10.1016/j.jclepro.2020.121725

Cheng, D., He, M., Ran, J., Cai, G., Wu, J., & Wang, X. (2018). In situ reduction of TiO2 nanoparticles on cotton fabrics through polydopamine templates for photocatalysis and UV protection. *Cellulose (London, England), 25*(2), 1413–1424. doi:10.1007/s10570-017-1606-1

Chen, H., Shao, S., Yu, Y., Huang, Y., Zhu, X., Zhang, S., Fan, J., Yin, G. Y., Chi, B., Wan, M., & Mao, C. (2020). A dual responsive biosensor for blood lead detection. *Analytica Chimica Acta, 1093*, 131–141. doi:10.1016/j.aca.2019.09.062 PMID:31735206

Chen, L. X., Chen, Z. W., Jiang, M., Lu, Z., Gao, C., Cai, G., & Singh, C. V. (2021). Insights on the Dual Role of Two-Dimensional Materials as Catalysts and Supports for Energy and Environmental Catalysis. *Journal of Materials Chemistry. A, Materials for Energy and Sustainability, 9*(4), 2018–2042. doi:10.1039/D0TA08649E

Chen, M., He, Y., Ye, Q., Wang, X., & Hu, Y. (2019). Shape-dependent solar thermal conversion properties of plasmonic Au nanoparticles under different light filter conditions. *Solar Energy, 182*, 340–347. doi:10.1016/j.solener.2019.02.070

Cheriyamundath, S., & Vavilala, S. L. (2021). Nanotechnology-based wastewater treatment. *Water and Environment Journal : the Journal / the Chartered Institution of Water and Environmental Management, 35*(1), 123–132. doi:10.1111/wej.12610

Chinaglia, S., Tosin, M., & Degli-Innocenti, F. (2018). Biodegradation rate of biodegradable plastics at molecular level. *Polymer Degradation & Stability, 147*, 237–244. doi:10.1016/j.polymdegradstab.2017.12.011

Chirayil, C. J., Abraham, J., Mishra, R. K., George, S. C., & Thomas, S. (2017). Instrumental techniques for the characterization of nanoparticles. In S. Thomas, R. Thomas, A. K. Zachariah, & R. Kumar Mishra (Eds.), *Thermal and rheological measurement techniques for nanomaterials characterization* (pp. 1–36). Elsevier. doi:10.1016/B978-0-323-46139-9.00001-3

Chis, A. A., Dobrea, C., Morgovan, C., Arseniu, A. M., Rus, L. L., Butuca, A., Juncan, A. M., Totan, M., Vonica-Tincu, A. L., Cormos, G., Muntean, A. C., Muresan, M. L., Gligor, F. G., & Frum, A. (2020). Applications and Limitations of Dendrimers in Biomedicine. *Molecules (Basel, Switzerland), 25*(17), 3982. doi:10.3390/molecules25173982 PMID:32882920

Chisenga, S. M., Tolesa, G. N., & Workneh, T. S. (2020). Biodegradable Food Packaging Materials and Prospects of the Fourth Industrial Revolution for Tomato Fruit and Product Handling. *International Journal of Food Sciences, 8879101*, 1–17. doi:10.1155/2020/8879101 PMID:33299850

Choi, J., Oh, H., Han, S. W., Ahn, S., Noh, J., & Park, J. B. (2017). Preparation and characterization of graphene oxide supported Cu, Cu2O, and CuO nanocomposites and their high photocatalytic activity for organic dye molecule. *Current Applied Physics, 17*(2), 137–145. doi:10.1016/j.cap.2016.11.020

Chu, S., Shi, X., Tian, Y., & Gao, F. (2022). pH-responsive polymer nanomaterials for tumor therapy. *Frontiers in Oncology, 12*, 855019. doi:10.3389/fonc.2022.855019 PMID:35392227

Cieslinski, J. T., & Kaczmarczyk, T. Z. (2011). Pool boiling of water-Al_2O_3 and water-Cu nanofluids on horizontal smooth tubes. *Nanoscale Research Letters, 6*(1), 1–9. doi:10.1186/1556-276X-6-220 PMID:21711741

Costa, S. S., Miranda, A. L., de Morais, M. G., Costa, J. A. V., & Druzian, J. I. (2019). Microalgae as source of polyhydroxyalkanoates (PHAs) — A review. *International Journal of Biological Macromolecules, 131*, 536–547. doi:10.1016/j.ijbiomac.2019.03.099 PMID:30885732

Crisan, M. C., Teodora, M., & Lucian, M. (2022). Copper Nanoparticles: Synthesis and Characterization, Physiology, Toxicity and Antimicrobial Applications. *Applied Sciences (Basel, Switzerland), 12*(1), 141. doi:10.3390/app12010141

Cui, M., Song, Z., Wu, Y., Guo, B., Fan, X., & Luo, X. (2016). A highly sensitive biosensor for tumor maker alpha fetoprotein based on poly(ethylene glycol) doped conducting polymer PEDOT. *Biosensors & Bioelectronics, 79*, 736–741. doi:10.1016/j.bios.2016.01.012 PMID:26774088

Curulli, A. (2023). Functional Nanomaterials Enhancing Electrochemical Biosensors as Smart Tools for Detecting Infectious Viral Diseases. In Molecules (Vol. 28, Issue 9). MDPI. doi:10.3390/molecules28093777

D'Souza, A. A., Kumari, D., & Banerjee, R. (2017). *Nanocomposite biosensors for point-of-care—evaluation of food quality and safety*. Nanobiosensors., doi:10.1016/B978-0-12-804301-1.00015-1

D'Ugo, E., Bruno, M., Mukherjee, A., Chattopadhyay, D., Giuseppetti, R., De Pace, R., & Magurano, F. (2021). Characterization of microbial response to petroleum hydrocarbon contamination in a lacustrine ecosystem. *Environmental Science and Pollution Research International, 28*(20), 26187–26196. doi:10.1007/s11356-021-13885-8 PMID:33871774

Da'ana, D. A., Zouari, N., Ashfaq, M. Y., Abu-Dieyeh, M., Khraisheh, M., Hijji, Y. M., & Al-Ghouti, M. A. (2021). Removal of Toxic Elements and Microbial Contaminants from Groundwater Using Low-Cost Treatment Options. *Current Pollution Reports*, *7*(3), 300–324. doi:10.1007/s40726-021-00187-3

Dadi, S., Celik, C., Mandal, A. K., & Ocsoy, I. (2021). Dopamine and norepinephrine assistant-synthesized nanoflowers immobilized membrane with peroxidase mimic activity for efficient detection of model substrates. *Applied Nanoscience*, *11*(1), 117–125. doi:10.1007/s13204-020-01577-7

Dai, L., Wang, Y., Zou, X., Chen, Z., Liu, H., & Ni, Y. (2020). Ultrasensitive Physical, Bio, and Chemical Sensors Derived from 1-, 2-, and 3-D Nanocellulosic Materials. *Small*, *16*(13), 1906567. doi:10.1002/smll.201906567 PMID:32049432

Damodharan, J. (2021). Nanomaterials in medicine–An overview. *Materials Today: Proceedings*, *37*, 383–385. doi:10.1016/j.matpr.2020.05.380

Dandia, A., Bansal, S., Sharma, R., Rathore, K. S., & Parewa, V. (2018). Microwave-assisted nanocatalysis: A CuO NPs/rGO composite as an efficient and recyclable catalyst for the Petasisborono–Mannich reaction. *RSC Advances*, *8*(53), 30280–30288. doi:10.1039/C8RA05203D PMID:35546842

Danisman, H., Çelebi, F., Danışman, Ş., & Bıçakçı, A. (2022). Effects of diamond-like carbon coating on frictional and mechanical properties of orthodontic brackets: An invitro study. *Apos Trends In Orthodontıcs*, *12*(1), 13–19. doi:10.25259/APOS_79_2021

Danushika, C. M., Rohan, S. D., & Renuka, N. L. (2021). Novel Acumens into Biodegradation: Impact of Nanomaterials and Their Contribution. In M. Kamila Cabral (Ed.), *M Kassio Ferreira, S Rodrigo Nogueira de*. Biodegradation Technology of Organic and Inorganic Pollutants. IntechOpen.

Daraee, H., Eatemadi, A., Abbasi, E., Aval, S. F., Kouhi, M., & Akbarzadeh, A. (2016). Application of gold nanoparticles in biomedical and drug delivery. *Artificial Cells, Nanomedicine, and Biotechnology*, *44*(1), 410–422. doi:10.3109/21691401.2014.955107 PMID:25229833

Das, I., Chattopadhyay, S., Mahato, A., Kundu, B., & De, G. (2016). Fabrication of a cubic zirconia nanocoating on a titanium dental implant with excellent adhesion, hardness andbiocompatibility. *RSC Advances*, *6*(64), 59030–59038. doi:10.1039/C6RA10661G

Dastjerdi, Z., Cranston, E. D., Berry, R., Fraschini, C., & Dubé, M. A. (2019). Polymer Nanocomposites for Emulsion-Based Coatings and Adhesives. *Macromolecular Reaction Engineering*, *13*(2), 1800050. doi:10.1002/mren.201800050

Dastjerdi, Z., Cranston, E. D., & Dubé, M. A. (2018). Pressure sensitive adhesive property modification using cellulose nanocrystals. *International Journal of Adhesion and Adhesives*, *81*, 36–42. doi:10.1016/j.ijadhadh.2017.11.009

Datsyuk, V., Trotsenko, S., Peikert, K., Höflich, K., Wedel, N., Allar, C., Sick, T., Deinhart, V., Reich, S., & Krcmar, W. (2019). Polystyrene nanofibers for nonwoven porous building insulation materials. *Engineering Reports*, *1*(2), e12037. doi:10.1002/eng2.12037

Daud, J., & Lee, K.-Y. (2017). Surface Modification of Nanocellulose. In Hanieh Kargarzadeh, I. Ahmad, S. Thomas, & A. Dufresne (Eds.), Handbook of Nanocellulose and Cellulose (pp. 101–122). Wiley-VCH. doi:10.1002/9783527689972.ch3

De Silva, R. T., Mantilaka, M. M. M. G. P. G., Ratnayake, S. P., Amaratunga, G. A. J., & de Silva, K. M. N. (2017). Nano-MgO reinforced chitosan nanocomposites for high performance packaging applications with improved mechanical, thermal and barrier properties. *Carbohydrate Polymers*, *157*, 739–747. doi:10.1016/j.carbpol.2016.10.038 PMID:27987986

de Souza, G. L., Almeida, A. C., Danlas, N. O., Paula, A., Turrioni, S., & Moura, C. C. G. (2020). Cytotoxicity and Effects of a New Cacium Hydroxide Nanoparticle Material on Production of Reactive Oxygen Species by LPS-Stimulated Dental Pulp Cells. *Iranian Endodontic Journal*, *15*(4), 227–235. doi:10.22037/iej.v15i4.28942 PMID:36704109

Dehghani, S., Hosseini, S. V., & Regenstein, J. M. (2018). Edible films and coatings in seafood preservation: A review. *Food Chemistry*, *240*, 505–513. doi:10.1016/j.foodchem.2017.07.034 PMID:28946304

Demirbas, A., Büyükbezirci, K., Celik, C., Kislakci, E., Karaagac, Z., Gokturk, E., Kati, A., Cimen, B., Yilmaz, V., & Ocsoy, I. (2019). Synthesis of long-term stable gold nanoparticles benefiting from red raspberry (Rubus idaeus), strawberry (Fragaria ananassa), and blackberry (Rubus fruticosus) extracts–gold ion complexation and investigation of reaction conditions. *ACS Omega*, *4*(20), 18637–18644. doi:10.1021/acsomega.9b02469 PMID:31737823

Demirbas, A., Karaytuğ, T., Arabaci, N., Yilmaz, E. S., & Ocsoy, I. (2020). Synthesis of metallic and metal oxide nanomaterials. In N. Srivastava, M. Srivastava, P. K. Mishra, & V. K. Gupta (Eds.), *Green Synthesis of Nanomaterials for Bioenergy Applications* (pp. 99–123). Wiley., doi:10.1002/9781119576785.ch4

Deng, C., Pi, X., Qian, P., Chen, X., Wu, W., & Xiang, J. (2017). High-performance ratiometric electrochemical method based on the combination of signal probe and in Devi, R., Yadav, S., Nehra, R., Yadav, S., & Pundir, C. S. (2013). Electrochemical biosensor based on gold-coated iron nanoparticles/chitosan composite bound xanthine oxidase for detection of xanthine in fish meat. *Journal of Food Engineering*, *115*(2), 207–214. doi:10.1016/j.jfoodeng.2012.10.014

Deng, Z., Zhao, L., Zhou, H., Xu, X., & Zheng, W. (2022). Recent advances in electrochemical analysis of hydrogen peroxide towards in vivo detection. *Process Biochemistry (Barking, London, England)*, *115*, 57–69. doi:10.1016/j.procbio.2022.01.025

Dey, A., Abdul Moyez, Sk., Kumar Mandal, M., & Roy, S. (2016). Fabrication of solar cells using extracted biomolecules from tea leaves and hybrid perovskites. Materials Today: Proceedings. 3(10),part A, 3498-3504,. http://dx.doi.org/ doi:10.1016/j.matpr.2016.10.033

Dey, A., Das, B. C., Biswas, A. B., Biswas, P., Dhar, A., Roy, S., & Moyez Sk, A. (2017). Graphene Co-Doped TiO2 Nanocomposites for Photocatalysis and Photovoltaics Applications. *Indian Journal of Science and Technology*, *10*(31), 1–6. doi:10.17485/ijst/2017/v10i31/113878

Dhaka, A., Mali, S. C., Sharma, S., & Trivedi, R. (2023). A review on biological synthesis of silver nanoparticles and their potential applications. *Results in Chemistry*, *6*, 101108. doi:10.1016/j.rechem.2023.101108

Diamanti-Kipioti, A., Gusberti, F. A., & Lang, N. P. (1987). Clinical and microbiological effects of fixed orthodontic appliances. *Journal of Clinical Periodontology*, *14*(6), 326–333. doi:10.1111/j.1600-051X.1987.tb00979.x PMID:3509967

Díez-Pascual, A. M., & Díez-Vicente, A. L. (2014). ZnO-reinforced poly(3-hydroxybutyrate-co-3-hydroxyvalerate) bionanocomposites with antimicrobial function for food packaging. *ACS Applied Materials & Interfaces*, *6*(12), 9822–9834. doi:10.1021/am502261e PMID:24846876

Ding, Y., Sun, Y., Liu, C., Jiang, Q. Y., Chen, F., & Cao, Y. (2023). SERS-Based Biosensors Combined with Machine Learning for Medical Application. In ChemistryOpen (Vol. 12, Issue 1). John Wiley and Sons Inc. doi:10.1002/open.202200192

Din, M. I., Ghaffar, T., Najeeb, J., Hussain, Z., Khalid, R., & Zahid, H. (2020). Potential perspectives of biodegradable plastics for food packaging application-review of properties and recent developments. *Food Additives & Contaminants. Part A, Chemistry, Analysis, Control, Exposure & Risk Assessment*, *37*(4), 665–680. doi:10.1080/19440049.2020.1718219 PMID:32049609

Din, S. H. (2019). Nano-Composites and their Applications: A review. *Characterization and Application of Nanomaterials*, *2*(1). doi:10.24294/can.v2i1.875

Dobson, P. King. Stephen and Jarvie. Helen (2019, May 14). Nanoparticle. *Encyclopedia Britannica.* https://www.britannica.com/science/nanoparticle

Dolai, J., Mandal, K., & Jana, N. R. (2021). Nanoparticle size effects in biomedical applications. *ACS Applied Nano Materials*, *4*(7), 6471–6496. doi:10.1021/acsanm.1c00987

dos Santos de Abreu, D., Strauss, M., & Santhiago, M. (2020). Flexible cellulose-based devices for monitoring physical parameters. In *Comprehensive Analytical Chemistry* (1st ed., Vol. 89). Elsevier B.V., doi:10.1016/bs.coac.2020.01.002

Dosio, F., Huang, Y.-F., Kumar Prabhakar, P., Khurana, N., Vyas, M., Sharma, V., El-Saber Batiha, G., Kaur, H., Singh, J., Kumar, D., Sharma, N., Kaushik, A., & Kumar, R. (2023). Aspects of Nanotechnology for COVID-19 Vaccine Development and Its Delivery Applications. *Pharmaceuticals.* doi:10.3390/pharmaceutics15020451

Do, T. T. A., Grijalvo, S., Imae, T., Garcia-Celma, M. J., & Rodríguez-Abreu, C. (2021). A nanocellulose-based platform towards targeted chemo-photodynamic/photothermal cancer therapy. *Carbohydrate Polymers*, *270*, 118366. doi:10.1016/j.carbpol.2021.118366 PMID:34364611

Dubey, A., Irudhayaraj, S., & Jaiswal, A. (2023). Bio-nanocomposites: A Next Generation Food Packaging Materials. In *Synthesis and Applications of Nanomaterials and Nanocomposites* (pp. 197–231). Springer Nature Singapore. doi:10.1007/978-981-99-1350-3_8

Dubey, S. K., Dey, A., Singhvi, G., Pandey, M. M., Singh, V., & Kesharwani, P. (2022). Emerging trends of nanotechnology in advanced cosmetics. *Colloids and Surfaces. B, Biointerfaces*, *214*, 112440. doi:10.1016/j.colsurfb.2022.112440 PMID:35344873

Dubey, S. P., Lahtinen, M., Särkkä, H., & Sillanpää, M. (2010). Bioprospective of Sorbus aucuparia leaf extract in development of silver and gold nanocolloids. *Colloids and Surfaces. B, Biointerfaces*, *80*(1), 26–33. doi:10.1016/j.colsurfb.2010.05.024 PMID:20620889

Ducheyne, P. (2015). 1.109. Bioactive Ceramics. *Comprehensive Biomaterials*, *1*, 157–179. doi:10.1016/B978-0-08-055294-1.00021-0

Dujardin, E., Peet, C., Stubbs, G., Culver, J. N., & Mann, S. (2003). Organization of metallic nanoparticles using tobacco mosaic virus templates. *Nano Letters*, *3*(3), 413–417. doi:10.1021/nl034004o

Dykman, L. A., & Khlebtsov, N. G. (2014). Uptake of engineered gold nanoparticles into mammalian cells. In Chemical Reviews, 114(2), 1258–1288. doi:10.1021/cr300441a

Eddy, N. O., Edet, U. E., Oladele, J. O., Kelle, H. I., Ogoko. E. C., Odiongenyi, A. O., Ameh, P., Ukpe, R. A., Ogbodo, R., Garg, R. & Garg, R. (2023b). Synthesis and application of novel microporous framework of nanocomposite from trona for photocatalysed degradation of methyl orange dye. *Journal of Environmental Monitoring and Assessment*. doi:10.1007/s10661-023-12014-x

Eddy, N. O., Garg, R., Garg, R., Eze, S. I., Ogoko, E. C., Kelle, H. I., Ukpe, R. A., Ogbodo, R. and Chijoke, F. (2023a). Sol-gel synthesis, computational chemistry, and applications of CaO nanoparticles for the remediation of methyl orange contaminated water. *Advances in Nano Research*. https://doi.org/. doi:10.12989/anr.2023.15.1.000

Eddy, N. O., Odiongenyi, A., Garg, R., Ukpe, R. A., & Garg, R. ChimaOgoko, E., Kelle, H. I., & Ogbodo, R. (2023). Ecotoxicological aspects of Nanotechnology. In Nanotechnology for Sustainable Agriculture, Food and Environment (pp. 211–223). CRC Press. doi:10.1201/9781003397861-16

Eddy, N. O., Garg, R., Garg, R., Aikoye, A., & Ita, B. I. (2022a). Waste to resource recovery: Mesoporous adsorbent from orange peel for the removal of trypan blue dye from aqueous solution. *Biomass Conversion and Biorefinery*. doi:10.1007/s13399-022-02571-5

Eddy, N. O., Garg, R., Garg, R., Eze, S. I., Ogoko, E. C., Kelle, H. I., Ukpe, R. A., Ogbodo, R., & Chijoke, F. (2023). Sol-gel synthesis, computational chemistry, and applications of Cao nanoparticles for the remediation of methyl orange contaminated water. *Advances in Nano Research*, *15*(1), 35–48. doi:10.12989/anr.2023.15.1.035

Eddy, N. O., Odiongenyi, A. O., Garg, R., Ukpe, R. A., Garg, R., El Nemr, A., Ngwu, C. M., & Okop, I. J. (2023). Quantum and experimental investigation of the application of Crassostrea gasar (mangrove oyster) shell–based CaO nanoparticles as adsorbent and photocatalyst for the removal of procaine penicillin from aqueous solution. *Environmental Science and Pollution Research International, 30*(23), 64036–64057. doi:10.1007/s11356-023-26868-8 PMID:37059957

Eddy, N. O., Ukpe, R. A., Ameh, P., Ogbodo, R., Garg, R., & Garg, R. (2022). Theoretical and experimental studies on photocatalytic removal of methylene blue (MetB) from aqueous solution using oyster shell synthesized CaO nanoparticles (CaONP-O). *Environmental Science and Pollution Research International, 30*(34), 81417–81432. doi:10.1007/s11356-022-22747-w PMID:36057067

Eddy, N. O., Ukpe, R. A., Garg, R., Garg, R., Odiongenyi, A. O., Ameh, P., Akpet, I. N., & Udo, E. S. (2023c). Review of in-depth knowledge on the application of oxides nanoparticles and nanocomposites of Al, Si and Ca as photocatalyst and antimicrobial agents in the treatment of contaminants in water. *Clean Technologies and Environmental Policy*. doi:10.1007/s10098-023-02603-2

Editorial, G., Cheon, J., Chan, W., Zuhorn, I., & Authors, C. (2019). The Future of Nanotechnology: Cross-disciplined Progress to Improve Health and Medicine. *Accounts of Chemical Research, 52*(9), 2405. doi:10.1021/acs.accounts.9b00423 PMID:31525881

Effiong, D. E., Uwah, T. O., Jumbo, E. U., & Akpabio, A. E. (2020). Nanotechnology in Cosmetics: Basics, Current Trends and Safety Concerns—A Review. *Advances in Nanoparticles, 9*, 1–22. doi:10.4236/anp.2020.91001

Egbosiuba, T. C., Egwunyenga, M. C., Tijani, J. O., Mustapha, S., Abdulkareem, A. S., Kovo, A. S., Krikstolaityte, V., Veksha, A., Wagner, M., & Lisak, G. (2022). Activated multi-walled carbon nanotubes decorated with zero valent nickel nanoparticles for arsenic, cadmium and lead adsorption from wastewater in a batch and continuous flow modes. *Journal of Hazardous Materials, 423*(PB), 126993. doi:10.1016/j.jhazmat.2021.126993

Eguílaz, M., Gutierrez, F., González-Domínguez, J. M., Martínez, M. T., & Rivas, G. (2016). Single-walled carbon nanotubes covalently functionalized with polytyrosine: A new material for the development of NADH-based biosensors. *Biosensors & Bioelectronics, 86*, 308–314. doi:10.1016/j.bios.2016.06.003 PMID:27387261

Eichhorn, S. J., Dufresne, A., Aranguren, M., Marcovich, N. E., Capadona, J. R., Rowan, S. J., Weder, C., Thielemans, W., Roman, M., Renneckar, S., Gindl, W., Veigel, S., Keckes, J., Yano, H., Abe, K., Nogi, M., Nakagaito, A. N., Mangalam, A., Simonsen, J., & Peijs, T. (2010). Current international research into cellulose nanofibres and nanocomposites. *Journal of Materials Science, 45*(1), 1–33. doi:10.1007/s10853-009-3874-0 PMID:32836382

Ekrikaya, S., Yilmaz, E., Arslan, S., Karaaslan, R., Ildiz, N., Celik, C., & Ocsoy, I. (2023). Dentin bond strength and antimicrobial activities of universal adhesives containing silver nanoparticles synthesized with Rosa canina extract. *Clinical Oral Investigations, 27*(11), 1–12. doi:10.1007/s00784-023-05306-6 PMID:37821654

Ekrikaya, S., Yilmaz, E., Celik, C., Demirbuga, S., Ildiz, N., Demirbas, A., & Ocsoy, I. (2021). Investigation of ellagic acid rich-berry extracts directed silver nanoparticles synthesis and their antimicrobial properties with potential mechanisms towards Enterococcus faecalis and Candida albicans. *Journal of Biotechnology, 341*, 155–162. doi:10.1016/j.jbiotec.2021.09.020 PMID:34601019

Elbony, F. A. (2020). Nanotechnology and Energy Efficiency In Buildings. JL. *EGYPTIAN SOCIETY OF ENGNEERS, 59*(1), 29–35.

Elcock, D. (2007). *Potential impacts of nanotechnology on energy transmission applications and needs (No. ANL/EVS/TM/08-3). Argonne National Lab.* ANL.

Elgrishi, N., Rountree, K. J., McCarthy, B. D., Rountree, E. S., Eisenhart, T. T., & Dempsey, J. L. (2018). A Practical Beginner's Guide to Cyclic Voltammetry. *Journal of Chemical Education, 95*(2), 197–206. doi:10.1021/acs.jchemed.7b00361

Elshafie, M., Younis, S. A., Serp, P., & Gad, E. A. M. (2020). Preparation characterization and non-isothermal decomposition kinetics of different carbon nitride sheets. *Egypt. J. Pet., 29*(1), 21–29. doi:10.1016/j.ejpe.2019.09.003

Erhardt, D. (2003). NEWS & VIEWS. *Nature*. www.nature.com/naturematerials

Esmaeili, A. (2011, December). Applications of nanotechnology in oil and gas industry. In AIP conference proceedings (Vol. 1414, No. 1, pp. 133-136). American Institute of Physics. doi:10.1063/1.3669944

Ezati, P., & Rhim, J.-W. (2020). pH-responsive chitosan-based film incorporated with alizarin for intelligent packaging applications. *Food Hydrocolloids, 102*, 105629. doi:10.1016/j.foodhyd.2019.105629

Ezhil Vilian, A. T. (2017). Krishnan Giribabu, Sang Rak Choe, Rethinasabapathy Muruganantham, Hoomin Lee, Changhyun Roh, Yun Suk Huh, Young-Kyu Han. A spick-and-span approach to the immobilization of horseradish peroxidase on Au nanospheres incorporated with a methionine/graphene biomatrix for the determination of endocrine disruptor bisphenol A. *Sensors and Actuators. B, Chemical, 251*, 804–812. doi:10.1016/j.snb.2017.05.122

Fahmy, S. A., Fawzy, I. M., Saleh, B. M., Issa, M. Y., Bakowsky, U., & Azzazy, H. M. E. S. (2021). Green synthesis of platinum and palladium nanoparticles using Peganum harmala L. seed alkaloids: Biological and computational studies. *Nanomaterials (Basel, Switzerland), 11*(4), 965. doi:10.3390/nano11040965 PMID:33918743

Farooq, A., Al-jowder, R., Narayanaswamy, R., Azzawi, M., Roche, P. J. R., & Whitehead, D. E. (2013). Gas detection using quenching fluorescence of dye-immobilized silica nanoparticles. *Sensors and Actuators. B, Chemical, 183*, 230–238. doi:10.1016/j.snb.2013.03.058

Farooq, A., Patoary, M. K., Zhang, M., Mussana, H., Li, M., Naeem, M. A., Mushtaq, M., Farooq, A., & Liu, L. (2020). Cellulose from sources to nanocellulose and an overview of synthesis and properties of nanocellulose/zinc oxide nanocomposite materials. *International Journal of Biological Macromolecules*, *154*, 1050–1073. doi:10.1016/j.ijbiomac.2020.03.163 PMID:32201207

Favier, V., Canova, G. R., Cavaillé, J. Y., Chanzy, H., Dufresne, A., & Gauthier, C. (1995). Nanocomposite materials from latex and cellulose whiskers. *Polymers for Advanced Technologies*, *6*(5), 351–355. doi:10.1002/pat.1995.220060514

Fawcett, D., Verduin, J. J., Shah, M., Sharma, S. B., & Poinern, G. E. J. (2017). A review of current research into the biogenic synthesis of metal and metal oxide nanoparticles via marine algae and seagrasses. *Journal of Nanoscience*, *2017*(4), 1–15. doi:10.1155/2017/8013850

Fayemi, O. E., Adekunle, A. S., Kumara Swamy, B. E., & Ebenso, E. E. (2018). Electrochemical sensor for the detection of dopamine in real samples using polyaniline/NiO, ZnO, and Fe3O4 nanocomposites on glassy carbon electrode. *Journal of Electroanalytical Chemistry (Lausanne, Switzerland)*, *818*, 236–249. doi:10.1016/j.jelechem.2018.02.027

Fei, A., Liu, Q., Huan, J., Qian, J., Dong, X., Qiu, B., Mao, H., & Wang, K. (2015). Label-free impedimetric apta-sensor for detection of femtomole level acetamiprid using gold nanoparticles decorated multiwalled carbon nanotube-reduced graphene oxide nanoribbon composites. *Biosensors & Bioelectronics*, *70*, 122–129. doi:10.1016/j.bios.2015.03.028 PMID:25797851

Feizpoor, S., & Habibi-Yangjeh, A. (2018). Ternary TiO2/Fe3O4/CoWO4 nanocomposites: Novel magnetic visible-light-driven photocatalysts with substantially enhanced activity through p-n heterojunction. *Journal of Colloid and Interface Science*, *524*, 325–336. doi:10.1016/j.jcis.2018.03.069 PMID:29656067

Ferdous, Z., & Nemmar, A. (2020). Health Impact of Silver Nanoparticles: A Review of the Biodistribution and Toxicity Following Various Routes of Exposure. *International Journal of Molecular Sciences*, *21*(7), 2375. doi:10.3390/ijms21072375 PMID:32235542

Fernández-Pampín, N., González Plaza, J. J., García-Gómez, A., Peña, E., Rumbo, C., Barros, R., Martel-Martín, S., Aparicio, S., & Tamayo-Ramos, J. A. (2022). Toxicology assessment of manganese oxide nanomaterials with enhanced electrochemical properties using human in vitro models representing different exposure routes. *Scientific Reports*, *12*(1), 20991. doi:10.1038/s41598-022-25483-w PMID:36471154

Ferrando-Magraner, E., Bellot-Arcís, C., Paredes-Gallardo, V., Almerich-Silla, J. M., García-Sanz, V., Fernández-Alonso, M., & Montiel-Company, J. M. (2020). Antibacterial propertiesof nanoparticles in dental restorative materials. A systematic review and meta-analysis. *Medicina*, *56*(2), 55. doi:10.3390/medicina56020055 PMID:32013103

Ferreira, M. A. M., & Filipe, J. A. (2018). Nanotechnology Applications – The Future Arrived Suddenly. In Computational Approaches in Biomedical Nano-Engineering (pp. 23–41). Wiley. doi:10.1002/9783527344758.ch2

Ferreira, F. V., Dufresne, A., Pinheiro, I. F., Souza, D. H. S., Gouveia, R. F., Mei, L. H. I., & Lona, L. M. F. (2018). How do cellulose nanocrystals affect the overall properties of biodegradable polymer nanocomposites: A comprehensive review. *European Polymer Journal*, *108*(July), 274–285. doi:10.1016/j.eurpolymj.2018.08.045

Ferrier, D. C., & Honeychurch, K. C. (2021). Carbon nanotube (CNT)-based biosensors. In Biosensors (Vol. 11, Issue 12). MDPI. doi:10.3390/bios11120486

Fisher, C. H. (1989). In R. B. Seymour, H. F. Mark, L. Pauling, C. H. Fisher, G. A. Stahl, L. H. Sperling, C. S. Marvel, C. E. Carraher, & R. B. Seymour (Eds.), *Anselm Payen Pioneer in Natural Polymers and Industrial Chemistry BT - Pioneers in Polymer Science* (pp. 47–61). Springer Netherlands., doi:10.1007/978-94-009-2407-9_5

Fitriani, F., Bilad, M. R., Aprilia, S., & Arahman, N. (2023). Biodegradable Hybrid Polymer Film for Packaging: A Review. *Journal of Natural Fibers*, *20*(1), 2159606. doi:10.1080/15440478.2022.2159606

Fontaine, M., Bartolami, E., Prono, M., Béal, D., Blosi, M., Costa, A. L., Ravagli, C., Baldi, G., Sprio, S., Tampieri, A., Fenoglio, I., Tran, L., Fadeel, B., & Carriere, M. (2023). Nanomaterial genotoxicity evaluation using the high-throughput p53-binding protein 1 (53BP1) assay. *PLoS One*, *18*(9), e0288737. doi:10.1371/journal.pone.0288737 PMID:37713377

Fortunati, E., Peltzer, M., Armentano, I., Torre, L., Jiménez, A., & Kenny, J. M. (2012). Effects of modified cellulose nanocrystals on the barrier and migration properties of PLA nano-biocomposites. *Carbohydrate Polymers*, *90*(2), 948–956. doi:10.1016/j.carbpol.2012.06.025 PMID:22840025

Freitas, R. A. (2005). Nanotechnology, nanomedicine and nanosurgery. In International Journal of Surgery, 3(4). doi:10.1016/j.ijsu.2005.10.007

Frigerio, C., Ribeiro, D. S. M., Rodrigues, S. S. M., Abreu, V. L. R. G., Barbosa, J. A. C., Prior, J. A. V., Marques, K. L., & Santos, J. L. M. (2012). Application of quantum dots as analytical tools in automated chemical analysis: A review. *Analytica Chimica Acta*, *735*, 9–22. doi:10.1016/j.aca.2012.04.042 PMID:22713912

Fu, C., Liu, C., Li, Y., Guo, Y., Luo, F., Wang, P., Guo, L., Qiu, B., & Lin, Z. (2016). Homogeneous electrochemical biosensor for melamine based on DNA triplex structure and exonuclease III-assisted recycling amplification. *Analytical Chemistry*, *88*(20), 10176–10182. doi:10.1021/acs.analchem.6b02753 PMID:27665764

Fuertes, G., Soto, I., Vargas, M., Valencia, A., Sabattin, J., & Carrasco, R. (2016). Nanosensors for a Monitoring System in Intelligent and Active Packaging. *Journal of Sensors*, *7980476*, 1–8. doi:10.1155/2016/7980476

Fujimori, Y., Sato, T., Hayata, T., Nagao, T., Nakayama, M., Nakayama, T., & Suzuki, K. (2012). Novel antiviral characteristics of nanosized copper (I) iodide particles showing inactivation activity against 2009 pandemic H1N1 influenza virus. *Applied and Environmental Microbiology*, *78*(4), 951–955. doi:10.1128/AEM.06284-11 PMID:22156433

Gadhave, R. V., Das, A., Mahanwar, P. A., & Gadekar, P. T. (2018). Starch based bio-plastics: The future of sustainable packaging. *Open Journal of Polymer Chemistry*, *8*(2), 21–33. doi:10.4236/ojpchem.2018.82003

Gaikwad, K. K., Singh, S., & Ajji, A. (2019). Moisture absorbers for food packaging applications. *Environmental Chemistry Letters*, *17*(2), 609–628. doi:10.1007/s10311-018-0810-z

Gaikwad, K. K., Singh, S., & Lee, Y. S. (2018). Oxygen scavenging films in food packaging. *Environmental Chemistry Letters*, *16*(2), 523–538. doi:10.1007/s10311-018-0705-z

Gajanan, K., & Tijare, S. N. (2018). Applications of nanomaterials. *Materials Today: Proceedings*, *5*(1), 1093–1096. doi:10.1016/j.matpr.2017.11.187

Garg, R., Garg, R., Sharma, R., Anjum, A., & Eddy, N. O. (2023). Introduction to Nanotechnology and Nanomaterials. In Nanotechnology for Sustainable Agriculture, Food and Environment (pp. 1–15). CRC Press. doi:10.1201/9781003397861-1

Garg, R., Garg, R., Khan, M. A., Bansal, M., & Garg, V. K. (2023). Utilization of biosynthesized silica-supported iron oxide nanocomposites for the adsorptive removal of heavy metal ions from aqueous solutions. *Environmental Science and Pollution Research International*, *30*(34), 81319–81332. doi:10.1007/s11356-022-21111-2 PMID:35672639

Garg, R., Garg, R., Okon Eddy, N., Ibrahim Almohana, A., Fahad Almojil, S., Amir Khan, M., & Ho Hong, S. (2022). Biosynthesized silica-based zinc oxide nanocomposites for the sequestration of heavy metal ions from aqueous solutions. *Journal of King Saud University. Science*, *34*(4), 101996. doi:10.1016/j.jksus.2022.101996

Garg, R., Garg, R., Sillanpää, M., Alimuddin, Khan, M. A., Mubarak, N. M., & Tan, Y. H. (2023). Rapid adsorptive removal of chromium from wastewater using walnut-derived biosorbents. *Scientific Reports*, *13*(1), 6859. doi:10.1038/s41598-023-33843-3 PMID:37100812

Garrick, B. J. (2008). Examples of Risks Having the Potential for Catastrophic Consequences. B. John Garrick, Quantifying and Controlling Catastrophic Risks. Academic Press. doi:10.1016/B978-0-12-374601-6.00007-8

Ge, Y., Liu, P., Xu, L., Qu, M., Hao, W., Liang, H., Sheng, Y., Zhu, Y., & Wen, Y. (2022). A portable wireless intelligent electrochemical sensor based on layer-by-layer sandwiched nanohybrid for terbutaline in meat products. *Food Chemistry*, *371*, 131140. doi:10.1016/j.foodchem.2021.131140 PMID:34583185

Geyter, N. (2021). Biological activity and antimicrobial property of Cu/aC: Hnanocomposites and nanolayered coatings on titanium substrates. *Materials Science and Engineering C*, *119*, 111513. doi:10.1016/j.msec.2020.111513 PMID:33321606

Ghaedi, M. (2021). Adsorption: Fundamental Processes and Applications. In *Adsorption: Fundamental Processes and Applications*, *33*(1). https://linkinghub.elsevier.com/retrieve/pii/B97801281880570 00011

Ghafar, H., Khan, M. I., Sarwar, H. S., Yaqoob, S., Hussain, S. Z., Tariq, I., & Sohail, M. F. (2020). Development and characterization of bioadhesive film embedded with lignocaine and calcium fluoride nanoparticles. *AAPS PharmSciTech*, *21*(2), 1–12. doi:10.1208/s12249-019-1615-5 PMID:31912272

Ghanbari, Kh., & Babaei, Z. (2016). Fabrication and characterization of non-enzymatic glucose sensor based on ternary NiO/CuO/polyaniline nanocomposite. *Analytical Biochemistry*, *498*, 37–46. doi:10.1016/j.ab.2016.01.006 PMID:26778527

Ghanbarizadeh, P., Parivazh, M. M., Abbasi, M., Osfouri, S., Dianat, M. J., Rostami, A., Dibaj, M., & Akrami, M. (2022). Performance Enhancement of Specific Adsorbents for Hardness Reduction of Drinking Water and Groundwater. *Water (Basel)*, *14*(17), 2749. doi:10.3390/w14172749

Gharpure, S., & Ankamwar, B. (123 C.E.). Use of nanotechnology in combating coronavirus. *3 Biotech*, *11*, 358. doi:10.1007/s13205-021-02905-6

Giridharan, T., Masi, C., Sindhu, S., & Arumugam, P. (2014). Studies on green synthesis, characterization and anti-proliferative potential of silver nano particle using Dodonaea viscosa and Capparis decidua. *Biosciences Biotechnology Research Asia*, *11*(2), 665–673. doi:10.13005/bbra/1320

Girigoswami, A., Thirumalai, A., Karthick, H., Pallavi, P., Gowtham, P., & Girigoswami, A. (2022). Nanotechnology driven improvement of smart food packaging. *Materials Research Innovations*. doi:10.1080/14328917.2022.2114667

GLOBE NEWSWIRE. (2023). Research and markets. Global Nanotechnology Clothing Market Report 2023: Rising Use of Nanotechnology Clothing in the Medical Industry Owing to its Anti-Microbial Properties Drives Growth. *Nanotechnology Clothing Global Market Report 2023*. https://www.researchandmarkets.com/

Golgoon, A., Aliofkhazraei, M., Toorani, M., Moradi, M. H., & Rouhaghdam, A. S. (2015). Corrosion and wear properties of nanoclay-polyester nanocomposite coatings fabricated by electrostatic method. *Procedia Materials Science*, *11*, 536–541. doi:10.1016/j.mspro.2015.11.042

Gomez, H., Ram, M. K., Alvi, F., Stefanakos, E., & Kumar, A. (2010). Novel synthesis, characterization, and corrosion inhibition properties of nanodiamond– polyaniline films. *The Journal of Physical Chemistry. C, Nanomaterials and Interfaces*, *114*(44), 18797–18804. doi:10.1021/jp106379e

González-Pedroza, M. G., Benítez, A. R. T., Navarro-Marchal, S. A., Martínez-Martínez, E., Marchal, J. A., Boulaiz, H., & Morales-Luckie, R. A. (2023). Biogeneration of silver nanoparticles from Cuphea procumbens for biomedical and environmental applications. *Scientific Reports*, *13*(1), 790. doi:10.1038/s41598-022-26818-3 PMID:36646714

Gooding, J. J., & Gaus, K. (2016). Single-Molecule Sensors: Challenges and Opportunities for Quantitative Analysis. *Angewandte Chemie International Edition*, *55*(38), 11354–11366. doi:10.1002/anie.201600495 PMID:27444661

Gopakumar, D. A., Thomas, S., & Grohens, Y. (2016). Nanocelluloses as Innovative Polymers for Membrane Applications. In D Puglia, E. Fortunati, & J. M. Kenny (Eds.), Multifunctional Polymeric Nanocomposites Based on Cellulosic Reinforcements (Issue 1, pp. 253–275). William Andrew Publishing. doi:10.1016/B978-0-323-44248-0.00008-0

Gopinath, V., MubarakAli, D., Priyadarshini, S., Priyadharsshini, N. M., Thajuddin, N., & Velusamy, P. (2012). Biosynthesis of silver nanoparticles from Tribulus terrestris and itsantimicrobial activity: A novel biological approach. *Colloids and Surfaces. B, Biointerfaces*, *96*, 69–74. doi:10.1016/j.colsurfb.2012.03.023 PMID:22521683

Gosai, A., Khondakar, K. R., Ma, X., & Ali, M. A. (2021). Application of functionalized graphene oxide based biosensors for health monitoring: Simple graphene derivatives to 3D printed platforms. In Biosensors (Vol. 11, Issue 10). MDPI. doi:10.3390/bios11100384

Gourlay, S. J., Rice, R. M., Hegyeli, A. F., Wade, C. W. R., Dillon, J. G., Jaffe, H., & Kulkarni, R. K. (1978). Biocompatibility Testing of Polymers: In Vivo Implantation Studies. *Journal of Biomedical Materials Research*, *12*(2), 219–232. doi:10.1002/jbm.820120207 PMID:649628

Govindappa, M., Hemashekhar, B., Arthikala, M. K., Rai, V. R., & Ramachandra, Y. L. (2018). Characterization, antibacterial, antioxidant, antidiabetic, anti-inflammatory and antityrosinase activity of green synthesized silver nanoparticles using Calophyllum tomentosum leaves extract. *Results in Physics*, *9*, 400–408. doi:10.1016/j.rinp.2018.02.049

Goyal, R., Sharma, M., & Amberiya, U. K. (2014). Innovative nano composite materials and applications in automobiles. *International Journal of Engineering Research & Technology (Ahmedabad)*, *3*(1), 3001–3009.

Grand View Research. (2023). *Base Metal Mining Market Size, Share & Trends Analysis Report By Product (Aluminum, Copper, Zinc, Lead), By End-Use (Construction, Automotive), By Region, And Segment Forecasts*. GVR. htttps://www.grandviewresearch.com/industry-analysis/base-metal-minin g-market

Grand View Research. (2023). *Nanomaterials Market Size, Share & Trends Analysis Report By Material (Gold, Silver, Iron, Copper), By Application (Aerospace, Automotive, Medical), By Region, And Segment Forecasts, 2023 – 2030*. https://www.google.com/search?q=Global+market+value+for+nano particles

Grand View Research. (2023). *Polymer Foam Market Size, Share & Trends Analysis Report By Type (Polystyrene, Polyurethane, Polyolefin, Melamine, Phenolic, PVC), By Application, By Region, And Segment Forecasts, 2022 – 2030*. https://www.grandviewresearch.com/industry-analysis/polymer-foam-market

Grass, G., Rensing, C., & Solioz, M. (2011). Metallic copper as an antimicrobial surface. *Applied and Environmental Microbiology*, *77*(5), 1541–1547. doi:10.1128/AEM.02766-10 PMID:21193661

Guan, X., Lin, S., Lan, J., Shang, J., Li, W., Zhan, Y., Xiao, H., & Song, Q. (2019). Fabrication of Ag/AgCl/ZIF-8/TiO2 decorated cotton fabric as a highly efficient photocatalyst for degradation of organic dyes under visible light. *Cellulose (London, England)*, *26*(12), 7437–7450. doi:10.1007/s10570-019-02621-8

Gudimalla, A., Jose, J., Varghese, R. J., & Thomas, S. (2021). Green synthesis of silver nanoparticles using Nymphae odorata extract incorporated films and antimicrobial activity. *Journal of Polymers and the Environment*, *29*(5), 1412–1423. doi:10.1007/s10924-020-01959-6

Gudkov, S. V., Burmistrov, D. E., Smirnova, V. V., Semenova, A. A. & Lisitsyn, A. B. (2022). A mini review of antibacterial properties of Al_2O_3 nanoparticles. *Nanomaterials (Bassel), 12*(15), 2635, 30. . doi:10.3390/nano12152635

Guillard, V., Gaucel, S., Fornaciari, C., Angellier-Coussy, H., Buche, P., & Gontard, N. (2018). The next generation of sustainable food packaging to preserve our environment in a circular economy context. *Frontiers in Nutrition*, *5*, 121. doi:10.3389/fnut.2018.00121 PMID:30564581

Gündüz, M. G., İşli, F., El-Khouly, A., Yıldırım, Ş., Fincan, G. S., & Şimşek, R. (2014). Microwave-assisted synthesis and myorelaxant activity of 9-indolyl-1, 8-acridinedione derivatives. *European Journal of Medicinal Chemistry*, *75*, 258–266. doi:10.1016/j.ejmech.2014.01.059 PMID:24534541

Guo, Z., Martucci, N. J., Liu, Y., Yoo, E., Tako, E. and Mahler, G. J. Silicon dioxide nanoparticles exposure affects small intestine function in an in vitro model. *Nanotoxicology, 12*(5), 485-508. . doi:10.1080/17435390.2018.1463407

Guo, L., Jackman, J. A., Yang, H. H., Chen, P., Cho, N. J., & Kim, D. H. (2015). Strategies for enhancing the sensitivity of plasmonic nanosensors. *Nano Today*, *10*(2), 213–239. doi:10.1016/j.nantod.2015.02.007

Guo, Q., Sun, H., Zhang, L., & Li, D. (2021). Cotton fabric-based rGO/BiVO4 recyclable photocatalytic nanocomposites for dye degradation under visible light. *Composites Communications*, *27*, 100846. doi:10.1016/j.coco.2021.100846

Guo, Q., Zhang, X., Zhao, F., Song, Q., Su, G., Tan, Y., Tao, Q., Zhou, T., Yu, Y., Zhou, Z., & Lu, C. (2020). Protein-Inspired Self-Healable Ti3C2MXenes/Rubber-Based Supramolecular Elastomer for Intelligent Sensing. *ACS Nano*, *14*(3), 2788–2797. doi:10.1021/acsnano.9b09802 PMID:32045216

Guo, S., & Wang, E. (2007). Synthesis and electrochemical applications of gold nanoparticles. *Analytica Chimica Acta*, *598*(2), 181–192. doi:10.1016/j.aca.2007.07.054 PMID:17719891

Gupta, V., Biswas, D., & Roy, S. (2022). A Comprehensive Review of Biodegradable Polymer-Based Films and Coatings and Their Food Packaging Applications. *Materials (Basel)*, *15*(17), 5899. doi:10.3390/ma15175899 PMID:36079280

Gurgur, E., Oluyamo, S. S., Adetuyi, A. O., Omotunde, O. I., & Okoronkwo, A. E. (2020). Green synthesis of zinc oxide nanoparticles and zinc oxide–silver, zinc oxide–copper nanocomposites using Bridelia ferruginea as biotemplate. *SN Applied Sciences*, *2*(5), 1–12. doi:10.1007/s42452-020-2269-3

Gutiérrez-Blanco, A., Peris, E., & Poyatos, M. (2018). Pyrene-connected tetraimidazolylidene complexes of iridium and rhodium. Structural features and catalytic applications. *Organometallics*, *37*(21), 4070–4076. doi:10.1021/acs.organomet.8b00633

Gutowski, V., Yang, W., Li, S., Dean, K., & Zhang, X. (2017). Lightweight Nanocomposite Materials, In book: Lightweight and Sustainable Materials for Automotive Applications, 277-302, Haleem, A., Javaid, M., Singh, R. P., Rab, S., & Suman, R. (2023). Applications of Nanotechnology in Medical field. *Global Health Journal (Amsterdam, Netherlands)*, *7*(2), 70–77.

Gwande, M. B., Goswami, A., Felpin, F., Asefa, T., Huang, X., Silva, R., Zou, X., Zboril, R., & Varma, R. S. (2016). Cu and Cu-based nanoparticles: Synthesis and applications in catalysis. *Chemical Reviews*, *116*(6), 3722–3811. doi:10.1021/acs.chemrev.5b00482 PMID:26935812

Haleem, A., Javaid, M., Singh, R. P., Suman, R., & Rab, S. (2020). Biosensors applications in the medical field: A brief review. *Sensors International*, *2*, 100100. doi:10.1016/j.sintl.2021.100100

Han, J. W., Ruiz-Garcia, L., Qian, J. P., & Yang, X. T. (2018). Food Packaging: A Comprehensive Review and Future Trends. *Comprehensive Reviews in Food Science and Food Safety*, *17*(4), 860–877. doi:10.1111/1541-4337.12343 PMID:33350114

Han, J., Liu, Y., Tian, Z., Yu, F., Sun, X., & Wang, W. (2021). Microtubular carbonized cotton fiber modified g-C3N4 for the enhancement of visible-light-driven photocatalytic activity. *Materials Today. Communications*, *29*, 102926. doi:10.1016/j.mtcomm.2021.102926

Haripriya, S., & Ajitha, P. (2017). Antimicrobial efficacy of silver nanoparticles of Aloe vera. *Journal of Advanced Pharmacy Education and Research*, *7*(2), 2249–3379.

Hasan, K. M. F., Xiaoyi, L., Shaoqin, Z., Horváth, P. G., Bak, M., Bejó, L., Sipos, G., & Alpár, T. (2022). Functional silver nanoparticles synthesis from sustainable point of view: 2000 to 2023 – A review on game changing materials. *Heliyon*, *8*(12), e12322. doi:10.1016/j.heliyon.2022.e12322 PMID:36590481

Hasan, M. R., Ahommed, M. S., Daizy, M., Bacchu, M. S., Ali, M. R., Al-Mamun, M. R., Saad Aly, M. A., Khan, M. Z. H., & Hossain, S. I. (2021). Recent development in electrochemical biosensors for cancer biomarkers detection. *Biosensors & Bioelectronics: X*, *8*, 100075. doi:10.1016/j.biosx.2021.100075

Hasheminya, S. M., Mokarram, R. R., Ghanbarzadeh, B., Hamishekar, H., & Kafil, H. S. (2018). Physicochemical, mechanical, optical, microstructural and antimicrobial properties ofnovel kefiran-carboxymethyl cellulose biocomposite films as influenced by copper oxide nanoparticles (CuONPs). *Food Packaging and Shelf Life*, *17*, 196–204. doi:10.1016/j.fpsl.2018.07.003

Hassanpour, P., Panahi, Y., Ebrahimi-Kalan, A., Akbarzadeh, A., Davaran, S., Nasibova, A. N., Khallov, R., & Kavetskyy, T. (2018). Biomedical applications of aluminium oxide nanoparticles. *Micro & Nano Letters*, *13*(9), 1227–1231. doi:10.1049/mnl.2018.5070

Hassan, S. S., Al-Ghouti, M. A., Abu-Dieyeh, M., & McKay, G. (2020). Novel bioadsorbents based on date pits for organophosphorus pesticide remediation from water. *Journal of Environmental Chemical Engineering*, *8*(1), 103593. doi:10.1016/j.jece.2019.103593

Haug, H., & Koch, S. W. (2009). *Quantum Theory of the Optical and Electronic Properties of Semiconductors*. WORLD SCIENTIFIC., doi:10.1142/7184

Heine, E., Knops, H. G., Schaefer, K., Vangeyte, P., & Moeller, M. (2007). Antimicrobial functionalisation of textile materials. *Multifunctional Barriers for Flexible Structure: Textile, Leather and Paper,* 23-38.

Heinze, T. (2015). Cellulose: Structure and properties. *Advances in Polymer Science*, *271*, 1–52. doi:10.1007/12_2015_319

Heo, D. N., Ko, W. K., Lee, H. R., Lee, S. J., Lee, D., Um, S. H., & Kwon, I. K. (2016). Titanium dental implants surface-immobilized with gold nanoparticles as osteoinductive agents for rapid osseointegration. *Journal of Colloid and Interface Science*, *469*, 129–137. doi:10.1016/j.jcis.2016.02.022 PMID:26874978

Hergt, R., Dutz, S., Müller, R., & Zeisberger, M. (2006). Magnetic particle hyperthermia: Nanoparticle magnetism and materials development for cancer therapy. *Journal of Physics Condensed Matter*, *18*(38), S2919–S2934. doi:10.1088/0953-8984/18/38/S26

Hermas, A. E. A., Salam, M. A., & Al-Juaid, S. S. (2013). In situ electrochemical preparation of multi-walled carbon nanotubes/polyaniline composite on the stainless steel. *Progress in Organic Coatings*, *76*(12), 1810–1813. doi:10.1016/j.porgcoat.2013.05.021

Herrmann, J. (2017). Photocatalysis. In *Kirk-Othmer Encyclopedia of Chemical Technology* (pp. 1–44). Wiley. doi:10.1002/0471238961.1608152019051816.a01.pub3

He, S., Lu, C., & Zhang, S. (2011). Facile and efficient route to polyimide-TiO2 nanocomposite coating onto carbon fiber. *ACS Applied Materials & Interfaces*, *3*(12), 4744–4750. doi:10.1021/am201239h PMID:22081943

Hilger, I., Hergt, R., & Kaiser, W. A. (2005, February). Use of magnetic nanoparticle heating in the treatment of breast cancer. [). IET Digital Library.]. *IEE Proceedings. Nanobiotechnology*, *152*(1), 33–39. doi:10.1049/ip-nbt:20055018 PMID:16441156

Hiremath, N., & Bhat, G. (2017). High-performance carbon nanofibers and nanotubes. In *Structure and properties of high-performance fibers* (pp. 79–109). Woodhead Publishing. doi:10.1016/B978-0-08-100550-7.00004-8

Hnana, K., Barhoun, A., & Draoui, K. (2019). Improvement of glyphosate adsorption using new composites based on ghassoul and chitosan: Kinetics and equilibrium study. *Mediterranean Journal of Chemistry*, *9*(4), 266–279. doi:10.13171/mjc9419111121087kd

Holmannova, D., Borsky, P., Svadlakova, T., Borska, L., & Fiala, Z. (2022). Carbon Nanoparticles and Their Biomedical Applications. *Applied Sciences (Basel, Switzerland)*, *12*(15), 7865. doi:10.3390/app12157865

Honarvar, Z., Hadian, Z., & Mashayekh, M. (2016). Nanocomposites in food packaging applications and their risk assessment for health. *Electronic Physician*, *8*(6), 2531–2538. doi:10.19082/2531 PMID:27504168

Hon, D. (1994). Cellulose: A random walk along its historical path. *Cellulose (London, England)*, *1*(1), 1–25. doi:10.1007/BF00818796

Hong, L. G., Yuhana, N. Y., & Zawawi, E. Z. E. (2021). Review of bioplastics as food packaging materials. *AIMS Materials Science*, *8*(2), 166–184. doi:10.3934/matersci.2021012

Hosne Asif, A. K. M. A., & Hasan, M. Z. (2018). Application of Nanotechnology in Modern Textiles: A Review. *International Journal of Current Engineering and Technology*, *8*(02). doi:10.14741/ijcet/v.8.2.5

Hossain, N., Islam, M. A., Chowdhury, M. A., & Alam, A. (2022). Advances of nanoparticles employment in dental implant applications. *Applied Surface Science Advances*, *12*, 100341. doi:10.1016/j.apsadv.2022.100341

Hosseini-Hosseinabad, S. M., Minaeian, S., Tavakoli, A., Sabaei, M., Yousefi Zoshk, M., Laripour, R., Ramezani, S., Hoseini, M., & Chamanara, M. (2022). Development of g-C3N4/ZnO nanocomposite as a novel, highly effective and durable photocatalytic antibacterial coating for cotton fabric. *Ceramics International*. doi:10.1016/j.ceramint.2022.12.080

Hou, W., & Wang, Q. (2009). UV-driven reversible switching of a polystyrene/titania nanocomposite coating between superhydrophobicity and superhydrophilicity. *Langmuir*, *25*(12), 6875–6879. doi:10.1021/la900151y PMID:19388630

Huang, J., Yang, B., Peng, Y., Huang, J., Wong, S. H. D., Bian, L., Zhu, K., Shuai, X., & Han, S. (2021). Nanomedicine-Boosting Tumor Immunogenicity for Enhanced Immunotherapy. In Advanced Functional Materials, 31(21). John Wiley and Sons Inc. doi:10.1002/adfm.202011171

Huang, X., Zhu, Y., & Kianfar, E. (2021). Nano Biosensors: Properties, applications and electrochemical techniques. In Journal of Materials Research and Technology (Vol. 12, pp. 1649–1672). Elsevier Editora Ltda. doi:10.1016/j.jmrt.2021.03.048

Huang, J., Li, Q., Sun, D., Lu, Y., Su, Y., Yang, X., Wang, H., Wang, Y., Shao, W., He, N., Hong, J., & Chen, C. (2007). Biosynthesis of silver and gold nanoparticles by novel sundried Cinnamomum camphora leaf. *Nanotechnology*, *18*(10), 105104. doi:10.1088/0957-4484/18/10/105104

Huang, Y., Cambre, M., & Lee, H. (2017). The Toxicity of Nanoparticles Depends on Multiple Molecular and Physicochemical Mechanisms. *International Journal of Molecular Sciences*, *18*(12), 2702. doi:10.3390/ijms18122702 PMID:29236059

Huang, Z., Raghuwanshi, V. S., & Garnier, G. (2017). Functionality of immunoglobulin G and immunoglobulin M antibody physisorbed on cellulosic films. *Frontiers in Bioengineering and Biotechnology*, *5*(AUG), 41. doi:10.3389/fbioe.2017.00041 PMID:28770196

Hua, Z., Yu, T., Liu, D., & Xianyu, Y. (2021). Recent advances in gold nanoparticle-based biosensors for food safety detection. *Biosensors & Bioelectronics*, *179*, 113076. Advance online publication. doi:10.1016/j.bios.2021.113076 PMID:33601132

Hu, J., Zhang, G., & Liu, S. (2012). Enzyme-responsive polymeric assemblies, nanoparticles and hydrogels. *Chemical Society Reviews*, *41*(18), 5933–5949. doi:10.1039/c2cs35103j PMID:22695880

Hulkoti, N. I., & Taranath, T. C. (2014). Biosynthesis of nanoparticles using microbes—A review. *Colloids and Surfaces. B, Biointerfaces*, *121*, 474–483. doi:10.1016/j.colsurfb.2014.05.027 PMID:25001188

Huo, K., Zhang, X., Wang, H., Zhao, L., Liu, X., & Chu, P. K. (2013). Osteogenic activity and antibacterial effects on titanium surfaces modified with Zn-incorporated nanotube arrays. *Biomaterials*, *34*(13), 3467–3478. doi:10.1016/j.biomaterials.2013.01.071 PMID:23439134

Hu, Q., Katti, P. S., & Gu, Z. (2014). Enzyme-responsive nanomaterials for controlled drug delivery. *Nanoscale*, *6*(21), 12273–12286. doi:10.1039/C4NR04249B PMID:25251024

Hussain, A., Attique, F., Naqvi, S. A. R., Ali, A., Ibrahim, M., Hussain, H., Zafar, F., Iqbal, R. S., Ayub, M. A., Assiri, M. A., Imran, M., & Ullah, S. (2022). Nanoformulation of Curcuma longa Root Extract and Evaluation of Its Dissolution Potential. *ACS Omega*. doi:10.1021/acsomega.2c06258 PMID:36643543

Hu, Y., Zhao, Z., & Wan, Q. (2011). Facile preparation of carbon nanotube-conducting polymer network for sensitive electrochemical immunoassay of Hepatitis B surface antigen in serum. *Bioelectrochemistry (Amsterdam, Netherlands)*, *81*(2), 59–64. doi:10.1016/j.bioelechem.2011.01.005 PMID:21458390

Iavicoli, I., Fontana, L., & Nordberg, G. (2016). The effects of nanoparticles on the renal system. *Critical Reviews in Toxicology*, *46*(6), 490–560. doi:10.1080/10408444.2016.1181047 PMID:27195425

Ibănescu, M., Muşat, V., Textor, T., Badilita, V., & Mahltig, B. (2014). Photocatalytic and antimicrobial Ag/ZnO nanocomposites for functionalization of textile fabrics. *Journal of Alloys and Compounds*, *610*, 244–249. doi:10.1016/j.jallcom.2014.04.138

Iderawumi, A. M., & Yusuff, M. A. (2021). Effects of nanoparticles on improvement in quality and shelf life of fruits and vegetables. *J Plant Biol Crop Res*, *4*, 2–7.

Iftekhar, S., Ramasamy, D. L., Srivastava, V., Asif, M. B., & Sillanpää, M. (2018). Understanding the factors affecting the adsorption of Lanthanum using different adsorbents: A critical review. *Chemosphere*, *204*, 413–430. doi:10.1016/j.chemosphere.2018.04.053 PMID:29677649

Ikeda, M., Ochi, R., & Hamachi, I. (2010). Supramolecular hydrogel-based protein and chemosensor array. *Lab on a Chip*, *10*(24), 3325–3334. doi:10.1039/c004908e PMID:20862441

Ildiz, N., Baldemir, A., Altinkaynak, C., Özdemir, N., Yilmaz, V., & Ocsoy, I. (2017). Self assembled snowball-like hybrid nanostructures comprising Viburnum opulus L. extract and metal ions for antimicrobial and catalytic applications. *Enzyme and Microbial Technology*, *102*, 60–66. doi:10.1016/j.enzmictec.2017.04.003 PMID:28465062

Ioelovich, M. (2016). Nanocellulose—fabrication, structure, properties, and application in the area of care and cure. In A. Grumezescu (Ed.), *Fabrication and Self-Assembly of Nanobiomaterials* (pp. 243–288). William Andrew Publishing. doi:10.1016/B978-0-323-41533-0.00009-X

Iravani, M., Khalilnezhad, Z., & Khalilnezhad, A. (2023). A review on application of nanoparticles for EOR purposes: History and current challenges. *Journal of Petroleum Exploration and Production Technology*, *13*(4), 959–994. doi:10.1007/s13202-022-01606-x PMID:36644438

Irshad, M. A., Nawaz, R., Rehman, M. Z., Adrees, M., Rizwan, M., Ali, S., Ahmad, S., & Tasleem, S. (2021). Synthesis, characterization and advanced sustainable applications of titanium dioxide nanoparticles: A review. *Ecotoxicology and Environmental Safety*, *212*, 111978. doi:10.1016/j.ecoenv.2021.111978 PMID:33561774

Islam, M. N., & Rahman, F. (2018). Production and modification of nanofibrillated cellulose composites and potential applications. In G. Koronis & A. Silva (Eds.), *Green Composites for Automotive Applications* (pp. 115–141). Elsevier Ltd. doi:10.1016/B978-0-08-102177-4.00006-9

Islam, N. U., Amin, R., Shahid, M., Amin, M., Zaib, S., & Iqbal, J. (2017). A multi-target therapeutic potential of Prunus domestica gum stabilized nanoparticles exhibited prospective anticancer, antibacterial, urease-inhibition, anti-inflammatory and analgesic properties. *BMC Complementary and Alternative Medicine*, *17*(1), 1–17. doi:10.1186/s12906-017-1791-3 PMID:28535789

Iyer, K. S., & Chavan, S. S. (2017). Nanotechnology and Its Application in Cosmetic Industry A Concise Review. *International Journal of Innovative Science and Research Technology*, *2*(9), 263–267.

Jabeen, F. (2018). Chapter 34 - Polymer–ceramic nanocomposites for controlled drug delivery, Editor(s): Inamuddin, Abdullah M. Asiri, Ali Mohammad, In Woodhead Publishing Series in Biomaterials, Applications of Nanocomposite Materials in Drug Delivery. Woodhead Publishing. doi:10.1016/B978-0-12-813741-3.00035-2

Jackson, T. C., Patani, B. O., & Ekpa, D. E. (2017). Nanotechnology in Diagnosis: A Review. *Advances in Nanoparticles*, *06*(03), 93–102. doi:10.4236/anp.2017.63008

Jadhav, K., Deore, S., Dhamecha, D., Hr, R., Jagwani, S., Jalalpure, S., & Bohara, R. (2018). Phytosynthesis of silver nanoparticles: Characterization, biocompatibility studies, and anticancer activity. *ACS Biomaterials Science & Engineering*, *4*(3), 892–899. doi:10.1021/acsbiomaterials.7b00707 PMID:33418773

Jadhav, K., Rajeshwari, H. R., Deshpande, S., Jagwani, S., Dhamecha, D., Jalalpure, S., & Baheti, D. (2018). Phytosynthesis of gold nanoparticles: Characterization, biocompatibility, and evaluation of its osteoinductive potential for application in implant dentistry. *Materials Science and Engineering C*, *93*, 664–670. doi:10.1016/j.msec.2018.08.028 PMID:30274099

Jain, P., & Pradeep, T. (2005). Potential of silver nanoparticle-coated polyurethane foam as an antibacterial water filter. *Biotechnology and Bioengineering*, *90*(1), 59–63. doi:10.1002/bit.20368 PMID:15723325

Jalilinejad, N., Rabiee, M., Baheiraei, N., Ghahremanzadeh, R., Salarian, R., Rabiee, N., Akhavan, O., Zarrintaj, P., Hejna, A., Saeb, M. R., Zarrabi, A., Sharifi, E., Yousefiasl, S., & Zare, E. N. (2023). Electrically conductive carbon-based (bio)-nanomaterials for cardiac tissue engineering. *Bioengineering & Translational Medicine*, *8*(1), e10347. doi:10.1002/btm2.10347 PMID:36684103

Jamaluddin, N. A., Talib, N., & Sani, A. S. A. (2021). Performance comparative of modified jatropha based nanofluids in orthogonal cutting process. *Evergreen*, *8*(2), 461–468. doi:10.5109/4480729

Janaki, A. C., Sailatha, E., & Gunasekaran, S. (2015). Synthesis, characteristics and antimicrobial activity of ZnO nanoparticles. *Spectrochimica Acta. Part A: Molecular and Biomolecular Spectroscopy*, *144*, 17–22. doi:10.1016/j.saa.2015.02.041 PMID:25748589

Jancikova, S., Jamróz, E., Kulawik, P., Tkaczewska, J., & Dordevic, D. (2019). Furcellaran/gelatin hydrolysate/rosemary extract composite films as active and intelligent packaging materials. *International Journal of Biological Macromolecules*, *131*, 19–28. doi:10.1016/j.ijbiomac.2019.03.050 PMID:30853584

Jandt, K. D., & Watts, D. C. (2020). Nanotechnology in dentistry: Present and future perspectives on dental nanomaterials. *Dental Materials*, *36*(11), 1365–1378. doi:10.1016/j.dental.2020.08.006 PMID:32981749

Jaswal, T. and Gupta, J. (2023). A review on the toxicity of silver nanoparticles on human health,Materials Today: *Proceedings, 81*(2), 859-863. . doi:10.1016/j.matpr.2021.04.266

Javaid, M., Haleem, A., Singh, R. P., Rab, S., & Suman, R. (2021). Exploring the potential of nanosensors: A brief overview. *Sensors International*, *2*(October), 100130. doi:10.1016/j.sintl.2021.100130

Jayakaran, P., Nirmala, G. S., & Govindarajan, L. (2019). Qualitative and Quantitative Analysis of Graphene-Based Adsorbents in Wastewater Treatment. *International Journal of Chemical Engineering*, *2019*, 1–17. doi:10.1155/2019/9872502

Jedrzejczak-Krzepkowska, M., Kubiak, K., Ludwicka, K., & Bielecki, S. (2016). Bacterial NanoCellulose Synthesis, Recent Findings. In M. Gama, F. Dourado, & S. Bielecki (Eds.), *Bacterial Nanocellulose: From Biotechnology to Bio-Economy* (pp. 19–46). Elsevier B.V. doi:10.1016/B978-0-444-63458-0.00002-0

Jeevanandam, J., Ling, J. K. U., Barhoum, A., San Chan, Y., & Danquah, M. K. (2022). Bionanomaterials: Definitions, sources, types, properties, toxicity, and regulations. In Fundamentals of Bionanomaterials (pp. 1-29). Elsevier.

Jesús De La, J., Bayo, T., Ángeles Martín-Lara, M., Calero Hoces, M., Miguel, P., Castillo, S., Pula, H. J., & Zamorano, M. (2023). *Management of Used COVID-19 Personal Protective Equipment: A Bibliometric Analysis and Literature Review*. doi:10.3390/app13042377

Jiang, Y., & Wu, J. (2019). Recent development in chitosan nanocomposites for surface-based biosensor applications. In ELECTROPHORESIS (Vol. 40, Issues 16–17, pp. 2084–2097). Wiley. https://doi.org/ doi:10.1002/elps.201900066

Jiang, W., Huang, L., Zhang, D., Wang, Y., & Pan, G. (2021). Graphene-Based Composites for Electrochemical Sensor Fabrication and Their Application for Drug Detection. *International Journal of Electrochemical Science*, *16*(3), 1–13. doi:10.20964/2021.03.69

Jiang, W., Wang, Y., Wargo, J. A., Lang, F. F., & Kim, B. Y. S. (2021). Considerations for designing preclinical cancer immune nanomedicine studies. *Nature Nanotechnology*, *16*(1), 6–15. doi:10.1038/s41565-020-00817-9 PMID:33349682

Jia, Y., Jiang, Y., He, Y., Zhang, W., Zou, J., Magar, K. T., Boucetta, H., Teng, C., & He, W. (2023). Approved Nanomedicine against Diseases. In Pharmaceutics, 15(3). MDPI. doi:10.3390/pharmaceutics15030774

Joghee, S., Ganeshan, P., Vincent, A., & Hong, S. I.Leaf Extract and Their Antimicrobial Activity. (2019). Ecofriendly biosynthesis of zinc oxide and magnesium oxide particles from medicinal plant Pisonia grandis R. Br. leaf extract and their antimicrobial activity. *BioNanoScience*, *9*(1), 141–154. doi:10.1007/s12668-018-0573-9

Jokar, M., Safaralizadeh, M. H., Hadizadeh, F., Rahmani, F., & Kalani, M. R. (2016). Design and evaluation of an APTA-nanosensor to detect Acetamiprid in vitro and silico. *Journal of Biomolecular Structure & Dynamics*, *34*(11), 2505–2517. doi:10.1080/07391102.2015.1123188 PMID:26609886

Joseph, T. M., Kar Mahapatra, D., Esmaeili, A., Piszczyk, Ł., Hasanin, M. S., Kattali, M., Haponiuk, J., & Thomas, S. (2023). Nanoparticles: Taking a unique position in medicine. *Nanomaterials (Basel, Switzerland)*, *13*(3), 574. doi:10.3390/nano13030574 PMID:36770535

Juanola-Feliu, E., Miribel-Català, P., Avilés, C., Colomer-Farrarons, J., González-Piñero, M., & Samitier, J. (2014). Design of a Customized Multipurpose Nano-Enabled Implantable System for In-Vivo Theranostics. *Sensors (Basel)*, *14*(10), 19275–19306. doi:10.3390/s141019275 PMID:25325336

Jung, Y., Min, J., Choi, J., Bang, J., Jeong, S., Pyun, K. R., Ahn, J., Cho, Y., Hong, S., Hong, S., Lee, J., & Ko, S. H. (2022). Smart paper electronics by laser-induced graphene for biodegradable real-time food spoilage monitoring. *Applied Materials Today*, *29*, 101589. doi:10.1016/j.apmt.2022.101589

Justino, C., Duarte, A., & Rocha-Santos, T. (2017). Recent Progress in Biosensors for Environmental Monitoring: A Review. *Sensors (Basel), 17*(12), 2918. doi:10.3390/s17122918 PMID:29244756

Kai, K. (2008). *Nanotechnology and Medical Robotics*. Legal and Ethical Responsibility.

Kalaivani, M., & Ravi, S. (2023). Green Synthesis of ZnO NPs and CdO-ZnO Nanocomposites using aqueous Extract of Water Hyacinth (Eichhorniacrassipes) Characterization, Structural and Nano-fertilizer using Application. *Indian Journal of Science and Technology, 16*(25), 1918–1926. doi:10.17485/IJST/v16i25.180

Kalakonda, S. N., Bammidi, R., Edubilli, H., Medapati, S., Boyina, S. L. D., & Prasad, V. V. S. R. (2023). Biosensors – An Insight into the Electrochemical and Optical Biosensors. *International Journal of Pharmaceutical Investigation, 13*(3), 402–412. doi:10.5530/ijpi.13.3.051

Kalantari, K., Ahmad, M. B., Fard Masoumi, H. R., Shameli, K., Basri, M., & Khandanlou, R. (2015). Rapid and high capacity adsorption of heavy metals by Fe3O4/montmorillonite nanocomposite using response surface methodology: Preparation, characterization, optimization, equilibrium isotherms, and adsorption kinetics study. *Journal of the Taiwan Institute of Chemical Engineers, 49*, 192–198. doi:10.1016/j.jtice.2014.10.025

Kandeepan, G. (2021). Biodegradable nanocomposite packaging films for meat and meat products: A review. *Journal of Packaging Technology and Research, 5*(3), 143–166. doi:10.1007/s41783-021-00123-2

Kannan, P., & Maduraiveeran, G. (2023). Metal Oxides Nanomaterials and Nanocomposite-Based Electrochemical Sensors for Healthcare Applications. In Biosensors (Vol. 13, Issue 5). MDPI. doi:10.3390/bios13050542

Kannan, P., & Maduraiveeran, G. (2021, December 31). Bimetallic Nanomaterials-Based Electrochemical Biosensor Platforms for Clinical Applications. *Micromachines, 13*(1), 76. doi:10.3390/mi13010076 PMID:35056240

Kapoor, V., Phan, D., & Pasha, A. B. M. T. (2018). Effects of metal oxide nanoparticles on nitrification in wastewater treatment systems: A systematic review. *Journal of Environmental Science and Health. Part A, Toxic/Hazardous Substances & Environmental Engineering, 53*(7), 659–668. doi:10.1080/10934529.2018.1438825 PMID:29469639

Karaagac, Z., Dadi, S., Ekrikaya, S., & Ocsoy, I. (2021). Plant Extract Guided Nanomaterials and Plant Disease Management. In M. A. Mallick, M. K. Solanki, B. Kumari, & S. K. Verma (Eds.), *In Nanotechnology in Sustainable Agriculture* (pp. 271–290). CRC Press.

Kargarzadeh, H., Huang, J., Lin, N., Ahmad, I., Mariano, M., Dufresne, A., Thomas, S., & Gałęski, A. (2018). Recent developments in nanocellulose-based biodegradable polymers, thermoplastic polymers, and porous nanocomposites. *Progress in Polymer Science, 87*, 197–227. doi:10.1016/j.progpolymsci.2018.07.008

Kargarzadeh, H., Mariano, M., Huang, J., Lin, N., Ahmad, I., Dufresne, A., & Thomas, S. (2017). Recent developments on nanocellulose reinforced polymer nanocomposites: A review. *Polymer*, *132*, 368–393. doi:10.1016/j.polymer.2017.09.043

Karimian, A., Yousefi, B., Sadeghi, F., Feizi, F., Najafzadehvarzi, H., & Parsian, H. (2022). Synthesis of biocompatible nanocrystalline cellulose against folate receptors as a novel carrier for targeted delivery of doxorubicin. *Chemico-Biological Interactions*, *351*, 109731. doi:10.1016/j.cbi.2021.109731 PMID:34728188

Karimi, L., Yazdanshenas, M. E., Khajavi, R., Rashidi, A., & Mirjalili, M. (2016). Functional finishing of cotton fabrics using graphene oxide nanosheets decorated with titanium dioxide nanoparticles. *Journal of the Textile Institute*, *107*(9), 1122–1134. doi:10.1080/00405000.2015.1093311

Karnani, R. L., & Chowdhary, A. (2013). Biosynthesis of silver nanoparticle by eco-friendly method. *Indian journal of Nanoscience*, 1(1), 25-31.

Karthikeyan, P., Nikitha, M., Pandi, K., Meenakshi, S., & Park, C. M. (2021). Effective and selective removal of organic pollutants from aqueous solutions using 1D hydroxyapatite-decorated 2D reduced graphene oxide nanocomposite. *Journal of Molecular Liquids*, *331*, 115795. doi:10.1016/j.molliq.2021.115795

Kasry, A., Ardakani, A. A., Tulevski, G. S., Menges, B., Copel, M., & Vyklicky, L. (2012). Highly efficient fluorescence quenching with graphene. In Journal of Physical Chemistry C, 116(4), 2858–2862. doi:10.1021/jp207972f

Kataria, N., & Garg, V. K. (2018). Green synthesis of Fe3O4 nanoparticles loaded sawdust carbon for cadmium (II) removal from water: Regeneration and mechanism. *Chemosphere*, *208*, 818–828. doi:10.1016/j.chemosphere.2018.06.022 PMID:29906756

Katz, L. M., Dewan, K., & Bronaugh, R. L. (2015). Nanotechnology in cosmetics. *Food and Chemical Toxicology*, *85*, 127–137. doi:10.1016/j.fct.2015.06.020 PMID:26159063

Kausar, A., Ahmad, I., Zhao, T., Aldaghri, O., Ibnaouf, K. H., & Eisa, M. H. (2023). Multifunctional Polymeric Nanocomposites for Sensing Applications—Design, Features, and Technical Advancements. In Crystals (Vol. 13, Issue 7). Multidisciplinary Digital Publishing Institute (MDPI). doi:10.3390/cryst13071144

Kausar, A. (2022). Potential of Polymer/Fullerene Nanocomposites for Anticorrosion Applications in the Biomedical Field. *Journal of Composites Science*, *6*(12), 394. doi:10.3390/jcs6120394

Kawadkar, J., Chauhan, M. K., & Maharana, M. (2011). Nanobiotechnology: Application of nanotechnology in diagnosis, drug discovery and drug development. *Asian Journal of Pharmaceutical and Clinical Research*, *4*(1), 23–25.

Kayal, S. (2021). Thermoresponsive magnetic/polymer composite nanoparticles for biomedical applications. *Materials Today: Proceedings*, *41*, 1116–1119. doi:10.1016/j.matpr.2020.08.605

Kazemi, F., Naghib, S. M., Zare, Y., & Rhee, K. Y. (2021). Biosensing Applications of Polyaniline (PANI)-Based Nanocomposites: A Review. In Polymer Reviews (Vol. 61, Issue 3, pp. 553–597). Taylor and Francis Ltd. doi:10.1080/15583724.2020.1858871

Kelle, H. I., Ogoko, E. C., Akintola O & Eddy, N. O. (2023). Quantum and experimental studies on the adsorption efficiency of oyster shell–based CaO nanoparticles (CaONPO) towards the removal of methylene blue dye (MBD) from aqueous solution. *Journal: Biomass Conversion and Biorefinery*. . doi:10.1007/s13399-023-04947-7

Keller, K. H. (2007). Nanotechnology and society. *Journal of Nanoparticle Research*, *9*(1), 5–10. doi:10.1007/s11051-006-9193-3

Keskinbora, K. H., & Jameel, M. A. (2018). Nanotechnology applications and approaches in medicine: A review. *J NanosciNanotechnol Res*, *2*(2), 6.

Ketelsen, M., Janssen, M., & Hamm, U. (2020). Consumers' response to environmentally-friendly food packaging - A systematic review. *Journal of Cleaner Production*, *254*, 120123. doi:10.1016/j.jclepro.2020.120123

Khalid, M. Y., Kamal, A., Otabil, A., Mamoun, O., & Liao, K. (2023). Graphene/epoxy nanocomposites for improved fracture toughness: A focused review on toughening mechanism. *Chemical Engineering Journal Advances*, *16*, 100537. doi:10.1016/j.ceja.2023.100537

Khan, A., Khan, R. A., Salmieri, S., Le Tien, C., Riedl, B., Bouchard, J., Chauve, G., Tan, V., Kamal, M. R., & Lacroix, M. (2012). Mechanical and barrier properties of nanocrystalline cellulose reinforced chitosan based nanocomposite films. *Carbohydrate Polymers*, *90*(4), 1601–1608. doi:10.1016/j.carbpol.2012.07.037 PMID:22944422

Khandelwal, A., Joshi, R., & Shrivastava, M. (2022). Application of Nanomaterials in Agriculture and their Impact on Environment. *International Journal of Agriculture Environment and Biotechnology*, *15*(2), 195–204.

Khan, I., Saeed, K., & Khan, I. (2019). Nanoparticles: Properties, applications and toxicities. *Arabian Journal of Chemistry*, *12*(7), 908–931. doi:10.1016/j.arabjc.2017.05.011

Khan, M., Cao, W., Li, J., Zaman, M. I., & Manan, A. (2014). Density functional theory calculations for the investigation of (Ag, N) codoping effect on the electronic and optical properties of anatase TiO 2. *International Journal of Modern Physics B*, *28*(18), 1450112. doi:10.1142/S0217979214501124

Khan, Y., Sadia, H., Ali Shah, S. Z., Khan, M. N., Shah, A. A., Ullah, N., Ullah, M. F., Bibi, H., Bafakeeh, O. T., Khedher, N. B., Eldin, S. M., Fadhl, B. M., & Khan, M. I. (2022). Classification, Synthetic, and Characterization Approaches to Nanoparticles, and Their Applications in Various Fields of Nanotechnology: A Review. *Catalysts*, *12*(11), 1386. doi:10.3390/catal12111386

Khatoon, A., Khan, F., Ahmad, N., Shaikh, S., Rizvi, S. M. D., Shakil, S., Al-Qahtani, M. H., Abuzenadah, A. M., Tabrez, S., Ahmed, A. B. F., Alafnan, A., Islam, H., Iqbal, D., & Dutta, R. (2018). Silver nanoparticles from leaf extract of Mentha piperita: Eco-friendly synthesis andeffect on acetylcholinesterase activity. *Life Sciences*, *209*, 430–434. doi:10.1016/j.lfs.2018.08.046 PMID:30138593

Khatoon, N., Alam, H., Manzoor, N., & Sardar, M. (2018). Removal of toxic contaminants from water by sustainable green synthesised non-toxic silver nanoparticles. *IET Nanobiotechnology / IET*, *12*(8), 1090–1096. doi:10.1049/iet-nbt.2018.5075 PMID:30964019

Khurshid, Z., Zafar, M., Qasim, S., Shahab, S., Naseem, M., & AbuReqaiba, A. (2015). Advances in nanotechnology for restorative dentistry. *Materials (Basel)*, *8*(2), 717–731. doi:10.3390/ma8020717 PMID:28787967

Kiessling, F., Liu, Z., & Gätjens, J. (2010). Advanced nanomaterials in multimodal imaging: Design, functionalization, and biomedical applications. *Journal of Nanomaterials*, *2010*, 1–15. doi:10.1155/2010/894303

Kim, J.-H., Shim, B. S., Kim, H. S., Lee, Y.-J., Min, S.-K., Jang, D., Abas, Z., & Kim, J. (2015). Review of nanocellulose for sustainable future materials. *International Journal of Precision Engineering and Manufacturing-Green Technology*, *2*(2), 197–213. doi:10.1007/s40684-015-0024-9

Kim, M.-Y., & Lee, K. H. (2022). Electrochemical sensors for sustainable precision agriculture—A review. *Frontiers in Chemistry*, *10*, 848320. doi:10.3389/fchem.2022.848320 PMID:35615311

Kinnear, C., Moore, T. L., Rodriguez-Lorenzo, L., Rothen-Rutishauser, B., & Petri-Fink, A. (2017). Form Follows Function: Nanoparticle Shape and Its Implications for Nanomedicine. In Chemical Reviews, 117(17), 11476–11521. American Chemical Society. doi:10.1021/acs.chemrev.7b00194

Kirk, K. A., Othman, A., & Andreescu, S. (2018). Nanomaterial-functionalized Cellulose: Design, Characterization and Analytical Applications. *Analytical Sciences*, *34*(1), 19–31. doi:10.2116/analsci.34.19 PMID:29321453

Kirson, E. D., & Yaari, Y. (2000). A novel technique for micro-dissection of neuronal processes. *Journal of Neuroscience Methods*, *98*(2), 119–122. doi:10.1016/S0165-0270(00)00194-1 PMID:10880825

Kneipp, J., Kneipp, H., & Kneipp, K. (2008). SERS—A single-molecule and nanoscale tool for bioanalytics. *Chemical Society Reviews*, *37*(5), 1052–1060. doi:10.1039/b708459p PMID:18443689

Koca, F. D., Yilmaz, D. D., Onmaz, N. E., & Ocsoy, I. (2020). Peroxidase-like activity and antimicrobial properties of curcumin-inorganic hybrid nanostructure. *Saudi Journal of Biological Sciences*, *27*(10), 2574–2579. doi:10.1016/j.sjbs.2020.05.025 PMID:32994713

Kohl, Y., Rundén-Pran, E., Mariussen, E., Hesler, M., El Yamani, N., Longhin, E. M., & Dusinska, M. (2020). Genotoxicity of Nanomaterials: Advanced In Vitro Models and High Throughput Methods for Human Hazard Assessment—A Review. *Nanomaterials (Basel, Switzerland)*, *10*(10), 1911. doi:10.3390/nano10101911 PMID:32992722

Komeily-Nia, Z., Montazer, M., Heidarian, P., & Nasri-Nasrabadi, B. (2019). Smart photoactive soft materials for environmental cleaning and energy production through incorporation of nanophotocatalyst on polymers and textiles. *Polymers for Advanced Technologies*, *30*(2), 235–253. doi:10.1002/pat.4480

Kontogiannis, O., Selianitis, D., Lagopati, N., Pippa, N., Pispas, S., & Gazouli, M. (2023). Surfactant and Block Copolymer Nanostructures: From Design and Development to Nanomedicine Preclinical Studies. In Pharmaceutics (Vol. 15, Issue 2). MDPI. doi:10.3390/pharmaceutics15020501

Kora, A. J., & Rastogi, L. (2018). Green synthesis of palladium nanoparticles using gum ghatti (Anogeissus latifolia) and its application as an antioxidant and catalyst. *Arabian Journal of Chemistry*, *11*(7), 1097–1106. doi:10.1016/j.arabjc.2015.06.024

Kori, A. H., Jagirani, M. S., & Soylak, M. (2023). Graphene-Based Nanomaterials: A Sustainable Material for Solid-Phase Microextraction (SPME) for Environmental Applications. *Analytical Letters*, *56*(15), 1–16. doi:10.1080/00032719.2023.2173221

Korzeniewska, E., Szczęsny, A., & Kasprowicz, D. (2020). IDEs structures created in the physical vacuum deposition process on textile substrates. *Journal of Physics: Conference Series*, *1534*(1), 012004. doi:10.1088/1742-6596/1534/1/012004

Kotsiri, Z., Vidic, J., & Vantarakis, A. (2022). Applications of biosensors for bacteria and virus detection in food and water–A systematic review. *Journal of Environmental Sciences (China)*, *111*, 367–379. doi:10.1016/j.jes.2021.04.009 PMID:34949365

Kottegoda, N., Sandaruwan, C., Priyadarshana, G., Siriwardhana, A., Rathnayake, U. A., Berugoda Arachchige, D. M., Kumarasinghe, A. R., Dahanayake, D., Karunaratne, V., & Amaratunga, G. A. J. (2017). Urea-hydroxyapatite nanohybrids for slow release of nitrogen. *ACS Nano*, *11*(2), 1214–1221. doi:10.1021/acsnano.6b07781 PMID:28121129

Kovačević, Z., Flinčec Grgac, S., & Bischof, S. (2021). Progress in Biodegradable Flame Retardant Nano-Biocomposites. *Polymers*. PMID:33673607

Kozawa, D., Cho, S. Y., Gong, X., Nguyen, F. T., Jin, X., Lee, M. A., Lee, H., Zeng, A., Xue, G., Schacherl, J., Gibson, S., Vega, L., & Strano, M. S. (2020). A Fiber Optic Interface Coupled to Nanosensors: Applications to Protein Aggregation and Organic Molecule Quantification. *ACS Nano*, *14*(8), 10141–10152. doi:10.1021/acsnano.0c03417 PMID:32667777

Kreyling, W. G., Semmler-Behnke, M., & Chaudhry, Q. (2010). A complementary definition of nanomaterial. *Nano Today*, *5*(3), 165–168. doi:10.1016/j.nantod.2010.03.004

Krishnani, S., Yadav, R., Rishi, N., & Goel, A. (2023). Microbial nanobionics: future perspectives and innovative approach to nanotechnology. *Environmental Applications of Microbial Nanotechnology: Emerging Trends in Environmental Remediation*, 17–32. doi:10.1016/B978-0-323-91744-5.00022-9

Krishna, V. M. (2016). Heat Transfer Enhancement by using ZnO Water Nanofluid in a Concentric Tube Heat Exchanger under Forced Convection Conditions. *International Journal of Innovations in Engineering and Technology*, *7*(4), 177–184.

Kulal, P., & Badalamoole, V. (2020). Efficient removal of dyes and heavy metal ions from waste water using Gum ghatti - graft - poly(4-acryloylmorpholine) hydrogel incorporated with magnetite nanoparticles. *Journal of Environmental Chemical Engineering*, *8*(5), 104207. doi:10.1016/j.jece.2020.104207

Kulkarni, S. J. (2017). Summary of Research and Studies on Heat Transfer Enhancements by Nanofluids. *International Journal of Engineering Research and General Science*, *5*(2), 175–181.

Kulkarni, S., Patil, P., Mujumdar, A., & Naik, J. (2018). Synthesis and evaluation of gas sensing properties of PANI, PANI/SnO2 and PANI/SnO2/rGO nanocomposites at room temperature. *Inorganic Chemistry Communications*, *96*, 90–96. doi:10.1016/j.inoche.2018.08.008

Kumah, E. A., Fopa, R. D., Harati, S., Boadu, P., Zohoori, F. V., & Pak, T. (2023). Human and environmental impacts of nanoparticles: A scoping review of the current literature. *BMC Public Health*, *23*(1), 1059. doi:10.1186/s12889-023-15958-4 PMID:37268899

Kumar, A., Sharma, M., Amari, A., & Vaish, R. (2024). Sonocatalytic induced dye degradation and antibacterial performance of SrTiO3 nanoparticles embedded cotton fabric. *Environmental Research*, *240*, 117541. doi:10.1016/j.envres.2023.117541 PMID:37914006

Kumar, A., Sharma, M., & Vaish, R. (2022a). Durable antibacterial cotton fabric via spray-coating of photocatalytic MoS2. *Materials Chemistry and Physics*, *126658*, 126658. doi:10.1016/j.matchemphys.2022.126658

Kumar, A., Sharma, M., & Vaish, R. (2022b). Screen printed calcium fluoride nanoparticles embedded antibacterial cotton fabric. *Materials Chemistry and Physics*, *288*, 126449. doi:10.1016/j.matchemphys.2022.126449

Kumar, A., Sharma, M., & Vaish, R. (2022c). BaTiO 3 Nanoparticles Embedded Antibacterial Cotton Fabric with UV Protection Characteristics. *Journal of Natural Fibers*, 1–14. doi:10.1080/15440478.2022.2139325

Kumar, A., Sharma, M., & Vaish, R. (2023). CaCu3Ti4O12 nanoparticle-loaded cotton fabric for dual photocatalytic antibacterial and dye degradation applications. *Environmental Science and Pollution Research International*, *30*(55), 117011–117021. Advance online publication. doi:10.1007/s11356-023-26835-3 PMID:37046162

Kumar, A., Sharma, M., & Vaish, R., & Ahmed, S. (2022). Poling effect on piezocatalytic antibacterial and dye degradation activities of BaTiO3 nanoparticles embedded cotton fabric. *Journal of Alloys and Compounds, 168530*. doi:10.1016/j.jallcom.2022.168530

Kumaran, N. S., & Vijayaraj, R. (2017). Biosynthesis of silver nano particles from Leucas aspera (Willd.) link and its anti-inflammatory potential against carrageen induced paw edema in rats. *International Journal of Pharmaceutical Sciences and Research*, *8*(6), 2588–2593. doi:10.13040/IJPSR.0975-8232.8(6).2588-93

Kumari, M., Sarkar, B., & Mukherjee, K. (2023). Nanoscale calcium oxide and its biomedical applications: A comprehensive review. *Biocatalysis and Agricultural Biotechnology*, *47*, 102506. doi:10.1016/j.bcab.2022.102506

Kumari, R., Suman, K., Karmakar, S., Lakra, S. G., Saurav, G. K., & Mahto, B. K. (2023). Regulation and safety measures for nanotechnology-based agri-products. *Frontiers in Genome Editing*, *5*, 1200987. doi:10.3389/fgeed.2023.1200987 PMID:37415849

Kumar, M. (2019). Applications of Nano science in sports: An analytical study. International Journal of Physiology. *Nutrition and Physical Education*, *4*(1), 2642–2646.

Kumar, N., & Dixit, A. (2019a). Nanotechnology-Driven Explosives and Propellants. In *Nanotechnology for Defence Applications*. Springer. doi:10.1007/978-3-030-29880-7_3

Kumar, N., & Dixit, A. (2019b). Nanotechnology-Enabled Management of Chemical, Biological, Radiological, and Nuclear Threats. In *Nanotechnology for Defence Applications*. Springer. doi:10.1007/978-3-030-29880-7_4

Kumar, N., Sangma, S. N., Ray, D. P., Ghosh, R. K., Ammayappan, L., & Chattopadhyay, S. N. (2020). Nanomaterial regulation and its applications in textile sector: A review. *International Journal of Bioresource Science*, *7*(2), 75–81. doi:10.30954/2347-9655.02.2020.5

Kumar, P. S. M., Francis, A. P., & Devasena, T. (2014). Biosynthesized and chemically synthesized titania nanoparticles: Comparative analysis of antibacterial activity. *Journal of Environmental Nanotechnology*, *3*(3), 73–81. doi:10.13074/jent.2014.09.143098

Kumar, P. V., Kala, S. M. J., & Prakash, K. S. (2019). Green synthesis of gold nanoparticles using Croton Caudatus Geisel leaf extract and their biological studies. *Materials Letters*, *236*, 19–22. doi:10.1016/j.matlet.2018.10.025

Kumar, R., Kumar, M., & Luthra, G. (2023). Fundamental approaches and applications of nanotechnology: A mini review. *Materials Today: Proceedings*. doi:10.1016/j.matpr.2022.12.172

Kumar, S., Ngasainao, M. R., Sharma, D., Sengar, M., Gahlot, A. P. S., Shukla, S., & Kumari, P. (2022). Contemporary nanocellulose-composites: A new paradigm for sensing applications. *Carbohydrate Polymers*, *298*, 120052. doi:10.1016/j.carbpol.2022.120052 PMID:36241259

Kuspanov, Z., Bakbolat, B., Baimenov, A., Issadykov, A., Yeleuov, M., & Daulbayev, C. (2023). Photocatalysts for a sustainable future: Innovations in large-scale environmental and energy applications. *The Science of the Total Environment*, *885*, 163914. doi:10.1016/j.scitotenv.2023.163914 PMID:37149164

Kuswandi, B. (2017). Environmental friendly food nano-packaging. *Environmental Chemistry Letters*, *15*(2), 205–221. doi:10.1007/s10311-017-0613-7

Kuswandi, B., Wicaksono, Y., Jayus, Abdullah, A., Heng, L. Y., & Ahmad, M. (2011). Smart packaging: Sensors for monitoring of food quality and safety. *Sensing and Instrumentation for Food Quality and Safety*, *5*(3–4), 137–146. doi:10.1007/s11694-011-9120-x

Lakshmipriya, T., & Gopinath, S. C. B. (2021). 1 - Introduction to nanoparticles and analytical devices, Editor(s): Subash C.B. Gopinath, Fang Gang, Nanoparticles in Analytical and Medical Devices, Elsevier. doi:10.1016/B978-0-12-821163-2.00001-7

Lanzalaco, S., & Molina, B. G. (2020, May 24). Polymers and Plastics Modified Electrodes for Biosensors: A Review. *Molecules (Basel, Switzerland)*, *25*(10), 2446. doi:10.3390/molecules25102446 PMID:32456314

Larson, K. K., He, M., Teichert, J. F., Naganawa, A., & Bode, J. W. (2012). Chemical sensing with shapeshifting organic molecules. *Chemical Science (Cambridge)*, *3*(6), 1825–1828. doi:10.1039/c2sc20238g

Lazcka, O., Del Campo, F. J., & Muñoz, F. X. (2007). Pathogen detection: A perspective of traditional methods and biosensors. In Biosensors and Bioelectronics (Vol. 22, Issue 7, pp. 1205–1217). doi:10.1016/j.bios.2006.06.036

Lee, K. Y., & Bismarck, A. (2016). Bacterial NanoCellulose as Reinforcement for Polymer Matrices. In Miguel Gama, F. Dourado, & S. Bielecki (Eds.), Bacterial Nanocellulose: From Biotechnology to Bio-Economy (pp. 109–122). Elsevier B.V. doi:10.1016/B978-0-444-63458-0.00006-8

Lee, C. W., Suh, J. M., & Jang, H. W. (2019). Chemical Sensors Based on Two-Dimensional (2D) Materials for Selective Detection of Ions and Molecules in Liquid. *Frontiers in Chemistry*, *7*, 708. doi:10.3389/fchem.2019.00708 PMID:31803712

Lee, J. H., Jo, J. K., Kim, D. A., Patel, K. D., Kim, H. W., & Lee, H. H. (2018). Nano-graphene oxide incorporated into PMMA resin to prevent microbial adhesion. *Dental Materials*, *34*(4), e63–e72. doi:10.1016/j.dental.2018.01.019 PMID:29402540

Lee, P. C., Kang, D., Oh, J. T., Seo, J. Y., Shin, D., Jung, J. U., Ko, Y. K., Ha, J. U., & Kim, M. G. (2023). Reducing Moisture Absorption in Polypropylene Nanocomposites for Automotive Headlamps Using Hydrophobicity-Modified Graphene/Montmorillonite. *Nanomaterials (Basel, Switzerland)*, *13*(9), 1439. doi:10.3390/nano13091439 PMID:37176984

Lee, Y., Garcia, M. A., Frey Huls, N. A. F., & Sun, S. (2010). Synthetic Tuning of the Catalytic Properties of Au-Fe_3O_4 Nanoparticles. *Angewandte Chemie International Edition*, *49*(7), 1271–1274. doi:10.1002/anie.200906130 PMID:20077449

LewisOscar, F., MubarakAli, D., Nithya, C., Priyanka, R., Gopinath, V., Alharbi, N. S., & Thajuddin, N. (2015). One pot synthesis and anti-biofilm potential of copper nanoparticles(CuNPs) against clinical strains of Pseudomonas aeruginosa. *Biofouling*, *31*(4), 379–391. doi:10.1080/08927014.2015.1048686 PMID:26057498

Li, Y., He, R., Niu, Y., & Li, F. (2019). Paper-Based Electrochemical Biosensors for Point-of-Care Testing of Neurotransmitters. In Journal of Analysis and Testing (Vol. 3, Issue 1, pp. 19–36). Nonferrous Metals Society of China. doi:10.1007/s41664-019-00085-0

Liang, J., Wei, M., Wang, Q., Zhao, Z., Liu, A., Yu, Z., & Tian, Y. (2018a). Sensitive Electrochemical Determination of Hydrogen Peroxide Using Copper Nanoparticles in a Polyaniline Film on a Glassy Carbon Electrode. *Analytical Letters*, *51*(4), 512–522. doi:10.1080/00032719.2017.1343832

Lian, Z., Zhang, Y., & Zhao, Y. (2016). Nano-TiO2 particles and high hydrostatic pressure treatment for improving functionality of polyvinyl alcohol and chitosan composite films and nano-TiO2 migration from film matrix in food simulants. *Innovative Food Science & Emerging Technologies*, *33*, 145–153. doi:10.1016/j.ifset.2015.10.008

Li, J., Wang, J., Guo, X., Zheng, Q., Peng, J., Tang, H., & Yao, S. (2015). Carbon nanotubes labeled with aptamer and horseradish peroxidase as a probe for highly sensitive protein biosensing by post=electropolymerization of insoluble precipitates on electrodes. *Analytical Chemistry*, *87*(15), 7610–7617. doi:10.1021/acs.analchem.5b00640 PMID:26176595

Lindh, H., Williams, H., Olsson, A., & Wikström, F. (2016). Elucidating the indirect contributions of packaging to sustainable development: A terminology of packaging functions and features. *Packaging Technology & Science*, *29*(4-5), 225–246. doi:10.1002/pts.2197

Li, S., Chen, H., Liu, X., Li, P., & Wu, W. (2022). Nanocellulose as a promising substrate for advanced sensors and their applications. *International Journal of Biological Macromolecules*, *218*, 473–487. doi:10.1016/j.ijbiomac.2022.07.124 PMID:35870627

Liu, D., Zhang, X., & You, T. (2014). Electrochemical Performance of Electrospun Free-Standing Nitrogen-Doped Carbon Nanofibers and Their Application for Glucose Biosensing ACS Appl. *ACS Applied Materials & Interfaces*, *6*(9), 6275–6280. doi:10.1021/am501713g PMID:24707890

Liu, G., Feng, D. Q., Qian, Y., Wang, W., & Zhu, J. J. (2019). Construction of FRET biosensor for off-on detection of lead ions based on carbon dots and gold nanorods. *Talanta*, *201*, 90–95. doi:10.1016/j.talanta.2019.03.101 PMID:31122466

Liu, S., Hu, Q., Qiu, J., Wang, F., Lin, W., Zhu, F., Wei, C., Zhou, N., & Ouyang, G. (2017). Enhanced Photocatalytic Degradation of Environmental Pollutants under Visible Irradiation by a Composite Coating. *Environmental Science & Technology*, *51*(9), 5137–5145. doi:10.1021/acs.est.7b00350 PMID:28379014

Liu, W. X., Song, S., Ye, M. L., Zhu, Y., Zhao, Y. G., & Lu, Y. (2022). Nanomaterials with Excellent Adsorption Characteristics for Sample Pretreatment: A Review. *Nanomaterials (Basel, Switzerland)*, *12*(11), 1845. doi:10.3390/nano12111845 PMID:35683700

Liu, Y., He, K., Chen, G., Leow, W. R., & Chen, X. (2017). Nature-Inspired Structural Materials for Flexible Electronic Devices. *Chemical Reviews*, *117*(20), 12893–12941. doi:10.1021/acs.chemrev.7b00291 PMID:28991450

Li, Y., Yang, C., Yin, X., Sun, Y., Weng, J., Zhou, J., & Feng, B. (2019). Inflammatory responses to micro/nano-structured titanium surfaces with silver nanoparticles in vitro. *Journal of Materials Chemistry. B*, *7*(22), 3546–3559. doi:10.1039/C8TB03245A

Lohse, S. E., & Murphy, C. J. (2012). Applications of colloidal inorganic nanoparticles: From medicine to energy. *Journal of the American Chemical Society*, *134*(38), 15607–15620. doi:10.1021/ja307589n PMID:22934680

López, A. D. F., De-la-Torre, G. E., Fernández Severini, M. D., Prieto, G., Brugnoni, L. I., Colombo, C. V., Dioses-Salinas, D. C., Rimondino, G. N., & Spetter, C. V. (2023). Chemical-analytical characterization and leaching of heavy metals associated with nanoparticles and microplastics from commercial face masks and the abundance of personal protective equipment (PPE) waste in three metropolitan cities of South America. *Marine Pollution Bulletin*, *191*, 114997. doi:10.1016/j.marpolbul.2023.114997 PMID:37148588

Lowe, S., O'Brien-Simpson, N. M., & Connal, L. A. (2015). Antibiofouling polymer interfaces: Poly (ethylene glycol) and other promising candidates. *Polymer Chemistry*, *6*(2), 198–212. doi:10.1039/C4PY01356E

Lu, J., Guo, J., Song, S., Yu, G., Liu, H., Yang, X., & Lu, Z. (2020). Preparation of Ag nanoparticles by spark ablation in gas as catalysts for electrocatalytic hydrogen production. *RSC Advances*, *10*(63), 38583–38587. doi:10.1039/D0RA06682F PMID:35517560

Luo, X., Han, Y., Chen, X., Tang, W., Yue, T., & Li, Z. (2020). Carbon dots derived fluorescent nanosensors as versatile tools for food quality and safety assessment: A review. *Trends in Food Science & Technology*, *95*, 149–161. doi:10.1016/j.tifs.2019.11.017

Luo, X., Morrin, A., Killard, A. J., & Smyth, M. R. (2006). Application of nanoparticles in electrochemical sensors and biosensors. Electroanalysis. *An International Journal Devoted to Fundamental and Practical Aspects of Electroanalysis*, *18*(4), 319–326.

Ma, B. (2021). *Aluminum matrix nanocomposites for applications in automobile industry* [Doctoral dissertation, Sorbonne Université].

Madianos, L., Tsekenis, G., Skotadis, E., Patsiouras, L., & Tsoukalas, D. (2018). A highly sensitive impedimetric apta-sensor for the selective detection of acetamiprid and atrazine based on microwires formed by platinum nanoparticles. *Biosensors & Bioelectronics*, *101*, 268–274. doi:10.1016/j.bios.2017.10.034 PMID:29096365

Maduraiveerana, G., Sasidharana, M., & Ganesan, V. (2017). *Electrochemical sensor and biosensor platforms based on advanced nanomaterials for biological and biomedical applications, Biosensors and Bioelectronics*. Elsevier., doi:10.1016/j.bios.2017.12.031

Magaye, R., Zhao, J., Bowman, L., & Ding, M. (2012). Genotoxicity of cobalt, nickel and copper based nanoparticles. *Experimental and Therapeutic Medicine*, *4*(4), 551–561. doi:10.3892/etm.2012.656 PMID:23170105

Mahajan, G., Kaur, M., & Gupta, R. (2022). Green functionalized nanomaterials: Fundamentals and future opportunities. In *Green Functionalized Nanomaterials for Environmental Applications* (pp. 21–41). Elsevier. doi:10.1016/B978-0-12-823137-1.00003-8

Maheshwari, R., Joshi, G., Mishra, D. K., & Tekade, R. K. (2019). Bionanotechnology in pharmaceutical research. In *Basic fundamentals of drug delivery* (pp. 449–471). Academic Press. doi:10.1016/B978-0-12-817909-3.00012-1

Mahmoud, A., Ezgi, Ö., Merve, A., & Özhan, G. (2016). In Vitro Toxicological Assessment of Magnesium Oxide Nanoparticle Exposure in Several Mammalian Cell Types. Int J Toxicol, 35(4):429-37. doi:10.1177/1091581816648624

Malhotra, B. D., & Pandey, C. M. (2017). *Biosensors: Fundamentals and Applications*. https://www.ebsco.com/terms-of-use

Malhotra, B. D., & Ali, M. A. (2018). *Nanomaterials in Biosensors: Fundamentals and Applications*. Nanomaterials for Biosensors., doi:10.1016/B978-0-323-44923-6.00001-7

Malhotra, B. D., Ali, M. A., Malhotra, B., & Ali, M. (2018). In B. D. Malhotra & M. A. Ali (Eds.), *Nanocomposite Materials: Biomolecular Devices. Nanomaterials for Biosensors* (pp. 145–159). William Andrew Publishing., doi:10.1016/B978-0-323-44923-6.00005-4

Maliael, M. T., Jain, R. K., & Srirengalakshmi, M. (2022). Effect of nanoparticle coatings on frictional resistance of orthodontic archwires: A systematic review and meta-analysis. *World Journal of Dentistry*, *13*(4), 417–424. doi:10.5005/jp-journals-10015-2066

Malik, S., Muhammad, K., & Waheed, Y. (2023). Nanotechnology: A revolution in modern industry. *Molecules (Basel, Switzerland)*, *28*(2), 661. doi:10.3390/molecules28020661 PMID:36677717

Mallakpour, S., Behranvand, V., & Hussain, C. M. (2023). Worldwide fight against COVID-19 using nanotechnology, polymer science, and 3D printing technology. *Polymer Bulletin*, *80*(1), 165–183. doi:10.1007/s00289-021-04006-z PMID:35106016

Mandal, A. K., Katuwal, S., Tettey, F., Gupta, A., Bhattarai, S., Jaisi, S., Bhandari, D. P., Shah, A. K., Bhattarai, N., & Parajuli, N. (2022). Current Research on Zinc Oxide Nanoparticles: Synthesis, Characterization, and Biomedical Applications. *Nanomaterials (Basel, Switzerland)*, *12*(17), 3066. doi:10.3390/nano12173066 PMID:36080103

Mandal, A., Maitra, S., Roy, S., Hazra, B., Ray, K., & Kargupta, K. (2023). Selective photoreduction of CO2 to methanol using Cu-doped 1D-Bi2S3/rGO nanocomposites under visible light irradiation. New Journal of Chemistry. [RSC]. *New Journal of Chemistry*, *47*(3), 1422–1434. doi:10.1039/D2NJ03892G

Mandal, T. K., & Parvin, N. (2011). Rapid detection of bacteria by carbon quantum dots. *Journal of Biomedical Nanotechnology*, *7*(6), 846–848. doi:10.1166/jbn.2011.1344 PMID:22416585

Mangalampalli, B., Dumala, N., & Grover, P. (2017). Acute oral toxicity study of magnesium oxide nanoparticles and microparticles in female albino Wistar rats. *Regulatory Toxicology and Pharmacology*, *98*, 170–184. doi:10.1016/j.yrtph.2017.09.005 PMID:28899817

Mangaraj, S., Yadav, A., Bal, L. M., Dash, S. K., & Mahanti, N. K. (2019). Application of biodegradable polymers in food packaging industry: A comprehensive review. *Journal of Packaging Technology and Research*, *3*(1), 77–96. doi:10.1007/s41783-018-0049-y

Manna, S., Gopakumar, D. A., Roy, D., Saha, P., & Thomas, S. (2018). Nanobiomaterials for removal of fluoride and chlorophenols from water. In *New Polymer Nanocomposites for Environmental Remediation*. Elsevier Inc. doi:10.1016/B978-0-12-811033-1.00020-2

Maraveas, C. (2020). Production of Sustainable and Biodegradable Polymers from Agricultural Waste. *Polymers*, *12*(5), 1127. doi:10.3390/polym12051127 PMID:32423073

Marbawi, H., Arif, M., Ros, M., Khudri, S., Othman, A. R., Effendi, M. I., Gansau, J. A., & Yasid, N. A. (2019).. Kinetic Analysis of the Adsorption of Glyphosate onto Palm Oil Fronds Activated Carbon. *BIOREMEDIATION SCIENCE AND TECHNOLOGY, 7*(1), 29–33.

Marszewski, M., Cao, S., Yu, J., & Jaroniec, M. (2015). Semiconductor-based photocatalytic CO 2 conversion. *Materials Horizons*, *2*(3), 261–278. doi:10.1039/C4MH00176A

Mather, R., & Wilson, J. (2017). Fabrication of Photovoltaic Textiles. *Coatings*, *7*(5), 63. doi:10.3390/coatings7050063

Mazov, I., Krasnikov, D., Stadnichenko, A., Kuznetsov, V., Romanenko, A., Anikeeva, O., & Tkachev, E. (2012). Direct vapor-phase bromination of multiwall carbon nanotubes. *Journal of Nanotechnology*, *2012*, 2012. doi:10.1155/2012/954084

McGaw, L. J., Elgorashi, E. E., & Eloff, J. N. (2014). 8 - Cytotoxicity of African Medicinal Plants Against Normal Animal and Human Cells, Editor(s): Victor Kuete, Toxicological Survey of African Medicinal Plants, Elsevier. doi:10.1016/B978-0-12-800018-2.00008-X

Mehmandoust, M., Erk, N., Karaman, C., & Karaman, O. (2022). An electrochemical molecularly imprinted sensor based on CuBi2O4/rGO@MoS2 nanocomposite and its utilization for highly selective and sensitive for linagliptin assay. *Chemosphere*, *291*, 132807. doi:10.1016/j.chemosphere.2021.132807 PMID:34762887

Mei, L., Xie, R., Zhu, S., Deng, S., Xu, H., Fan, X., Yin, W., & Gu, Z. (2023). Neurotoxicity study of lead-based perovskite nanoparticles. *Nano Today*, *50*, 101830. doi:10.1016/j.nantod.2023.101830

Mendonça, G., Mendonça, D. B., Aragao, F. J., & Cooper, L. F. (2008). Advancing dental implant surface technology–from micron-to nanotopography. *Biomaterials*, *29*(28), 3822–3835. doi:10.1016/j.biomaterials.2008.05.012 PMID:18617258

Menossi, M., Casalongué, C., & Alvarez, V. A. (2021). Bio-nanocomposites for Modern Agricultural Applications. In *Handbook of Consumer Nanoproducts*. Springer., doi:10.1007/978-981-15-6453-6_68-1

Meskher, H., Mustansar, H. C., Thakur, A. K., Sathyamurthy, R., Lynch, I., Singh, P., Han, T. K., & Saidur, R. (2022). Recent trends in carbon nanotube (CNT)-based biosensors for the fast and sensitive detection of human viruses: a critical review. In Nanoscale Advances (Vol. 5, Issue 4, pp. 992–1010). Royal Society of Chemistry. doi:10.1039/D2NA00236A

Michalet, X., Pinaud, F. F., Bentolila, L. A., Tsay, J. M., Doose, S., Li, J. J., Sundaresan, G., Wu, A. M., Gambhir, S. S., & Weiss, S. (2005). Quantum Dots for Live Cells, in Vivo Imaging, and Diagnostics. *Science Mag.* www.sciencemag.org/cgi/content/full/307/5709/538/DC1

Mills, D. K., & Nestorova, G. G. (2023). Biosensor Development and Innovation in Healthcare and Medical Applications. In Sensors (Vol. 23, Issue 5). MDPI. doi:10.3390/s23052717

Mills, A. (2005). Oxygen indicators and intelligent inks for packaging food. *Chemical Society Reviews*, *34*(12), 1003–1011. doi:10.1039/b503997p PMID:16284666

Mir, S. A., Dar, B. N., Wani, A. A., & Shah, M. A. (2018). Effect of plant extracts on the techno-functional properties of biodegradable packaging films. *Trends in Food Science & Technology*, *80*, 141–154. doi:10.1016/j.tifs.2018.08.004

Mishra, R. K., Sabu, A., & Tiwari, S. K. (2018). Materials chemistry and the futurist eco-friendly applications of nanocellulose: Status and prospect. *Journal of Saudi Chemical Society*, *22*(8), 949–978. doi:10.1016/j.jscs.2018.02.005

Mittal, A. K., Chisti, Y., & Banerjee, U. C. (2013). Synthesis of metallic nanoparticles usingplant extracts. *Biotechnology Advances*, *31*(2), 346–356. doi:10.1016/j.biotechadv.2013.01.003 PMID:23318667

Mittal, A. K., Kaler, A., & Banerjee, U. C. (2012). Free Radical Scavenging and AntioxidantActivity of Silver Nanoparticles Synthesized from Flower Extract of Rhododendron dauricum. *Nano Biomedicine and Engineering*, *4*(3). doi:10.5101/nbe.v4i3.p118-124

MMR. (2023). Polym*eric Nanoparticles Market: Global Industry Analysis and Forecast (2022-2029).* MMR. www.maximizemarketresearch.com/market-report/global-polymeric-nanoparticles-market/104820/#

Moallem, S. A., Dehghani, N., Mehri, S., Shahsavand, S., Alibolandi, M., & Hadizadeh, F. (2015). Synthesis of novel 1, 8-acridinediones derivatives: Investigation of MDR r eversibility on breast cancer cell lines T47D and tamoxifenresistant T47D. *Research in Pharmaceutical Sciences*, *10*(3), 214. PMID:26600848

Modi, S., Yadav, V. K., Amari, A., Alyami, A. Y., Gacem, A., Harharah, H. N., & Fulekar, M. H. (2023). Photocatalytic Degradation of Methylene Blue Dye from Wastewater by Using Doped Zinc Oxide Nanoparticles. *Water (Basel)*, *15*(12), 2275. doi:10.3390/w15122275

Moghaddam, A., Namvar, F., Moniri, M., Md. Tahir, P., Azizi, S., & Mohamad, R. (2015). Nanoparticles biosynthesized by fungi and yeast: A review of their preparation, properties, and medical applications. *Molecules (Basel, Switzerland)*, *20*(9), 16540–16565. doi:10.3390/molecules200916540 PMID:26378513

Mohamed El-Rafie, H., & Hamed, M. A. A. (2014). Antioxidant and anti-inflammatory activities of silver nanoparticles biosynthesized from aqueous leaves extracts of four Terminalia species. *Adv. Nat. Sci. Nanosci. Nanotechnol*, *5*(3), 1–11. doi:10.1088/2043-6262/5/3/035008

Mohammadi, M. M., Kumar, A., Liu, J., Liu, Y., Thundat, T., & Swihart, M. T. (2020). Hydrogen Sensing at Room Temperature Using Flame-Synthesized Palladium-Decorated Crumpled Reduced Graphene Oxide Nanocomposites. *ACS Sensors*, *5*(8), 2344–2350. doi:10.1021/acssensors.0c01040 PMID:32786377

Mohammadi, Z., & Jafari, S. M. (2020). Detection of food spoilage and adulteration by novel nanomaterial-based sensors. *Advances in Colloid and Interface Science*, *286*, 102297. doi:10.1016/j.cis.2020.102297 PMID:33142210

Mohammad, Z. H., Ahmad, F., Ibrahim, S. A., & Zaidi, S. (2022). Application of nanotechnology in different aspects of the food industry. *Discover Food*, *2*(1), 12. doi:10.1007/s44187-022-00013-9

Moldovan, B., David, L., Vulcu, A., Olenic, L., Perde-Schrepler, M., Fischer-Fodor, E., Baldea, I., Clichici, S., & Filip, G. A. (2017). In vitro and in vivo anti-inflammatory properties of green synthesized silvernanoparticles using Viburnum opulus L. fruits extract. *Materials Science and Engineering C*, *79*, 720–727. doi:10.1016/j.msec.2017.05.122 PMID:28629073

Mondal, S. (2017). Preparation, properties and applications of nanocellulosic materials. *Carbohydrate Polymers*, *163*, 301–316. doi:10.1016/j.carbpol.2016.12.050 PMID:28267510

Morales-Lozoya, V., Espinoza-Gómez, H., Flores-López, L. Z., Sotelo-Barrera, E. L., Núñez-Rivera, A., Cadena-Nava, R. D., & Rivero, I. A. (2021). Study of the effect of the differentparts of Morinda citrifolia L.(noni) on the green synthesis of silver nanoparticles and theirantibacterial activity. *Applied Surface Science*, *537*, 147855. doi:10.1016/j.apsusc.2020.147855

Moshood, T. D., Nawanir, G., Mahmud, F., Mohamad, F., Ahmad, M. H., & AbdulGhani, A. (2022). Sustainability of biodegradable plastics: New problem or solution to solve the global plastic pollution? *Current Research in Green and Sustainable Chemistry*, *5*, 100273. doi:10.1016/j.crgsc.2022.100273

Motelica, L., Ficai, D., Ficai, A., Oprea, O. C., Kaya, D. A., & Andronescu, E. (2020). Biodegradable Antimicrobial Food Packaging: Trends and Perspectives. *Foods*, *9*(10), 1438. doi:10.3390/foods9101438 PMID:33050581

Moteriya, P., & Chanda, S. (2020). Green Synthesis of Silver Nanoparticles fromCaesalpinia pulcherrima Leaf Extract and Evaluation of Their Antimicrobial, Cytotoxic and Genotoxic Potential (3-in-1 System). *Journal of Inorganic and Organometallic Polymers and Materials*, *30*(10), 3920–3932. doi:10.1007/s10904-020-01532-7

Moulahoum, H., Ghorbanizamani, F., Guler Celik, E., & Timur, S. (2022). Nano-Scaled Materials and Polymer Integration in Biosensing Tools. *Biosensors (Basel)*, *12*(5), 301. doi:10.3390/bios12050301 PMID:35624602

Mousa, A. B., Moawad, R., & Abdallah, Y. (2021). Zinc oxide nanoparticles promise anticancer and antibacterial activity in ovarian cancer. *Pharmaceutical Research*. doi:10.1007/s11095-023-03505-0 PMID:37016170

Mukhopadhyay, R., Sarkar, B., Khan, E., Alessi, D. S., Biswas, J. K., Manjaiah, K. M., Eguchi, M., Wu, K. C. W., Yamauchi, Y., & Ok, Y. S. (2022). Nanomaterials for sustainable remediation of chemical contaminants in water and soil. *Critical Reviews in Environmental Science and Technology*, *52*(15), 2611–2660. doi:10.1080/10643389.2021.1886891

Mukhtar Iderawumi, A., Jiandong, H., Shakeel, A., Linze, L., & Syed Muhammad Zaigham Abbas, N. (2023). Advances in the Use of Organic and Organomineral Fertilizers in Sustainable Agricultural Production. In H. Khalid Rehman (Ed.), *Organic Fertilizers*. IntechOpen.

Muniyappan, N., Pandeeswaran, M., & Amalraj, A. (2021). Green synthesis of gold nanoparticles using Curcuma pseudomontana isolated curcumin: Its characterization, antimicrobial, antioxidant and anti-inflammatory activities. *Environmental Chemistry and Ecotoxicology*, *3*, 117–124. doi:10.1016/j.enceco.2021.01.002

Muñoz-Carrillo, M. G.,Garcidueñas-Piña, C., Pedroza-Jimenez,M. M., García, R. C., Carbajal-Hernández, A. L., Andrade,G. and Morales-Dominguez J. F. (2023). *Toxicity of Silver Nanoparticles Synthesized from Opuntia ficus-indica on Danio rerio and Arabidopsis thaliana and Their Antimicrobial Activity*. https://doi.org/ doi:10.1155/2023/6903787

Murad, U., Khan, S. A., Ibrar, M., Ullah, S., & Khattak, U. (2018). Synthesis of silver and gold nanoparticles from leaf of Litchi chinensis and its biological activities. *Asian Pacific Journal of Tropical Biomedicine*, *8*(3), 142. doi:10.4103/2221-1691.227995

Nabikhan, A., Kandasamy, K., Raj, A., & Alikunhi, N. M. (2010). Synthesis of antimicrobialsilver nanoparticles by callus and leaf extracts from saltmarsh plant, Sesuviumportulacastrum L. *Colloids and Surfaces. B, Biointerfaces*, *79*(2), 488–493. doi:10.1016/j.colsurfb.2010.05.018 PMID:20627485

Nagajyothi, P. C., Sreekanth, T. V. M., Tettey, C. O., Jun, Y. I., & Mook, S. H. (2014). Characterization, antibacterial, antioxidant, and cytotoxic activities of ZnO nanoparticles using Coptidis Rhizoma. *Bioorganic & Medicinal Chemistry Letters*, *24*(17), 4298–4303. doi:10.1016/j.bmcl.2014.07.023 PMID:25088397

Naghdi, T., Golmohammadi, H., Vosough, M., Atashi, M., Saeedi, I., & Maghsoudi, M. T. (2019). Lab-on-nanopaper: An optical sensing bioplatform based on curcumin embedded in bacterial nanocellulose as an albumin assay kit. *Analytica Chimica Acta*, *1070*, 104–111. doi:10.1016/j.aca.2019.04.037 PMID:31103163

Nair, R. R., Blake, P., Grigorenko, A. N., Novoselov, K. S., Booth, T. J., Stauber, T., & Geim, A. K. (2008). Fine structure constant defines visual transparency of graphene. *science, 320*(5881), 1308-1308. https://doi.org/ doi:10.1126/science.115696

Naqvi, S. M. Z. A., Haq, S. I. U., Iderawumi, A. R. M., Anwar, A., Mujtaba, G., Khan, N. M., Islam, A., Awais, M., Mustafa, T., & Iqbal, J. (2022). GREEN SYNTHESIS OF SILVER NANOPARTICLES AND ANTI-OXIDANT ACTIVITY IN PLANTS UNDER SEMIARID CONDITION—A REVIEW. *Pakistan Journal of Weed Science Research*, *28*(3), 319–330. doi:10.28941/pjwsr.v28i3.1064

Narayanan, M., Devi, P. G., Natarajan, D., Kandasamy, S., Devarayan, K., Alsehli, M., & Pugazhendhi, A. (2021). Green synthesis and characterization of titanium dioxide nanoparticles using leaf extract of Pouteria campechiana and larvicidal and pupicidal activity on Aedes aegypti. *Environmental Research*, *200*, 111333. doi:10.1016/j.envres.2021.111333 PMID:34051198

Naresh, V., & Lee, N. (2021). A Review on Biosensors and Recent Development of Nanostructured Materials-Enabled Biosensors. *Sensors (Basel)*, *21*(4), 1109. doi:10.3390/s21041109 PMID:33562639

Naseem, T., & Durrani, T. (2021). The role of some important metal oxide nanoparticles for wastewater and antibacterial applications: A review. *Environmental Chemistry and Ecotoxicology*, *3*, 59–75. doi:10.1016/j.enceco.2020.12.001

Naseri, M., Fotouhi, L., & Ehsani, A. (2018). Recent Progress in the Development of Conducting Polymer-Based Nanocomposites for Electrochemical Biosensors Applications: A Mini-Review. *Chemical Record (New York, N.Y.)*, *18*(6), 599–618. doi:10.1002/tcr.201700101 PMID:29460399

Nasrollahzadeh, M., Sajjadi, M., Iravani, S., & Varma, R. S. (2021). Starch, cellulose, pectin, gum, alginate, chitin and chitosan derived (nano) materials for sustainable water treatment: A review. *Carbohydrate Polymers*, *251*, 116986. doi:10.1016/j.carbpol.2020.116986 PMID:33142558

Nassar, M. Y., NourEldien, M. S., Ibrahim, I. M., & Aly, H. M. (2023). A Facile Hydrothermal Synthesis of S-VO2-Cellulose Nanocomposite for Photocatalytic Degradation of Methylene Blue Dye. *Processes (Basel, Switzerland)*, *11*(5), 1322. doi:10.3390/pr11051322

Nath, D., & Banerjee, P. (2013). Green nanotechnology–a new hope for medicalbiology. *Environmental Toxicology and Pharmacology*, *36*(3), 997–1014. doi:10.1016/j.etap.2013.09.002 PMID:24095717

Nay, A. A., Abd-Elhamid, A. I., Aly, A. A., & Brase, S. (2022). Recent progress in the applications of silica-based nanoparticles. *RSC Advances*, *12*(22), 13706–13726. doi:10.1039/D2RA01587K PMID:35530394

Ndaba, B., Roopnarain, A., Haripriya, R. A. M. A., & Maaza, M. (2022). Biosynthesized metallic nanoparticles as fertilizers: An emerging precision agriculture strategy. *Journal of Integrative Agriculture*, *21*(5), 1225–1242. doi:10.1016/S2095-3119(21)63751-6

Nechyporchuk, O., Belgacem, M. N., & Bras, J. (2016). Production of cellulose nanofibrils: A review of recent advances. *Industrial Crops and Products*, *93*, 2–25. doi:10.1016/j.indcrop.2016.02.016

Neelakantan, P. S. (2014). Nanotechnology In Dentistry - What Does The Future Hold In Store? *Dentistry (Sunnyvale, Calif.)*, *04*(02). doi:10.4172/2161-1122.1000198

Nesa, S. H. S., & Tarangini, K. (2023). A review on augmentation of natural fabric materials with novel bio/nanomaterials and their multifunctional perspectives. *Hybrid Advances, 100020.*

Nguyen, L. H., Naficy, S., Chandrawati, R., & Dehghani, F. (2019). Nanocellulose for Sensing Applications. *Advanced Materials Interfaces*, *6*(18), 30–33. doi:10.1002/admi.201900424

Nguyen, T. A., Nguyen, T. V., Thai, H., & Shi, X. (2016). Effect of nanoparticles on the thermal and mechanical properties of epoxy coatings. *Journal of Nanoscience and Nanotechnology*, *16*(9), 9874–9881. doi:10.1166/jnn.2016.12162

Nguyen-Tri, P., Nguyen, T. A., Carriere, P., & Ngo Xuan, C. (2018). Nanocomposite coatings: Preparation, characterization, properties, and applications. *International Journal of Corrosion*, *2018*, 2018. doi:10.1155/2018/4749501

Nicola, R., Costişor, O., Ciopec, M., Negrea, A., Lazău, R., Ianăşi, C., Picioruş, E. M., Len, A., Almásy, L., Szerb, E. I., & Putz, A. M. (2020). Silica-coated magnetic nanocomposites for Pb2+ removal from aqueous solution. *Applied Sciences (Basel, Switzerland)*, *10*(8), 2726. Advance online publication. doi:10.3390/app10082726

Nikalje, A. P. (2015). Nanotechnology and its Applications in Medicine. *Medicinal Chemistry (Sunnyvale, Calif.)*, *5*(2), 81–89. doi:10.4172/2161-0444.1000247

Nikazar, S., Barani, M., Rahdar, A., Zoghi, M., & Kyzas, G. Z. (2020). Photo-and magnetothermally responsive nanomaterials for therapy, controlled drug delivery and imaging applications. *ChemistrySelect*, *5*(40), 12590–12609. doi:10.1002/slct.202002978

Nilavukkarasi, M., Vijayakumar, S., & Kumar, S. P. (2020). Biological synthesis and characterization of silver nanoparticles with Capparis zeylanica L. leaf extract for potentantimicrobial and anti proliferation efficiency. *Materials Science for Energy Technologies*, *3*, 371–376. doi:10.1016/j.mset.2020.02.008

Nilsen-Nygaard, J., Fernández, E. N., Radusin, T., Rotabakk, B. T., Sarfraz, J., Sharmin, N., Sivertsvik, M., Sone, I., & Pettersen, M. K. (2021). Current status of biobased and biodegradable food packaging materials: Impact on food quality and effect of innovative processing technologies. *Comprehensive Reviews in Food Science and Food Safety*, *20*(2), 1333–1380. doi:10.1111/1541-4337.12715 PMID:33547765

Noah, N. M., & Ndangili, P. M. (2022). Green synthesis of nanomaterials from sustainable materials for biosensors and drug delivery. *Sensors International*, *3*, 100166. doi:10.1016/j.sintl.2022.100166

Noga, M., Milan, J., Frydrych, A., & Jurowski, K. (2023). Toxicological Aspects, Safety Assessment, and Green Toxicology of Silver Nanoparticles (AgNPs)—Critical Review: State of the Art. *International Journal of Molecular Sciences*, *24*(6), 5133. doi:10.3390/ijms24065133 PMID:36982206

Noreen, I., Hashmi, A., Iqbal, Y., Malik, F. P., Iqbal, M., & Amin, F. (2023). In-situ formation of Azo dye capped-silver nanoparticles and their nanocomposite with reduced graphene oxide for dye degradation. *Materials Chemistry and Physics*, *308*, 128299. doi:10.1016/j.matchemphys.2023.128299

Nunes, G. S., Lins, J. A. P., Silva, F. G. S., Araujo, L. C., Silva, F. E. P. S., Mendonça, C. D., Badea, M., Hayat, A., & Marty, J. L. (2014). Design of a macroalgae amperometric biosensor; application to the rapid monitoring of organophosphate insecticides in an agroecosystem. *Chemosphere*, *111*, 623–630. doi:10.1016/j.chemosphere.2014.05.028 PMID:24997974

Ocheje Ameh, P., Okon Eddy, N., Ameh, P. O., & Eddy, N. O. (2020). Theoretical and Experimental Investigations of the Corrosion Inhibition Action of Piliostigma Thonningii Extract on Mild Steel in Acidic Medium. *Communication in Physical Science, 2018*(1), 27–42. https://journalcps.com/

Ocsoy, I., Tasdemir, D., Mazicioglu, S., & Tan, W. (2018). Nanotechnology in plants. In R. Varshney, M. Pandey, & A. Chitikineni (Eds.), *Plant Genetics and Molecular Biology* (pp. 263–275). Springer., doi:10.1007/10_2017_53

Ocsoy, I., Temiz, M., Celik, C., Altinsoy, B., Yilmaz, V., & Duman, F. (2017). A green approach for formation of silver nanoparticles on magnetic graphene oxide and highlyeffective antimicrobial activity and reusability. *Journal of Molecular Liquids*, *227*, 147–152. doi:10.1016/j.molliq.2016.12.015

Odoemelam, S. A., Oji, E. O., Eddy, N. O., Garg, R., Garg, R., Islam, S., Khan, M. A., Khan, N. A., & Zahmatkesh, S. (2023). Zinc oxide nanoparticles adsorb emerging pollutants (glyphosate pesticide) from aqueous solutions. *Environmental Monitoring and Assessment*, *195*(6), 78195. doi:10.1007/s10661-023-11255-0 PMID:37166547

Ogoko, E. C., Kelle, H. I., Akintola, O., & Eddy, N. O. (2023). Experimental and theoretical investigation of *Crassostrea* gigas (gigas) shells based CaO nanoparticles as a photocatalyst for the degradation of bromocresol green dye (BCGD) in an aqueous solution. *Biomass Conversion and Biorefinery*. doi:10.1007/s13399-023-03742-8

Ohashi, T., & Dai, L. (2006). C60 and carbon nanotube sensors. Carbon Nanotechnology: *Recent Developments in Chemistry, Physics, Materials Science and Device Applications*, 525–575. doi:10.1016/B978-044451855-2/50018-8

Oh, S. T., Yoon, S. J., Choa, Y. H., Jeong, Y. K., & Niihara, K. (2006). Wear Behavior of Nano-Sized Metal Particle Dispersed Al_2O_3 Nanocomposites. *Key Engineering Materials*, *317–318*, 369–372. . doi:10.4028/www.scientific.net/KEM.317-318.369

Omietimi, H. B., Afolalu, S. A., Kayode, J. F., Monye, S. I., Lawal, S. L., & Emetere, M. E. (2023). An overview of nanotechnology and its application. *E3S Web of Conferences, 391*. doi:10.1051/e3sconf/202339101079

Onaizi, A. M., Huseien, G. F., Lim, N. H. A. S., Amran, M., & Samadi, M. (2021). Effect of nanomaterials inclusion on sustainability of cement-based concretes: A comprehensive review. *Construction & Building Materials*, *306*, 124850. doi:10.1016/j.conbuildmat.2021.124850

Onyeaka, H., Passaretti, P., Miri, T., & Al-Sharify, Z. T. (2022). The safety of nanomaterials in food production and packaging. *Current Research in Food Science*, *5*, 763–774. doi:10.1016/j.crfs.2022.04.005 PMID:35520272

Osman, R. B., & Swain, M. V. (2015). A critical review of dental implant materials with anemphasis on titanium versus zirconia. *Materials (Basel)*, *8*(3), 932–958. doi:10.3390/ma8030932 PMID:28787980

Othman, S. H. (2014). Bio-nanocomposite materials for food packaging applications: Types of biopolymer and nano-sized filler. *Agriculture and Agricultural Science Procedia*, *2*, 296–303. doi:10.1016/j.aaspro.2014.11.042

Oyewo, O. A., Agboola, O., Onyango, M. S., Popoola, P., & Bobape, M. F. (2018). Current Methods for the Remediation of Acid Mine Drainage Including Continuous Removal of Metals From Wastewater and Mine Dump. In M. N. V. Prasad, P. J. de C. Favas, & S. K. B. T.-B.-G. for M. S. R. Maiti (Eds.), Bio-Geotechnologies for Mine Site Rehabilitation (pp. 103–114). Elsevier. doi:10.1016/B978-0-12-812986-9.00006-3

Pacheco, I., & Buzea, C. (2021). Nanomaterials and Nanocomposites: Classification and Toxicity. In O. V. Kharissova, L. M. Torres-Martínez, & B. I. Kharisov (Eds.), *Handbook of Nanomaterials and Nanocomposites for Energy and Environmental Applications*. Springer. doi:10.1007/978-3-030-36268-3_

Pacquit, A., Frisby, J., Diamond, D., Lau, K. T., Farrell, A., Quilty, B., & Diamond, D. (2007). Development of a smart packaging for the monitoring of fish spoilage. *Food Chemistry*, *102*(2), 466–470. doi:10.1016/j.foodchem.2006.05.052

Padmanabhan, S. (2023). *No. 1* (Vol. 29). Nanotechnology in Orthodontics. In Seminars in Orthodontics. WB Saunders., doi:10.1053/j.sodo.2023.01.005

Pak, S. C., Penrose, W., & Hesketh, P. J. (2001). An ultrathin platinum film sensor to measure biomolecular binding. In *Biosensors & Bioelectronics, 16*. www.elsevier.com/locate/bios

Palani, G., Trilaksana, H., Sujatha, R. M., Kannan, K., Rajendran, S., Kornirjrnko, K., Nykiel, M., & Uthayakuma, M. (2023). Silver nanoparticles for waste water management. *Molecules (Basel, Switzerland)*, *28*(8), 3520. doi:10.3390/molecules28083520 PMID:37110755

Panacek, A., Prucek, R., Safarowa, D., Dittrich, M., Richtrova, J., Benickova, K., Zhoril, R. & Kvitek, L. (2011). Acute and chronic toxicity effects of silver nanoparticles (nps) on drosophila melanogaster. *environmental Science & Technology*, *45* (11), 4974-4979, DOI: doi:10.1021/es104216b

Panda, S. K. B. C., Sen, K., & Mukhopadhyay, S. (2023). Sustainable Photocatalytic Desizing Process for the Starch-Based Size. *ACS Omega*, *8*(21), 18726–18734. doi:10.1021/acsomega.3c00713 PMID:37273639

Pandey, S., Makhado, E., Kim, S., & Kang, M. (2023). Recent developments of polysaccharide based superabsorbent nanocomposite for organic dye contamination removal from wastewater — A review. *Environmental Research, 217*, 114909. doi:10.1016/j.envres.2022.114909 PMID:36455632

Panel, M. (2019). Byte by byte: Policy innovation for transforming Africa's food system with digital technologies, Dakar. Dakar, Senegal. *International Food Policy Research Institute of Bonn, 9780896296848.*

Panou, A., & Karabagias, I. K. (2023). Biodegradable Packaging Materials for Foods Preservation: Sources, Advantages, Limitations, and Future Perspectives. *Coatings, 13*(7), 1176. doi:10.3390/coatings13071176

Pan, S., Lin, H., Deng, J., Chen, P., Chen, X., Yang, Z., & Peng, H. (2015). Novel wearable energy devices based on aligned carbon nanotube fiber textiles. *Advanced Energy Materials, 5*(4), 1401438. doi:10.1002/aenm.201401438

Papi, M., De Spirito, M., & Palmieri, V. (2023). Nanotechnology in the COVID-19 era: Carbon-based nanomaterials as a promising solution. *Carbon, 210*, 118058. doi:10.1016/j.carbon.2023.118058 PMID:37151958

Park, M. V. D. Z., Neigh, A. M., Vermeulen, J. P., de la Fonteyne, L. J. J., Verharen, H., Briedé, van Loveren, H., & de Jong, W. H. (2011). The effect of particle size on the cytotoxicity, inflammation, developmental toxicity and genotoxicity of silver nanoparticles. *Biomaterials, 32*(36), 9810–9817. doi:10.1016/j.biomaterials.2011.08.085 PMID:21944826

Park, Y. W., Kim, S. M., Lee, J. Y., & Jang, W. (2015). Application of biosensors in smart packaging. *Molecular & Cellular Toxicology, 11*(3), 277–285. doi:10.1007/s13273-015-0027-1

Patel, G., Pillai, V., Bhatt, P., & Mohammad, S. (2020). Application of nanosensors in the food industry. In B. Han, V. K. Tomer, T. A. Nguyen, A. Farmani, & P. Kumar Singh (Eds.), *Nanosensors for Smart Cities*. Elsevier. doi:10.1016/B978-0-12-819870-4.00020-7

Patel, V., Kruse, P., & Selvaganapathy, P. R. (2020). Solid State Sensors for Hydrogen Peroxide Detection. *Biosensors (Basel), 11*(1), 9. doi:10.3390/bios11010009 PMID:33375685

Patra, J. K., & Baek, K. H. (2016). Biosynthesis of silver nanoparticles usingaqueous extract of silky hairs of corn and investigation of its antibacterial and anticandidal synergistic activity and antioxidant potential. *IET Nanobiotechnology / IET, 10*(5), 326–333. doi:10.1049/iet-nbt.2015.0102 PMID:27676382

Pawlak, K., & Kołodziejczak, M. (2020). The role of agriculture in ensuring food security in developing countries: Considerations in the context of the problem of sustainable food production. *Sustainability (Basel), 12*(13), 5488. doi:10.3390/su12135488

Payal, & Pandey, P. (2021). Role of Nanotechnology in Electronics: A Review of Recent Developments and Patents. *Recent Patents on Nanotechnology, 16*(1), 45–66. doi:10.2174/1872210515666210120114504

Peña-Bahamonde, J., Nguyen, H.N., Fanourakis, S.K. *et al.* Recent advances in graphene-based biosensor technology with applications in life sciences.

Peng, L., Dong, S., Wei, W., Yuan, X., & Huang, T. (2017). Synthesis of reticulated hollow spheres structure NiCo2S4 and its application in organophosphate pesticides biosensor. *Biosensors & Bioelectronics*, *92*, 563–569. doi:10.1016/j.bios.2016.10.059 PMID:27836591

Perera, K. Y., Jaiswal, A. K., & Jaiswal, S. (2023). *Biopolymer-Based Sustainable Food Packaging Materials: Challenges, Solutions, and Applications*. Foods.

Pérez-Lucas, G., El Aatik, A., Aliste, M., Navarro, G., Fenoll, J., & Navarro, S. (2023). Removal of Contaminants of Emerging Concern from a Wastewater Effluent by Solar-Driven Heterogeneous Photocatalysis: A Case Study of Pharmaceuticals. *Water, Air, and Soil Pollution*, *234*(1), 55. Advance online publication. doi:10.1007/s11270-023-06075-4

Petala, E., Baikousi, M., Karakassides, M. A., Zoppellaro, G., Filip, J., Tuček, J., Vasilopoulos, K. C., Pechoušek, J., & Zbořil, R. (2016). Synthesis, physical properties and application of the zero-valent iron/titanium dioxide heterocomposite having high activity for the sustainable photocatalytic removal of hexavalent chromium in water. *Physical Chemistry Chemical Physics*, *18*(15), 10637–10646. doi:10.1039/C6CP01013J PMID:27035846

Phanthong, P., Reubroycharoen, P., Hao, X., Xu, G., Abudula, A., & Guan, G. (2018). Nanocellulose: Extraction and application. *Carbon Resources Conversion*, *1*(1), 32–43. doi:10.1016/j.crcon.2018.05.004

Philip, D. (2010). Green synthesis of gold and silver nanoparticles using Hibiscus rosa sinensis. *Phys. E Low-Dimens. Physica E, Low-Dimensional Systems and Nanostructures*, *42*(5), 1417–1424. doi:10.1016/j.physe.2009.11.081

Pinto, R. J., Lucas, J. M., Silva, F. M., Girão, A. V., Oliveira, F. J., Marques, P. A., & Freire, C. S. (2019). Bio-based synthesis of oxidation resistant copper nanowires using an aqueous plant extract. *Journal of Cleaner Production*, *221*, 122–131. doi:10.1016/j.jclepro.2019.02.189

Pi, Y., Li, X., Xia, Q., Wu, J., Li, Y., Xiao, J., & Li, Z. (2018). Adsorptive and photocatalytic removal of Persistent Organic Pollutants (POPs) in water by metal-organic frameworks (MOFs). *Chemical Engineering Journal*, *337*, 351–371. doi:10.1016/j.cej.2017.12.092

Poggio, C., Lombardini, M., Colombo, M., Dagna, A., Saino, E., Arciola, C. R., & Visai, L. (2011). Antibacterial effects of six endodontic sealers. *The International Journal of Artificial Organs*, *34*(9), 908–913. doi:10.5301/ijao.5000055 PMID:22094573

Pospiskova, K., Safarik, I., Sebela, M., & Kuncova, G. (2013). Magnetic particle–based biosensor for biogenic amines using an optical oxygen sensor as a transducer. *Mikrochimica Acta*, *180*(3–4), 311–318. doi:10.1007/s00604-012-0932-0

Pourmadadi, M., Rajabzadeh-Khosroshahi, M., Saeidi Tabar, F., Ajalli, N., Samadi, A., Yazdani, M., Yazdian, F., Rahdar, A., & Díez-Pascual, A. M. (2022, October 26). Two-Dimensional Graphitic Carbon Nitride (g-C3N4) Nanosheets and Their Derivatives for Diagnosis and Detection Applications. *Journal of Functional Biomaterials*, *13*(4), 204. doi:10.3390/jfb13040204 PMID:36412845

Pouzesh, M., Nekouei, S., Ferdosi Zadeh, M. A., Keshtpour, F., Wang, S., & Nekouei, F. (2019). Fabrication of stable copper nanoparticles embedded in nanocellulose film as a bionanocomposite plasmonic sensor and thereof for optical sensing of cyanide ion in water samples. *Cellulose (London, England)*, *26*(8), 4945–4956. doi:10.1007/s10570-019-02405-0

Prabhu, B. M., Ali, S. F., Murdock, R. C., Hussain, S. M., & Srivatsan, M. (2010). Copper nanoparticles exert size and concentration dependent toxicity on somatosensory neurons of rat. *Nanotoxicology*, *4*(2), 150–160. doi:10.3109/17435390903337693 PMID:20543894

Prameela, K. L. (2017). Nanomaterial's applications in agriculture. *J Chem Pharm Sci*, *10*(1), 593–596.

Prasad, R., Bhattacharyya, A., & Nguyen, Q. D. (2017). Nanotechnology in sustainable agriculture: Recent developments, challenges, and perspectives. *Frontiers in Microbiology*, *8*, 1014. doi:10.3389/fmicb.2017.01014 PMID:28676790

Prasad, R., Pandey, R., & Barman, I. (2016). Engineering tailored nanoparticles withmicrobes: Quo vadis? *Wiley Interdisciplinary Reviews. Nanomedicine and Nanobiotechnology*, *8*(2), 316–330. doi:10.1002/wnan.1363 PMID:26271947

Primožič, M., Knez, Ž., & Leitgeb, M. (2021). (Bio)nanotechnology in Food Science-Food Packaging. *Nanomaterials (Basel, Switzerland)*, *11*(2), 292. doi:10.3390/nano11020292 PMID:33499415

Puglia, D. Fortunati, E., & Kenny, J. (2016). Multifunctional Polymeric Nanocomposites Based on Cellulosic Reinforcements (First Edit). William Andrew Publishing. doi:10.1016/C2014-0-04565-0

Pujar, A. S., & Bagawan, A. M. (2016). Applications of Nanotechnology: In Medicine. *International Journal of Scientific Research*, *7*(6), 1211–1215.

Pumera, M. (2011). Graphene in biosensing. *Materials Today*, *14*(7–8), 308–315. doi:10.1016/S1369-7021(11)70160-2

Qian, Z. S., Shan, X. Y., Chai, L. J., Ma, J. J., Chen, J. R., & Feng, H. (2014). DNA nanosensor based on biocompatible graphene quantum dots and carbon nanotubes. *Biosensors & Bioelectronics*, *60*, 64–70. doi:10.1016/j.bios.2014.04.006 PMID:24768864

Qin, M., Yuan, W., Zhang, X., Cheng, Y., Xu, M., Wei, Y., Chen, W., & Huang, D. (2022). Di Huang. Preparation of PAA/PAM/MXene/TA hydrogel with antioxidant, healable ability as a strain sensor. *Colloids and Surfaces. B, Biointerfaces*, *214*, 112482. doi:10.1016/j.colsurfb.2022.112482 PMID:35366577

Qiu, X., Zhang, Y., Zhu, Y., Long, C., Su, L., Liu, S., & Tang, Z. (2021). Applications of nanomaterials in asymmetric photocatalysis: Recent progress, challenges, and opportunities. *Advanced Materials, 33*(6), 2001731. doi:10.1002/adma.202001731 PMID:32672886

Qurashi, A., Rather, J. A., Yamazaki, T., Sohail, M., De Wael, K., Merzougui, B., & Hakeem, A. S. (2015). Swift electrochemical detection of paraben an endocrine disruptor by In2O3 nanobricks. *Sensors and Actuators. B, Chemical, 221*, 167–171. doi:10.1016/j.snb.2015.06.026

Qu, X., Alvarez, P. J., & Li, Q. (2013). Applications of nanotechnology in water and wastewater treatment. *Water Research, 47*(12), 3931–3946. doi:10.1016/j.watres.2012.09.058 PMID:23571110

Rabiee, N., Khatami, M., Soufi, G. J., Fatahi, Y., Iravani, S., & Varma, R. S. (2021). Diatoms with Invaluable Applications in Nanotechnology, Biotechnology, and Biomedicine: Recent Advances. *ACS Biomaterials Science & Engineering, 7*(7), 3053–3068. doi:10.1021/acsbiomaterials.1c00475 PMID:34152742

Rahman, K. T., Moin, T. S., Chowdhury, M. F. M., & Khan, M. N. (2023). Emerging Nano-Enable Materials in the Sports Industry. *Emerging Applications of Nanomaterials, 141*, 75–100. doi:10.21741/9781644902288-4

Rahman, M., Zahin, F., Saadi, M. A. S. R., Sharif, A., & Hoque, M. E. (2018). *Surface Modification of Advanced and Polymer Nanocomposites*., doi:10.1007/978-3-319-76090-2_6

Rajangam, B., Daniel, D. K., & Krastanov, A. I. (2018). Progress in enzyme inhibition based detection of pesticides. *Engineering in Life Sciences, 18*(1), 4–19. doi:10.1002/elsc.201700028 PMID:32624856

Rajarathinam, T., Kang, M., Hong, S., & Chang, S.-C. (2023). Nanocomposite-Based Electrochemical Sensors for Neurotransmitters Detection in Neurodegenerative Diseases. *Chemosensors (Basel, Switzerland), 11*(2), 103. doi:10.3390/chemosensors11020103

Rajendiran, R., Seelam, P. K., Patchaiyappan, A., Balla, P., Shankar, H., Ravi, B., Perupogu, V., & Lassi, U. (2023). Morphologically tailored facet dependent silver nanoparticles supported α-Al2O3 catalysts for chemoselective reduction of aromatic nitro compounds. *Chemical Engineering Journal, 451*, 138507. doi:10.1016/j.cej.2022.138507

Rajeshkumar, S., Bharath, L. V., & Geetha, R. (2019). Broad spectrum antibacterial silver nanoparticle green synthesis: Characterization, and mechanism of action. In A. K. Shukla & S. Iravani (Eds.), *In Micro and Nano Technologies, Green synthesis, characterization and applications of nanoparticles* (pp. 429–444). Elsevier. doi:10.1016/B978-0-08-102579-6.00018-6

Rajesh, U. C., Wang, J., Prescott, S., Tsuzuki, T., & Rawat, D. S. (2015). RGO/ZnO nanocomposite: An efficient, sustainable, heterogeneous, amphiphilic catalyst for synthesis of 3-substituted indoles in water. *ACS Sustainable Chemistry & Engineering, 3*(1), 9–18. doi:10.1021/sc500594w

Raj, S., Jose, S., Sumod, U. S., & Sabitha, M. (2012). Nanotechnology in Cosmetics: Opportunities and Challenges. *Journal of Pharmacy & Bioallied Sciences, 4*(3), 186–193. doi:10.4103/0975-7406.99016 PMID:22923959

Ram, B., Chauhan, G. S., Mehta, A., Gupta, R., & Chauhan, K. (2018). Spherical nanocellulose-based highly efficient and rapid multifunctional naked-eye Cr(VI) ion chemosensor and adsorbent with mild antimicrobial properties. *Chemical Engineering Journal, 349*(Vi), 146–155. doi:10.1016/j.cej.2018.05.085

Ramesh, M., Janani, R., Deepa, C., & Rajeshkumar, L. (2023a). Nanotechnology-Enabled Biosensors: A Review of Fundamentals, Design Principles, Materials, and Applications. In Biosensors, 13(1). MDPI. doi:10.3390/bios13010040

Ramezani Farani, M., Farsadrooh, M., Zare, I., Gholami, A., & Akhavan, O. (2023). Green Synthesis of Magnesium Oxide Nanoparticles and Nanocomposites for Photocatalytic Antimicrobial, Antibiofilm and Antifungal Applications. *Catalysts*, *13*(4), 642. doi:10.3390/catal13040642

Ramos, A. P., Cruz, M. A., Tovani, C. B., & Ciancaglini, P. (2017). Biomedical applications of nanotechnology. *Biophysical Reviews*, *9*(2), 79–89. doi:10.1007/s12551-016-0246-2 PMID:28510082

Ramos, M., Valdés, A., & Garrigós, M. C. (2016). Multifunctional Applications of Nanocellulose-Based Nanocomposites. In D. Puglia, E. Fortunati, & J. M. Kenny (Eds.), *Multifunctional Polymeric Nanocomposites Based on Cellulosic Reinforcements* (pp. 177–204). William Andrew Publishing. doi:10.1016/B978-0-323-44248-0.00006-7

Ramos-Zúñiga, J., Bruna, N., & Pérez-Donoso, J. M. (2023). Toxicity Mechanisms of Copper Nanoparticles and Copper Surfaces on Bacterial Cells and Viruses. *International Journal of Molecular Sciences*, *10503*(13), 10503. doi:10.3390/ijms241310503 PMID:37445681

Ramsden, J. J. (1990). What is nanotechnology? *Nature*, *344*, 524–526.

Rani, N., Kumari, K., Sangwan, P., Barala, P., Yadav, J., Vijeta, Rahul, & Hooda, V. (2022). Nano-Iron and Nano-Zinc Induced Growth and Metabolic Changes in Vigna radiata. *Sustainability (Basel)*, *14*(14), 8251. doi:10.3390/su14148251

Rani, S., Sharma, S., Bansal, M., Garg, R., & Garg, R. (2022). *Enhanced Zn(II) adsorption by chemically modified sawdust based biosorbents*. Environmental Science and Pollution Research, Ii., doi:10.1007/s11356-022-22963-4

Rao, C. N. R., Sood, A. K., Subrahmanyam, K. S., & Govindaraj, A. (2009). Graphene: The new two-dimensional nanomaterial. In Angewandte Chemie - International Editioni, 48(42), 7752–7777. doi:10.1002/anie.200901678

Rao, A., Bankar, A., Kumar, A. R., Gosavi, S., & Zinjarde, S. (2013). Removal of hexavalent chromium ions by Yarrowia lipolytica cells modified with phyto-inspired Fe0/Fe3O4 nanoparticles. *Journal of Contaminant Hydrology*, *146*, 63–73. doi:10.1016/j.jconhyd.2012.12.008 PMID:23422514

Raorane, D. V., Chaughule, R. S., Pednekar, S. R., & Lokur, A. (2019). Experimentalsynthesis of size-controlled TiO2 nanofillers and their possible use as composites in restorative dentistry. *The Saudi Dental Journal*, *31*(2), 194–203. doi:10.1016/j.sdentj.2019.01.008 PMID:30983829

Rasheed, A., & Khalid, F. A. (2014, June). Fabrication and properties of CNTs reinforced polymeric matrix nanocomposites for sports applications. []. IOP Publishing.]. *IOP Conference Series. Materials Science and Engineering*, *60*(1), 012009. doi:10.1088/1757-899X/60/1/012009

Rasheed, T. (2023). Carbon dots as potential greener and sustainable fluorescent nanomaterials in service of pollutants sensing. *Trends in Analytical Chemistry*, *158*, 116841. doi:10.1016/j.trac.2022.116841

Rastgoo, M., Montazer, M., Malek, R. M. A., Harifi, T., & Mahmoudi Rad, M. (2016). Ultrasound mediation for one-pot sonosynthesis and deposition of magnetite nanoparticles on cotton/polyester fabric as a novel magnetic, photocatalytic, sonocatalytic, antibacterial and antifungal textile. *Ultrasonics Sonochemistry*, *31*, 257–266. doi:10.1016/j.ultsonch.2016.01.008 PMID:26964948

Rastogi, A., Tripathi, D.K., Yadav, S. *et al.* A(2019). pplication of silicon nanoparticles in agriculture. *3 Biotech* **9**, 90. . doi:10.1007/s13205-019-1626-7

Rauf, A., Abu-Izneid, T., Ahmed Khalil, A., Hafeez, N., Olatunde, A., Mominur Rahman, M., Semwal, P., Saleh Al-Awthan, Y., Salem Bahattab, O., Khan, I. N., Arslan Khan, M., & Sharma, R. (2022). Nanoparticles in clinical trials of COVID-19: An update. *International Journal of Surgery*, *104*, 106818. doi:10.1016/j.ijsu.2022.106818 PMID:35953020

Raura, N., Garg, A., Arora, A., & Roma, M. (2020). Nanoparticle technology and its implications in endodontics: A review. *Biomaterials Research*, *24*(1), 1–8. doi:10.1186/s40824-020-00198-z PMID:33292702

Raza, A., Altaf, S., Ali, S., Ikram, M., & Li, G. (2022). Recent advances in carbonaceous sustainable nanomaterials for wastewater treatments. *Sustainable Materials and Technologies*, *32*, e00406. doi:10.1016/j.susmat.2022.e00406

Recordati, C., De Maglie, M., Bianchessi, S., Argentiere, S., Cella, C., Mattiello, S., Cubadda, F., Aureli, F., D'Amato, M., Raggi, A., Lenardi, C., Milani, P., & Scanziani, E. (2016). Tissue distribution and acute toxicity of silver after single intravenous administration in mice: Nano-specific and size-dependent effects. *Particle and Fibre Toxicology*, *13*(1), 12. doi:10.1186/s12989-016-0124-x PMID:26926244

Remya, R. R., Julius, A., Suman, T. Y., Mohanavel, V., Karthick, A., Pazhanimuthu, C., Samrot, A. V., & Muhibbullah, M. (2022). Role of Nanoparticles in Biodegradation and Their Importance in Environmental and Biomedical Applications. *Journal of Nanomaterials*, *6090846*, 1–15. doi:10.1155/2022/6090846

Riazi, H., Taghizadeh, G., & Soroush, M. (2021). MXene-Based Nanocomposite Sensors. In ACS Omega (Vol. 6, Issue 17, pp. 11103–11112). American Chemical Society (ACS). https://doi.org/ doi:10.1021/acsomega.0c05828

Rick, J., Tsai, M.-C., & Hwang, B. (2015). Biosensors Incorporating Bimetallic Nanoparticles. *Nanomaterials (Basel, Switzerland)*, *6*(1), 5. doi:10.3390/nano6010005 PMID:28344262

Riyanti, F., Hasanudin, H., Rachmat, A., Purwaningrum, W., & Hariani, P. L. (2023). Photocatalytic degradation of methylene blue and Congo red dyes from aqueous solutions by bentonite-Fe3O4 magnetic. *Communications in Science and Technology*, *8*(1), 1–9. doi:10.21924/cst.8.1.2023.1007

Rizi, H. A., Shin, D. H., & Rizi, S. Y. (2022). Polymeric Nanoparticles in Cancer Chemotherapy: A Narrative Review. *Iranian Journal of Public Health*, *51*(2), 226–239. doi:10.18502/ijph.v51i2.8677 PMID:35866132

Robles, E., Labidi, J., Halász, K., & Csóka, L. (2017). Key issues in reinforcement involving nanocellulose. In M. Jawadi, S. Boufi, & K. H.P.S (Eds.), Cellulose-Reinforced Nanofibre Composites: Production, Properties and Applications (pp. 401–425). Woodhead Publishing. doi:10.1016/B978-0-08-100957-4.00018-8

Rocchitta, G., Spanu, A., Babudieri, S., Latte, G., Madeddu, G., Galleri, G., Nuvoli, S., Bagella, P., Demartis, M. I., Fiore, V., Manetti, R., & Serra, P. A. (2016). Enzyme biosensors for biomedical applications: Strategies for safeguarding analytical performances in biological fluids. In Sensors (Switzerland), 16(6). MDPI AG. doi:10.3390/s16060780

Rocha-Santos, T. A. P. (2014). Sensors and biosensors based on magnetic nanoparticles. In *TrAC - Trends in Analytical Chemistry* (Vol. 62, pp. 28–36). Elsevier B.V., doi:10.1016/j.trac.2014.06.016

Rodrigues, M. C., Rolim, W. R., Viana, M. M., Souza, T. R., Gonçalves, F., Tanaka, C. J., Bueno-Silva, B., & Seabra, A. B. (2020). Biogenic synthesis and antimicrobial activity of silica-coated silvernanoparticles for esthetic dental applications. *Journal of Dentistry*, *96*, 103327. doi:10.1016/j.jdent.2020.103327 PMID:32229160

Rodrigues, S. S. M., Ribeiro, D. S. M., Soares, J. X., Passos, M. L. C., Saraiva, M. L. M. F. S., & Santos, J. L. M. (2017). Application of nanocrystalline CdTe quantum dots in chemical analysis: Implementation of chemo-sensing schemes based on analyte-triggered photoluminescence modulation. *Coordination Chemistry Reviews*, *330*, 127–143. doi:10.1016/j.ccr.2016.10.001

Rodriguez, J. L., Dalia, A. B., & Weiser, J. N. (2012). Increased chain length promotes pneumococcal adherence and colonization. *Infection and Immunity*, *80*(10), 3454–3459. doi:10.1128/IAI.00587-12 PMID:22825449

Rodriguez-Perez, L., Herranz, M. Á., & Martín, N. (2013). The chemistry of pristine graphene. *Chemical Communications (Cambridge)*, *49*(36), 3721–3735. doi:10.1039/c3cc38950b PMID:23539514

Roduner, E. (2006). Size matters: Why nanomaterials are different. *Chemical Society Reviews*, *35*(7), 583–592. doi:10.1039/b502142c PMID:16791330

Rojas-Velázquez, Á. N., Guillén-Castillo, O. I., Alcalá-Jauregui, J. A., Loredo-Osti, C., Ramírez-Tobías, H. M., Romero-Méndez, M. J., Méndez-Cortés, H., & Hernández-Montoya, A. (2023). Effect of a nitrogenous nanocomposite on leaching and N content in lettuce in soil columns. *Discover Nano*, *18*(1), 98. doi:10.1186/s11671-023-03874-w PMID:37523022

Roman, M. (2015). Toxicity of cellulose nanocrystals: A review. *Industrial Biotechnology (New Rochelle, N.Y.)*, *11*(1), 25–33. doi:10.1089/ind.2014.0024

Rosário, F., Creylman, J., Verheyen, G., Van Miert, S., Santos, C., Hoet, P., & Oliveira, H. (2022). Impact of Particle Size on Toxicity, Tissue Distribution and Excretion Kinetics of Subchronic Intratracheal Instilled Silver Nanoparticles in Mice. *Toxics*, *10*(5), 260. doi:10.3390/toxics10050260 PMID:35622673

Roy, A., & Bhattacharya, J. (2015). *Nanotechnology in industrial wastewater treatment.* IWA Publishing. https://library.oapen.org/handle/20.500.12657/25811

Roy, A., Dwari, D., Ram, M. K., & Datta, P. (2022). Piezoelectric nanomaterials for biomedical applications. In Food, Medical, and Environmental Applications of Nanomaterials (pp. 355-377). Elsevier. doi:10.1016/B978-0-12-822858-6.00012-1

Roy, S., Han, G. S., Shin, H., Lee, J. W., Mun, J., Shin, H., & Jung, H. S. (2015). Low Temperature Synthesis of Rutile TiO2 Nanocrystals and Their Photovoltaic and Photocatalytic Properties. Journal of Nanoscience and Nanotechnology. *American Scientific Publishers.*, *6*, 4516–4521. doi:10.1166/jnn.2015.9778 PMID:26369074

Ruiz, C., Soriano, L., & Valcárcel, M. (2017). Nanocellulose as analyte and analytical tool: Opportunities and challenges. *Trends in Analytical Chemistry*, *87*, 1–18. doi:10.1016/j.trac.2016.11.007

Saba, N., Jawaid, M., & Sultan, M. T. H. (2017). Thermal properties of oil palm biomass based composites. In M. Jawadi, P. Tahir, & N. Saba (Eds.), *Lignocellulosic Fibre and Biomass-Based Composite Materials: Processing, Properties and Applications* (Vol. 1, pp. 95–122). Woodhead Publishing., doi:10.1016/B978-0-08-100959-8.00006-8

Saba, N., Paridah, M. T., Abdan, K., & Ibrahim, N. A. (2016). Effect of oil palm nano filler on mechanical and morphological properties of kenaf reinforced epoxy composites. *Construction & Building Materials*, *123*, 15–26. doi:10.1016/j.conbuildmat.2016.06.131

Sachdeva, H. (2020). Recent advances in the catalytic applications of GO/rGO for green organic synthesis. *Green Process Synth*, *9*(1), 515–537. doi:10.1515/gps-2020-0055

Sadeghi, K., Lee, Y., & Seo, J. (2021). Ethylene Scavenging Systems in Packaging of Fresh Produce: A Review. *Food Reviews International*, *37*(2), 155–176. doi:10.1080/87559129.2019.1695836

Saha, K., Agasti, S. S., Kim, C., Li, X., & Rotello, V. M. (2012). Gold nanoparticles in chemical and biological sensing. In Chemical Reviews (Vol. 112, Issue 5, pp. 2739–2779). doi:10.1021/cr2001178

Sahu, A. K., Acharya, P. P., Kar, B. R., Nemani, R., & Gowda, M. (2019). Nanotechnology In Water Treatment, Case Study In Bangalore. *International Journal of Advanced Research*, *7*(2), 142–149. doi:10.21474/IJAR01/8470

Saleem, H., & Zaidi, S. (2020). Sustainable Use of Nanomaterials in Textiles and Their Environmental Impact. *Materials (Basel)*, *13*(22), 5134. doi:10.3390/ma13225134 PMID:33203051

Salehabadi, A., & Enhessari, M. (2018). Application of (mixed) metal oxides-based nanocomposites for biosensors. *Materials for Biomedical Engineering*, 357-396. https://doi.org/ doi:10.1016/B978-0-08-102814-8.00013-5

Sandra, V., Stojanovic, B., Ivanović, L., Miladinovic, S., & Milojević, S. (2018). *Application of nanocomposites in the automotive industry. Mobility and Vehicle Mechanics*. MVM.

Santhosh, C., Malathi, A., Daneshvar, E., Kollu, P., & Bhatnagar, A. (2018). Photocatalytic degradation of toxic aquatic pollutants by novel magnetic 3D-TiO2@HPGA nanocomposite. *Scientific Reports*, *8*(1), 1–15. doi:10.1038/s41598-018-33818-9 PMID:29311619

Sanuja, S., Agalya, A., & Umapathy, M. J. (2015). Synthesis and characterization of zinc oxide–neem oil–chitosan bionanocomposite for food packaging application. *International Journal of Biological Macromolecules*, *74*, 76–84. doi:10.1016/j.ijbiomac.2014.11.036 PMID:25499891

Sarfraz, J., Gulin-Sarfraz, T., Nilsen-Nygaard, J., & Pettersen, M. K. (2020). Nanocomposites for Food Packaging Applications: An Overview. *Nanomaterials (Basel, Switzerland)*, *11*(1), 10. doi:10.3390/nano11010010 PMID:33374563

Sarwar, A., Wang, J., Khan, M. S., Farooq, U., Riaz, N., Nazir, A., Mahmood, Q., Hashem, A., Al-Arjani, A. B. F., & Alqarawi, A. A., & Abd_allah, E. F. (. (2021). Iron oxide (Fe3o4)-supported sio2 magnetic nanocomposites for efficient adsorption of fluoride from drinking water: Synthesis, characterization, and adsorption isotherm analysis. *Water (Switzerland)*, *13*(11), 1514. doi:10.3390/w13111514

Schaefer, D., & Cheung, W. M. (2018). Smart Packaging: Opportunities and Challenges. *Procedia CIRP*, *72*, 1022–1027. doi:10.1016/j.procir.2018.03.240

Scholz, S., Kondakov, D., Lüssem, B., & Leo, K. (2015). Degradation Mechanisms and Reactions in Organic Light-Emitting Devices. *Chemical Reviews*, *115*(16), 8449–8503. doi:10.1021/cr400704v PMID:26230864

Schrand, A. M., Rahman, M. F., Hussain, S. M., Schlager, J. J., Smith, D. A., & Syed, A. F. (2010). Metal-based nanoparticles and their toxicity assessment. In Wiley Interdisciplinary Reviews: Nanomedicine and Nanobiotechnology (Vol. 2, Issue 5, pp. 544–568). doi:10.1002/wnan.103

Schulte, P. A., & Salamanca-Buentello, F. (2007). Ethical and Scientific Issues of Nanotechnology in the Workplace. *Environmental Health Perspectives*, *115*(1), 5–1. doi:10.1289/ehp.9456 PMID:17366812

Sciau, P. (2012). Nanoparticles in Ancient Materials: The Metallic Lustre Decorations of Medieval Ceramics. In The Delivery of Nanoparticles. InTech. doi:10.5772/34080

Scientific Committee on Consumer Safety. (2016). *Hydroxypatite (nano)*. European Commission. https://ec.europa.eu/ health/scientific_committees /consumer _safety /docs /sccs_ o_191.pdf

Selvaraj, A., George, A. M., & Rajeshkumar, S. (2021). Efficacy of zirconium oxidenanoparticles coated on stainless steel and nickel titanium wires in orthodontictreatment. *Bioinformation*, *17*(8), 760. doi:10.6026/97320630017760 PMID:35540697

Sen, M. (2020). *Nanocomposite Materials*. www.intechopen.com

Senthilkumar, R., Vaidyanathan, S., & Sivaraman, B. (2013). Effect of copper nanofluid concentration on thermal performance of heat pipes. *Frontiers in Heat Pipes*, *4*, 1–5. doi:10.5098/fhp.v4.1.3004

Seo, G., Lee, G., Kim, M. J., Baek, S. H., Choi, M., Ku, K. B., Lee, C. S., Jun, S., Park, D., Kim, H. G., Kim, S. J., Lee, J. O., Kim, B. T., Park, E. C., & Kim, S. (2020). Rapid Detection of COVID-19 Causative Virus (SARS-CoV-2) in Human Nasopharyngeal Swab Specimens Using Field-Effect Transistor-Based Biosensor. *ACS Nano*, *14*(4), 5135–5142. doi:10.1021/acsnano.0c02823 PMID:32293168

Sethy, N. K., Arif, Z., Mishra, P. K., & Kumar, P. (2020). Green synthesis of TiO2 nanoparticles from Syzygium cumini extract for photo-catalytic removal of lead (Pb) in explosive industrial wastewater. *Green Processing and Synthesis*, *9*(1), 171–181. doi:10.1515/gps-2020-0018

Shah, M., Fawcett, D., Sharma, S., Tripathy, S. K., & Poinern, G. E. J. (2015). Greensynthesis of metallic nanoparticles via biological entities. *Materials (Basel)*, *8*(11), 7278–7308. doi:10.3390/ma8115377 PMID:28793638

Shah, R. K. (2023). Efficient photocatalytic degradation of methyl orange dye using facilely synthesized α-Fe2O3 nanoparticles. *Arabian Journal of Chemistry*, *16*(2), 104444. doi:10.1016/j.arabjc.2022.104444

Shaikh, S., Yaqoob, M., & Aggarwal, P. (2021). An overview of biodegradable packaging in food industry. *Current Research in Food Science*, *4*, 503–520. doi:10.1016/j.crfs.2021.07.005 PMID:34401747

Shalaby, M. N., & Saad, M. M. (2020). Advanced material engineering and nanotechnology for improving sports performance and equipment. *International Journal of Psychosocial Rehabilitation*, *24*(10), 2314–2322.

Shalauddin, M., Akhter, S., Basirun, W. J., Bagheri, S., Anuar, N. S., & Johan, M. R. (2019). Hybrid nanocellulose/f-MWCNTs nanocomposite for the electrochemical sensing of diclofenac sodium in pharmaceutical drugs and biological fluids. *Electrochimica Acta*, *304*, 323–333. doi:10.1016/j.electacta.2019.03.003

Shankar, S. S., Rai, A., Ahmad, A., & Sastry, M. (2004). Rapid synthesis of Au, Ag, and bimetallic Au core–Ag shell nanoparticles using Neem (Azadirachta indica) leaf broth. *Journal of Colloid and Interface Science*, *275*(2), 496–502. doi:10.1016/j.jcis.2004.03.003 PMID:15178278

Sharma, P. (2023). Chapter 6 - Genotoxicity of the nanoparticles, Editor(s): Nar Singh Chauhan, Sarvajeet Singh Gill. In Nanomaterial-Plant Interactions, The Impact of Nanoparticles on Agriculture and Soil. Academic Press. doi:10.1016/B978-0-323-91703-2.00017-8

Sharma, A., Thakur, M., Bhattacharya, M., Mandal, T., & Goswami, S. (2019). Commercial application of cellulose nano-composites – A review. *Biotechnology Reports (Amsterdam, Netherlands)*, *21*, e00316. doi:10.1016/j.btre.2019.e00316 PMID:30847286

Sharma, C., & Bhardwaj, N. K. (2019). Bacterial nanocellulose: Present status, biomedical applications and future perspectives. *Materials Science and Engineering C*, *104*, 109963. doi:10.1016/j.msec.2019.109963 PMID:31499992

Sharma, D., Kanchi, S., Sabela, M. I., & Bisetty, K. (2016). Insight into the biosensing of graphene oxide: Present and future prospects. *Arabian Journal of Chemistry*, *9*(2), 238–261. doi:10.1016/j.arabjc.2015.07.015

Sharma, N. K., Vishwakarma, J., Rai, S., Alomar, T. S., AlMasoud, N., & Bhattarai, A. (2022). Green route synthesis and characterization techniques of silver nanoparticles and their biological adeptness. *ACS Omega*, *7*(31), 27004–27020. doi:10.1021/acsomega.2c01400 PMID:35967040

Sharma, R., Garg, R., Bali, M., & Eddy, N. O. (2023a). Biogenic synthesis of iron oxide nanoparticles using leaf extract of Spilanthes acmella: Antioxidation potential and adsorptive removal of heavy metal ions. *Environmental Monitoring and Assessment*, *195*(11), 1–17. doi:10.1007/s10661-023-11860-z PMID:37857875

Sharma, R., Garg, R., Bali, M., & Eddy, N. O. (2023b). Potential applications of green-synthesized iron oxide NPs for environmental remediation. *Environmental Monitoring and Assessment*, *195*(11), 1397. Advance online publication. doi:10.1007/s10661-023-12035-6 PMID:37910248

Sharma, V. K., Yngard, R. A., & Lin, Y. (2009). Silver nanoparticles: Green synthesis and theirantimicrobial activities. *Advances in Colloid and Interface Science*, *145*(1-2), 83–96. doi:10.1016/j.cis.2008.09.002 PMID:18945421

Sharma, V., Tripathi, A. K., & Mittal, H. (2022). Technological revolutions in smart farming: Current trends, challenges & future directions. *Computers and Electronics in Agriculture*, *201*, 107217. doi:10.1016/j.compag.2022.107217

Sharmila, G., Fathima, M. F., Haries, S., Geetha, S., Kumar, N. M., & Muthukumaran, C. (2017). Green synthesis, characterization and antibacterial efficacy of palladium nanoparticlessynthesized using Filicium decipiens leaf extract. *Journal of Molecular Structure*, *1138*, 35–40. doi:10.1016/j.molstruc.2017.02.097

Shekarian, E., Jafari Nasr, M. R., Tarighaleslami, A. H., Walmsley, T. G., Atkins, M. J., Sahebjamee, N., & Alaghehbandan, M. (2016). Impact of hybrid heat transfer enhancement techniques in shell and tube heat exchanger design. *Chemical Engineering Transactions*, *52*, 1159–1164. doi:10.3303/CET1652194

Sheng, Y., Li, Y., Peng, Y., Ma, L., Gao, W., Yan, C., & Liu, X. (2023). Effect of hydrophilic silica nanoparticles on foam properties of mixtures of silicone and hydrocarbon surfactants. *Chemical Engineering Science*, *278*, 118769. doi:10.1016/j.ces.2023.118769

Shen, X., Hu, Y., Xu, G., Chen, W., Xu, K., Ran, Q., Ma, P., Zhang, Y., Li, J., & Cai, K. (2014). Regulation of the biological functions of osteoblasts and bone formation by Zn-incorporated coating on microrough titanium. *ACS Applied Materials & Interfaces*, *6*(18), 16426–16440. doi:10.1021/am5049338 PMID:25148131

Sherlala, A. I. A., Raman, A. A. A., Bello, M. M., & Buthiyappan, A. (2019). Adsorption of arsenic using chitosan magnetic graphene oxide nanocomposite. *Journal of Environmental Management*, *246*, 547–556. doi:10.1016/j.jenvman.2019.05.117 PMID:31202019

Shi, J., Votruba, A. R., Farokhzad, O. C., & Langer, R. (2010). Nanotechnology in drug delivery and tissue engineering: From discovery to applications. In Nano Letters (Vol. 10, Issue 9, pp. 3223–3230). doi:10.1021/nl102184c

Shilar, F. A., Ganachari, S. V., & Patil, V. B. (2022). Advancement of nano-based construction materials-A review. *Construction & Building Materials*, *359*, 129535. doi:10.1016/j.conbuildmat.2022.129535

Shreyash, N., Bajpai, S., Khan, M. A., Vijay, Y., Tiwary, S. K., & Sonker, M. (2021). Green synthesis of nanoparticles and their biomedical applications: A review. *ACS Applied Nano Materials*, *4*(11), 11428–11457. doi:10.1021/acsanm.1c02946

Shrivastava, S., & Dash, D. (2009). Applying nanotechnology to human health: Revolution in biomedical sciences. *Journal of Nanotechnology*, *2009*, 2009. doi:10.1155/2009/184702

Shukla, D., & Vankar, P. S. (2012). Synthesis of plant parts mediated gold nanoparticles. *International Journal of Green Nanotechnology*, *4*(3), 277–288. doi:10.1080/19430892.2012.706175

Sid, S., Mor, R. S., Kishore, A., & Sharanagat, V. S. (2021). Bio-sourced polymers as alternatives to conventional food packaging materials: A review. *Trends in Food Science & Technology*, *115*, 87–104. doi:10.1016/j.tifs.2021.06.026

Silva, M. M. S., Dias, A. C. M. S., Silva, B. V. M., Gomes-Filho, S. L. R., Kubota, L. T., Goulart, M. O. F., & Dutra, R. F. (2015). Electrochemical detection of dengue virus NS1 protein with a poly(allylamine)/carbon nanotube layered immunoelectrode. *Journal of Chemical Technology and Biotechnology*, *90*(1), 194–200. doi:10.1002/jctb.4305

Silvestre, O. F., Rao, A., & Liz-Marzán, L. M. (2023). Self-assembled colloidal gold nanoparticles as substrates for plasmon enhanced fluorescence. *European Journal of Materials*, *3*(1), 1. doi: 10.1080/26889277.2023.2202676

Singh, R., & Kumar, S. (2017). Green technologies and environmental sustainability. *Green Technologies and Environmental Sustainability,* 1–492. doi:10.1007/978-3-319-50654-8

Singh, A., Dasauni, K., Kumar Nailwal, T., & Nenavathu, B. P. (2023). Formulation of dual functional gCN/TeO2-ZnO nanocomposites as a controlled release nanofertilizer and antibacterial agent. *Nanotechnology*, *34*(15), 155602. doi:10.1088/1361-6528/acb2d1 PMID:36715368

Singh, H., Sharma, A., Bhardwaj, S. K., Arya, S. K., Bhardwaj, N., & Khatri, M. (2021). Recent advances in the applications of nano-agrochemicals for sustainable agricultural development. *Environmental Science. Processes & Impacts*, *23*(2), 213–239. https://pubs.rsc.org/en/content/articlelanding/2021/em/d0em00404a/unauth. doi:10.1039/D0EM00404A PMID:33447834

Singh, P., Kushwaha, C. S., Singh, V. K., Dubey, G. C., & Shukla, S. (2021). Chemiresistive sensing of volatile ammonia over zinc oxide encapsulated polypyrrole based nanocomposite. *Sensors and Actuators. B, Chemical*, *342*, 130042. doi:10.1016/j.snb.2021.130042

Singh, R. P., Singh, P., & Singh, K. R. B. (2020a). Introduction to Composite Materials. In *Composite Materials* (pp. 1–28). CRC Press. doi:10.1201/9781003080633-1

Singh, R., & Dutta, S. (2018). A review on H2 production through photocatalytic reactions using TiO2/TiO2-assisted catalysts. *Fuel*, *220*, 607–620. doi:10.1016/j.fuel.2018.02.068

Sivarajasekar, N., Dharshni, Ra., & Dharshini, Rd. (2019). Liquid-liquid extraction for wastewater treatment View project THERMAL BIOSENSORS AND THEIR APPLICATIONS. *American International Journal of Research in Science*, *50*. http://www.iasir.net

Sivaram, N. M., Gopal, P. M., & Barik, D. (2019). Toxic Waste From Textile Industries. In *Energy from Toxic Organic Waste for Heat and Power Generation* (pp. 43–54). Elsevier. doi:10.1016/B978-0-08-102528-4.00004-3

Skrabalak, S. E., Chen, J., Sun, Y., Lu, X., Au, L., Cobley, C. M., & Xia, Y. (2008). Gold nanocages: Synthesis, properties, and applications. *Accounts of Chemical Research*, *41*(12), 1587–1595. doi:10.1021/ar800018v PMID:18570442

Slavin, Y. N., Asnis, J., Hńfeli, U. O., & Bach, H. (2017). Metal nanoparticles: Understanding the mechanisms behind antibacterial activity. *Journal of Nanobiotechnology*, *15*(1), 1–20. doi:10.1186/s12951-017-0308-z PMID:28974225

Smiddy, M., Papkovskaia, N., Papkovsky, D. B., & Kerry, J. P. (2002). Use of oxygen sensors for the non-destructive measurement of the oxygen content in a modified atmosphere and vacuum packs of cooked chicken patties; the impact of oxygen content on lipid oxidation. *Food Research International*, *35*(6), 577–584. doi:10.1016/S0963-9969(01)00160-0

Sodano, V. (2023). Nanotechnology and food system: Assessing the european union regulatory system. *European Journal of Development Studies*, *3*(5), 25–31. doi:10.24018/ejdevelop.2023.3.5.294

Solanki, R., Shankar, A., Modi, U., & Patel, S. (2023). New insights from nanotechnology in SARS-CoV-2 detection, treatment strategy, and prevention. In *Materials Today Chemistry* (Vol. 29). Elsevier Ltd., doi:10.1016/j.mtchem.2023.101478

Soltani, S., Gacem, A., Choudhary, N., Yadav, V. K., Alsaeedi, H., Modi, S., Patel, A., Khan, S. H., Cabral-Pinto, M. M. S., Yadav, K. K., & Patel, A. (2023). Scallion Peel Mediated Synthesis of Zinc Oxide Nanoparticles and Their Applications as Nano fertilizer and Photocatalyst for Removal of Organic Pollutants from Wastewater. *Water (Basel)*, *15*(9), 1672. doi:10.3390/w15091672

Some, S., Sen, I. K., Mandal, A., Aslan, T., Ustun, Y., Yilmaz, E. Ş., & Ocsoy, I. (2018).Biosynthesis of silver nanoparticles and their versatile antimicrobialproperties. *Materials Research Express*, *6*(1), 01200. https://doi.org/ doi:10.1088/2053-1591/aae23e

Song, M., Lin, X., Peng, Z., Xu, S., Jin, L., Zheng, X., & Luo, H. (2021). Materials and Methods of Biosensor Interfaces With Stability. In Frontiers in Materials (Vol. 7). Frontiers Media SA. doi:10.3389/fmats.2020.583739

Song, J. Y., & Kim, B. S. (2009). Rapid biological synthesis of silver nanoparticles using plant leaf extracts. *Bioprocess and Biosystems Engineering*, *32*(1), 79–84. doi:10.1007/s00449-008-0224-6 PMID:18438688

Sonneville-Aubrun, O., Simonnet, J. T., & L'alloret, F. (2004). Nanoemulsions: A new vehicle for skincare products. *Advances in Colloid and Interface Science*, *108*, 145–149. doi:10.1016/j.cis.2003.10.026 PMID:15072937

Sorrentino, A., Gorrasi, G., & Vittoria, V. (2007). Potential perspectives of bio-nanocomposites for food packaging applications. *Trends in Food Science & Technology*, *18*(2), 84–95. doi:10.1016/j.tifs.2006.09.004

Souza, V. G. L., & Fernando, A. L. (2016). Nanoparticles in food packaging: Biodegradability and potential migration to food—A review. *Food Packaging and Shelf Life*, *8*, 63–70. doi:10.1016/j.fpsl.2016.04.001

Sperling, R. A., Gil, P. R., Zhang, F., Zanella, M., & Parak, W. J. (2008). Biological applications of gold nanoparticles. *Chemical Society Reviews*, *37*(9), 1896–1908. doi:10.1039/b712170a PMID:18762838

Sridharan, R., Monisha, B., Kumar, P. S., & Gayathri, K. V. (2022). Carbon nanomaterials and its applications in pharmaceuticals: A brief review. *Chemosphere*, *294*, 133731. doi:10.1016/j.chemosphere.2022.133731 PMID:35090848

Srivastava, R., Varghese, R., & Abraham, A. R. (2023). Bio-nanomaterials: applications and utility in agricultural and medical field. In Applications of Multifunctional Nanomaterials (pp. 693-707). Elsevier.

Stefaniuk, M., Oleszczuk, P., & Ok, Y. S. (2016). Review on nano zerovalent iron (nZVI): From synthesis to environmental applications. *Chemical Engineering Journal*, *287*, 618–632. doi:10.1016/j.cej.2015.11.046

Stephanie, R., Kim, M. W., Kim, S. H., Kim, J. K., Park, C. Y., & Park, T. J. (2021, February). Recent advances in bimetallic nanomaterials and their nanocomposites for biosensing applications. *Trends in Analytical Chemistry*, *135*, 116159. doi:10.1016/j.trac.2020.116159

Stoimenov, P. K., Klinger, R. L., Marchin, G. L., & Klabunde, K. J. (2002). Metal oxide nanoparticles as bactericidal agents. *Langmuir, 18*(17), 6679–6686. doi:10.1021/la0202374

Sundaram, M. U., & Mahamani, A. (2015). Development of Carbon Nanotube Reinforced Aluminium Matrix Composite Brake Drum for Automotive Applications. Research and Innovation in Carbon Nanotube-Based Composites, Dr. BrahimAttaf (Ed.), Chapter, 2015.

Sundrarajan, M., & Gowri, S. (2011). Green synthesis of titanium dioxide nanoparticles by Nyctanthes arbor-tristis leaves extract. *Chalcogenide Letters, 8*(8), 447–451.

Sun, J., Shen, J., Chen, S., Cooper, M. A., Fu, H., Wu, D., & Yang, Z. (2018). Nanofiller Reinforced Biodegradable PLA/PHA Composites: Current Status and Future Trends. *Polymers, 10*(5), 505. doi:10.3390/polym10050505 PMID:30966540

Sun, L., Fang, K., Chen, W., Liu, K., Zhu, J., & Zhang, C. (2022). Fabrication of a novel superhydrophobic cotton by HDTMS with TiO2 supported activated carbon nanocomposites for photocatalysis and oil/water separation. *Industrial Crops and Products, 189*, 115836. doi:10.1016/j.indcrop.2022.115836

Sun, Y. P., Zhou, B., Lin, Y., Wang, W., Fernando, K. A. S., Pathak, P., Meziani, M. J., Harruff, B. A., Wang, X., Wang, H., Luo, P. G., Yang, H., Kose, M. E., Chen, B., Veca, L. M., & Xie, S. Y. (2006). Quantum-sized carbon dots for bright and colorful photoluminescence. *Journal of the American Chemical Society, 128*(24), 7756–7757. doi:10.1021/ja062677d PMID:16771487

Su, P.-G., & Lu, P.-H. (2020). Electrical and Humidity-Sensing Properties of Impedance-Type Humidity Sensors that Were Made of Ag Microwires/PPy/SnO2 Ternary Composites. *Chemosensors (Basel, Switzerland), 8*(4), 92. doi:10.3390/chemosensors8040092

Suppan, S. (2017). Applying nanotechnology to fertilizer: rationales, research, risks and regulatory challenges. The Institute for Agriculture and Trade Policy works locally and globally, Brazil.

Suresh, S., Mathan, P., Sagadevan, S., & Periasamy, M. (2014). RECENT TRENDS IN NANOBIOSENSORS AND THEIR APPLICATIONS-. *RE:view.*

Su, S., & Kang, P. M. (2020). Systemic Review of Biodegradable Nanomaterials in Nanomedicine. *Nanomaterials (Basel, Switzerland), 10*(4), 656. doi:10.3390/nano10040656 PMID:32244653

Sutunkova, M. P., Solovyeva, S. N., Chemyshov, I. N., Kilnova, S. V., Gurvich, V. B., Shur, V. Y., Shishkine, E. V., Zubarev, I. V., Privalova, L. I., & Katsneison, B. A. (2020). Manifestation of Systemic Toxicity in Rats after a Short-Time Inhalation of Lead Oxide Nanoparticles. *International Journal of Molecular Sciences, 21*(3), 690. doi:10.3390/ijms21030690 PMID:31973040

Suwannahong, K., Wongcharee, S., Kreetachart, T., Sirilamduan, C., Rioyo, J., & Wongphat, A. (2021). Evaluation of the microsoft excel solver spreadsheet-based program for nonlinear expressions of adsorption isotherm models onto magnetic nanosorbent. *Applied Sciences (Basel, Switzerland), 11*(16), 7432. doi:10.3390/app11167432

Suwan, T., Khongkhunthian, S., & Okonogi, S. (2018). Green synthesis and inhibitory effects against oral pathogens of silver nanoparticles mediated by rice extracts. *Drug Discoveries & Therapeutics*, *12*(4), 189–196. doi:10.5582/ddt.2018.01034 PMID:30224591

Su, X., Chen, W., Han, Y., Wang, D., & Yao, J. (2021). In-situ synthesis of Cu2O on cotton fibers with antibacterial properties and reusable photocatalytic degradation of dyes. *Applied Surface Science*, *536*, 147945. doi:10.1016/j.apsusc.2020.147945 PMID:33012933

Svagan, A. J., Åkesson, A., Cárdenas, M., Bulut, S., Knudsen, J. C., Risbo, J., & Plackett, D. (2012). Transparent films based on PLA and montmorillonite with tunable oxygen barrier properties. *Biomacromolecules*, *13*(2), 397–405. doi:10.1021/bm201438m PMID:22229499

Tahir, M. B., Nawaz, T., Nabi, G., Sagir, M., Khan, M. I., & Malik, N. (2022). Role of nanophotocatalysts for the treatment of hazardous organic and inorganic pollutants in wastewater. *International Journal of Environmental Analytical Chemistry*, *102*(2), 491–515. doi:10.1080/03067319.2020.1723570

Taib, N.-A. A. B., Rahman, M. R., Huda, D., Kuok, K. K., Hamdan, S., Bakri, M. K. B., Julaihi, M. R. M. B., & Khan, A. (2023). A review on poly lactic acid (PLA) as a biodegradable polymer. *Polymer Bulletin*, *80*(2), 1179–1213. doi:10.1007/s00289-022-04160-y

Tamang, S., Rai, S., Bhujel, R., Bhattacharyya, N. K., Swain, B. P., & Biswas, J. (2023). A concise review on GO, rGO and metal oxide/rGO composites: Fabrication and their supercapacitor and catalytic applications. *Journal of Alloys and Compounds*, *947*, 169588. doi:10.1016/j.jallcom.2023.169588

Tan, S. J., Campolongo, M. J., Luo, D., & Cheng, W. (2011). Building plasmonic nanostructures with DNA. In Nature Nanotechnology, 6(5), 268–276. Nature Publishing Group. doi:10.1038/nnano.2011.49

Tan, C., Han, F., Zhang, S., Li, P., & Shang, N. (2021). Novel bio-based materials and applications in antimicrobial food packaging: Recent advances and future trends. *International Journal of Molecular Sciences*, *22*(18), 9663. doi:10.3390/ijms22189663 PMID:34575828

Tan, K., Heo, S., Foo, M., Chew, I. M., & Yoo, C. (2019). An insight into nanocellulose as soft condensed matter: Challenge and future prospective toward environmental sustainability. *The Science of the Total Environment*, *650*(1), 1309–1326. doi:10.1016/j.scitotenv.2018.08.402 PMID:30308818

Tariq, W., Ali, F., Arslan, C., Nasir, A., Gillani, S. H., & Rehman, A. (2022). Synthesis and applications of graphene and graphenebased nanocomposites: Conventional to artificial intelligence approaches. *Frontiers in Environmental Chemistry*, *3*, 890408. doi:10.3389/fenvc.2022.890408

Tayeb, A. H., Amini, E., Ghasemi, S., & Tajvidi, M. (2018). Cellulose Nanomaterials-Binding Properties and Applications: A Review. *Molecules (Basel, Switzerland)*, *23*(10), 2684. doi:10.3390/molecules23102684 PMID:30340374

Teiri, H., Hajizadeh, Y., Samaei, M. R., Pourzamani, H., & Mohammadi, F. (2020). Modelling the phytoremediation of formaldehyde from indoor air by Chamaedorea Elegans using artificial intelligence, genetic algorithm and response surface methodology. *Journal of Environmental Chemical Engineering*, *8*(4), 103985. Advance online publication. doi:10.1016/j.jece.2020.103985

Teng, W.-Y., Jeng, S.-C., Kuo, C.-W., Lin, Y.-R., Liao, C.-C., & Chin, W.-K. (2008). Nanoparticles-doped guest-host liquid crystal displays (*Vol. 33*, Issue 15).

TERI ENVIS Resource Partner on Renewable Energy and Climate Change. (2021). *Application of Nanotechnology in Solar Energy*. 1-32. http://www.envis.nic.in/ WriteReadData /userfile s/file/n ew sletters/19- TERI.pdf

Thakkar, K. N., Mhatre, S. S., & Parikh, R. Y. (2010). Biological synthesis of metallicnanoparticles. *Nanomedicine; Nanotechnology, Biology, and Medicine*, *6*(2), 257–262. doi:10.1016/j.nano.2009.07.002 PMID:19616126

Thakur, V., Guleria, A., Kumar, S., Sharma, S., & Singh, K. (2021). Recent advances in nanocellulose processing, functionalization and applications: A review. In Materials Advances (Vol. 2, Issue 6, pp. 1872–1895). Royal Society of Chemistry. doi:10.1039/D1MA00049G

Thakur, M., Wang, B., & Verma, M. L. (2022). Development and applications of nano biosensors for sustainable agricultural and food industries: Recent developments, challenges, and perspectives. *Environmental Technology & Innovation*, *26*, 102371. doi:10.1016/j.eti.2022.102371

Thakur, V. K., & Thakur, M. K. (2014). Processing and characterization of natural cellulose fibers/thermoset polymer composites. *Carbohydrate Polymers*, *109*, 102–117. doi:10.1016/j.carbpol.2014.03.039 PMID:24815407

Thomas, P., Duolikun, T., Rumjit, N. P., Moosavi, S., Lai, C. W., Bin Johan, M. R., & Fen, L. B. (2020). Comprehensive review on nanocellulose: Recent developments, challenges and future prospects. *Journal of the Mechanical Behavior of Biomedical Materials*, *110*(May), 103884. doi:10.1016/j.jmbbm.2020.103884 PMID:32957191

Thomas, R., Snigdha, S., Bhavitha, K. B., Babu, S., Ajith, A., & Radhakrishnan, E. K. (2018). Biofabricated silver nanoparticles incorporated polymethyl methacrylate as a dental adhesive material with antibacterial and antibiofilm activity against Streptococcus mutans. *Biotech*, *8*(9), 1–10. doi:10.1007/s13205-018-1420-y PMID:30221117

ThukkaramM.VaidulychM.KyliánO.RigoleP.AliakbarshiraziS.AsadianM., & De

Tinucci, L., Borgarello, E., Minero, C., & Pelizzetti, E. (1993). Treatment of industrial wastewaters by photocatalytic oxidation on TiO2. In *Proceedings of the 1st International Conference on TiO2 Photocatalytic Purification and treatment of Water and Air* (pp. 585-594). Ontario, Canada.

Tiruvenkadam, N., Thyla, P. R., Senthilkumar, M., Bharathiraja, M., & Murugesan, A. (2015). Synthesis of new aluminum nano hybrid composite liner for energy saving in diesel engines. *Energy Conversion and Management*, *98*, 440–448. doi:10.1016/j.enconman.2015.04.017

Titus, D., Samuel, E. J. J., & Roopan, S. M. (2019). Nanoparticle characterization techniques. In A. K. Shukla & S. Iravani (Eds.), *In Micro and Nano Technologies,Green synthesis, characterization and applications of nanoparticles* (pp. 303–319). Elsevier., doi:10.1016/B978-0-08-102579-6.00012-5

Tiwari, D. K., Behari, J., & Sen, P. (2008). Application of nanoparticles in waste water treatment. *World Applied Sciences Journal*, *3*(3), 417–433.

Tizro, S., & Baseri, H. (2017). Removal of Cobalt Ions from Contaminated Water Using Magnetite Based Nanocomposites: Effects of Various Parameters on the Removal Efficiency. *J. Water Environ. Nanotechnol.*, *2*(23), 174–185. 10.22090

Trovatti, E., Oliveira, L., Freire, C. S., Silvestre, A. J., Neto, C. P., Pinto, J. J. C., & Gandini, A. (2010). Novel bacterial cellulose–acrylic resin nanocomposites. *Composites Science and Technology*, *70*(7), 1148–1153. doi:10.1016/j.compscitech.2010.02.031

Tshikovhi, A., Mishra, S. B., & Mishra, A. K. (2020). Nanocellulose-based composites for the removal of contaminants from wastewater. *International Journal of Biological Macromolecules*, *152*, 616–632. doi:10.1016/j.ijbiomac.2020.02.221 PMID:32097743

Tsujimura, S., Murata, K., & Akatsuka, W. (2014). Exceptionally High Glucose Current on a Hierarchically Structured Porous Carbon Electrode with "Wired" Flavin Adenine Dinucleotide-Dependent Glucose Dehydrogenase. *Journal of the American Chemical Society*, *136*(41), 14432–14437. doi:10.1021/ja5053736 PMID:25244161

Tümay, S. O., Şanko, V., Şenocak, A., & Demirbas, E. (2021). A hybrid nanosensor based on novel fluorescent iron oxide nanoparticles for highly selective determination of Hg2+ ions in environmental samples. *New Journal of Chemistry*, *45*(32), 14495–14507. doi:10.1039/D1NJ02908H

Uchugonova, A., Lessel, M., Nietzsche, S., Zeitz, C., Jacobs, K., Lemke, C., & König, K. (2012). Nanosurgery of cells and chromosomes using near-infrared twelve-femtosecond laser pulses. doi:10.1117/1.JBO.17.10.101502

Uko, C. A., Tijani, J. O., Abdulkareem, S. A., Mustapha, S., Egbosiuba, T. C., & Muzenda, E. (2022). Adsorptive properties of MgO/WO3 nanoadsorbent for selected heavy metals removal from indigenous dyeing wastewater. *Process Safety and Environmental Protection*, *162*, 775–794. doi:10.1016/j.psep.2022.04.057

Ulus, R., Yeşildağ, İ., Elmastaş, M., & Kaya, M. (2015). Rapid synthesis of novel 1,8-dioxoacridine carboxylic acid derivatives by microwave irradiation and their free radical scavenging activity. *Medicinal Chemistry Research*, *24*(10), 3752–3759. doi:10.1007/s00044-015-1417-6

Urnukhsaikhan, E., Bold, B. E., Gunbileg, A., Sukhbaatar, N., & Mishig-Ochir, T. (2021). Antibacterial activity and characteristics of silver nanoparticles biosynthesized from *Carduus crispus*. *Scientific Reports*, *11*(1), 21047. doi:10.1038/s41598-021-00520-2 PMID:34702916

Vahdat, A., Ghasemi, B., & Yousefpour, M. (2020). Mechanical properties of the hydroxyapatite and magnetic nanocomposite of hydroxyapatite adsorbents. *South African Journal of Chemical Engineering*, *33*, 90–94. doi:10.1016/j.sajce.2020.05.007

Vaidiglesias, V., Alba-Gonzalez, A., Fernandez-Bertolez, N., Touzani, A., Laffon, B., & Folgueira, M. (2023). Effects of zinc oxide nanoparticle exposure on human glial cells and zebra embryos. *International Journal of Molecular Sciences*, *24*(15), 12297. doi:10.3390/ijms241512297 PMID:37569675

Valasek, J. (1921). Piezo-Electric and Allied Phenomena in Rochelle Salt. *Physical Review*, *17*(4), 475–481. doi:10.1103/PhysRev.17.475

Valença, D. P., Alves, K. G. B., Melo, C. P. D., & Bouchonneau, N. (2015). Study of the efficiency of polypyrrole/ZnO nanocomposites as additives in anticorrosion coatings. *Materials Research*, *18*(suppl 2), 273–278. doi:10.1590/1516-1439.371614

Valsalam, S., Agastian, P., Arasu, M. V., Al-Dhabi, N. A., Ghilan, A. K. M., Kaviyarasu, K., & Arokiyaraj, S. (2019). Rapid biosynthesis and characterization of silver nanoparticles fromthe leaf extract of Tropaeolum majus L. and its enhanced in-vitro antibacterial, antifungal,antioxidant and anticancer properties. *Journal of Photochemistry and Photobiology. B, Biology*, *191*, 65–74. doi:10.1016/j.jphotobiol.2018.12.010 PMID:30594044

Van Gerwen, P., Laureyn, W., Laureys, W., Huyberechts, G., Op, M., Beeck, D., Baert, K., Suls, J., Sansen, W., Jacobs, P., Hermans, L., & Mertens, R. (1998). Nanoscaled interdigitated electrode arrays for biochemical sensors. In Sensors and Actuators (Vol. 49). doi:10.1016/S0925-4005(98)00128-2

Varadarajan, G., & Venkatachalam, P. (2016). Sustainable textile dyeing processes. *Environmental Chemistry Letters*, *14*(1), 113–122. doi:10.1007/s10311-015-0533-3

Varghese, E., & George, M. (2015) Green synthesis of zincoxide nanoparticles. *International Journal of Advance Research In Science And Engineering*.

Vasile, C. (2018). Polymeric Nanocomposites and Nanocoatings for Food Packaging: A Review. *Materials (Basel)*, *11*(10), 1834. doi:10.3390/ma11101834 PMID:30261658

Vasile, C., & Baican, M. (2021). Progresses in food packaging, food quality, and safety—Controlled-release antioxidant and/or antimicrobial packaging. *Molecules (Basel, Switzerland)*, *26*(5), 1263. doi:10.3390/molecules26051263 PMID:33652755

Venkatesh, R., Sakthivel, P., Vivekanandan, M., Kannan, C. R., Krishna, J. P., Dhanabalan, S., Thirugnanasambandham, T., & Majora, M. (2023). Synthesis and Thermal Adsorption Characteristics of Silver-Based Hybrid Nanocomposites for Automotive Friction Material Application. *Adsorption Science and Technology*, *2023*, 2023. doi:10.1155/2023/1003492

Verma, Y., Rani, V., & Rana, S. V. S. (2020). Assessment of cadmium sulphide nanoparticles toxicity in the gills of a fresh water fish. *Environmental Nanotechnology, Monitoring & Management*, *13*, 100280. doi:10.1016/j.enmm.2019.100280

Verran, J., & Maryan, C. J. (1997). Retention of Candida albicans on acrylic resin and silicone of different surface topography. *The Journal of Prosthetic Dentistry*, *77*(5), 535–539. doi:10.1016/S0022-3913(97)70148-3 PMID:9151275

Vidotti, M., Carvalhal, R. F., Mendes, R. K., Ferreira, D. C. M., & Kubota, L. T. (2011). Biosensors based on gold nanostructures. *Journal of the Brazilian Chemical Society*, *22*(1), 3–20. doi:10.1590/S0103-50532011000100002

Vidyasagar, P., Patel, R. R., Singh, S. K., & Singh, M. (2023). Grenn synthesis of silver nanoparticles: Methods, biological applications, delivery and toxicity. *Materials Advances*, *4*(8), 1831–1849. doi:10.1039/D2MA01105K

Vilas-Boas, V., & Vinken, M. (2021, January). Hepatotoxicity induced by nanomaterials: Mechanisms and in vitro models. *Archives of Toxicology*, *95*(1), 27–52. doi:10.1007/s00204-020-02940-x PMID:33155068

Villalba-Rodríguez, A. M., Martínez-Zamudio, L. Y., Martínez, S. A. H., Rodríguez-Hernández, J. A., Melchor-Martínez, E. M., Flores-Contreras, E. A., González-González, R. B., & Parra-Saldívar, R. (2023). Nanomaterial Constructs for Catalytic Applications in Biomedicine: Nanobiocatalysts and Nanozymes. *Topics in Catalysis*, *66*(9-12), 707–722. doi:10.1007/s11244-022-01766-4 PMID:36597435

Vinoth, S., Shalini Devi, K., & Pandikumar, A. (2021). A comprehensive review on graphitic carbon nitride-based electrochemical and biosensors for environmental and healthcare applications. *Trends in Analytical Chemistry*, *140*, 116274. doi:10.1016/j.trac.2021.116274

Vo-Dinh, T., Liu, Y., Fales, A. M., Ngo, H., Wang, H. N., Register, J. K., Yuan, H., Norton, S. J., & Griffin, G. D. (2015). SERS Nanosensors and Nanoreporters: Golden Opportunities in Biomedical Applications. *Wiley Interdisciplinary Reviews. Nanomedicine and Nanobiotechnology*, *7*(1), 17–33. doi:10.1002/wnan.1283 PMID:25316579

Vollath, D. (2008). Nanomaterials an introduction to synthesis, properties and application. *Environmental Engineering and Management Journal*, *7*(6), 865–870.

Vroman, I., & Tighzert, L. (2009). Biodegradable polymers. In Materials, 2(2), 307–344. doi:10.3390/ma2020307

Vuoriluoto, M., Orelma, H., Zhu, B., Johansson, L.-S., & Rojas, O. J. (2016). Control of Protein Affinity of Bioactive Nanocellulose and Passivation Using Engineered Block and Random Copolymers. *ACS Applied Materials & Interfaces*, *8*(8), 5668–5678. doi:10.1021/acsami.5b11737 PMID:26844956

Waghmode, M. S., Gunjal, A. B., Mulla, J. A., Patil, N. N., & Nawani, N. N. (2019). Studies on the titanium dioxide nanoparticles: Biosynthesis, applications and remediation. *SN Applied Sciences*, *1*(4), 310. doi:10.1007/s42452-019-0337-3

Wahab, A., Munir, A., Saleem, M. H., AbdulRaheem, M. I., Aziz, H., Mfarrej, M. F. B., & Abdi, G. (2023). Interactions of Metal-Based Engineered Nanoparticles with Plants: An Overview of the State of Current Knowledge, Research Progress, and Prospects. *Journal of Plant Growth Regulation*, *42*(9), 5396–5416. doi:10.1007/s00344-023-10972-7

Wan, C., Zhang, L., Yong, K.-T., Li, J., & Wu, Y. (2021). Recent progress in flexible nanocellulosic structures for wearable piezoresistive strain sensors. *Journal of Materials Chemistry. C, Materials for Optical and Electronic Devices*, *9*(34), 11001–11029. doi:10.1039/D1TC02360H

Wang, B., Chen, Y., Wu, Y., Weng, B., Liu, Y., Lu, Z., Li, C. M., & Yu, C. (2016). Aptamer induced assembly of fluorescent nitrogen-doped carbon dots on gold nanoparticles for sensitive detection of AFB1. *Biosensors & Bioelectronics*, *78*, 23–30. doi:10.1016/j.bios.2015.11.015 PMID:26584079

Wang, C., Luo, H., Zhang, Z., Wu, Y., Zhang, J., & Chen, S. (2014). Removal of As(III) and As(V) from aqueous solutions using nanoscale zero valent iron-reduced graphite oxide modified composites. *Journal of Hazardous Materials*, *268*, 124–131. doi:10.1016/j.jhazmat.2014.01.009 PMID:24480523

Wang, D., Huang, B., Liu, J., Guo, X., Abudukeyoumu, G., Zhang, Y., Ye, B.-C., & Li, Y. (2018). A novel electrochemical sensor based on Cu@Ni/MWCNTs nanocomposite for simultaneous determination of guanine and adenine. *Biosensors & Bioelectronics*, *102*, 389–395. doi:10.1016/j.bios.2017.11.051 PMID:29174972

Wanger, G., Onstott, T. C., & Southam, G. (2008). Stars of the terrestrial deep subsurface: A novel "star-shaped" bacterial morphotype from a South African platinum mine. *Geobiology*, *6*(3), 325–330. doi:10.1111/j.1472-4669.2008.00163.x PMID:18498531

Wang, F., Lu, Y., Chen, Y., Sun, J., & Liu, Y. (2018). Colorimetric Nanosensor Based on the Aggregation of AuNP Triggered by Carbon Quantum Dots for Detection of Ag+ Ions. *ACS Sustainable Chemistry & Engineering*, *6*(3), 3706–3713. doi:10.1021/acssuschemeng.7b04067

Wang, H., Meng, J., Li, F., & Li, T. (2023). Graphitic carbon nitride/ metal-organic framework composite functionalized cotton for efficient oil-water separation and dye degradation. *Journal of Cleaner Production*, *385*, 135758. doi:10.1016/j.jclepro.2022.135758

Wang, J. L., & Meng, L. J. (2014). Influence of carbon nano-fiber on mechanical property of PALC. *Applied Mechanics and Materials*, *535*, 785–787. . doi:10.4028/www.scientific.net/AMM.535.785

Wang, J., Euring, M., Ostendorf, K., & Zhang, K. (2022). Biobased materials for food packaging. *Journal of Bioresources and Bioproducts*, *7*(1), 1–13. doi:10.1016/j.jobab.2021.11.004

Wang, J., Shen, B., Lan, M., Kang, D., & Wu, C. (2020). Carbon nanotubes (CNTs) production from catalytic pyrolysis of waste plastics: The influence of catalyst and reaction pressure. *Catalysis Today*, *351*, 50–57. doi:10.1016/j.cattod.2019.01.058

Wang, L., Teles, M. P. R., Arabkoohsar, A., Yu, H., Ismail, K. A. R., Mahian, O., & Wongwises, S. (2022). A holistic and state-of-the-art review of nanotechnology in solar cells. In *Sustainable Energy Technologies and Assessments* (Vol. 54). Elsevier Ltd., doi:10.1016/j.seta.2022.102864

Wang, L., Wang, J., Tang, M., Wang, C., Gao, D., & Zhou, Y. (2023). Developing a Z-scheme Ag2CO3/ZIF-8 heterojunction for the surface decoration of cotton fabric toward repeatable photocatalytic dye degradation. *Applied Surface Science*, *610*, 155605. doi:10.1016/j.apsusc.2022.155605

Wang, L., Zhang, Y., Wu, A., & Wei, G. (2017). Designed graphene-peptide nanocomposites for biosensor applications: A review. *Analytica Chimica Acta*, *985*, 24–40. doi:10.1016/j.aca.2017.06.054 PMID:28864192

Wang, M., Zhang, M., Zhang, M., Aizezi, M., Zhang, Y., Hu, J., & Wu, G. (2019). In-situ mineralized robust polysiloxane–Ag@ZnO on cotton for enhanced photocatalytic and antibacterial activities. *Carbohydrate Polymers*, *217*, 15–25. doi:10.1016/j.carbpol.2019.04.042 PMID:31079671

Wang, N., Yu, J., Yan, J., & Hua, F. (2023). Recent advances in antibacterial coatings for orthodontic appliances. *Frontiers in Bioengineering and Biotechnology*, *11*, 1093926. doi:10.3389/fbioe.2023.1093926 PMID:36815889

Wang, P., Fu, F., & Liu, T. (2021). A review of the new multifunctional nano zero-valent iron composites for wastewater treatment: Emergence, preparation, optimization and mechanism. *Chemosphere*, *285*(July), 131435. doi:10.1016/j.chemosphere.2021.131435 PMID:34256206

Wang, Q., Han, N., Shen, Z., Li, X., Chen, Z., Cao, Y., Si, W., Wang, F., Ni, B.-J., & Thakur, V. K. (2023). MXene-based electrochemical (bio) sensors for sustainable applications: Roadmap for future advanced materials. *Nano Materials Science*, *5*(1), 39–52. doi:10.1016/j.nanoms.2022.07.003

Wang, R., Xu, Y., Zhang, T., & Jiang, Y. (2015). Rapid and sensitive detection of Salmonella typhimurium using aptamer-conjugated carbon dots as fluorescence probe. *Analytical Methods*, *7*(5), 1701–1706. doi:10.1039/C4AY02880E

Wang, S., Gao, B., Li, Y., Creamer, A. E., & He, F. (2017). Adsorptive removal of arsenate from aqueous solutions by biochar supported zero-valent iron nanocomposite: Batch and continuous flow tests. *Journal of Hazardous Materials*, *322*, 172–181. doi:10.1016/j.jhazmat.2016.01.052 PMID:26852252

Wang, T., Zhu, H., Zhuo, J., Zhu, Z., Papakonstantinou, P., Lubarsky, G., Lin, J., & Li, M. (2013). Biosensor based on ultrasmall MoS2 nanoparticles for electrochemical detection of H2O2 released by cells at the nanomolar level. *Analytical Chemistry*, *85*(21), 10289–10295. doi:10.1021/ac402114c PMID:24067077

Wang, W., Xu, G., Cui, X. T., Sheng, G., & Luo, X. (2014). Enhanced catalytic and dopamine sensing properties of electrochemically reduced conducting polymer nanocomposite doped with pure graphene oxide. *Biosensors & Bioelectronics*, *58*, 153–156. doi:10.1016/j.bios.2014.02.055 PMID:24632460

Wang, X., Li, F., & Guo, Y. (2020). Recent Trends in Nanomaterial-Based Biosensors for Point-of-Care Testing. *Frontiers in Chemistry, 8*, 586702. doi:10.3389/fchem.2020.586702 PMID:33195085

Wang, X., Xu, C., Yi, M., & Zhang, H. (2011). Effects of Nano-ZrO2 on the microstructure and mechanical properties of Ti(C, N)-based cermet die materials. *Advanced Materials Research, 154–155*, 1319–1323. . doi:10.4028/www.scientific.net/AMR.403-408.1319

Wang, Y., Qing, X., Zhou, Q., Zhang, Y., Liu, Q., Liu, K., Wang, W., Li, M., Lu, Z., Chen, Y., & Wang, D. (2017). The woven fiber organic electrochemical transistors based on polypyrrole nanowires/reduced graphene oxide composites for glucose sensing. *Biosensors & Bioelectronics, 95*, 138–145. doi:10.1016/j.bios.2017.04.018 PMID:28437640

Wang, Y., Yang, L., Zou, R., Lan, J., Yao, A., Xiao, H., & Lin, S. (2022). Z-scheme CeO2/Ag/CdS heterojunctions functionalized cotton fibers as highly recyclable and efficient visible light-driven photocatalysts for the degradation of dyes. *Journal of Cleaner Production, 380*, 135012. doi:10.1016/j.jclepro.2022.135012

Wan, Q., Han, L., Guo, Y., Yu, H., Tan, L., Zheng, A., & Chen, Y. (2021). Graphene-based biosensors with high sensitivity for detection of Ovarian cancer cells. *Molecules (Basel, Switzerland), 26*(23), 7265. doi:10.3390/molecules26237265 PMID:34885851

Wen, L., Wang, Y., Lu, D., Shengyang, H., & Heyou, H. (2010). Preparation of KF/CaO nanocatalyst and its application in biodiesel production from Chinese tallow seed oil. *Fuel, 89*(9), 2267–2271. doi:10.1016/j.fuel.2010.01.028

Westlake, J. R., Tran, M. W., Jiang, Y., Zhang, X., Burrows, A. D., & Xie, M. (2023). Biodegradable biopolymers for active packaging: Demand, development and directions. *Sustainable Food Technology, 1*(1), 50–72. doi:10.1039/D2FB00004K

Wiechers, J. W., & Musee, N. (2010). Engineered inorganic nanoparticles and cosmetics: Facts, issues, knowledge gaps and challenges. In Journal of Biomedical Nanotechnology (Vol. 6, Issue 5, pp. 408–431). doi:10.1166/jbn.2010.1143

Williford, J. M., Santos, J. L., Shyam, R., & Mao, H. Q. (2015). Shape control in engineering of polymeric nanoparticles for therapeutic delivery. *Biomaterials Science, 3*(7), 894–907. doi:10.1039/C5BM00006H PMID:26146550

Wu, Q., Liu, J., Wang, X., Feng, L., Wu, J., Zhu, X., Wen, W., & Gong, X. (2020). Organ-on-a-chip: Recent breakthroughs and future prospects. In BioMedical Engineering Online (Vol. 19, Issue 1). BioMed Central Ltd. doi:10.1186/s12938-020-0752-0

Wu, G., Lee, D., Cha, Y. L., Kim, W., Kim, D.-J., Du, H., & Zhang, X. (2022). *Polyaniline/Graphene-Functionalized Flexible Waste Mask Sensors for Ammonia and Volatile Sulfur Compound Monitoring*. American Chemical Society. doi:10.1021/acsami.2c15443

Wu, L., Wen, W., Wang, X., Huang, D., Cao, J., Qi, X., & Shen, S. (2022). Ultrasmall iron oxide nanoparticles cause significant toxicity by specifically inducing acute oxidative stress to multiple organs. *Particle and Fibre Toxicology*, *19*(1), 24. doi:10.1186/s12989-022-00465-y PMID:35351185

Wu, X., Han, X., Guo, Y., Liu, Q., Sun, R., Wen, Z., & Dai, C. (2023). Application prospect of calcium peroxide nanoparticles in biomedical field. *Reviews on Advanced Materials Science*, *62*(1), 20220308. doi:10.1515/rams-2022-0308

Wu, X., Xing, Y., Zeng, K., Huber, K., & Zhao, J. X. (2018). Study of Fluorescence Quenching Ability of Graphene Oxide with a Layer of Rigid and Tunable Silica Spacer. *Langmuir*, *34*(2), 603–611. doi:10.1021/acs.langmuir.7b03465 PMID:29275632

Wu, X., Yuan, X., Zhao, J., Ji, D., Guo, H., Yao, W., Li, X., & Zhang, L. (2023). Study on the effects of different pectinase/cellulase ratios and pretreatment times on the preparation of nanocellulose by ultrasound-assisted bio-enzyme heat treatment. *RSC Advances*, *13*(8), 5149–5157. doi:10.1039/D2RA08172E PMID:36777933

Wu, Z., Deng, W., Luo, J., & Deng, D. (2019). Multifunctional nano-cellulose composite films with grape seed extracts and immobilized silver nanoparticles. *Carbohydrate Polymers*, *205*, 447–455. doi:10.1016/j.carbpol.2018.10.060 PMID:30446127

Wu, Z., Fang, K., Chen, W., Zhao, Y., Xu, Y., & Zhang, C. (2021). Durable superhydrophobic and photocatalytic cotton modified by PDMS with TiO2 supported bamboo charcoal nanocomposites. *Industrial Crops and Products*, *171*, 113896. doi:10.1016/j.indcrop.2021.113896

Wyser, Y., Adams, M., Avella, M., Carlander, D., Garcia, L., Pieper, G., Rennen, M., Schuermans, J., & Weiss, J. (2016). Outlook and Challenges of Nanotechnologies for Food Packaging. *Packaging Technology & Science*, *29*(12), 615–648. doi:10.1002/pts.2221

Xia, Y., Xiong, Y., Lim, B., & Skrabalak, S. E. (2009). Shape-controlled synthesis of metal nanocrystals: Simple chemistry meets complex physics? In Angewandte Chemie - International Edition (Vol. 48, Issue 1, pp. 60–103). doi:10.1002/anie.200802248

Xie, K., Wang, J., Xu, S., Hao, W., Zhao, L., Huang, L., & Wei, Z. (2023). Application of Two-Dimensional MXene materials in sensors. *Materials & Design*, *228*, 111867. doi:10.1016/j.matdes.2023.111867

Xuan, L., Ju, Z., Skonieczna, M., Zhou, P., & Huang, R. (2020). Nanoparticles-induced potential toxicity on humans health: Application, toxicity mechanisms and evaluation models. *MedComm*, *4*(4), e327. doi:10.1002/mco2.327 PMID:37457660

Xu, Q., Xie, L., Diao, H., Li, F., Zhang, Y., Fu, F., & Liu, X. (2017). Antibacterial cotton fabric with enhanced durability prepared using silver nanoparticles and carboxymethyl chitosan. *Carbohydrate Polymers*, *177*, 187–193. doi:10.1016/j.carbpol.2017.08.129 PMID:28962757

Xu, X., Ray, R., Gu, Y., Ploehn, H. J., Gearheart, L., Raker, K., & Scrivens, W. A. (2004). Electrophoretic analysis and purification of fluorescent single-walled carbon nanotube fragments. *Journal of the American Chemical Society*, *126*(40), 12736–12737. doi:10.1021/ja040082h PMID:15469243

Yadav, A., Kumar, H., Sharma, R., & Kumari, R. (2023). Synthesis, processing, and applications of 2D (nano) materials: A sustainable approach. *Surfaces and Interfaces*, *39*, 102925. doi:10.1016/j.surfin.2023.102925

Yadav, A., Yadav, K., & Abd-Elsalam, K. A. (2023). Nanofertilizers: Types, Delivery and Advantages in Agricultural Sustainability. *Agrochemicals*, *2*(2), 296–336. doi:10.3390/agrochemicals2020019

Yadav, D., Kwak, M., Chauhan, P. S., Puranik, N., Lee, P. C., & Jin, J. O. (2022, November). Cancer immunotherapy by immune checkpoint blockade and its advanced application using bio-nanomaterials. []. Academic Press.]. *Seminars in Cancer Biology*, *86*, 909–922. doi:10.1016/j.semcancer.2022.02.016 PMID:35181474

Yadav, K. K., Kumar, S., Pham, Q. B., Gupta, N., Rezania, S., Kamyab, H., Yadav, S., Vymazal, J., Kumar, V., Tri, D. Q., Talaiekhozani, A., Prasad, S., Reece, L. M., Singh, N., Maurya, P. K., & Cho, J. (2019). Fluoride contamination, health problems and remediation methods in Asian groundwater: A comprehensive review. *Ecotoxicology and Environmental Safety*, *182*(January), 109362. doi:10.1016/j.ecoenv.2019.06.045 PMID:31254856

Yakoubi, A., El, C., & Dhafer, B. (2023). Advanced Plasmonic Nanoparticle-Based Techniques for the Prevention, Detection, and Treatment of Current COVID-19. *Plasmonics*, *19*(18), 311–347. doi:10.1007/s11468-022-01754-0 PMID:36588744

Yamada, K., Kim, C. T., Kim, J. H., Chung, J. H., Lee, H. G., & Jun, S. (2014). Single walled carbon nanotube-based junction biosensor for detection of escherichia coli. *PLoS One*, *9*(9), e105767. Advance online publication. doi:10.1371/journal.pone.0105767 PMID:25233366

Yang, D. (2021). Application of Nanotechnology in the COVID-19 Pandemic This article was published in the following Dove Press journal. *International Journal of Nanomedicine*, *16*, 623–649. Advance online publication. doi:10.2147/IJN.S296383 PMID:33531805

Yang, H., Zhu, S., & Pan, N. (2004). Studying the mechanisms of titanium dioxide as ultraviolet-blocking additive for films and fabrics by an improved scheme. *Journal of Applied Polymer Science*, *92*(5), 3201–3210. doi:10.1002/app.20327

Yang, L. (2023). Challenges and recent progress in carbon-based nanocomposites for sportswear and sensing applications. *Materials Research Express*, *10*(5), 052001. doi:10.1088/2053-1591/accf64

Yang, X., & Wang, Z. (2022). Effect of Particle Size and Concentration of Nanocellulose on Thermal Properties of Nanocellulose Aerogel. *Journal of Physics: Conference Series*, *2437*(1), 012061. doi:10.1088/1742-6596/2437/1/012061

Yang, Y.-N., Lu, K.-Y., Wang, P., Ho, Y.-C., Tsai, M.-L., & Mi, F.-L. (2020). Development of bacterial cellulose/chitin multi-nanofibers based smart films containing natural active microspheres and nanoparticles formed in situ. *Carbohydrate Polymers*, *228*, 115370. doi:10.1016/j.carbpol.2019.115370 PMID:31635728

Yan, L., Kasal, B., & Huang, L. (2016). A review of recent research on the use of cellulosic fibres, their fibre fabric reinforced cementitious, geo-polymer and polymer composites in civil engineering. *Composites. Part B, Engineering*, *92*, 94–132. doi:10.1016/j.compositesb.2016.02.002

Yao, B., Yao, J., Fan, Z., Zhao, J., Zhang, K., & Huang, W. (2022). Rapid Advances of Versatile MXenes for Electrochemical Enzyme-Based Biosensors, Immunosensors, and Nucleic Acid-Based Biosensors. *ChemElectroChem*, *9*(11), e202200103. Advance online publication. doi:10.1002/celc.202200103

Yao, L., Bojic, D., & Liu, M. (2023). Applications and safety of gold nanoparticles as therapeutic devices in clinical trials. *Journal of Pharmaceutical Analysis*, *13*(9), 960–967. doi:10.1016/j.jpha.2023.06.001 PMID:37842655

Yasmine, R., Ahmad, J., Qamar, S., & Qureshi, M. I. (2023). Engineered nanomaterials for sustainable agricultural production, soil improvement, and stress management: An overview. *Engineered Nanomaterials for Sustainable Agricultural Production, Soil Improvement and Stress Management*, 1-23. doi:10.1016/B978-0-323-91933-3.00022-2

Yazdanian, M., Rostamzadeh, P., Rahbar, M., Alam, M., Abbasi, K., Tahmasebi, E., & Yazdanian, A. (2022). The potential application of green-synthesized metal nanoparticles indentistry: A comprehensive review. *Bioinorganic Chemistry and Applications*, *2311910*, 1–27. Advance online publication. doi:10.1155/2022/2311910 PMID:35281331

Ye, J.-S., Hong, B.-D., Wu, Y.-S., Chen, H.-R., & Lee, C.-L. (2016). Heterostructured Palladium-Platinum Core-Shell Nanocubes for Use in a Nonenzymatic Amperometric Glucose Sensor Microchim. *Mikrochimica Acta*, *183*(12), 3311–3320. doi:10.1007/s00604-016-1976-3

Yetisen, A. K., Qu, H., Manbachi, A., Butt, H., Dokmeci, M. R., Hinestroza, J. P., Skorobogatiy, M., Khademhosseini, A., & Yun, S. H. (2016). Nanotechnology in Textiles. *ACS Nano*, *10*(3), 3042–3068. doi:10.1021/acsnano.5b08176 PMID:26918485

Yilmaz, O., & Yorgancioglu, A. (2019). Nanocoatings: preparation, properties, and biomedical applications. In C. Vasile (Ed.), *In Micro and Nano Technologies, Polymeric Nanomaterials In Nanotherapeutics* (pp. 299–331). Elsevier. doi:10.1016/B978-0-12-813932-5.00008-X

Ying, S., Guan, Z., Ofoegbu, P. C., Clubb, P., Rico, C., He, F., & Hong, J. (2022). Green synthesis of nanoparticles: Current developments and limitations. *Environmental Technology & Innovation*, *26*, 102336. doi:10.1016/j.eti.2022.102336

Yoon, J., Shin, M., Lim, J., Lee, J.-Y., & Choi, J.-W. (2020). Recent Advances in MXene Nanocomposite-Based Biosensors. *Biosensors (Basel)*, *10*(11), 185. doi:10.3390/bios10110185 PMID:33233574

Yougbare, S., Chang, T. K., Tan, S. H., Kuo, J. C., Hsu, P. H., Su, C. Y., & Kuo, T. R. (2019). Antimicrobial gold nanoclusters: Recent developments and future perspectives. *International journal of molecular sciences, 20*(12), 2924.n doi:10.3390/ijms20122924

Yousef, H.A., Fahmy, H.M., Arafa, F. N. *et al.* (2023). Nanotechnology in pest management: advantages, applications, and challenges. *Int J Trop Insect Sci, 43,* 1387–1399. https://. doi:10.1007/s42690-023-01053-z

Yousefiasl, S., Zare, I., Ghovvati, M., & Ghomi, M. (2023). Enzyme-Responsive Materials: Properties, Design, and Applications. In Stimuli-Responsive Materials for Biomedical Applications (pp. 203-229). American Chemical Society.

Yu, L., & Li, N. (2019). Noble metal nanoparticles-based colorimetric biosensor for visual quantification: A mini review. In Chemosensors, 7(4). MDPI Multidisciplinary Digital Publishing Institute. doi:10.3390/chemosensors7040053

Yukuyama, M. N., Ghisleni, D. D., Pinto, T. J., & Bou-Chacra, N. A. (2016). Nanoemulsion: Process selection and application in cosmetics–a review. *International Journal of Cosmetic Science, 38*(1), 13–24. doi:10.1111/ics.12260 PMID:26171789

Yuvakkumar, R., Suresh, J., & Hong, S. I. (2014). Green synthesis of zincoxide nanoparticles. *Advanced Materials Research, 952*, 137–140. . doi:10.4028/www.scientific.net/AMR.952.137

Zaib, S., & Iqbal, J. (2019). Nanotechnology: Applications, techniques, approaches, & the advancement in toxicology and environmental impact of engineered nanomaterials. *Importance & Applications of Nanotechnology, 8.*

Zaid, M. H. M., Abdullah, J., Yusof, N. A., Wasoh, H., Sulaiman, Y., Noh, M. F. M., & Issa, R. (2020). Reduced Graphene Oxide/TEMPO-Nanocellulose Nanohybrid-Based Electrochemical Biosensor for the Determination of Mycobacterium tuberculosis. *Journal of Sensors, 4051474*, 1–11. Advance online publication. doi:10.1155/2020/4051474

Zaitseva, N. V., Zemlyanova, M. A., Stepankov, M. S., & Ignatova, A. M. (2019). Studying and Assessing the Toxicity of Calcium Oxide Nanoparticles under One-Time Inhalation Exposure. *Nanotechnologies in Russia, 14*(9-10), 497–503. doi:10.1134/S1995078019050148

Zakaria, M. Y., Abd El-Halim, S. M., Beshay, B. Y., Zaki, I., & Abourehab, M. A. S. (2023). 'Poly phenolic phytoceutical loaded nano-bilosomes for enhanced caco-2 cell permeability and SARS-CoV 2 antiviral activity': In-vitro and insilico studies. *Drug Delivery, 30*(1), 1. doi:10.1080/10717544.2022.2162157 PMID:36587813

Zankana, M. M., Al-dalawy, S. M., & Barzinjy, A. A. (2023). Synthesis and characterization of bio-nanocomposites: Functionalization of graphene oxide with a biocompatible amino acid. *Hybrid Advances, 3*, 100070. doi:10.1016/j.hybadv.2023.100070

Zanna, H. K., Gunda, H. J., Tikau, F. W., & Gero, S. M. (2018).. . *Assessment of Heavy Metals in Trona from Yobe State, Nigeria., 6*(1), 10–15.

Zanoni, I., Blosi, M., Fiorini, V., Crosera, M., Ortelli, S., Stagni, S., Stefan, A., Psilodimitrakopoulos, S., Stratakis, E., Filon, F. L., & Luisa Costa, A. (2020). Use of Cotton Textiles Coated by Ir(III) Tetrazole Complexes within Ceramic Silica Nanophases for Photo-Induced Self-Marker and Antibacterial Application. *Nanomaterials (Basel, Switzerland)*, *10*(6), 1020. doi:10.3390/nano10061020 PMID:32471044

Zeng, M., Echols, I., Wang, P., Lei, S., Luo, J., Peng, B., He, L., Zhang, L., Huang, D., Mejia, C., Wang, L., Mannan, M. S., & Cheng, Z. (2018). Highly Biocompatible, Underwater Superhydrophilic and Multifunctional Biopolymer Membrane for Efficient Oil–Water Separation and Aqueous Pollutant Removal. *ACS Sustainable Chemistry & Engineering*, *6*(3), 3879–3887. doi:10.1021/acssuschemeng.7b04219

Zhang, C., Gholipour, G., & Mousavi, A. A. (2021). State-of-the-Art Review on Responses of RC Structures Subjected to Lateral Impact Loads. *Archives of Computational Methods in Engineering*, *28*(4), 2477–2507. doi:10.1007/s11831-020-09467-5

Zhang, H., Ma, X., Liu, Y., Duan, N., Wu, S., Wang, Z., & Xu, B. (2015). Gold nanoparticles enhanced SERS aptasensor for the simultaneous detection of Salmonella typhimurium and Staphylococcus aureus. *Biosensors & Bioelectronics*, *74*, 872–877. doi:10.1016/j.bios.2015.07.033 PMID:26241735

Zhang, H., & Sablani, S. (2021). Biodegradable packaging reinforced with plant-based food waste and by-products. *Current Opinion in Food Science*, *42*, 61–68. doi:10.1016/j.cofs.2021.05.003

Zhang, L., Li, Q., Zhou, J., & Zhang, L. (2012). Synthesis and photophysical behavior of pyrene-bearing cellulose nanocrystals for Fe 3+ sensing. *Macromolecular Chemistry and Physics*, *213*(15), 1612–1617. doi:10.1002/macp.201200233

Zhang, L., Qian, M., Cui, H., Zeng, S., Wang, J., & Chen, Q. (2021). Spatiotemporal Concurrent Liberation of Cytotoxins from Dual-Prodrug Nanomedicine for Synergistic Antitumor Therapy. *ACS Applied Materials & Interfaces*, *13*(5), 6053–6068. doi:10.1021/acsami.0c21422 PMID:33525873

Zhang, M., Liu, X., Xie, Y., Zhang, Q., Zhang, W., Jiang, X., & Lin, J. (2020). Biological safe gold nanoparticle-modified dental aligner prevents the Porphyromonas gingivalisbiofilm formation. *ACS Omega*, *5*(30), 18685–18692. doi:10.1021/acsomega.0c01532 PMID:32775870

Zhang, S., Jiang, W., Liu, G., Liu, S., Chen, H., Lyu, G., Yang, G., Liu, Y., & Ni, Y. (2023). Preparation of ultrafine and highly loaded silver nanoparticle composites and their highly efficient applications as reductive catalysts and antibacterial agents. *Journal of Colloid and Interface Science*, *629*, 766–777. doi:10.1016/j.jcis.2022.09.018 PMID:36099844

Zhang, X., Tan, X., Wang, P., & Qin, J. (2023). Application of Polypyrrole-Based Electrochemical Biosensor for the Early Diagnosis of Colorectal Cancer. *Nanomaterials (Basel, Switzerland)*, *13*(4), 674. doi:10.3390/nano13040674 PMID:36839042

Zhang, Y., Zhang, K., & Ma, H. (2009). Electrochemical DNA Biosensors Based on Gold Nanoparticles / Cysteamine / Poly(glutamic acid). *American Journal of Biomedical Sciences*, *1*(2), 115–125. doi:10.5099/aj090200115

Zhao, Y., Ye, M., Chao, Q., Jia, N., Ge, Y., & Shen, H. (2009). Simultaneous detection of multifood-borne pathogenic bacteria based on functionalized quantum dots coupled with immunomagnetic separation in food samples. *Journal of Agricultural and Food Chemistry, 57*(2), 517–524. doi:10.1021/jf802817y PMID:19154162

Zhao, Y., Zhang, W., Lin, Y., & Du, D. (2013). The vital function of Fe3O4@Au nanocomposites for hydrolase biosensor design and its application in detection of methyl parathion. *Nanoscale, 5*(3), 1121–1126. doi:10.1039/c2nr33107a PMID:23280070

Zha, R., Shi, T., He, L., & Zhang, M. (2022). Nanoengineering and green chemistry-oriented strategies toward nanocelluloses for protein sensing. *Advances in Colloid and Interface Science, 308*, 102758. doi:10.1016/j.cis.2022.102758 PMID:36037672

Zheng, J., Wang, C. G., Zhou, H., Ye, E., Xu, J., Li, Z., & Loh, X. J. (2021). *Current research trends and perspectives on solid-state nanomaterials in hydrogen storage*. Research. doi:10.34133/2021/3750689

Zhou, X. Q., Hayat, Z., Zhang, D. D., Li, M. Y., Hu, S., Wu, Q., Cao, Y. F., & Yuan, Y. (2023). Zinc oxide nanoparticles: Synthesis, characterization, modification, and applications in food and agriculture. *Processes (Basel, Switzerland), 11*(4), 1193. doi:10.3390/pr11041193

Zhou, X., Huang, X., Qi, X., Wu, S., Xue, C., Boey, F. Y. C., Yan, Q., Chen, P., & Zhang, H. (2009). In Situ Synthesis of Metal Nanoparticles on Single-Layer Graphene Oxide and Reduced Graphene Oxide Surfaces. *The Journal of Physical Chemistry. C, Nanomaterials and Interfaces, 113*(25), 10842–10846. doi:10.1021/jp903821n

Zhou, Z., Zhang, X., Wu, X., & Lu, C. (2016). Self-stabilized polyaniline@ graphene aqueous colloids for the construction of assembled conductive network in rubber matrix and its chemical sensing application. *Composites Science and Technology, 125*, 1–8. doi:10.1016/j.compscitech.2016.01.016

Zhu, J., Fang, K., Chen, W., Liu, K., Sun, L., & Zhang, C. (2022). Preparation of superhydrophobic, antibacterial and photocatalytic cotton by the synergistic effect of dual nanoparticles of rGO-TiO2/QAS-SiO2. *Industrial Crops and Products, 189*, 115801. doi:10.1016/j.indcrop.2022.115801

Zikalala, N., Matshetshe, K., Parani, S., & Oluwafemi, O. S. (2018). Biosynthesis protocols for colloidal metal oxide nanoparticles. *Nano-Structures & Nano-Objects, 16*, 288–299. doi:10.1016/j.nanoso.2018.07.010

Zinge, C., & Kandasubramanian, B. (2020). Nanocellulose based biodegradable polymers. *European Polymer Journal, 133*(May), 109758. doi:10.1016/j.eurpolymj.2020.109758

Zou, B., Huang, C. Z., Wang, J., & Liu, B. Q. (2006). Effect of Nano-Scale TiN on the Mechanical Properties and Microstructure of Si_3N_4 Based Ceramic Tool Materials. *Key Engineering Materials, 315–316*, 154–158. . doi:10.4028/www.scientific.net/KEM.315-316.154

Zou, R., Teng, X., Lin, Y., & Lu, C. (2020). Graphitic carbon nitride-based nanocomposites electrochemiluminescence systems and their applications in biosensors. *Trends in Analytical Chemistry, 132*, 116054. doi:10.1016/j.trac.2020.116054

About the Contributors

Rajni Garg is working as Associate Professor (Chemistry) in the Department of Applied Science and Humanities at Galgotias College of Engineering and Technology, Greater Noida (Uttar Pradesh), India. She received her B.Sc. (Medical) at the D.A.V. College, Abohar (Punjab), India; M.Sc. (H.S.) in Chemistry, Panjab University, India; M.Phil. in Chemistry at CDLU, India and Ph.D. (Chemistry) at the GKU, India. Her areas of interest include nanochemistry, green chemistry and supramolecular chemistry. She has authored and co-authored many scientific papers published in international and national journals, and proceedings of international and national scientific conferences. She has also authored three textbooks and edited two reference books. She has published five patents on the potential applications of nanomaterials. Currently she is working on green synthesis and potential applications of nanoparticles in environmental remediation.

Ansar Anjum is the Dean Student Welfare and Professor in the Department of Applied Sciences at Galgotias College of Engineering and Technology, Greater Noida, India. With a robust tenure of 14 years dedicated to the realms of teaching and research, she has cultivated a profound expertise. Her scholarly pursuits center around the intricate domains of water remediation, nanochemistry, green chemistry, and materials science. She has authored numerous scientific papers, proceedings, and a patent, demonstrating her expertise and commitment to environmental sustainability. She has authored and co-authored numerous scientific papers published in renowned international and national journals, as well as in the prestigious international and national scientific conferences. Beyond her research endeavors, she has also served as a reviewer for many scientific journals. Currently, Dr. Anjum is actively engaged in research on the green synthesis and potential applications of nanoparticles, striving to develop eco-friendly and sustainable nanotechnologies.

***

Gholamreza Abdi did PhD in medicinal, aromatic and spice plant breeding and physiology at the Department of Agricultural Botany, Biotechnology Centre from university Tehran, Iran and presently working as assistant Scientist at Deptt. of biotechnology, Persian Gulf University. He has a vast experience in the Molecular Breeding and Biotechnological aspects. He led his service for the eminent and esteem organizations seed company, Iran. Dr. Abdi is a member of the Society for Medicinal plant Research and Society for Advancement of Human and Nature. He has published approximately 5 articles in Magazine and Newspapers and having total 58 National and International publications with topics spanning a wide range from DNA profiling methodology to plant molecular systematic, phytochemistry, entomological aspects, Nanotechnology, Plant Breeding, Molecular Breeding and Mutational Breeding areas. He has three book chapters in Springer, Netherland. He contributed a number of pioneering studies at the early stage of plant DNA fingerprinting. He is awarded with national Golden Awards, Young Scientist Award for his outstanding contribution in the agriculture sector. His current research interests focus on the Marker Assisted Selection in plants and its study on the genomic and post-genomic level.

Mukhtar Iderawumi Abdulraheem is a member of the Centre for Applied Agriculture Optoelectronics Research (CAAOR), Henan Agricultural University, Zhengzhou, China, where he conducted various research works and experiments. Abdulraheem attended the Federal College of Education (Special), Oyo; Federal College of Agriculture, Ibadan; and Federal University of Technology, Akure. He further attended the Ladoke Akintola University of Technology, Ogbomoso, Nigeria for a Certificate in Management Information Systems (MIS)/Data Processing. He is a renowned researcher, teacher, adviser, scientist, and academician with over sixty (60) published articles in reputed journals. Abdulraheem made presentations at seminars and Conferences and completed six (8) conference paper proceedings, Four (4) chapters in edited books; Three (3) seminar papers, and attended over forty-five (45) workshops/conference/webinars. He is a member of several professional bodies and an Editorial Board Member of more than 10 journals. He is a multidisciplinary researcher with publications cutting across the agricultural field.

Hillary Onyeka Abugu is a senior lecturer in the Department of Pure and Industrial Chemistry at the University of Nigeria, Nsukka. He possesses a Ph.D in Analytical Chemistry. His current research interest is in the development or modification of analytical methods for materials conversion and applications in environmental monitoring, assessment and remediation. He has a total of 288 citations, i-10 and h-index of 11 respectively. He has supervised several undergraduate and postgraduate students and is actively involved in lecturing, research and development.

Nyeneime William Akpanudo is a Lecturer 1 at the Akwa Ibom State University and her her current research interest is in the development of nanoparticles of novel metals for remediation of hydrocarbon-related contaminants in the environment. She has also worked on the applications of some natural-based adsorbents for the remediation of some classes of environmental-based contaminants. Her current Google Scholar metrics are total citation, h-index and i10-index of 56, 5 and 2 respectively.

Nawal Alghamdi is an academic and researcher in the Physics Department at the University of Tabuk, Saudi Arabia, since 2021. She obtained her Master's degree from the Department of Electronic and Electrical Engineering at the University of Sheffield, United Kingdom. Also, she obtained her PhD degree from the Physics Department at the University of Sheffield in 2021. She has wide experience in semiconductor materials and devices research, especially in semiconductor-based sensors.

Soley Arslan completed her doctorate (2005-2009) studies at the University of Erciyes. Dr. Arslan was promoted to Assoc. Prof and Full Professor in 2015 and 2021, respectively.

Anand Balu is an Associate professor at the Sir Theagaraya College,Chennai, India with a major research interest in materials science, especially, the formulation of green corrosion inhibitors for metals. He has supervised and graduated several scholars and is currently mentoring a good number of academicians. He has an average h-index value of 7 and an i-10 index of 6.

Seda Baktir completed her undergraduate education (2012-2017) at Baskent University (2012-2017) and is currently continuing her doctorate at Erciyes University.

Jyoti Bhattacharjee is currently B. Tech student at the Department of Chemical Engineering, University of Calcutta. Miss Bhattacharjee has passed her class 10 from Pratt Memorial School with a grade of A++ and class 12 from Hariyana Vidya Mandir with a grade of A++. She has been carrying out research work on the synthesis of thermoelectric materials for the Last one year. She has done a certified course on Machine Learning for chemical industries to boost profit, C and Python, and an online Training Participation in the Program on Overview of Space Science, ISRO. She has received a certificate from the Institute of Self- Reliance, and a Certificate of Recognition for the phenomenal and worthy article in the 2023 issue of "THE MAGNIFICENCE OF HUMBLES". Miss Bhattacharjee is an ABACUS second runner-up and was awarded a Certificate of Appreciation for excellent board results from Coal India Limited.She has published one review paper.

Hikmetnur Danisman completed her undergraduate education (2009-2014) at Erciyes University and completed her doctorate at Gaziosmanpasa University (2016-2020). Dr. Danisman has been working as a Assistant Professor in the Nuh Naci Yazgan University Faculty of Dentistry Department of Orthodontics since 2020.

Maumita Das Mukherjee is an Assistant Professor at Amity Institute of Applied Science, Amity University, Noida, India. She has her research interest work in the area "Studies on Nano-Structured Zirconia Based DNA Biosensors for M. tuberculosis Detection". She has many research publications in International and National journals to her credit. She is the life member of Indian Association for the Cultivation of Sciences, Kolkata, India. Throughout her illustrious career, Dr. Mukherjee has made significant contributions to the field of chemistry. She has authored and co-authored numerous scientific papers published in renowned international and national journals, as well as in the proceedings of prestigious international and national scientific conferences.

Nnabuk Okon Eddy graduated from the University of Calabar with a BSc Degree in Chemistry (1995) and PhD in Physical Chemistry (2010) and was appointed Professor of Physical Chemistry in 2015. He currently serves as a Professor of Physical/Computational Chemistry at the University of Nigeria, Nsukka. He had previously served at ABU, Zaria, and Federal University, Lokoja, before his present appointment at the UNN. Prof. Eddy has led several research teams as Principal Investigator at both National and International levels and is the current Chairman of the Federal Government's Technical Committee on Agrochemicals. His pioneering contributions in this role have led to the development of more than forty Gazetted standards for fertilizers, pesticides, and soil stimulants. He is also the Representative/ Team Leader of Nigeria in the African Regional Standard Organization (ARSO) and in the International Standard Organization, where he has also contributed to the development of International standards in Agrochemicals. He has also contributed to policies on the prohibition of chemical weapons, having received training in this respect at the Pacific Northwest National Laboratory at the expense of the Federal Government of Nigeria. His contribution to waste reduction through recycling and resource recovery has received remarkable global impacts with an average h-index of 49 and i10 index of 130. He has led different accreditation teams to the Chemistry Departments of some Nigerian universities and has, over the years, successfully supervised many postgraduate students at both the MSc and PhD levels. He is a consultant to the University of Zambia and NAEC on Chemistry and Nuclear Science curricula. He is among the top 2% most cited scientists and the first Physical Sciences in Nigeria (ADI ranking).

Rishav Garg is working as Professor and Head, Department of Civil Engg. at Galgotias College of Engg and Tech, Greater Noida (UP), India. He received his B.Tech. in Civil Engg. at the REC, Kurukshetra India; M.Tech in Structural Engg. at PTU, Kapurthala, India and Ph.D. in Civil Engg. at PTU, Kapurthala, India. His research interests include applications of nanotechnology and composite materials in sustainable construction. He has authored and co-authored several technical papers in international and national journals. He is actively engaged in developing techniques for strength and durability enhancement of cementitious materials. He has also presented his work in many international and national scientific conferences. He has published patents on sustainable construction practices. He is actively engaged in strategic development as a member of various academic and professional bodies with his rich industrial, research, and academic experience. In addition to his academic and administrative contributions, he is actively providing his services in consultation for innovative projects.

Akshay Gaur is presently working as Research fellow at School of Engineering, IIT Mandi (H.P). His research area includes Energy Harvesting, Dye Degradation and Environment Cleaning.

Franci Nathalie Gómez Jaimes is a Chemistry Professor at the University of Santander, Bucaramanga, Colombia. Her research work includes obtaining and chemical transformation of natural products.

Kasahun Gudeta is a lecturer at the Department of Applied Biology, Adama Science and Technology University, Adama, Ethiopia.

Oluwadamilola Oluwatoyin Hazzan is a multidisciplinary researcher in environmental science and engineering at the Institute of Urban Environment, Chinese Academy of Sciences, Xiamen, China. She has expertise in Climate Change Adaptation, Disaster Resilience, Agriculture and Food Security, GIS and Modelling and Vulnerability Assessment. Currently, her research focuses on microbial communities in bioelectrochemical systems and the mechanism of electron transfer between microorganisms and electron acceptors. She is open to collaborations and potential opportunities with other researchers.

Inés Hernández Celi is a Microbiology Professor at the University Santo Tomás, Bucaramanga, Colombia. Her research work includes antifungal activity measurement and biomass analysis.

Priya Jaswal is presently working as Assistant Professor in Textile Engineering Department at Jawaharlal Nehru Government Engineering College, Sundernagar Distt. Mandi (H.P). She has completed her Ph.D and M.Tech from NIT Jalandhar. Her research area includes Fibre Spinning, Testing, Smart Textiles and Modelling & Simulation.

Nafeesa Farooq Khan is a distinguished scientist with a Ph.D. in Botany from the prestigious University of Kashmir. Her academic journey has been marked by a strong commitment to advancing the field of molecular ecology. With a profound understanding of plant biology at the molecular level, Dr. Khan has made significant contributions to the scientific community. Her research focus lies in the intricate realm of molecular ecology, where she delves into the genetic and molecular mechanisms that govern plant interactions with their environment. In addition to her research and publications, Dr. Khan may have engaged in collaborative projects, attended international conferences, and played a role in mentoring the next generation of scientists. Her influence may extend beyond the academic realm, possibly involving outreach activities, policy advocacy, or other initiatives aimed at promoting the significance of molecular ecology in addressing broader ecological challenges.

Sunil J. Kulkarni has done his PhD in Chemical Engineering with focus on adsorption for wastewater treatment. He has done his Masters in chemical engineering from Tatyasaheb Kore Institute of Engineering and Technology, Warananagar. He has worked as an Assistant Professor in Chemical Engineering Department of Datta Meghe College of Engineering, Airoli, Navi Mumbai, India for 24 years. He has 25 years of teaching and research experience. Dr. Sunil is involved with different academic activities across the universities in Maharashtra. His area of investigation in doctoral studies is phenol and heavy metal removal from waste water. He has also worked on kinetics and isotherms of various sorption operations. His area of research includes adsorption, clean technology and environmental biotechnology. He is on the reviewer board of many international journals and reviewed many international papers. He is also governing body member, promotional editor and executive member of many reputed international journals. Currently he is working as Associate Professor in chemical engineering at GIT, Lavel, India.

Amit Kumar is presently working as Assistant Professor in Textile Engineering Department at Jawaharlal Nehru Government Engineering College, Sundernagar, Mandi (H.P). He has received Ph.D Degree from IIT Mandi (H.P) and M.Tech in Energetic Materials and Polymers from Defence Institute of Advanced Technology Pune. His research area includes Antibacterial Textiles, Dye Degradation, Environment Cleaning, Phase Change Polymers and Smart Textiles.

Carlos Andrés Martínez Bonilla is a Chemistry Professor at the University Santo Tomás, Bucaramanga, Colombia. His research interests include nanomaterials, glycopolymers, chemo/bio sensors, chemical analysis of organic (VOCs) and inorganic (heavy metals) pollutants, DOSY experiments, and functional nanocomposites.

Ismail Ocsoy (GOOGLE SCHOLAR: H-index:42, Citation:5238) completed his master (2009-2011) and doctorate (2011-2014) studies at the University of Florida. Dr. Ocsoy joined Erciyes University, Faculty of Pharmacy as an independent researcher in Turkey in 2015. Dr. Ocsoy was promoted to Assoc. Prof and Full Professor in 2017 and 2023, respectively. The most important contribution of Dr. Ocsoy's scientific work to universal science is in the field of Nanotechnology and Nano-biotechnology. His publications are in high impact factor (EF) journals (Advanced Materials (EF:32.086)), Journal of American Chemical Society (EF: 16.383), ACS Nano (EF: 18.027) and Analytical Chemistry (EF: 8.008) including especially in the most respected journals of the world. Dr. Ocsoy was selected as a National Leading Researcher in 2022 within the scope of the TUBITAK 2247-A National Leading Researchers.

Paula Peña González works in environmental chemistry at the Universidad Santo Tomás of Bucaramanga, Colombia. Her research interests include the development of chemosensors and nanomaterials for the detection and removal of environmental contaminants for the design of environmental remediation strategies.

Chirag Porwal is presently working as Research fellow at School of Engineering, IIT Mandi (H.P). His research area includes Energy Harvesting, Dye Degradation and Environment Cleaning.

Subhasis Roy is working as Assistant Professor at Department of Chemical Engineering, University of Calcutta, India. Dr. Roy received Ph.D. from IIT Kharagpur, India. He worked as a Postdoctoral Researcher at Sungkyunkwan University, South Korea, Research Scientist at IISER, Kolkata, and worked as a visiting researcher Raman Fellow, Ohio University, USA. Dr. Roy has published more than 65 scientific research papers in peer-reviewed journals and 22 conference proceedings. He has 18 book chapters to his credit and acted as editor for 2 books and filed several patents. Dr. Roy was awarded DST-Fast Track Young Scientist, TARE award by SERB and Mission innovation DBT-DST joint funding. He holds membership of many professional societies such as Royal Society of Chemistry (MRSC), National Academy of Sciences, India (MNASc), Fellow of Indian Chemical Society (FICS), Linnean Society of London (FLS)and reviewers as well as editorial board member of many international journals. His expertise is in waste management and recycling, perovskite solar cells, multiferroic composite thin films, photocatalysts for water splitting, fuel cells, and hydrogen energy.

Vanshika Sharma is presently perusing B.Tech in Textile Engineering from Jawaharlal Nehru Government Engineering College, Sundernagar Distt. Mandi (H.P). She has qualified GATE 2023 and has got admitted to dual M.Tech + PhD program at IIT Mandi.

Chandan Singh obtained M.Sc. (2017) and Ph.D. (2023) from Department of Botany under the supervision of Prof. Deepak Vyas, Dr. Harisingh Gour Vishwavidyalaya, Sagar. Dr. Singh is providing consultancy to different NGOs and farmer for use of Biochar and spent mushroom compost (SMC) also working as promoter of Science to the School children. Dr. Singh has published articles in national and international Journal and several book chapters in National and International publishers. He works as a reviewer in various International and national Journals. He has research interest in Organic farming, Biochar, Mushroom biology, Landscaping and Garden Management, Roof gardening, and Secondary Metabolites of Mushroom and their role in enhancing resistance in crop plant.

Richard Alexis Ukpe holds a Ph.D in Environmental Chemistry and he is employed as a senior lecturer in the Department of Chemistry at the Federal University, Otuoke and is actively involved in environmental research including search and application of green corrosion inhibitors, adsorption and photocatalytic removal of environmental contaminants, environmental monitoring and nano synthesis. His current Google Scholar metrics are a total number of citations, h-index and i10-index of 176, 8 and 7 respectively. Dr. Richard Alexis Ukpe is primarily involved in teaching, research and development.

Rahul Vaish is presently working as Professor in School of Engineering at IIT Mandi (H.P). He has received Ph.D Degree from IISC Bangalore and M.Tech in Material Science and Engineering from Motilal Nehru National Institute of Technology (MNNIT), Allahabad, U.P. His research area includes Glasses for Electrical Applications, Energy Harvesting, Smart Polymers, and Environment Cleaning. He has authored and co-authored more than 300 scientific papers published in international and national journals.

Abdul Wahab is an enthusiastic and hard-working researcher seeking a challenging position. In addition to the strong communication and collaboration skills, the author dedicated to contributing to innovative projects within interdisciplinary teams. Aside from being hardworking and enthusiastic, author is also passionate about biochemistry and molecular biology.

Index

P

R

S

T

IGI Global
Publishing Tomorrow's Research Today
www.igi-global.com

Publishing Tomorrow's Research Today

Uncover Current Insights and Future Trends in
Scientific, Technical, & Medical (STM)
with IGI Global's Cutting-Edge Recommended Books

Print Only, E-Book Only, or Print + E-Book.
Order direct through IGI Global's Online Bookstore at **www.igi-global.com** or through your preferred provider.

ISBN: 9798369303689
© 2024; 299 pp.
List Price: US$ **300**

ISBN: 9798369314791
© 2024; 287 pp.
List Price: US$ **330**

ISBN: 9798369300442
© 2023; 542 pp.
List Price: US$ **270**

ISBN: 9798369308929
© 2024; 426 pp.
List Price: US$ **265**

ISBN: 9781668489383
© 2023; 299 pp.
List Price: US$ **325**

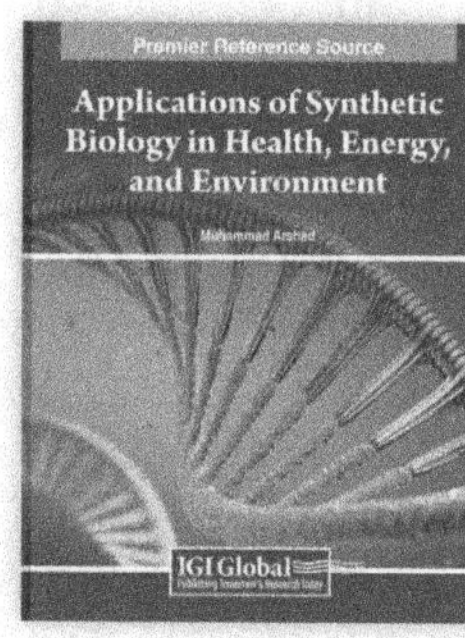

ISBN: 9781668465776
© 2023; 454 pp.
List Price: US$ **325**

Do you want to stay current on the latest research trends, product announcements, news, and special offers?
Join IGI Global's mailing list to receive customized recommendations, exclusive discounts, and more.
Sign up at: www.igi-global.com/newsletters.

Scan the QR Code here to view more related titles in STM.

www.igi-global.com Sign up at www.igi-global.com/newsletters facebook.com/igiglobal twitter.com/igiglobal linkedin.com/igiglobal

Ensure Quality Research is Introduced to the Academic Community

Become a Reviewer for IGI Global Authored Book Projects

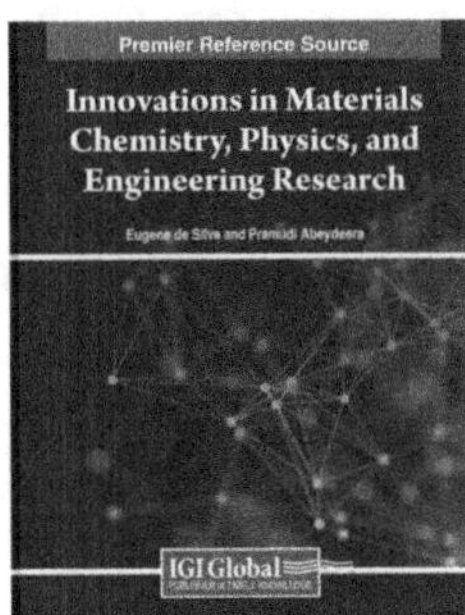

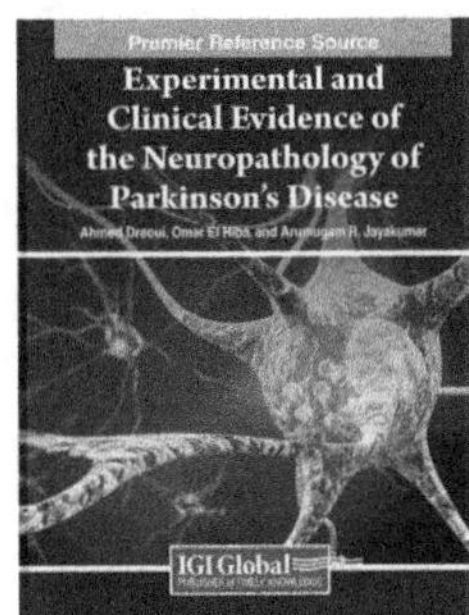

The overall success of an authored book project is dependent on quality and timely manuscript evaluations.

Applications and Inquiries may be sent to:
development@igi-global.com

Applicants must have a doctorate (or equivalent degree) as well as publishing, research, and reviewing experience. Authored Book Evaluators are appointed for one-year terms and are expected to complete at least three evaluations per term. Upon successful completion of this term, evaluators can be considered for an additional term.

If you have a colleague that may be interested in this opportunity, we encourage you to share this information with them.

IGI Global
Publishing Tomorrow's Research Today
www.igi-global.com

IGI Global Open Access Journal Program

Publishing Tomorrow's Research Today

IGI Global's Open Access Journal Program

Including Nearly 200 Peer-Reviewed, Gold (Full) Open Access Journals across IGI Global's Three Academic Subject Areas: Business & Management; Scientific, Technical, and Medical (STM); and Education

Consider Submitting Your Manuscript to One of These Nearly 200 Open Access Journals for to Increase Their Discoverability & Citation Impact

Web of Science Impact Factor 6.5 — Journal of Organizational and End User Computing

Web of Science Impact Factor 4.7 — Journal of Global Information Management

Web of Science Impact Factor 3.2 — International Journal on Semantic Web and Information Systems

Web of Science Impact Factor 2.6 — Journal of Database Management

Choosing IGI Global's Open Access Journal Program Can Greatly Increase the Reach of Your Research

Higher Usage
Open access papers are 2-3 times more likely to be read than non-open access papers.

Higher Download Rates
Open access papers benefit from 89% higher download rates than non-open access papers.

Higher Citation Rates
Open access papers are 47% more likely to be cited than non-open access papers.

Submitting an article to a journal offers an invaluable opportunity for you to share your work with the broader academic community, fostering knowledge dissemination and constructive feedback.

Submit an Article and Browse the IGI Global Call for Papers Pages

We can work with you to find the journal most well-suited for your next research manuscript.
For open access publishing support, contact: journaleditor@igi-global.com

Publishing Tomorrow's Research Today

IGI Global

e-Book Collection

Including Essential Reference Books Within Three Fundamental Academic Areas

Business & Management
Scientific, Technical, & Medical (STM)
Education

- Acquisition options include Perpetual, Subscription, and Read & Publish
- No Additional Charge for Multi-User Licensing
- No Maintenance, Hosting, or Archiving Fees
- Continually Enhanced Accessibility Compliance Features (WCAG)

Over **150,000+** Chapters | Contributions From **200,000+** Scholars Worldwide | More Than **1,000,000+** Citations | Majority of e-Books Indexed in Web of Science & Scopus | Consists of Tomorrow's Research Available Today!

Recommended Titles from our e-Book Collection

Innovation Capabilities and Entrepreneurial Opportunities of Smart Working
ISBN: 9781799887973

Advanced Applications of Generative AI and Natural Language Processing Models
ISBN: 9798369305027

Using Influencer Marketing as a Digital Business Strategy
ISBN: 9798369305515

Human-Centered Approaches in Industry 5.0
ISBN: 9798369326473

Modeling and Monitoring Extreme Hydrometeorological Events
ISBN: 9781668487716

Data-Driven Intelligent Business Sustainability
ISBN: 9798369300497

Information Logistics for Organizational Empowerment and Effective Supply Chain Management
ISBN: 9798369301593

Data Envelopment Analysis (DEA) Methods for Maximizing Efficiency
ISBN: 9798369302552

Request More Information, or Recommend the IGI Global e-Book Collection to Your Institution's Librarian

For More Information or to Request a Free Trial, Contact IGI Global's e-Collections Team: eresources@igi-global.com | 1-866-342-6657 ext. 100 | 717-533-8845 ext. 100

Are You Ready to **Publish Your Research?**

IGI Global
Publishing Tomorrow's Research Today

IGI Global offers book authorship and editorship opportunities across three major subject areas, including Business, STM, and Education.

Benefits of Publishing with IGI Global:

- Free one-on-one editorial and promotional support.
- Expedited publishing timelines that can take your book from start to finish in less than one (1) year.
- Choose from a variety of formats, including Edited and Authored References, Handbooks of Research, Encyclopedias, and Research Insights.
- Utilize IGI Global's eEditorial Discovery® submission system in support of conducting the submission and double-blind peer review process.
- IGI Global maintains a strict adherence to ethical practices due in part to our full membership with the Committee on Publication Ethics (COPE).
- Indexing potential in prestigious indices such as Scopus®, Web of Science™, PsycINFO®, and ERIC – Education Resources Information Center.
- Ability to connect your ORCID iD to your IGI Global publications.
- Earn honorariums and royalties on your full book publications as well as complimentary content and exclusive discounts.

Join Your Colleagues from Prestigious Institutions, Including:

Australian National University
Massachusetts Institute of Technology
Johns Hopkins University
Tsinghua University 1911
Harvard University
Columbia University in the City of New York

Learn More at: www.igi-global.com/publish

or by Contacting the Acquisitions Department at: acquisition@igi-global.com

Individual Article & Chapter Downloads

US$ 37.50/each

Easily Identify, Acquire, and Utilize Published Peer-Reviewed Findings in Support of Your Current Research

- Browse Over ***170,000+ Articles & Chapters***
- ***Accurate & Advanced*** Search
- Affordably Acquire ***International Research***
- ***Instantly Access*** Your Content
- Benefit from the ***InfoSci® Platform Features***

THE UNIVERSITY *of* NORTH CAROLINA *at* CHAPEL HILL

It really provides an excellent entry into the research literature of the field. It presents a manageable number of highly relevant sources on topics of interest to a wide range of researchers. The sources are scholarly, but also accessible to 'practitioners'.

- Ms. Lisa Stimatz, MLS, University of North Carolina at Chapel Hill, USA

www.ingramcontent.com/pod-product-compliance
Lightning Source LLC
LaVergne TN
LVHW080342011026
849161LV00009B/2162
* 9 7 9 8 3 6 9 3 1 0 9 4 6 *